CALIFORNIA STATE U
This b

The Biology and Conservation

THE SYMPOSIA OF THE NATIONAL ZOOLOGICAL PARK

of the CALLITRICHIDAE

DEVRA G. KLEIMAN, *Editor*

A symposium held at the
Conservation and Research Center,
National Zoological Park,
Smithsonian Institution
August 18–20, 1975

Smithsonian Institution Press
Washington, D.C.
1977

Copyright © 1978 Smithsonian Institution. All rights reserved.

Library of Congress Cataloging in Publication Data

Main entry under title:

The Biology and conservation of the Callitrichidae.

(The symposia of the National Zoological Park)
Bibliography: p.
Includes index.
1. Callitrichidae—Congresses. 2. Wildlife conservation—Congresses. I. Kleiman, Devra G.
II. Washington, D.C. National Zoological Park.
III. Series: Washington, D.C. National Zoological Park. The symposia of the National Zoological Park.
QL737.P92B56 599'.82 78-2428
ISBN 0-87474-586-1
ISBN 0-87474-587-X pbk.
Panel Discussion, August 20, 1975:
Reproduction of Small Colonies of
Endangered Callitrichids, Panelists W.

Contents

Preface

Our knowledge of the biology of the South American Callitrichidae—marmosets and tamarins—has lagged behind our knowledge of other primate families. Reasons for the gap are numerous. Field research has been hampered by the species' small size, rapid movements, and forested habitat, even though all callitrichids are diurnal. Moreover, South American primates, in general, have been less studied than Asian and African forms. Captive observations have been hindered by poor reproduction and disease problems, some of which resulted from inadequate nutrition of captive specimens.

During the past decade, there has been a developing interest in this family, but the following facts indicate the recency of this interest and the extent of our ignorance. There have been only two long-term field studies of callitrichids, both initiated in 1973 and reported on in this volume. Gestations are known for only a handful of species, and preliminary data on the estrous cycle length for one genus were first published in 1973. Several species have never been maintained in captivity, and few species are bred in sufficient numbers that they can be considered as viable self-sustaining populations. The ultimate indication of our ignorance of this family is the tendency to refer to all forms as "the marmoset," regardless of genus or species. Whereas a researcher would never publish a study where the subject species was called only "the macaque," or "the baboon," or even "the lemur," the literature is replete with studies on "the marmoset." The assumption that species are alike arises only because we know so little of the differences. As will be seen in this volume, there are important differences in the biology and behavior of the callitrichid species.

This conference was an outgrowth of a meeting held in February 1972 at the National Zoological Park entitled *Saving the Lion Marmoset,*[1] which had as its purpose the sharing of information concerning callitrichid biology in an effort to improve the plight of the endangered golden lion tamarin, *Leontopithecus r. rosalia.* In the intervening years, new studies of marmosets and tamarins were initiated, and it became apparent that a second larger meeting might prove fruitful for a variety of reasons. First, numerous South American countries had ceased exporting New World primates and were actively developing plans for large-scale captive breeding programs that would provide animals for export. Secondly, the biomedical community was also beginning to see the need for long-term propagation programs, but had had little interaction with either field biologists or ethologists studying callitrichids. Thirdly, field studies were beginning to provide much new information on the status, ecology, and habits of callitrichids. Also, the status of the endangered forms, such as the lion tamarin, was worsening, and one other species, used extensively in biomedical research, the cotton-top tamarin, *Saguinus oedipus,* was found to be rapidly diminishing in numbers. Finally, the first studies on the basic reproductive cycles of callitrichids were appearing. With all of this new information and the knowledge that for both the biomedical community and the conservation-oriented biologists, the ultimate concern had to be with conservation, I organized the meeting, deliberately including both the "users" and "conservers" of callitrichids. The result can be seen in the contributed papers and the panel discussions during which biologists from different disciplines and with a diversity of interests shared their knowledge of callitrichids. Discussions were lively and disagreements occasionally occurred about means and ends; however, a resolution was drawn up by a committee and endorsed wholeheartedly by conference participants which expressed continued concern for the worsening status of the endangered forms and requested action toward the regulation of species most commonly used in biomedical research, by the development of captive propagation programs and the preservation of habitat.

Participants in this symposium did not always agree on callitrichid taxonomy. In some cases, an Editor's note has been added for clarification. The Panamanian rufous-naped or Geoffroy's tamarin is occasionally referred to as *Saguinus* (*oedipus*) *geoffroyi* in text to distinguish it from the closely related cotton-top tamarin, *Saguinus oedipus oedipus.*

The conference was supported by the Smithsonian Research Foundation and the proceedings are being published through a generous grant from the Friends of the National Zoo. The Pan American Health Organization and National Institutes of Health kindly helped with travel funds and travel arrangements for some South American participants. Ms. Wyotta Holden and Mr. Joe Reed helped with numerous aspects of the organization of the conference, and I thank them for their efforts. Other persons who worked to make the conference a success and to whom I am grateful are P. R. Chu, L. Collins, C. A. Dorsey, P. Kibbee, D. Mack, M. McComas, N. A. Muckenhirn, and C. Wemmer. Special thanks are due to my husband, John F. Eisenberg, for his thoughtfulness, patience, and support.

Devra G. Kleiman
National Zoological Park
Washington, D.C., 20008

September 1, 1976

[1] Bridgwater, D. D., ed. *Saving the Lion Marmoset.* Wheeling, West Virginia: Wild Animal Propagation Trust, 1972.

List of Participants

Robert C. Bailey
Matthews Hall, Harvard University, Cambridge, Massachusetts, 02138

†J. S. Batson
Medical Division, Oak Ridge Associated Universities, P.O. Box 117, Oak Ridge, Tennessee, 37830

*Benjamin Blood
Animal Resources Branch, National Institutes of Health, Bethesda, Maryland, 20014

Hilary O. Box
Department of Psychology, University of Reading, Bldg. 3, Earley Gate, Whitenights, Reading RG6 2AL, England

Rogerio Castro
Proyecto Primates, P.O. Box 621, Iquitos, Loreto, Peru

†Napoleon Castro R.
Ministerio de Agricultura, Lima, Peru

*Mary-Scott Cebul
Division of Health Science Resources, Section of Comparative Medicine, Yale University, School of Medicine, 375 Congress Avenue, New Haven, Connecticut, 06510

John L. Cicmanec
Litton Bionetics, Kensington, Maryland, 20795

Adelmar F. Coimbra-Filho
Departamento de Conservação Ambiental, FEEMA, Rio de Janeiro, Brazil

*Robert Cooper
Department of Zoology, San Diego State College, San Diego, California, 92182

Gary A. Dawson
The Museum, Michigan State University, East Lansing, Michigan, 48824

*Carolyn Dorsey
National Zoological Park, Washington, D.C., 20008

*Nicole Duplaix
New York Zoological Society, Bronx, New York, 10460

John F. Eisenberg
National Zoological Park, Washington, D.C., 20008

Gisela Epple
Monell Chemical Senses Center, 3500 Market Street, Philadelphia, Pennsylvania, 19104

†Curtis H. Freese[1]
Department of Pathobiology, Johns Hopkins University, Baltimore, Maryland, 21205

Margaret A. Freese[1]
Department of Medicine, Johns Hopkins University, Baltimore, Maryland, 21205

Nazareth Gengozian
Medical Division, Oak Ridge Associated Universities, P.O. Box 117, Oak Ridge, Tennessee, 37830

Kenneth M. Green
National Zoological Park, Washington, D.C., 20008

†John K. Hampton, Jr.
California Polytechnic State University, San Luis Obispo, California 93407

Suzanne H. Hampton
KinetoMatic, Route 1, Box 63C Templeton, California 93465

John P. Hearn
MRC Reproductive Biology Unit, 2 Forrest Road, Edinburgh EH1 2QW, Scotland

Robert J. Hoage
National Zoological Park, Washington, D.C., 20008

*Keith Hobbs
OLAC House, Blackthorn, Bicester, Oxon OX6 OTP, England

Jennifer C. Ingram
Department of Psychology, 8–10 Berkeley Square, Bristol B58 1HH, England

†Marvin Jones
San Diego Zoological Garden, San Diego, California, 92112

W. R. Kingston[2]
Hillcrest House, Belton, Loughborough, Leics. LE12 9TE, England

Devra G. Kleiman
National Zoological Park, Washington, D.C., 20008

*Rainer Lorenz
Lehrstuhl für Anthropologie der Georg-August Universität, Burgerstr. 50, D-34 Göttingen, West Germany

*David Mack
National Zoological Park, Washington, D.C., 20008

†Alceo Magnanini
Fundação Estadual de Engenharia do Meio Ambiente, Departamento de Conservação Ambiental, Rio de Janeiro, Brazil.

Jeremy J. C. Mallinson
Jersey Wildlife Preservation Trust, Les Augres Manor, Jersey, Channel Islands

†Elizabeth McLanahan
National Zoological Park, Washington, D.C., 20008

*Luis Melendez
PAHO, 525 23rd Street, N.W., Washington, D.C., 20037

Russell A. Mittermeier
Museum of Comparative Zoology, Harvard University, Cambridge, Massachusetts, 02138

Patricia F. Neyman
Department of Zoology, University of California, Berkeley, California, 94720

*Carol Nicoll
Department of Psychology, University of California, Berkeley, California, 94720

*John Perry
National Zoological Park, Washington, D.C., 20008

A. G. Pook
Jersey Wildlife Preservation Trust, Les Augres Manor, Jersey, Channel Islands

James A. Porter
South American Primates, 10525 S. W. 185th Terrace, Miami, Florida, 33157

*Toby Pyle
World Wildlife Fund, 1319 Eighteenth Street, N.W., Washington, D.C., 20036

Marleni Flores Ramirez[3]
Puerto Rico Nuclear Center, Tropical Agro-Sciences Division, Mayaguez, Puerto Rico, 00708

*Theodore H. Reed
National Zoological Park, Washington, D.C., 20008

†Juan Revilla C.
Universidad Nacional Mayor de San Marcos, Lima, Peru

Harmut Rothe
Institute of Anthropology, University of Göttingen, Büsenweg 3, D-34 Göttingen, West Germany

*George B. Schaller
New York Zoological Society, Bronx, New York, 10460

*Patricia Scollay
Department of Psychology, San Diego State University, San Diego, California, 92110

†T. A. Smith
Medical Division, Oak Ridge Associated Universities, P. O. Box 117, Oak Ridge, Tennessee, 37830

Pekka Soini
Proyecto Primates, P. O. Box 621, Iquitos, Loreto, Peru

*Warren Thomas
Los Angeles Zoo, 5333 Zoo Drive, Los Angeles, California, 90027

*Richard W. Thorington
Department of Vertebrate Zoology, Museum of Natural History, Smithsonian Institution, Washington, D.C., 20560

*Robert Whitney
Animal Resources Branch, National Institutes of Health, Bethesda, Maryland, 20014

Charles G. Wilson
Overton Park Zoo, Memphis, Tennessee, 38112

Karen Mitchell
Interpreter Translator, 4103 Byrd Court, Kensington, Maryland, 20795

[1] Present address: Department of Biology, Southwestern at Memphis, 2000 North Parkway, Memphis, Tennessee 38112

[2] Present address: Proyecto Primates, P.O. Box 621, Iquitos, Peru

[3] Present address: Department of Biology, University of Puerto Rico, RUM, Mayaguez, P.R., 00708

*Attended, but did not present a paper.

†Did not attend, but author of a paper.

Joint Resolution by Participants in Conference on the "Biology and Conservation of the Callitrichidae," August 20, 1975

Sixty scientists from the United States, South America, and Europe have participated in an International Conference on the Biology and Conservation of Callitrichids (marmosets and tamarins). The meeting was convened by the National Zoological Park, Smithsonian Institution, Washington, D.C., and held at Front Royal, Virginia, from August 18-20. The speakers presented most of the known biology of these monkeys, drawn from studies both in the field and in the laboratory. This statement conveys our deep concern about the present situation. We examine here the immediate threats to these monkeys and suggest methods to repair the situation and to avoid future problems.

---

1. Endangered Species

Several species of callitrichids are threatened so severely that they are near to extinction:

Lion tamarin, the 3 subspecies of *Leontopithecus rosalia*
Buff-headed marmoset, *Callithrix flaviceps*
White-eared marmoset, *Callithrix aurita*
Cotton-top tamarin, *Saguinus oedipus oedipus*

Threats

The clearance of natural homelands. The lion tamarins are now dependent on only 2 percent of their original natural range and not more than 600 are thought to survive in the wild.

Suggestions

a. A total ban on the export, shooting of or interference with these animals.
b. Urgent study by census and observations.
c. Close management of the remaining home environments by reforestation where possible and by a supply of the known limiting factors. These animals depend on particular types of trees and on shelter holes, which might be created in such trees for them. A supply of additional foods may also be necessary to assist their survival in the present unnaturally reduced environments to which they are restricted.

The plight of these callitrichids is extremely serious and any delay in the application of effective measures may well prove disastrous. Conservation organizations and the governments of source countries are requested to assist.

---

2. Species Used in Research

There are other species used by pharmacological interests and by biomedical research. These species are not immediately threatened, but may become so as the growing needs of research outstrip the replacement of the natural wild stocks. This problem will arise over the next ten years until breeding programs within the user countries make excessive demand on the wild unnecessary.

The needs for these animals in research are very real where they allow the study of naturally occurring human diseases such as colonic cancer and hepatitis, or where they form excellent models in human-oriented research in immunology, virology, reproduction, contraception, and teratology. The potential of marmosets and tamarins in such studies is only starting to be recognized. The main species involved are:

Common marmoset, *Callithrix jacchus*
Saddle-backed tamarin, *Saguinus fuscicollis*
Moustache tamarin, *Saguinus mystax*

Threats

a. Appalling losses at present during shipment from the wild to the user countries, which result in an unreal escalation in demand.
b. Accelerated rate of deforestation in developing South American countries.

Suggestions

a. Development of breeding centers and research within the home countries.
b. Development within the user countries of breeding centers that apply rigorous genetic management of stocks.
c. Rigorous control of export quotas from the homelands.
d. Control of the conditions at export centers and during shipments.
e. Assistance from scientific organizations and the governments of user countries is requested.

On behalf of the combined interests of conservation and scientific research, we request the urgent application of practical measures to change the present unacceptable situation.

SECTION I:

THE EVOLUTION AND ECOLOGY OF THE CALLITRICHIDAE

Introduction

The cebid monkeys of South America have radiated to fit ecological niches similar to those seen in the Old World. The Callitrichidae, however, as Eisenberg points out, exist in a squirrel-like niche without parallel in the Old World. They have also opted for a high reproductive output, by bearing twins and triplets and, at least in the laboratory, exhibiting little seasonality. As indicated by both Eisenberg and Dawson, they can be considered as *r* selected, relative to other primate species.

Based on captive studies, it has been suggested that marmosets and tamarins have a monogamous social structure, with nuclear family groups exhibiting territoriality. The in-depth field studies by Dawson and Neyman on *Saguinus* spp. have complicated this simple picture, and raised numerous questions. Although group size is typically within the limits of what one might expect for a nuclear family, there are observations of larger groups that persist in time and are not mere feeding aggregations. Moreover, both authors report that it is not uncommon to find some transient individuals of all ages and reproductive status (although Dawson noted that juveniles and young adults were more often transients) which seem to change their group affiliation with relative impunity. These findings challenge the notions of nuclear family group integrity and strict territoriality.

Dawson has proposed the term age-dependent, one male, one female, dominance system which retains the concept that the core of a *S. (oedipus) geoffroyi* group is an adult breeding pair and their dependent offspring, but suggests that nonbreeding subordinate individuals may be unrelated to the alpha pair. Neyman suggests that *S. oedipus* pairs may not be permanent, but mainly bond for the duration of reproduction. Support for either proposal can only come from future studies of the kin relationships of individuals in several groups within a restricted locality. If a transient individual is accepted by a group without aggression, we must know whether that group contains a close relative, e.g., older sibling, of the intruder before eliminating the term family from our callitrichid vocabulary. Within a small population, neighboring groups are certainly genetically related, and transients may well be uncles, nieces, or cousins of one or both members of the alpha pair. It is clear that some sexually mature members of the population do not breed and thus, they may perpetuate their own genes by helping close relatives to rear offspring.

Group size and stability appear to be dependent on the quality of the habitat. Dawson has shown that *S. (oedipus) geoffroyi* groups are less stable in their composition in drier upland forest (although group size was constant) than in lowland wet forest. The high turnover in group members and changes in group size observed by Neyman in *S. oedipus* in Colombia may have resulted from both the fact that her study area was semidisturbed and the habitat was extremely seasonal and quite dry, with 60 percent of the trees losing their leaves during the dry season. Coimbra also reports that habitat destruction in the range of *L. r. rosalia* has seriously affected the social structure of the remaining population. More studies of species in optimal and suboptimal habits (both natural and human-induced) where individuals and their relationships are identifiable will clarify how and why environmental parameters affect social structure. Moreover, species-specific differences in social organization and lability will undoubtedly arise as more field studies on the various callitrichid forms are conducted. There seems to be little disagreement concerning the fact that *Cebuella pygmaea* lives in small stable nuclear family groups, but its dependence on exudates from tree species, a nonseasonal food source, may promote a more stable social structure than in *Saguinus.*

The papers by Ramirez et al., Mittermeier and Coimbra-Filho, and Coimbra-Filho detail species-specific differences in feeding habits and shelter-seeking behavior which could influence a species' spatial distribution, territoriality, and therefore its social organization. In many of the papers in this section, it is implied that different species have habitat preferences, a not unexpected suggestion; however, with this implication come many potential differences, e.g., in the mobility of groups. *Cebuella,* because of its apparent dependency on exudate source trees, has a very small home range and is not very mobile.

Studies of the ecology of callitrichids are in their infancy, but the following chapters suggest that future research will provide exciting answers to some very intriguing questions. Especially interesting will be studies relating feeding ecology and habitat requirements to social structure in different callitrichid species.

JOHN F. EISENBERG
Office of Zoological Research
National Zoological Park
Smithsonian Institution
Washington, D.C. 20008

Comparative Ecology and Reproduction of New World Monkeys

ABSTRACT

In the New World, the primates have had a long evolutionary history. Fossil primates in South America date from the Oligocene. Primate evolution in South America probably began during the Eocene and the callitrichids have been separated from the cebids for a period dating from at least the Oligocene. Callitrichid primates are unique for a variety of reasons and appear to occupy a niche in South America which has not been exploited by primates in the Paleotropics. In the New World, the marmosets and tamarins are diurnal, frugivore-insectivores; they are small in size and forage in second-growth areas as well as in the open forest areas in riverine situations. In many respects, the New World callitrichids appear to occupy ecological niches which are exploited in the Paleotropics by rodents of the family Sciuridae. It is proposed that the callitrichids are secondarily small, having evolved from an ancestral form somewhat larger than the contemporary species. Thus, their small size should not be taken as a "primitive" attribute. The callitrichids do possess numerous morphological features which are conservative or plesiomorph. Such plesiomorph characters of the callitrichids include extensive gland fields in the anogenital area and in the sternal area of both sexes, the possession of a functional Jacobson's organ, and the possession of claws rather than nails on the digits.

The small size of the callitrichids evidently increased selection pressures favoring twinning and a shortened reproductive cycle relative to the larger Cebidae, thus resulting in the potential for increased numbers of offspring from females during their reproductive life span. In this sense, the callitrichids can be said to have undergone *r* selection. The increased number of young in the Callitrichidae presents certain problems to the female in that the combined weight of the young at birth may exceed 10 percent of the mother's body weight. In extreme cases, the combined litter weight may approximate 28 percent of the mother's weight. This is in excess of values for most of the Cebidae. The increased energetic drain on the lactating mother probably created a positive selection for increased parental care on the part of the male and her older offspring. The parental care manifests itself in the transport of the young at varying ages after birth. Participation by the male and older siblings in the transport of young is uniquely developed in the Callitrichidae, but parallel trends are shown in *Aotus, Callicebus* and *Pithecia.* The integrity of the pair bond would appear to be important in the rearing cycle of the callitrichids and, for this reason, males and females exhibit pronounced hostility toward strange same-sexed adults. It would appear that the natural social unit of the marmoset is the nuclear family. Cognizance of the rearing pattern is a necessary prerequisite to captive management.

In contradistinction to the Callitrichidae, the larger Cebidae of the subfamily Atelinae show an extremely slow development schedule. The intrinsic rate of natural increase for the genus *Ateles* is approximately half of that for the Old World genus *Macaca.* In the life table strategy for *Lagothrix* and *Ateles*, they show a strong resemblance toward many of the smaller anthropoids, such as the gibbon, *Hylobates.*

Evolution and Adaptation

In the New World, the primates have had a long evolutionary history with fossil primates in North America dating from the Eocene. In South America, the earliest extant fossils may be dated from the Oligocene. Not only has the primate fauna of South America had a long history, but it has been exceedingly rich in the number of forms which coexisted during any single geological epoch. To quote from Hershkovitz (1974), "the history of South American primates is longer and much more complex than was generally supposed. . . .

Platyrrhines are nearly related to catarrhines . . . the more primitive organization of platyrrhines in general indicates that, if not ancestral to catarrhines, they are almost certainly early offshoots of the haplorhine stock from which catarrhines arose. In the latter event, the ancestral stock may have lived in the Afro-South American mass of Gondwanaland." Thus, the history of primate evolution in South America probably began early in the Eocene after separation from Africa, and the callitrichids have been separated from the cebids for a period dating from at least the Oligocene, and perhaps even earlier.

The living forms of some South American primates exploit ecological niches similar to those niches occupied by primates in the Paleotropics (Eisenberg et al., 1972). There are, however some notable differences. Nocturnal, morphologically conservative primates, such as the Old World galagos and lorises, are not represented in the Neotropics by similarly adapted primate forms. The only nocturnal neotropical primate, *Aotus*, differs in its niche occupancy from either lorisoids or galagoids. This niche is, in fact, partly occupied by didelphine marsupials in the Neotropics (Eisenberg and Thorington, 1973).

With regard to the diurnal neotropical primates, *Alouatta* occupies a niche similar to that of some species of Old World colobine primates; however, *Alouatta*'s degree of adaptation to a folivorous diet has been accomplished differently from the colobines. Rather than having an enlarged and ensacculated stomach for the digestion of leaves, in *Alouatta* the large intestine and caecum have been lengthened (Hladik, 1967). *Cebus* would appear to occupy a niche similar to *some* species of *Cercopithecus* and *Macaca* in the Paleotropics while *Ateles* occupies a feeding niche similar to the hylobatids of the Old World tropics. The niches of some of the smaller neotropical primates of the genera *Callicebus, Pithecia, Cacajao,* and *Saimiri* do not completely parallel any primates of the Old World tropics, except possibly for the West African *Miopithecus* which in some respects resembles *Saimiri* in its feeding ecology (Gautier-Hion, 1966).

The primate radiation on the island of Madagascar was accomplished with a very morphologically conservative stock and although niche occupancy by this taxon on Madagascar has replicated trends in the Old World tropics, there are many more nocturnal forms in Madagascar. Also, the concordance between niche occupancy by the Madagascan lemuriforms and the cercopithecine primates in the Paleotropics is only complementary in certain restricted cases (see Eisenberg et al., 1972).

Callitrichid primates are unique for a variety of reasons, and from the standpoint of their inclusive ecological niche, they do not have any strict primate parallels in the Paleotropics (see Moynihan, 1976). In the New World, the marmosets and tamarins are diurnal, frugivore-insectivores. They are small in size and forage in second-growth areas as well as in open forest areas in riverine habitat. In many respects, the New World callitrichids appear to occupy ecological niches which are exploited in the Paleotropics by members of the family Sciuridae (see Medway, 1969; Kingdon, 1975; Rosevear, 1969). Sciurid rodents probably did not enter the South American continent until the Pliocene, and to this day there has not been any vast radiation of tropical squirrels in the New World tropics comparable to the forested regions of Africa and Southeast Asia (Hershkovitz, 1972). Of course, the caviomorph rodents invaded the South American continent in the Miocene or perhaps the late Oligocene (Patterson and Pasqual, 1972), but the caviomorphs did not evolve a diurnal squirrel-like form. They have, however, occupied several nocturnal, arboreal, frugivorous niches and thus would seem to have avoided outright competition with callitrichids.

Primitive and Advanced Characters

The small size of the callitrichids is probably a secondarily derived characteristic. I believe that the ancestral form was somewhat larger than recent callitrichids, perhaps within the size range of the squirrel monkey, *Saimiri.* For this reason, I consider their small size as a specialized attribute and not primitive. However, the callitrichids do possess numerous morphological features which may be considered primitive or "plesiomorph," including extensive sternal glands on the ventrum and gland fields associated with the labia of the female and the scrotum of the male. The possession of these glands together with a functional Jacobson's organ in the adult clearly defines the callitrichids as plesiomorph or conservative, and *Callimico* then serves as a conservative "bridge" between callitrichids and cebids (see Table 1).

Within the cebids, the genera *Lagothrix*, *Ateles*, and *Brachyteles* (the Atelinae) show many apomorph or advanced characteristics including (1) reduction of the sternal gland field to a small localized point in the case of *Ateles*, (2) reduction of the labial gland field in the female, (3) absence of a large scrotal gland field in the male, (4) lack of a functional Jacobson's organ in the adult (5) lack of a baculum in the penis of the adult male, (6) extraordinary enlargement of the clitoris and labia in the female, and (7) possession of a fully prehensile tail. Such an enlargment of the female genitalia may also be noted as a parallel trend in the distantly related *Callicebus* and *Pithecia* (see Table 1).

Thus, the callitrichids represent a taxonomic unit with several plesiomorph characters. In the Cebidae, the subfamilies Alouattinae, Pithecinae, Aotinae, and Callicebinae represent intermediate grades and the Cebinae and Atelinae exhibit the most apomorph or advanced specializations.

Reproductive Trends

With the exception of *Callimico*, the Callitrichidae are characterized by the production of twins. The interbirth interval in most callitrichids is either 6 (captive) or 12 (captive and field) months and the estrous cycle approximates 15 to 16 days. Sexual maturity in the female marmoset or tamarin may be attained at 18 months or in the male at around 24 months. On the other hand, the family Cebidae is characterized by (a) a modal litter size of one, and (b) ages at sexual maturity of 3½ or more years in the female and 4 or more years in the male. The estrous cycle is highly variable in the Cebidae, determined at 10.9 days for *Saimiri*, 16 to 20 days for *Cebus*, and 24 to 27 days for *Ateles* (see Table 2 for a summary of reproductive parameters).

Thus it may be seen that the marmosets and tamarins have opted for a reproductive strategy which involves a short estrous cycle, twinning, and a rapid maturation of the young. The evolution of small size in the callitrichids evidently increased selective pressures favoring twinning and a shortened reproductive cycle relative to the larger Cebidae, thus offering the potential for increasing the number of offspring during the reproductive span of a given adult female. In this sense then, the callitrichids can be said to have undergone *r*-selection.

The increased number of young in the Callitrichidae creates certain problems for the female, however, in that the combined weight of the young at birth may exceed 10 percent of the mother's body weight. In extreme cases, the combined litter weight may approximate 28 percent of the mother's weight. This is far in excess of values for most of the Cebidae which generally produce young at birth weighing less than 10 percent of the mother's weight and as low as 5 percent (Kleiman, 1977a). The increased energetic drain on the lactating mother probably created a positive selection for increased parental care on the part of the male and older offspring. The parental care manifests itself in the transport of young at varying ages after birth, but rather early in most callitrichids. Participation by the male and older siblings in the transport of young is uniquely developed in the Callitrichidae, but similar tendencies are also shown in *Aotus*, *Callimico*, *Callicebus*, and perhaps *Pithecia*.

It is worth mentioning that male transport of older young, pair maintenance through a rearing cycle, and small troop size (i.e., restricted to a nuclear family) may be a phylogenetically old syndrome in the Ceboidea. This sociological pattern of rearing appears in *Pithecia*, *Aotus*, and *Callicebus*, all to some extent conservative cebids of intermediate size and diverse niche occupancy (see also Moynihan, 1967 and 1976).

The integrity of the pair bond appears to be very important in the callitrichid rearing cycle and, for this reason, female hostility toward a strange adult female and male hostility toward strange adult males is very pronounced, especially in captivity. The natural social unit of marmosets and tamarins during rearing is probably an extended family, and cognizance of the rearing pattern is a necessary prerequisite to sound captive management.

Field data for *Saguinus* (Neyman, 1977; Dawson, 1977) suggest that in the field tamarins may emigrate and enter new groups. Thus, pair alliances may not persist beyond a given rearing cycle in the wild. Of course, in captivity with movement into or out of a family under total restriction, only the founding pair will reproduce even after their offspring become sexually mature in the home cage (Epple, 1975).

Table 2. Reproductive data comparisons for the Ceboidea.

Species	*Age at Full Sexual Maturity** ♂	♀	*Gestation (Days)*	*Interbirth Interval***	*Number of Young (Mode)*	*Estrous Cycle (Days)*	*Reference*
Cacajao rubicundus	>6 yrs	>5 yrs	?	~3 yrs	1	?	Fontaine, 1974
Pithecia pithecia	>4 yrs	?	163	?	1	?	Hick, 1973
Ateles fusciceps	~5 yrs	>4 yrs	226-232	~3 yrs	1	24-27	Eisenberg, 1973
Ateles geoffroyi	~5 yrs	4yrs	226-232	~3yrs	1	24-27	Eisenberg, 1973
Lagothrix lagotricha	>5 yrs	4 yrs	225	?	1	?	Williams, 1968
Cebus apella	4-5 yrs	3 1/2 yrs	180	?	1	16-20	Asdell, 1964
Saimiri sciureus	30+ mos	?	157	~12 mos	1	10.9	Travis and Holmes, 1974; Rosenblum and Cooper, 1968
Aotus trivirgatus					1		Renquist (pers. comm.)
Alouatta villosa			180-194		1	13-24	Glander, 1975
Callimico goeldii	18 mos?	12 mos?	154	~12 mos	1	?	Lorenz, 1972
Callithrix jacchus	9-13 mos	20-24 mos	140-148	6 or 12 mos	2	16.4	Hearn and Lunn, 1975; Hearn, 1977
Leontopithecus rosalia	≥24 mos	18 mos	125-130	6 or 12 mos	2	14-21	Kleiman, 1977b
Saguinus oedipus	≥24 mos	18 mos	~140	6 or 12 mos	2	~15	Hampton and Hampton, 1977

*Full sexual maturity in the male implies "sociological" maturity, i.e., full growth and not only the onset of spermatogenesis.

**Full-term rearing.

Mating Patterns

Copulation in the Callitrichidae is of a standard primate form. The female stands quadrupedally with the legs flexed and the male mounts while she is so crouched, intromitting for less than half a minute. Generally, several intromissions precede an ejaculation. This type of pattern is similar for *Leontopithecus, Saimiri,* and *Cebus;* however, in the larger cebid primates, where the tree limb itself may be quite small relative to the size of the animal, the female may recline forward supporting herself, in part, on her chest. The male will intromit while crouched and generally a multiple series of mounts can precede ejaculation. In the case of the howler monkey, *Alouatta,* ejaculation apparently can occur on a single mount and the average mount duration is slightly less than a minute. This contrasts with most *Macaca* spp.,where mounts are extremely brief, the male usually grips the female's hind legs with his hind feet, and many mounts are necessary before an ejaculation.

Long mounts occur in *Lagothrix* and *Ateles,* and the female's posture is even further modified. The female may initially recline forward while the male mounts; the male always (*Ateles*) or frequently (*Lagothrix*) locks his legs over the female's thighs while intromitting, and at the same time grips a branch with his prehensile tail. The female may then sit up vertically while the male sits in a near vertical position, and she may even turn around to face the male during intromission. Intromission may be as long as 10 minutes in *Lagothrix;* in *Ateles,* intromission with thrusting may last for over 30 minutes (16–35 min, N = 4) (Eisenberg, 1976).

The assumption of a forward reclining posture by the female with full structural support from tree limbs appears to correlate with a larger body size and arboreal copulation. The female's sitting

Table 1. Anatomical comparisons for the Ceboidea.*

Taxon	*Males Ano-genital Area*				
	Baculum	*Glans Prominent*	*Glans with Spines*	*Scrotum with Enlarged Gland Field*	*Gland Field Around Perineum*
Callitrichidae					
Callithrix	Present	−	+	++	−
Cebuella	Present	−	−	++	−
Leontopithecus	Present	−	?	++	?
Saguinus	Present	−	−	++	+
Callimico	?	++	−	++	−
Cebidae					
Aotus	Absent	−	−	−	+
Callicebus	Present	−	−	++	−
Pithecia	Present	−	−	++	+
Cacajao	Present	+	−	++	++
Alouatta	Present	+	−	−	−
Saimiri	?	++	++	−	−
Cebus	Small	++	+	+	−
Lagothrix	Absent	++	+	−	−
Ateles	Absent	++	+	−	−

− = Not conspicuously differentiated.

+ = Conspicuous.

++++ = Hypertrophied.

* All anatomical data from Hill (1957, 1960, 1962).

Gland fields defined for this paper correspond to Epple and Lorenz (1967) as follows: Type A = sternal gland complex; Type B = sternal field; Type C = gular field; Type D = gular/sternal field; Type E = sternal/epigastric field; Type F = sternal nodule.

Both Sexes							*Females Ano-genital Area*		
Jacobson's Organ	*Sternal Gland Complex* *Type A*	*Type B*	*Type C*	*Type D*	*Type E*	*Type F*	*Clitoris Enlarged*	*Lateral Labial Gland Field*	*Perineal Gland Field*
Present in adult	+						–	+++	++
Present in adult	+						–	+++	+
Present in adult		+					–	?	?
Present in adult	+						–	+++	+
Present in adult	+						–	?	?
?	+						–	–	+
?				+			+++	++	–
?			+				++	–	++
?					+		+	++	++
?			+				–	++	–
Foetal				+			+	++	++
Present in adult					+		+	+	–
Foetal		+					+++	–	–
Foetal						+	++++	–	–

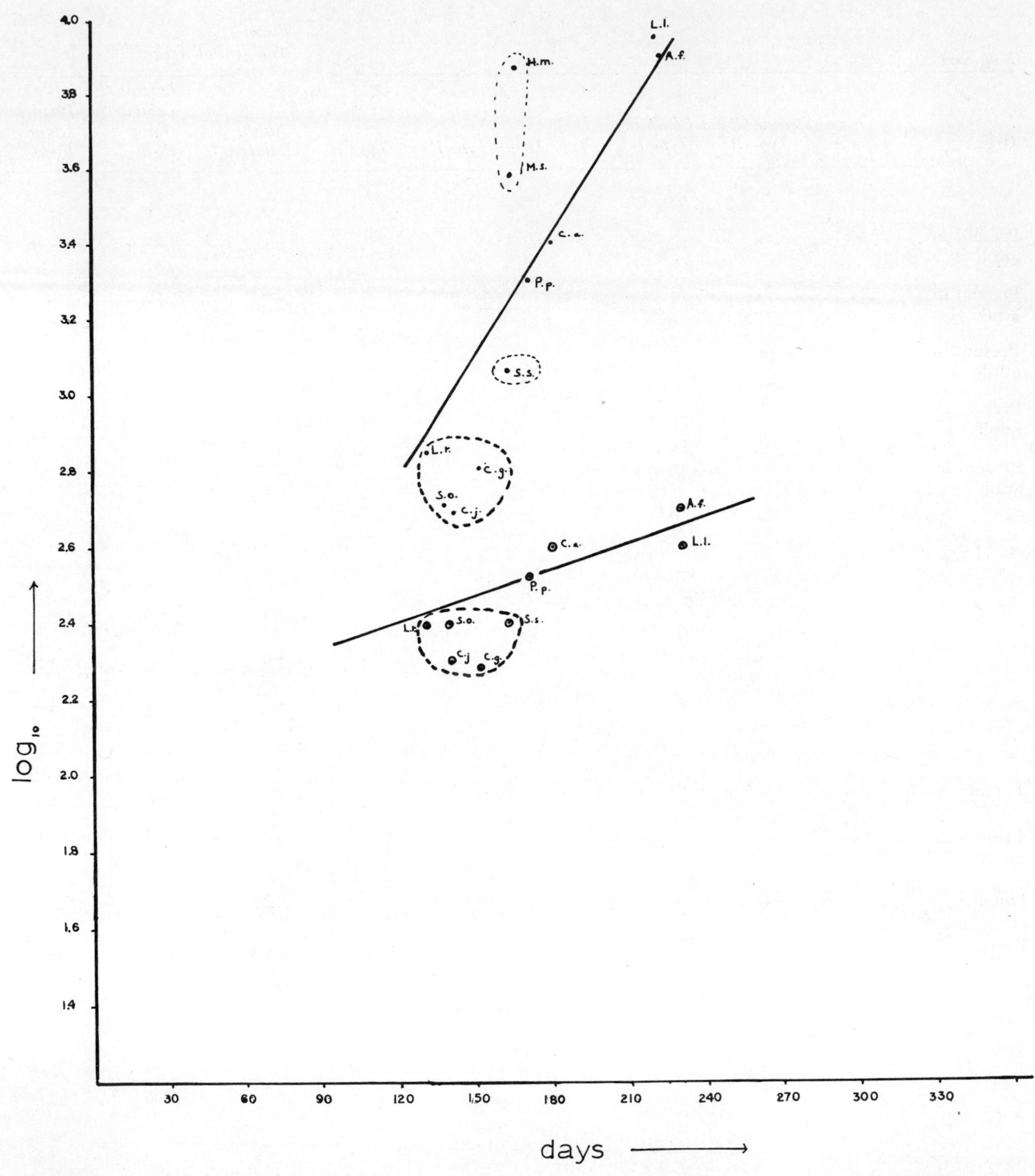

Figure 1. Correlation of gestation length and size for ceboid primates. Abscissa = gestation length in days. Ordinate = $\log^{10}$ of weight in grams (upper curve), head and body length in millimeters (lower curve). Values for the Callitrichidae and *Saimiri* are compared to the major axis for the Cebidae. Weights and lengths from Napier and Napier (1967); Ceboid gestations are listed in Table 2. Gestation for *Macaca* from Napier and Napier (1967).

M. m = *Macaca mulatta*
M. s = *Macaca sinica*
L. l = *Lagothrix lagotricha*
A. f = *Ateles fusciceps*
C. a = *Cebus apella*
P. p = *Pithecia pithecia*
L. r = *Leontopithecus rosalia*
S. o = *Saguinus oedipus*
S. s = *Saimiri sciureus*
C. g = *Callimico goeldii*
C. j = *Callithrix jacchus*

upright on the male's lap with his legs hooked across her thighs appears to correlate with a prolonged mount and also occurs only in species where a fully prehensile tail permits the male to be anchored during intromission. This prolonged intromission is typically accompanied by considerable foreplay in *Lagothrix* and *Ateles* and it also is noteworthy that both of these genera lack a baculum. Perhaps attainment of erection in an animal dependent on vasodilation and vasoconstriction for the maintenance of turgidity necessitates a longer continuous intromission rather than a series of short intromissions attainable by those species possessing a baculum. The nature of intromission in *Lagothrix* and *Ateles* is reminiscent of some ungulates having a vascular penis, including the horse (*Equus*), and the rhinoceros (*Rhinocerus*), as well as man and, as such, represents an interesting convergence in behavioral evolution.

Gestation Trends in the Ceboidea

A regression analysis of gestation length against body size, using (a) weight of nonpregnant adult females, and (b) head-body length of females, indicates that for both callitrichids and cebids the gestation is slightly longer than cercopithecid primates in the same size class. If I assume that the callitrichids are secondarily small and use the Cebidae as a major regression axis, then it can be seen that the callitrichid and squirrel monkey gestations are slightly longer than would have been predicted on the basis of their size alone (see Figure 1). This would indicate that the young *Saimiri* are more precocial at birth, thus allowing a rapid attainment of motor skills for independent locomotion. The added energetic drain on a lactating mother carrying a large baby is ameliorated in two ways. In the case of *Saimiri*, the mother is apparently helped in transporting her young by "aunties." In the example of the Callitrichidae, the adult male carries the young.

Gestation in *Ateles* and *Lagothrix* has been determined at 225 to 230 days (Eisenberg, 1973; Williams, 1967). Interbirth intervals for females which have successfully raised their young for both wild and captive populations vary from 630-980 days. Thus, the subfamily Atelinae shows an extremely slow developmental schedule. The intrinsic rate of natural increase for the genus *Ateles* is approximately half that of the Old World genus, *Macaca* (Eisenberg, 1976). In the life table strategy for *Lagothrix* and *Ateles*, they show a strong convergence toward many of the smaller anthropoids, such as the gibbon, *Hylobates*. What life table data will look like for other species of Cebidae can at this point only be guessed at; however, a perusal of captive records as they exist would indicate that the uakari, *Cacajao rubicundus*, has a low recruitment similar to that demonstrated for the Atelinae (Fontaine, 1974).

Literature Cited

Asdell, S. A.
1964. *Patterns of Mammalian Reproduction.* Ithaca: Cornell University Press.

Dawson, G. A.
1977. Composition and stability of social groups of the tamarin, *Saguinus oedipus geoffroyi*, in Panama: Ecological and behavioral implications. In *The Biology and Conservation of the Callitrichidae*, edited by Devra G. Kleiman, pp. 23–37. Washington, D.C.: Smithsonian Institution Press.

Eisenberg, J. F.
1973. Reproduction in two species of spider monkeys, *Ateles fusciceps* and *A. geoffroyi*. *J. Mammal.*, 54:955–957.

1976 Communication mechanisms and social integration in the black spider monkey, *Ateles fusciceps robustus*, and related species. *Smithson. Contribs. Zool.* No. 213:1–108.

Eisenberg, J. F.; N. Muckenhirn; and R. Rudran
1972 The relationship between ecology and social structure in primates. *Science*, 176:863–874.

Eisenberg, J. F., and R. W. Thorington, Jr.
1973 A preliminary analysis of a neotropical mammal fauna. *Biotropica*, 5:150–161.

Epple, G
1975. The behavior of marmoset monkeys (Callithricidae) In *Primate Behavior: Developments in Field and Laboratory Research, Volume 4*, edited by L. A. Rosenblum, pp. 195–239. New York: Academic Press.

Epple, G., and R. Lorenz
1967. Vorkommen, Morphologie und Funktion der Sternaldrüse bei den Platyrrhini. *Folia Primat.*, 7:98–126.

Fontaine, R.
1974. The individual nonsocial behavior of *Cacajao rubicundus* in a nonsocial environment. Master's thesis, Bucknell University, Lewiston, Pennsylvania.

Gautier-Hion, A.
1966. L'ecologie et l'ethologie du talapoin, *Miopithecus talapoin talapoin*. *Biol. gabonica*, 2:311–329.

Glander, K. E.
1975. Habitat and resource utilization: An ecological

view of social organization in mantled howling monkeys. Ph.D. dissertation, University of Chicago, Chicago, Illinois.

Hampton, S. H., and J. K. Hampton, Jr.
1977. **Detection of reproductive cycles and pregnancy in tamarins (*Saguinus* spp.). In *The Biology and*** *Conservation of the Callitrichidae* , edited by Devra G. Kleiman, pp. 173–179. Washington, D.C.: Smithsonian Institution Press.

Hearn, J. P.
1977. The endocrinology of reproduction in the common marmoset, *Callithrix jacchus.* In *The Biology and Conservation of the Callitrichidae,* edited by Devra G. Kleiman, pp. 163–171. Washington, D.C.: Smithsonian Institution Press.

Hearn, J. P. and S. F. Lunn
1975. The reproductive biology of the marmoset monkey, *Callithrix jacchus. Breeding Simians for Developmental Biology,* edited by F. T. Perkins and P. N. O'Donoghue, pp. 191–202. Laboratory Animal Handbooks, vol. 6.

Hershkovitz, P.
1972. The recent mammals of the neotropical region: a zoogeographic and ecological review. In *Evolution, Mammals, and Southern Continents,* edited by A. Keast, F. Erk, and B. Glass, pp. 311–341. Albany: State University of New York Press.
1974. A new genus of late Oligocene monkey (Cebidae, Platyrrhini) with notes on postorbital closure and Platyrrhine evolution. *Folia Primat.*, 21:1–35.

Hick, U.
1973. Wir sind umgezogen. *Zeitschrift der Kölner Zoo,* 16(4):127–145.

Hill, W. C. O.
1957. *Primates: Comparative Anatomy and Taxonomy. Volume III. Pithecoidea, Platyrrhini.* Edinburgh: Edinburgh University Press.
1960. *Primates: Comparative Anatomy and Taxonomy. Volume IV. Cebidae, Part A.* Edinburgh: Edinburgh University Press.
1962. *Primates: Comparative Anatomy and Taxonomy. Volume V. Cebidae, Part B.* Edinburgh: Edinburgh University Press.

Hladik, C. M.
1967. Surface relative du Tractus digestif de quelques Primates. *Mammalia,* 31:120–147.

Kingdon, J.
1975. *East African Mammals, Volume II, Part B.* New York: Academic Press.

Kleiman, D. G.
1977a. Monogamy in mammals. *Quarterly Review of Biology*, 52:39–69.
1977b. Characteristics of reproduction and sociosexual interactions in pairs of lion tamarins *(Leontopithecus rosalia)* during the reproductive cycle. In *The Biology and Conservation of the Callitrichidae,* edited by Devra G. Kleiman, pp. 181–190. Washington, D.C.: Smithsonian Institution Press.

Lorenz, R.
1972. Management and reproduction of the Goeldi's monkey, *Callimico goeldii* (Thomas 1904) Callimiconidae, Primates. In *Saving the Lion Marmoset,* edited by D. Bridgwater, pp. 92–110. Oglebay Park, Wheeling, West Virginia: The Wild Animal Propagation Trust.

Medway, Lord.
1969. *The Wild Mammals of Malaya and Offshore Islands Including Singapore.* London: Oxford University Press.

Moynihan, M.
1967. Comparative aspects of communication in New World primates. In *Primate Ethology,* edited by D. Morris, pp. 236–266. Chicago: Aldine Publishing Company.
1976. *The New World Primates.* Princeton: Princeton University Press.

Napier, J. R. and Napier, P. H.
1967. *A Handbook of Living Primates.* London: Academic Press.

Neyman, P.
1977. Aspects of the ecology and social organization of free-ranging cotton-top tamarins (*Saguinus oedipus*) and the conservation status of the species. In *The Biology and Conservation of the Callitrichidae,* edited by Devra G. Kleiman, pp. 39–71. Washington, D.C.: Smithsonian Institution Press.

Patterson, B. and R. Pasqual
1972. The fossil mammal fauna of South America. In *Evolution, Mammals, and Southern Continents.* edited by A. Keast, F. Erk, and B. Glass, pp. 247–310. Albany: State University of New York Press.

Rosenblum, L. A. and R. W. Cooper, editors
1968. *The Squirrel Monkey.* New York and London: Academic Press.

Rosevear, D. R.
1969. *The Rodents of West Africa.* London: British Museum (Natural History).

Travis, J. C. and W. N. Holmes
1974. Some physiological and behavioural changes associated with oestrus and pregnancy in the squirrel monkey (*Saimiri sciureus*). *J. Zool., London,* 174:41–66.

Williams, L.
1967. Breeding Humboldt's woolly monkey at Murrayton Woolly Monkey Sanctuary. *Internat. Zoo Yearb.*, 7:86–89.
1968. *Man and Monkey.* Philadelphia and New York: J. B. Lippincott Company.

GARY A. DAWSON
The Museum and
Department of Fisheries and Wildlife
Michigan State University

Composition and Stability of Social Groups of the Tamarin, *Saguinus oedipus geoffroyi*, in Panama: Ecological and Behavioral Implications

ABSTRACT

Data were collected on group size, composition, and stability of *Saguinus oedipus geoffroyi* inhabiting second-growth, tropical dry forest in the Panama Canal Zone. Group size, determined for 71 groups, did not vary seasonally. Sex-age class composition for five groups was determined by live-trapping and marking. Subsequent retrapping and observations of these groups revealed high turnover rates of group constituents, while group size and sex-age class proportions within groups remained relatively stable. Median stability did not vary significantly among sex-age classes, although emigration and disappearance rates were somewhat higher for juveniles than adults. Differences in stability and patterns of home-range usage between lowland and upland groups were correlated with availability of moisture and degree of deciduousness of their home ranges during the dry season. Data documenting the lack of sexual dimorphism in this species are presented.

The absence of pronounced sexual dimorphism, duality of male-female roles, evidence for a one male, one female dominance system, high incidence of intragroup aggression, intraspecific communication system, foraging strategy, and distribution of resources are considered in defining the species' social structure and its adaptive significance in a seasonal environment.

Introduction

Despite the difficulties in studying arboreal primates (Aldrich-Blake, 1970), many field investigations of New World monkeys (all of which are arboreal) have been conducted in recent years. The howler monkey, *Alouatta,* which was the subject of the first comprehensive field investigation of a New World primate (Carpenter, 1934), has been the object of most studies. Some field investigations have been undertaken also on the species of *Ateles, Saimiri,* and *Cebus.* While the quantity and quality of the data regarding the ecology and behavior of these species cannot compare with that for terrestrial Catarrhines, it is substantial. These primates form large social groups (Mason, 1966) and are characterized by uni-male or age-graded, multi-male social organizations (Eisenberg et al., 1972). In contrast, those Platyrrhines which form small social units, e.g., *Aotus, Callicebus,* and all of the Callitrichidae, have not been well studied. Information regarding their social organization is fragmentary and often anecdotal.

Mason's (1966, 1968) studies of *Callicebus moloch* may be considered the pioneer work on the social organization of a New World species occurring in small groups. Moynihan (1964, 1970) discussed the behavior patterns of *Aotus trivirgatus* and *Saguinus oedipus geoffroyi,* but his primary emphasis lay in laboratory study rather than in field observation. While contributions such as those made by Coimbra-Filho and Mittermeier (1973) and Thorington (1968) add to our knowledge of callitrichid ecology, they result from casual observations rather than from comprehensive field study.

The dearth of field data regarding the social organization and ecology of the Callitrichidae, and of the genus *Saguinus,* in particular, was a strongly-motivating factor in my choice of a subject for field study.

Range and Description

The Panamanian tamarin, *Saguinus oedipus geoffroyi* (see Hershkovitz, 1966) is the only member of the family Callitrichidae endemic to Central America. The ancestors of the present form apparently immigrated from South America during the late Pliocene (Hershkovitz, 1969). The modern range extends from the Colombian province of Chocó north to at least central Panamá (Hershkovitz, 1949; Handley, 1966; Moynihan, 1970). Moynihan (1970) indicates that *S. o. geoffroyi* is characteristic of areas of moderate humidity, and is almost absent from the humid Atlantic coast; however, the trapping records of Telford et al. (1972) indicate that it is extremely abundant in a strip of disturbed vegetation about 5 km wide along the Atlantic coast in the province of San Blas. Furthermore, *S. o. geoffroyi* is reported to be abundant on edges of riverain forest in the humid province of Darién (P. Galindo, pers. comm.). It appears then, that a lack of habitat in the appropriate seral stages, rather than a climatologically determined habitat type, restricts the range of this tamarin in the humid lowlands of Panamá.

Adult *S. o. geoffroyi* weigh about 500 g; adult body length is approximately 235 mm. A detailed description of the pelage may be found in Hershkovitz (1966) and Moynihan (1970). The pelage of the female does not differ from that of the male. Indeed, the only readily apparent sexually dimorphic characters, aside from the external sexual organs, are the suprapubic and circumgenital glands, which are larger and more highly developed in the female (Wislocki, 1930).

Study Site

This investigation took place on the Pacific slope of Panamá in the Canal Zone. The climate there is markedly seasonal, with a mid-December to mid-April dry season. Rainfall during the dry season is infrequent, patchy in distribution, and usually well below 7.6 mm per month. Monthly rainfall during the wet season averages between 10.3 mm and 27.8 mm (records, Panama Canal Company).

The extant forest has been designated as Dry Tropical Forest (Holdridge and Budowski, 1956; Holdridge, 1967). Physiognomically, this forest is composed of second-growth trees and shrubs, 2 m to 18 m in height, with associated vines and lianas. The shrub layer is well developed; various species of tropical grasses grow in numerous openings. Together, the shrubs and grasses often form dense tangles that are impenetrable without the aid of a machete. Taller trees (to about 23 m), primarily *Anacardium excelsum,* form a gallery forest along both permanent and intermittent streams. A more complete description of the area may be found in Fleming (1971).

The focal point of the study was the former Rodman Ammunition Depot, a 1015 ha reserve located west of the Panama Canal (8°57′N, 79°37′ W, elev. 18 m to 168 m). The Rodman Depot (hereinafter referred to as "Rodman") is under lease to Gorgas Memorial Laboratory for biomedical purposes. The area proved to be ideal since it: (1) supported a tamarin population of about 20 to 30 animals per km^2 (Southwick, 1975), (2) was relatively inaccessible to the public, (3) contained vegetation that had not been seriously disturbed for over two decades, and (4) contained a dendritic network of hard-surfaced roads which was accessible at all seasons.

Materials and Methods

The data were obtained from a number of sources: (1) observations of five social groups with individually marked animals, (2) observations made while radio-tracking two social groups, (3) examination of 131 tamarins segregated by sex-age class, (4) measurements of museum specimens, and (5) general observations made while engaged in field study.

Tomahawk Live Traps (Tomahawk Live Trap Co., Box 323, Tomahawk, Wisc.) measuring 41 cm × 14 cm × 14 cm and baited with banana and mango were used in trapping live tamarins. Traps were wired to trees and vines 3 m to 10 m above the ground in areas frequented by tamarins. Traps were washed with stream water when necessary to remove accumulated bait and the odor of trapped animals.

Individuals were marked with leather collars wrapped with from one to three strips of brilliantly colored plastic tape according to a predetermined group color code. Longevity of the collars and readability of the tape proved excellent; only 3 of 68 collars required replacement during the study period, and all collar codes were readable throughout the study. Small mammal ear tags (Salt Lake Stamp Co., Salt Lake City, Utah) were attached to the left ear as an additional means of identification.

Live trapping of study group constituents began in Rodman in January 1973. Groups were selected whose estimated home ranges lay at fairly discrete intervals along a moisture-vegetation continuum which extended from the moist bottomlands with evergreen gallery forests to seasonally xeric uplands which support scrubby, highly deciduous forests. This sampling scheme was based on an hypothesis suggested by Janzen and Schoener (1968), which held that insectivorous vertebrates inhabiting seasonally xeric tropical uplands must employ different patterns of spatial utilization than their lowland counterparts in order to exploit the more seasonally variable insect resources characteristic of upland areas.

Initially, I wished to test this hypothesis by utilizing radio-location telemetry to compare the movements and activity patterns of a social group inhabiting the more equitable lowlands with those of a troop from the seasonally inhospitable uplands. Toward this end, I trapped and radio-marked an animal from a group which occupied a lowland, river-bottom site, and an animal from a group which occupied a seasonally xeric hillside.

The scope of the study broadened when Nelson (1975) discovered that an upland group which he was observing was unstable in its constituents, although group size remained constant. I then began monitoring the composition of the two radio-tagged groups in order to ascertain whether site differences might also influence site behavior. Three additional groups, Green and Red, which occupied habitats of intermediate deciduousness, and Yellow, which occupied a home range that was even drier than that of the upland group, were subsequently trapped and marked.

The five groups were retrapped and observed at irregular intervals from February to December 1973. From January through May 1974 they were observed at monthly intervals. Observations were made from blinds while the groups fed at prebaited platforms 3m to 5 m above the ground. Identification of individual animals was aided by the use of 7 × 14 binoculars. Each group was observed for two consecutive days during each of the observation periods in 1974. No changes in group composition were noted on any of the second day counts.

Movements of the lowland and upland groups were monitored through the use of radio-location telemetry equipment (AVM Instrument Co., 808 W. Springfield Ave., Champaign, Ill.) as described by Montgomery et al. (1973). The groups were followed for several days to determine their approximate home ranges; machetes were then used to construct trail systems within the home ranges. Where possible, trails were arranged in a grid pattern with each square measuring about 300 feet (91.5 m) on a side, although fallen trees and steep slopes made some deviations necessary. This system allowed precise (within 20 m) determinations of group locations. The lowland group's movements were monitored for about 500 hours following the completion of the trail system. This included 26 complete wet-season days and three complete dry-season days. The upland group proved more difficult to track because it often traveled rapidly, and inhabited a large, topographically rugged home range. The approximately 100 hours logged while tracking the upland group include location data from 14 wet-season days and 17 dry-season days.

One hundred thirty-one tamarins from areas peripheral to Rodman were examined at the rate of five animals every two weeks over the course of one year. Individuals were sampled without known bias from groups which had been under observation for 5 to 15 minutes in order to ascertain group size. The following data were collected: (1) Weight (to the nearest 0.1 g) and standard mammalogical measurements (in mm): (2) Sex and reproductive condition; (3) Degree of pigmentation of the suprapubic gland for purposes of aging.

Tamarins were divided into four sex-age classes: mature males, mature females, immature males, and immature females. The criterion used for separation of immature from mature animals was the degree of pigmentation of the suprapubic and circumgenital glands. These apocrine and holocrine structures are unpigmented or very slightly pigmented in prepubertal tamarins (Epple, 1967b). Tamarins reach puberty at approximately 18 months (Hampton et al., 1966); my observations on known-age animals indicate that the suprapubic gland becomes pigmented at 16 to 18 months. The change in pigmentation is striking, and appears to be a reliable method of determining age. Body weight can also be used as an aging criterion, but it is less proximally related to the physiological changes which occur at puberty than is the suprapubic gland. Only 3 of 38 juveniles examined exceeded 425 g; no mature tamarins weighed less than this.

Skeletal measurements presented herein were taken from preserved specimens housed in the Michigan State University Museum. Needle-pointed calipers were used to obtain measurements to .01 mm. Measurements of canine teeth follow Kinzey (1972).

All "F" values and their associated probability statements are the result of single classification analyses of variance with two-sided hypotheses (Sokal and Rohlf, 1969). Single classification Chi-square tests and Chi-square goodness of fit test also follow Sokal and Rohlf (1969). Differences in median stability among social groups and among sex-age classes were analyzed by formulae from Hays (1963). The use of the Bonferroni normal statistic in identifying significant components of the Chi-square analysis follows Neu et al. (1974).

Results

Group Size and Composition

TRAPPING SUCCESS: Trap success averaged 39.04 trap days/tamarin (68 captures, 70 recaptures over 5,388 trap days). Of those tagged animals, only 50 figured prominently in the study. Six infants in the five study groups died or disappeared before I could capture them. Juveniles were more susceptible to capture and recapture than adults ($\chi^2 = 10.356, p < .005$), with expected values for their capture based on the proportion of juveniles in the population.

GROUP SIZE: Mean group size was determined for 71 groups. Rounded mean size was used for 24 groups seen frequently in Rodman; the remainder of the data results from single observations. Mean size was found to be 6.39 ± .31 when carried infants were excluded from the count, and 6.93 ± .36 when the infants were included (Figure 1).

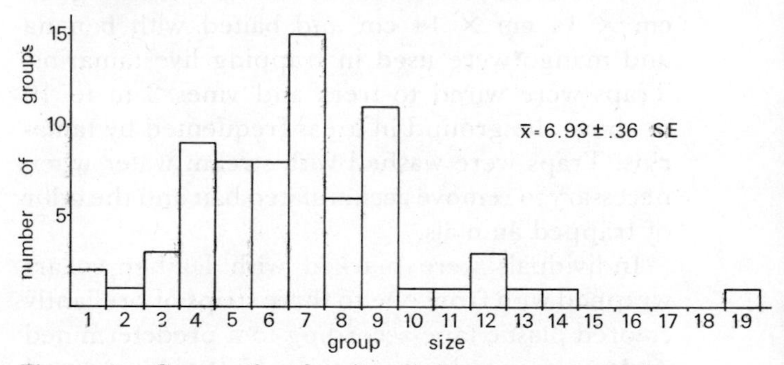

Figure 1. Group size for 71 distinct groups of the Panamanian tamarin (*S. oedipus*) encountered on the Pacific slope of the Panama Canal Zone from January 1973 to May 1974.

Individual tamarins and aggregates smaller than the mean were often observed during periods of casual observation; however, observations of such small groups over a longer time period (5 to 30 minutes) or when "mobbing" the investigator, revealed that these individuals and aggregates were almost invariably constituents of larger groups. No "floating" population of extra-group individuals was evident.

The above data must be further qualified by the observation that the two largest groups appeared to be casual groupings of tamarins at concentrated food sources rather than permanent social groups. Both "groups" were observed only once, although I often passed through that area.

Group size was compared over three trimesters: (1) November to February, the peak breeding season; (2) March to June, the peak season of parturition, and (3) July to October. Carried infants were excluded from the analysis since they are incapable of participating actively in intergroup movement, and, moreover, their numbers could be expected to increase group size in the third trimester. Analysis of variance indicated that no differences in group sizes existed over trimesters ($F = .249$, $p > .75$, d.f. = 2,68).

POPULATION SEX RATIO: The sex ratio for *S. o. geoffroyi* in this study did not differ significantly from 1:1. This held true for both adult and juvenile age classes ($\chi^2 = .582$, $p > .1$, 1 d.f. for adults; $\chi^2 = 2.400$, $p > .1$, 1 d.f. for juveniles). No sex-specific mortality factor appears to be affecting either age class.

COMPOSITION OF MONITORED GROUPS: A summary of the sex-age class compositions for the five monitored groups is given in Table 1. A definitive statement on either group or population sex-age class structure cannot be made on the basis of the composition of five groups; however, a more substantive statement can be made regarding the sex-age class structure of a population with the addition of sex-age class data from the extra-Rodman sample.

A Chi-square goodness of fit test was used to compare the sex-age class distribution of the five groups of known composition with the sex-age class distribution from the extra-Rodman sample. The two distributions did not differ significantly; in fact, they proved practically identical ($\chi^2 = 0.359$, $p > .9$, 3 d.f.). This supports the contention that the sex-age class structures found in both samples are representative of the population's sex-age class structure.

Table 1. Sex-age class composition of five social groups of the Panamanian tamarin, *S. o. geoffroyi*.[1]

Group (n)[2]	*Sex-age Classes* *Mature Males*		*Mature Females*		*Immature Males*		*Immature Females*		*Total*	
	$\bar{x}$	*SE*	$\bar{x}$	*SE*	$\bar{x}$	*SE*	$\bar{x}$	*SE*	$\bar{x}$	*SE*
Lowland (14)	2.0	0.1	2.6	0.1	1.3	0.3	0.3	0.1	6.1	0.3
Green (7)	2.1	0.5	2.3	0.2	1.1	0.5	1.4	0.3	7.0	0.4
Red (7)	1.9	0.1	2.3	0.8	1.0	0.0	0.0	0.0	5.1	0.3
Upland (9)	4.0	0.4	0.8	0.1	0.7	0.4	0.9	0.3	6.3	0.5
Yellow (8)	2.0	0.4	2.4	0.5	1.3	0.3	0.8	0.3	6.4	0.5
Total[3]	2.4	0.4	2.1	0.3	1.1	0.1	0.7	0.2	6.2	0.3

[1]Carried young excluded
[2]n = number of censuses
[3]Average of means as estimate of group composition

GROUP STABILITY: While the five monitored groups maintained relatively stable group sizes and sex-age class proportions (Table 1), their particular complements of individuals proved unstable. Forty-one immigrations and emigrations (considering emigration and subsequent immigration by the same individual as separate events), 14 disappearances, and 1 death were recorded while observing the 5 groups over a period of from 10 to 15 months per group. Since systematic, i.e. monthly counts, were made only during the last five months, additional changes in group composition may have gone unnoticed. Twenty immigrations and emigrations, and 6 disappearances were recorded during 19 observation periods during the wet season. Twenty-one immigrations and emigrations, 8 disappearances, and 1 death were recorded over 29 dry-season counts, 25 of which were made during the 1974 dry season. Rates of mobility between wet and dry seasons are not comparable due to the irregular intervals among wet-season censuses. The data demonstrate, however, that changes in group composition occurred frequently during both seasons.

Comparison of Group Stability

Two measures were used to compare stability among groups: (1) the percentage of individuals associated with a group above the mean for that group, and (2) comparisons of median stability, where group stability is based on the stability values of its components, and stability itself is defined as the frequency of occurrence of individuals over pertinent sampling periods. The first measure is crude in that it is sensitive only to gross changes in group association; the second is more finite since it also measures the tenure of individual constituents.

In Table 2, groups are organized along a gradient according to the relative availability of moisture and degree of deciduousness of the vegetation in their home ranges. It is apparent that the number of individuals associated with groups above and beyond the mean varied directly with the degree of deciduousness in their home range and inversely with the availability of moisture.

Median group stability was tested over the 1974 dry season, when each group was censused monthly for five months, and over the 1973 wet season. Median stability values for the 1974 dry season are listed in Table 2. No values of median stability are presented for the 1973 dry season since uneven, inadequate sampling precluded comparisons among all groups. For the eight of ten wet season comparisons which were possible, a separate index of median stability, based on common census periods, was developed for each paired comparison in order to correct for unbalanced sampling.

An a priori test to determine whether differences in median stability occurred among the five groups during the 1974 dry season indicated that significant differences did occur ($\chi^2 = 8.444$, $.1 > p > .05$, 4 d.f.). A posteriori paired comparisons indicated that two groups, the lowland and the upland, exhibited significantly higher dry-season stabilities than Yellow group ($\chi^2 = 4.549$, $p < .05$, 1 d.f., and $\chi^2 = 5.519$, $p < .025$, 1 d.f., respectively). All other comparisons were nonsignificant, with $p > .1$ or more for all comparisons.

Paired comparisons of group median stability during the wet season indicated that the lowland group was significantly more stable than Yellow group ($\chi^2 = 3.976$, $p < .05$, 1 d.f.). Median stabilities among other groups did not differ, with $p > .1$ or more for all comparisons.

Stability Among Sex-age Classes

The median test was also used to test for differences in stability among sex-age classes, both for the entire study period and the 1974 dry season. No significant differences in stability among sex-age classes were noted for either period ($\chi^2 = 4.402$, $p > .1$, 3 d.f., and $\chi^2 = .483$, $p > .975$, 3 d.f., respectively).

In a separate test, emigrations and disappearances for all sex-age classes over all counts were compared with the expected number of emigrations and disappearances per class based on the proportion of each sex-age class in the overall population (Table 3). The results ($\chi^2 = 9.245$, $p < .05$, 3 d.f.) indicate that the expected numbers of emigrations and disappearances per class differed significantly from the numbers observed. It can be seen from Table 3 that mature tamarins of both sexes disappeared and emigrated somewhat less than expected, while the immature tamarins emigrated and disappeared more frequently than their numbers in the population

Table 2. Comparison of mean group size and associated individuals over a moisture-vegetation gradient.

	Estimated Percent Deciduousness, Dry Season 1974	*Group*	*$\bar{x} \pm SE$*	*Associated Individuals*[1]	*% Associated Above Mean*	*Median Stability Dry Season 1974*
River Valley						
Moisture gradient	10	Lowland[4]	6.143 ± .254	10	62.79	100[2]
	20	Green	7.000 ± .436	12	71.43	80
	20	Red	5.143 ± .261	9	75.00	80
	30	Upland	6.334 ± .500	13	105.24	100[3]
Hillside	40	Yellow	6.375 ± .518	17	166.67	40

[1]Includes recruitment by birth
[2]Median dry-season stability differs from that of Yellow at $p < .05$
[3]Median dry-season stability differs from that of Yellow at $p < .025$
[4]Median wet-season stability differs from that of Yellow at $p < .05$

would predict. A comparison of the observed proportion of emigrations and disappearances per age class with the expected proportions was made using the Bonferroni Z statistic in order to determine whether or not the significant Chi-square value could be attributed to extreme differences between observed and expected values in one or more of the sex-age classes (See Table 3). The results indicate that while the differences between observed and expected proportions approached significance ($.2 > p > .1$) for each comparison, the significant Chi-square value was not attributable to significant differences in any sex-age class.

Table 3. Emigrations and disappearances among sex-age classes: expected values based on sex-age class proportions in the population.

Sex-age Class	*Proportion in Population (Expected Proportion)*	*Observed Emigration and Disappearance*	*Expected Emigration and Disappearance*	*Chi-Square Values*	*Proportions Observed*	*90% Family Confidence Coefficient on Observed Proportion*
Mature males	.405	8	12.947	1.890	.250	$.079 \leq p \leq .421$
Mature females	.351	9	11.325	.445	.281	$.103 \leq p \leq .459$
Immature males	.137	9	4.397	4.996	.281	$.103 \leq p \leq .459$
Immature females	.107	6	3.421	1.944	.188	$.038 \leq p \leq .338$
				$x^2 = 9.245$		

Immature tamarins, then, tend to emigrate and disappear more frequently than adults. This observation seems to be at odds with the stability comparison, which revealed no differences in stability among sex-age classes. This can be explained, however, by noting that the prolonged dependence of the young on the group early in life, which results in high stability scores, compensates for the higher rates of emigration and disappearance which occur among older immature tamarins.

Group losses by emigration or disappearance were recouped by immigration and natality. The sex-age class proportions of 18 immigrants entering the monitored groups for the first time did not differ significantly from expected proportions based on sex-age class proportions in the population ($\chi^2 = .992$, $p > .5$, 3 d.f.). A second source of recruitment, immigrating animals which had formerly emigrated, was found to be substantial (13 animals). Emigrants were found to reunite with their former groups in proportion to the numbers of each sex-age class which were known to have emigrated ($\chi^2 = 1.092$, $p > .5$, 3 d.f.).

While emigration to adjacent groups was a common phenomenon (11 observations), long-distance emigrations were also observed. One male about 18 months of age traveled 6.5 km from the home range of his original group. Several tamarins, including one female of about 12 months of age, were known to have traveled at least 2 km from their group of origin.

Home Range Data

An attempt was made to quantify both home range size and utilization patterns for two groups, the lowland group, on the humid extreme of the moisture gradient, and the upland group, on the dry side. The lowland group occupied a home range of about 26 ha. This home range was defended as a territory in the strictest sense—a defended area with a distinct geographical boundary (Bates, 1970). The approximately 13 percent of this range overlapped by three other groups was the site of frequent intergroup contact and territorial marking. I was unable to quantify the home range of the upland group due to its large size, the short range of my telemetry equipment in that densely foliated, eroded area, and the lack of time for cutting the needed trails. During the wet season, the upland group was known to occupy an area of 32 ha; its probable home range was much larger. The 32 ha portion of the home range was not defended as a geographical entity—at least 83 percent of it was overlapped by at least five other groups. The groups using this area apparently did so through separation from other groups in time and/or space. Although frequent aggressive encounters and agonistic displays between groups occurred, both offense and defense seemed directed toward defending the immediate area around the group rather than toward defending a geographic territory per se.

Importance of Group Components

Perhaps the most important observation to be made here is that practically no sexual dimorphism exists in this species. In Table 4, parameters of nine characters which are commonly sexually dimorphic in primates are listed and the intersexual differences tested. The only parameter which differed significantly between sexes was body weight, where females average 4.3 percent heavier than males. Mature males and females are roughly equal in size, weight, and dental armament, and are thus equally capable, in a physical sense, of overt participation in aggressive encounters, both inter and intragroup, and in predator defense.

On the basis of morphological similarities, one might expect a more equal apportionment of roles between sexes than is found in most primate societies. My field observations bear this out. Both males and females participated actively in the care of the infant, a fact which is already well substantiated in the literature (e.g. Hampton, 1964; Shadle et al., 1965; Epple 1967a, 1970a; Weber 1972). Both sexes were observed to actively participate in territorial defense. In the majority of instances, this consisted of uttering "long whistles" (Moynihan, 1970) at the other group, with perhaps a few rushes toward the opposing group being made by individuals of either group. However, six intergroup encounters were observed where opponents made physical contact; of these, three took place near artificial feeding sites and three took place in natural situations. Males tended to be more aggressive in the six altercations involving contact. Both sexes were seen to mark vegetation with their suprapubic and circumgenital glands following both the aggressive encounters and those involving only agonistic display. Females appeared to mark more frequently than males, and to initiate marking behavior following an encounter.* Both sexes ex-

Table 4. Comparison of body weight, body length, and selected cranial and dental measurements of adult male and female *S. o. geoffroyi.*

Characteristic	$\bar{x}$ *SE*	*n*	*F*	*P*
Body length (mm) males	232.43 ± 1.43	53	2.249	$p > .20$
Body length (mm) females	241.52 ± 3.47	48		
Body weight[1] males	486.42 ± 5.89	53	4.628	$p < .10$
Body weight[2] females	507.46 ± 8.11	41		
Skull length (mm)[3] males	51.16 ± 1.73	48	.913	$p > .50$
Skull length (mm) females	50.92 ± 1.08	44		
Zygomatic breadth (mm) males	35.38 ± 1.30	48	.031	$p > .50$
Zygomatic breadth (mm) females	35.34 ± 1.06	41		
Effective ($\bar{x}$) maxillary canine length males	6.12 ± .17	52	1.691	$p > .20$
Effective ($\bar{x}$) maxillary canine length females	6.39 ± .11	46		
Longer maxillary canine length[4] males	6.84 ± .09	48	.536	$p > .50$
Longer maxillary canine length females	6.93 ± .07	45		
Labiolingual breadth, maxillary canine males	2.38 ± .02	48	3.515	$p > .10$
Labiolingual breadth, maxillary canine females	2.33 ± .02	44		
Left mandibular canine length males	6.59 ± .09	48	1.523	$p > .20$
Left mandibular canine length females	6.49 ± .08	44		
Left mandibular canine breadth males	2.67 ± .02	48	2.741	$p > .20$
Left mandibular canine breadth females	2.61 ± .02	45		

[1]Individual weights to nearest 0.1 g

[2]Pregnant females excluded

[3]This and following measurements using needle-pointed calipers to .01 mm

[4]Canine measurements follow Kinzey, 1972

* *Editor's note:* Epple (this volume, page231) also reports higher scent-marking scores for female *S. fuscicollis* during and after encounters with strange conspecifics, regardless of whether the aliens are males or females.

hibit behavior indicative of "control" animals in the sense of Bernstein (1964, 1966).

The lack of sexual dimorphism and the duality of roles noted above might lead one to suspect the existence of a social structure with a one male, one female dominance system such as the one described by Epple (1967a, 1967b, 1972) for laboratory groups of *Callithrix jacchus.* Observations at feeding platforms during the dry season indicated that certain males and females occasionally dominated members of their own sex in competition for food. I also observed that a single female per group had priority to reproduction—i.e. only one infant or set of twins was found in any one group at a given time, although the group might contain two or more sexually mature females. Epple (1970a), who apparently observed the same phenomenon in laboratory groups of *S. o. geoffroyi,* indicated that only alpha females bore young.

A further test of Epple's hypothesis that one and only one alpha (reproductive) female occurs per group was made by comparing the expected number of reproductive females, based on one female per average group of 6.93 animals, with the observed number found in the collected sample. A Chi-square test ($\chi^2 = 1.820$, $p > .1$, 1 d.f.) indicated that the hypothesized number of reproductive females did not differ significantly from the number observed.

Reproductive females appear to be older than nonreproductive females as evidenced by a higher frequency of scarring, especially on the ears and the face, and the high incidence of damaged canine teeth ($\chi^2 = 5.604$, $p < .025$, 1 d.f. and $\chi^2 = 4.433$, $p < .05$, 1 d.f., respectively). (Hershkovitz, 1970, observed that dental caries and damaged teeth occurred only in the older marmoset age classes.) The incidence of a highly developed suprapubic gland was higher in reproductive females than in nonreproductive females (21/22 vs. 6/26, $\chi^2 = 11.112$, $p < .001$, 1 d.f.).

Males did not differ from females in regard to the number of torn ears and facial scars ($\chi^2 = .512$, $p > .1$, 1 d.f.) or broken canines ($\chi^2 = .570$, $p > .1$, 1 d.f.). When combined with observational data, these findings offer further support for the coexistence of a single dominant male per group and also the lack of sex-specific mortality.

Discussion

Carpenter (1952, 1954) initiated the synthesis of primate studies with the aim of presenting an integrated classification system for primate social structures. More recent reviews (e.g. Crook and Gartlan, 1966; Bernstein, 1970) attempted to order the various primate taxa according to somewhat arbitrary evolutionary grades based on social "complexity." Eisenberg et al. (1972) reassessed this system and developed a more workable system which relates social structure to the ecology of a given species. The following discussion has as its goal the redefinition of the position of *S. o. geoffroyi* on this continuum. Group size and structure of the Panamanian tamarins will be discussed in relation to habitat, diet, and its system of communication. The transitory nature of some group constituents, and its implications, will also be discussed. A new category of social organization for primates, the age-dependent, one male, one female dominance system will be presented. It is suggested that other *Saguinus* spp., and other callitrichids, occupy a similar position, although further studies are needed to confirm this hypothesis.

Epple (1972), in summarizing the accounts from older literature, reported that callitrichid group size varied from 2 to 12 animals. Thorington (1968) observed groups of *Saguinus midas* which varied from 2 to 6 individuals, and reported that researchers in Surinam had observed groups of *S. midas* numbering up to 20 animals. Hladik and Hladik (1969) and Moynihan (1970), working in Panamá with *S. o. geoffroyi,* observed groups numbering 2 to 9 tamarins, with single individuals also in evidence. The last author considered an account of group sizes of up to 12 animals (Chapman, 1929) to be grossly inaccurate.

My own observations of 71 groups yielded a mean of 6.93 ±.36, with a range from 1 to 19. The few single animals observed were almost always seen to join a group if watched for a half hour or more. The two largest groups, one of 14 and one of 19, were observed only once at localized food sources, and thus may represent casual groupings of tamarins rather than permanent social groups. Most of the groups observed during the study ranged from 3 to 9 animals.

The mean size of specific social groups is determined by a multitude of factors: e.g. phylogenetic history, population density, diet, foraging system, spatial and temporal distribution of resources in the environment, relative distribution and magnitude of population depressants such as predation and parasitism, and the recent history of epizootics, (Crook and Gartlan, 1966, Bernstein, 1970, Eisenberg et al., 1972). It would be pointless to pursue a discussion of *Saguinus* group size as determined by these factors, since so many of them remain unqualified; however, some general comments can be made.

For the most part, callitrichids inhabit "edge" vegetation in the broadest sense, which can generally be considered dense vegetation offering low levels of visibility (Thorington, 1968; Moynihan, 1970; Coimbra-Filho and Mittermeier, 1973; Dawson, 1976). Under these conditions, it is not surprising that the visual system of communication is rather primitive and rudimentary (Moynihan, 1970), and that the auditory and olfactory systems are highly developed (Epple, 1967a, 1967b, 1968, 1970a, 1972; Moynihan, 1970). High-frequency vocalizations, which form the bulk of the acoustical repertoire (Epple, 1968; Moynihan, 1970), have the probable advantage of being absorbed rapidly by the foliage, thus enabling the passage of specific information to group conspecifics at short range, without attracting predators, of which there are many, from longer range (Marler, 1965; Moynihan, 1967, 1970).

Tamarins, particularly *Saguinus*, are highly insectivorous and frugivorous (Hladik and Hladik, 1969; Dawson, 1976). The fruits most often used are small, and their seeds are generally dispersed by birds and small mammals (Dawson, 1976). These fruits are more uniformly distributed over time and space than are the large, fleshy fruits (Smythe, 1970), which are the mainstay of larger primate frugivores. The tamarin food source is more or less uniformly distributed rather than clumped.

Observations from this study indicated that little, if any, cooperation occurs in the capture of individual food items, although at times an insect flushed by one tamarin might be captured and eaten by another. Of more significance is the observation that individual group members acted in a sentinel capacity, warning those engaged in feeding of the approach of an aerial or terrestrial predator. Increased probabilities of individual or kin survival through group cooperation in predator detection, and through the mobbing of predators by the group are probably the primary advantages accrued to individuals in the *Saguinus* social system. The number of "alarm" vocalizations in the acoustical repertoire of these small, vulnerable primates (Epple, 1968; Moynihan, 1970) and the high frequency with which they are uttered at a variety of stimuli on any given day seem to support this view. Mobbing of natural predators was not observed. Tamarins, however, often rushed toward the investigator while he was removing live-trapped individuals from the trees in which they were captured. These rushes, which were both vigorous and accompanied by intense vocalizations, sometimes ended only a few feet from the observer. It thus seems probable that tamarins would mob a smaller predator. Other possible advantages of group behavior, such as thermoregulatory benefits which may be gained in nocturnal huddling behavior (Dawson, 1976), should not be ignored.

When taken together, the limited range of intragroup communication, the relatively uniform distribution of food resources within a given season, and the space required by foragers in a non-cooperative, individually oriented foraging system may also militate against the formation of large social groups. The high levels of aggression exhibited by group dominants (Epple, 1967a, 1970b, 1972) and territorial behavior noted by Thorington (1968) and Moynihan (1970), are probably instrumental in maintaining both individual and group foraging space required in the exploitation of dispersed, low-density food resources.

Carpenter (1952, 1954) observed that troop size tended to be standard for a given species, and that sex-age ratios also appeared to be stable. Eisenberg et al. (1972) indicated that habitat differences cause widespread variation in intraspecific social structure within some taxa and not others. One may infer from their discussion that, within a given habitat at a given point in time, the factors governing modal group size are fairly constant, although the habitat type itself may be quite "patchy" (as defined by MacArthur and Levins, 1964). This appeared to be true in this study where, although home range sites varied greatly in regard to the availability of food and moisture during the dry season, group size and group sex-age ratios remained relatively constant.

The mechanism necessary for the maintenance of stable group size and sex-age class proportions is obvious—immigration and natality must compensate for losses in proportion to the loss in each sex-age class. Natal recruitment alone is insufficient to compensate for these losses. The data demonstrate that these vacancies are indeed filled, and that immigration is the primary mechanism.

The causal mechanism underlying the high rates of change in social group constituents is unknown. One might postulate, on the basis of laboratory observations (e.g. Epple, 1967a, 1970b, 1972) that aggression on the part of the dominant pair toward subordinates is responsible for group instability. On the other hand, the fact that subordinate individuals move as frequently as they do suggests that a degree of tolerance exists between the dominant pair and subordinates, and that emigration from the group may be precipitated by the aggression of other subordinates. The situation is further obfuscated by the lack of knowledge regarding the degree of relatedness among social groups—viz., the acceptance of immigrating animals by a group may be facilitated if one of the dominant pair is related to the immigrant and has been associated with it in the past. The elucidation of the causal mechanism and the identification and quantification of those factors which influence this mechanism merit further research.

The source of immigrating animals appears to be other social groups rather than a population of extra-group individuals. Evidence for this is partly negative—single animals and small groups were almost always associated with larger groups and rarely remained isolated for extended periods. Moreover, baited feeding platforms were never visited by single animals. In contrast, ten immigrants, roughly one half of the total, were marked animals from other groups. It might be surmised that immigrants come from two sources: (1) groups which maintain their integrity while exhibiting a turnover of constituents, and (2) groups which for some reason prove to be nonviable and thus dissipate. The existence of the latter source is necessary if one is to explain the maintenance of group size with no existing extra-group population as a source of replacements.

Four of the five groups each contained a reproductive female which remained with its respective group for the course of the study. Epple (1970a), working with *S. o. geoffroyi,* observed that only alpha females bore young, although "inferior," but sexually mature females copulated with the fertile male (alpha male). It appears likely, then, that the reproductive females in these free-ranging groups are alpha animals. In addition, at least one adult male remained a permanent member of each group throughout the study. In three of five groups, the resident males were observed to supplant other males at feeding platforms and in some natural feeding situations. They were also observed to "lead" in chasing an intruding group from the feeding platform or territory. Evidence exists, then, for a stable group core of one reproductive or alpha female and one resident, probably dominant male. The other stable element of the group is the dependent young, usually the survivor of twins, which does not leave the group for extended periods until it is roughly a year or more of age. In light of their temporary existence, the remainder of the group constituents may be considered transients.

The above observations, when combined with those in the prior section regarding the relative roles of male and female tamarins, suggest that the typical *S. o. geoffroyi* group consists of an alpha male, an alpha female, sedentary young of the year, and a transient complement of presumably subordinate animals of both sexes, which appear to be younger than the alpha female, and probably the alpha male as well. Using terminology parallel to that used by Eisenberg et al. (1972), this configuration may be considered an age-related, male-female troop. The existence of the dual dominance system is plausible given the lack of sexual dimorphism and the duality of roles discussed previously. The statistically equivalent rates of intergroup movement by males and females add credibility to the concept—viz., unisexual dominance systems should generate higher rates of movement, and perhaps extra-group individuals, of the dominated sex.

As noted previously, the existence of the dual dominance system in free-ranging callitrichids was predicted by Epple (1967a, 1967b, 1970a, 1970b, 1972) on the basis of laboratory observations in which *Callithrix jacchus* was the primary model, although supportive evidence was garnered from less intensive observations of other callitrichids, including *S. o. geoffroyi* (which she referred to as *Oedipomidas spixi* after Hill, 1957). She thought of this system as existing within a family group consisting of one or two adult pairs and their offspring of successive years. This was in keeping with the traditional concept regarding the social structure of the Callitrichidae, and was strengthened by her observation that, while dual dominance systems also formed in groups of unrelated *C. jacchus* and *Saguinus fuscicollis,* they were rarely stable over long periods of time, whereas family groups remained peaceful over the three-year study. The social configuration which I observed appeared to be identical to that in Epple's "artificial," i.e., unrelated groups.

Epple (1967a, 1967b) indicated that the alpha animals "tolerated" juveniles. A "tolerance" for juveniles was also observed in an artificial feeding situation by Nelson (1975). In the current study, a female approximately 12 months of age was tagged in the upland group in June 1973. During the two subsequent weeks she was alternately a member of the upland group and adjacent Yellow group. She then disappeared until August, when she was seen with another group 400 m south of the upland group's home range. In September she became a member of Green group, whose home range was located 2 km from that of the upland group. It would appear then, that juveniles may move from group to group with a minimum of interference.

The integration of new group members, and the maintenance of dominance in the group may at times, however, require overt physical encounters. Epple (1972) observed that the introduction of a strange conspecific into the laboratory group precipitated aggressive behavior, particularly on the part of the alpha animal of the same sex. She also observed that serious fights between alpha animals and their subordinates occurred frequently, and often after the group had been stable for a year or more. In the current study, the long rasps (Moynihan, 1970) which accompany aggressive contact were heard frequently, although it was often difficult to determine whether the conflicts were intra- or intergroup in nature. The field observations of Moynihan (1970) indicate that intragroup altercations are more frequent, and, as evidenced by the vocalizations, the more intense. It is possible that high levels of intragroup aggression are the driving force behind the intergroup flow of transient animals.

A group social structure composed of a single dominant pair and a shifting complement of subordinates would have several probable advantages over the traditionally accepted "extended family" group with an identical dominance system:

1. Gene flow in the population would be greatly increased since individuals from other groups, rather than siblings, would ascend to dominant, i.e., breeding, positions. This would limit genetic drift and assist in maintaining the integrity of the species.

2. Immigrants ascending to breeding positions in an open social system would contribute a more varied genotype (i.e., heterozygous at more loci), tested over a wider variety of ecological conditions, to the offspring. It follows that offspring exhibiting more variable, more broadly tested genotypes would be genetically more "fit" for survival under a wider variety of environmental conditions than offspring from isolated social groups subject to a limited range of selective pressures. This consideration is especially important since callitrichids in general are *r* selected (as defined by Pianka, 1970) relative to other primate species, and may be thought of as colonizing forms which inhabit widely scattered, temporally discrete areas of second growth over a wide range of environmental conditions.

3. Replacement of individuals lost to mortality would be more rapid, thus facilitating the maintenance of group size at optimal levels.

4. The higher incidence of intragroup aggression in non-family groups would favor selection of more aggressive alpha animals. (It might well be argued to the contrary that high levels of aggression might not be adaptive since they might prove to be socially disruptive. However, Bernstein and Gordon (1974) suggested that "rather than disrupting social relationships, aggression serves to enforce regulated social interactions which maintain primate societies.") It is possible that tamarins which are highly aggressive in intragroup conflicts might also prove highly aggressive in intergroup conflicts. The capacity for high levels of intergroup aggression would be adaptive in maintaining the large foraging territory necessary to exploit dispersed food resources.

5. Long distance movements by transient individuals, and high levels of individual aggression fostered in non-family groups might aid in the discovery and colonization of small, widely separated areas of second growth such as those generated by slash and burn agriculture.

Gartlan (1968) and Eisenberg et al. (1972) discussed recent literature in which intraspecific group structure and function were found to vary over environmental gradients. Jolly (1972) found that lemur groups which had formerly defended fixed territories abandoned these territories and adopted a system of temporal and spatial segregation during a severe dry season. She concluded that "the change may reflect long-term population effects or seasonal food shortage." Social structure, and spatial utilization, then, may vary across both environmental gradients and time, and may even vary within groups over variable environmental conditions.

Anderson (1970), in a consideration of ecological structure and gene flow in small mammal populations, divided potentially suitable habitats into colonization habitats and survival habitats. In the former, populations capitalized on abundant, temporally discrete resources; in the latter, segments of the population survived in areas which were relatively salubrious at all times. Time-space segregation afforded the modus operandi for exploitation of resources in the colonization habitat, while strict territoriality proved to be the energetically feasible alternative in the stable survival habitat.

In Rodman, tamarins occupying lowland areas, which provide a relatively stable food resource throughout the dry season, may be thought of as occupying survival habitats. Their well-defined territories, relatively stable group compositions, and higher rates of infant survival (Dawson, 1976) support this contention. Those tamarins occupying upland sites utilize a seasonal resource, and may be thought of as inhabiting colonization habitats. The observed segregation of groups in space and time, the apparent inability of these groups to establish and defend a geographically fixed territory, and the low rates of infant survival are in agreement with Anderson's (1970) observations on colonizing animals.

At this point, it is necessary to note that a fundamental difference exists between the situation in Anderson's (1970) study and that involving *S. o. geoffroyi* in Panamá. Anderson (1970) found it necessary to invoke group selection in explaining the genetic differences found between survival and colonization components of the house mouse, *Mus*, since the colonizing element did not contribute genetic information to the survival populations (i.e., there was no direct genetic feedback). As indicated earlier, the colonization element of *S. o. geoffroyi* does have a feedback mechanism—immigration to survival habitats and the rise to sexual dominance by individuals from colonization habitats. Thus, the *S. o. geoffroyi* social system allows the transmission of genetic information in the traditional Darwinian manner.

Anderson (1970) characterized the function of the survival component of the population as "maintaining the ongoing population in both the ecological and genetical senses," and the function of the colonization component as "a test of the full range of ecological and genetic variation under all of the accessible variations of the fundamental niche of the species," while at the same time providing for the dissipation of excess animals. It is suggested that populations of *S. o. geoffroyi* follow this pattern, and that the accrued advantages are responsible for their success under environmental conditions which are less than equitable.

Acknowledgments

I undertook this study as a predoctoral student in the Department of Fisheries and Wildlife, Michigan State University. The research was supported by grants to the author from the Midwest Universities Consortium for International Activities (MUCIA), and the Lister Hill Fellowship granted by The Gorgas Memorial Institute. I wish to thank the people responsible for funding this project. I wish to express my appreciation to Drs. R. H. Baker, J. M. Hunter, M. D. Young, P. Galindo, and R. Beumer, without whose efforts and inspiration this study would not have been possible. I also wish to thank E. Mendez and J. Baglien for their constant assistance and encouragement. My thanks to J. Enders, D. Christian, and Dr. W. R. Dukelow for reviewing the manuscript. L. Jenkins and P. Stinson provided secretarial and artistic assistance. I was ably assisted in the field by G. W. Barrett, Jr. Statistical computations were facilitated by the Wang 600-14 calculator in the Michigan State University Museum.

Literature Cited

Aldrich-Blake, F. P. G.
1970. Problems of social structure in forest monkeys. In *Social Behavior in Birds and Mammals,* edited by J. H. Crook, pp. 79–101. New York: Academic Press.

Anderson, P. K.
1970. Ecological structure and gene flow in small mammals. *Symp. Zool. Soc. Lond.*, no. 26: 299–325.

Bates, B. C.
1970. Territorial behavior in primates: a review of recent field studies. *Primates,* 11: 271–284.

Bernstein, I. S.
1964. Role of the dominant male rhesus in response to external challenges to the group. *J. Comp. Physiol. Psychol.*, 57: 404–406.
1966. Analysis of a key role in a capuchin *(Cebus albifrons)* group. *Tulane Stud. Zool.*, 13(2): 49–54.
1970. Primate status hierarchies. In *Primate Behavior: Developments in Field and Laboratory Research* Volume I, edited by L. A. Rosenblum, pages 71–110. New York: Academic Press.

Bernstein, I. S., and T. P. Gordon
1974. The function of aggression in primate societies. *Amer. Sci.* 62: 304–311.

Carpenter, C. R.
1934. A field study of the behavior and social relations of howling monkeys (*Alouatta palliata*). *Comp. Psych. Monogr.*, 16(5): 1–212.
1952. Social behavior of non-human primates. Structure et physiologie des societés animals. *Colloques Internationaux du Centre National de la Recherche Scientifique,* 34: 227–246.
1954. Tentative generalizations on the grouping behavior of non-human primates. *Human Biology,* 26(3): 269-276.

Chapman, F. M.
1929. *My tropical air castle; nature studies in Panama.* New York: Appleton.

Coimbra-Filho, A. F., and R. A. Mittermeier
1973. Distribution and ecology of the genus *Leontopithecus* Lesson, 1840, in Brazil. *Primates,* 14(1): 47-66.

Crook, J. H., and J. S. Gartlan
1966. Evolution of primate societies. *Nature,* 210: 1200-1203.

Dawson, G. A.
1976. Behavioral ecology of the Panamanian tamarin, *Saguinus oedipus* (Callitrichidae, Primates). Ph.D. dissertation, Michigan State University, East Lansing, Michigan.

Eisenberg, J. F., N. A. Muckenhirn, and R. Rudran
1972. The relation between ecology and social structure in primates. *Science,* 176: 863-874.

Epple, G.
1967a. Soziale kommunikation bei *Callithrix jacchus* Erxleben, 1777. In *Progress in Primatology,* edited by D. Starck, R. Schneider, and H. J. Kuhn, pages 247-254. Stuttgart: Gustav Fischer Verlag.
1967b. Vergleichende Untersuchungen über Sexual- und Sozialverhalten der Krallenaffen (Hapalidae) *Folia Primat.,* 7(1): 37-65.
1968. Comparative studies on vocalization in marmoset monkeys (Hapalidae). *Folia Primat.,* 8:1-40.
1970a. Maintenance, breeding, and development of marmoset monkeys (Callithricidae) in captivity. *Folia Primat.,* 12(1): 56-76.
1970b. Quantitative studies on scent marking in the marmoset (*Callithrix jacchus*). *Folia Primat.,* 13: 48-62.
1972. Social communication by olfactory signals in marmosets. *International Zoo Yearb.* 12: 36-42.

Fleming, T. H.
1971. Population ecology of three species of Neotropical rodents. *Univ. of Michigan Mus. of Zool. Pub.,* no. 143.

Gartlan, J. S.
1968. Structure and function in primate society. *Folia Primat.* 8: 89-120.

Hampton, J. K. Jr.
1964. Laboratory requirements and observations of *Oedipomidas oedipus. Amer. J. Phys. Anthrop.,* 22: 239-244.

Hampton, J. K., S. H. Hampton, and B. T. Landwehr
1966. Observations on a successful breeding colony of the marmoset, *Oedipomidas oedipus. Folia Primat.,* 4: 265-287.

Handley, C. O. Jr.
1966. Checklist of the mammals of Panama. In *Ectoparasites of Panama,* edited by R. L. Wenzel and V. J. Tipton, pages 753-795. Chicago: Field Museum of Natural History.

Hays, W. L.
1963. *Statistics for Psychologists.* New York: Holt, Rinehart, and Winston.

Hershkovitz, P.
1949. Mammals of northern Columbia: Preliminary report number 4: monkeys (Primates) with taxonomic revision of some forms. *Proc. U. S. Nat. Mus.,* 98: 323-427.
1966. Taxonomic notes on the tamarins, genus *Saguinus,* (Callithricidae Primates), with descriptions of four new forms. *Folia Primat.,* 4: 381-395.
1969. The evolution of mammals on southern continents. VI The recent mammals of the Neotropical region: A zoogeographic and ecological review. *Quart. Rev. of Biol.,* 44: 1-70.
1970. Dental and periodontal diseases and abnormalities in wild-caught marmosets, (Primates, Callithricidae) *Amer. J. Phys. Anthrop.* 32(3): 377-394.

Hill, W. C. O.
1957. *Primates. Comparative anatomy and taxonomy. Vol. III Pithecoidea.* Edinburgh: University Press.

Hladik, A. and C. M. Hladik
1969. Rapports trophiques entre végétation et primates dans la forêt de Barro Colorado (Panama). *Terre Vie,* 23: 25-117.

Holdridge, L. R.
1967. *Life zone ecology.* San José, Costa Rica: Tropical Science Center.

Holdridge, L. R., and G. Budowski
1956. Report of an ecological survey of the Republic of Panama. *Caribbean Forester,* 17: 92-110.

Janzen, D. H., and T. W. Schoener
1968. Differences in insect abundance and diversity between wetter and drier sites during a tropical dry season. *Ecology,* 49: 98-110.

Jolly, A.
1972. Troop continuity and troop spacing in *Propithecus verreauxi* and *Lemur catta* at Berenty (Madagascar). *Folia Primat.,* 17(5): 335-362.

Kinzey, W. G.
1972. Canine teeth of the monkey, *Callicebus moloch:* lack of sexual dimorphism. *Primates,* 13(4): 365-369.

MacArthur, R., and R. Levins
1964. Competition, habitat selection, and character displacement in a patchy environment. *Proc. U.S. Nat. Acad. Sci.,* 51: 1207-1210.

Marler, P.
1965. Communication in monkeys and apes. In *Primate Behavior—Field Studies of Monkeys and Apes,* edited by I. De Vore, pp. 544-584. New York: Holt, Rinehart, and Winston.

Mason, W. A.
1966. Social organization of the South American monkey, *Callicebus moloch:* a preliminary report. *Tu-*

lane Stud. Zool., 13(1): 23-28.

1968. Use of space by *Callicebus* groups. In *Primates—Studies in Adaptation and Variability,* edited by P. C. Jay, pages 200-216. New York: Holt, Rinehart, and Winston.

Montgomery, G. G., W. W. Cochran, and M. W. Sunquist
1973. Radiolocating arboreal vertebrates in tropical forest. *J. Wildl. Manage.*, 37(3): 426-428.

Moynihan, M.
1964. Some behavior patterns of Platyrrhine monkeys. I. The Night Monkey (*Aotus trivirgatus*). *Smithsonian Misc. Coll.* 146 (5): 1-84.
1967. Comparative aspects of communication in New World primates. In *Primate Ethology,* edited by D. Morris, pages 236-266. London: Weidenfeld and Nicolson.
1970. Some behavior patterns of platyrrhine monkeys. II. *Saguinus geoffroyi* and some other tamarins. *Smithson. Contrib. Zool.*, no. 28: 1-77.

Nelson, T. W.
1975. Quantitative observations of feeding behavior in *Saguinus geoffroyi* (Callithricidae, Primates). *Primates* 16(2): 223-226.

Neu, C. W., C. R. Byers, and J. M. Peek
1974. A technique for analysis of utilization-availability data. *J. Wildl. Manage.*, 38(3): 541-545.

Pianka, E. R.
1970. On r and K-Selection. *Amer. Nat.:* 104: 592-597.

Shadle, A. R., E. A. Mirand, and J. T. Grace, Jr.
1965. Breeding responses in tamarins. *Lab. Anim. Care* 15: 1-10.

Smythe, N.
1970. Relationships between fruiting seasons and seed dispersal methods in a Neotropical forest. *Amer. Nat.*, 104: 25-35.

Sokal, R. R., and F. J. Rohlf
1969. *Biometry.* San Francisco: W. H. Freeman and Company.

Southwick, C. H. (Chairman)
1975. *Nonhuman primates: usage and availability for biomedical programs.* Washington D.C.: Natl. Acad. Sci., 122 pages.

Telford, S. R. Jr., A. Herrer, and H. A. Christensen
1972. Enzootic cutaneous leishmaniasis in eastern Panama III: ecological factors relating to the mammalian hosts. *Ann. Trop. Med. Parasitol.* 66(2): 173-179.

Thorington, R. W. Jr.
1968. Observations of the tamarin, *Saguinus midas. Folia Primat.* 9: 95-98.

Weber, E.
1972. Breeding cotton-headed tamarins at Melbourne Zoo. *Internat. Zoo. Yearb.* 12: 49-50.

Wislocki, G. B.
1930. A study of scent glands in the marmosets, especially *Oedipomidas geoffroyi. J. Mammal.* 11: 475-482.

PATRICIA F. NEYMAN[1]
Department of Zoology
University of California
Berkeley, California 94720

Aspects of the Ecology and Social Organization of Free-ranging Cotton-Top Tamarins (*Saguinus oedipus*) and the Conservation Status of the Species

ABSTRACT

Fifty-three *Saguinus oedipus* were live-trapped, marked, and released in one section of a relatively mature secondary forest located in the Western Carribean coastal lowlands of Colombia. In a total of over 2,500 active field hours, approximately 750 hours of contact were made with marked and unmarked groups of tamarins.

Captures were made with decoy-type live traps and several types of collars were used for marking. Trapping and marking methods are discussed in detail since they may be applicable to future callitrichid field studies.

Certain groups containing between three and thirteen members restricted their movements to a well-defined home range. Although some groups changed considerably in size and composition during the study, they continued to occupy the same areas. Contact with neighboring groups or intruding individuals could usually be characterized as agonistic, including frequent "Rasp" vocalizations, vocalizations associated with separation of members from the group ("Dips" and "Long Calls"), frequent short chases, and occasional body contact. Encounters occurred most often in the overlap areas entered frequently by both groups, and terminated with the two groups gradually drifting apart.

[1]Author's name previously Warner.

Groups of from one to five were present occasionally within the home ranges of "established" groups. Some "transient" groups contained individuals which previously had been observed with "established" groups. Usually "transients" were chased vigorously by the resident group on contact, but two strange individuals joined resident groups with no sign of antagonism. Individuals examined from "transient" groups were adults whose age (based on weight and tooth wear) ranged from young to rather old. Both sexes were included.

Home-range sizes for three groups were 7.8, 7.8, and 10.0 hectares, with the corresponding group size (maximum and minimum numbers) ranging from 13 to 5; 5 to 3; and 6 to 3 individuals. Adjoining group home-range overlap was 20 to 30 percent. Density based on these three groups ranged from 0.3 to 1.8 tamarins per hectare.

Movement patterns, daily routine, group cohesion, and relations with other species are briefly described. Most of the 25 observed sleeping sites were broad tree forks, but a few were dense branch masses. Sites were frequently reused. In all but two cases, all individuals of a group slept together.

Food items included fruits of trees, vines and epiphytes, insects, newly sprouting leaves or buds, leaves, leaf stems, and in one case a frog. Unidentified material was gathered from flowers, surfaces of certain fruits and tree branches or trunks, and the decayed parts of certain trees.

Variation in tooth wear indicated a probable spectrum of ages among the adults of the larger groups examined (eight members). Only one or a pair of infants or juveniles were observed in any group. Present data are not sufficient to support or reject the "extended family" social organization advanced by various workers (Epple, 1972b; Eisenberg, Muckenhirn, and Rudran, 1972); However, the changes in group size and composition observed in this and other studies on *Saguinus* (Dawson, 1976, 1977; Thorington, 1968; Durham and Durham, in press; Castro and Soini, 1977; Izawa, 1976) argue against the suggestion.

It is suggested that the availability of *Saguinus oedipus* to exporters is a misleading index of their abundance. The habitat destruction now occurring is the major threat to this species' future in the wild. The immediate establishment of adequately protected reserves can assure its long-term survival. The need for broad long-term policies on fauna preservation and financing of basic research, especially on callitrichids, is emphasized.

SUMARIO

Cincuenta y tres *Saguinus oedipus* fueron atrapados vivos, marcados y posteriormente liberados en un area de un bosque secundario relativamente maduro, ubicado en la costa Oeste del Caribe, en Colombia. De un total de 2,500 horas de trabajo activo en el terreno, 750 hrs fueron de contacto con grupos marcados y no marcados de dicha especie.

Las capturas fueron hechas con trampas de señuelo vivo fabricadas con alambre por la gente del lugar. Collares de cuero, envueltos en cinta plástica en colores, medallas numeradas hechas de "Formica", y pequeñas campanitas fueron utilizadas en el marcaje. El método de captura, marcaje y determinación de la edad relativa de los individuos se discute en detalle puesto que pueden ser utilizados en futuros trabajos similares con otros Callitrichidae.

Algunos grupos restringieron sus movimientos a 'home ranges' (área de actividad) bien definidos, conteniendo entre 3 a 13 miembros. A pesar de que algunos de estos grupos cambiaron considerablemente en tamaño y composición durante el estudio, permanecieron sin cambios en los límites de los 'home ranges'. Los contactos con grupos vecinos o individuos que pretendieron entrar al área pudieron ser, en general, caracterizados como agonísticos, que incluyeron pequeños enfrentamientos, aumento considerable en la cantidad de vocalizaciones emitidas, persecuciones, y ocasionales contactos cuerpo a cuerpo. Los encuentros ocurrieron en las zonas de sobreposición de las áreas de actividad de los grupos, terminando con una gradual separación de estas.

Grupos de uno a cinco individuos estuvieron temporariamente presentes en varias oportunidades dentro las áreas de los grupos 'establecidos' descritos más arriba. En varios casos, contenían individuos marcados previamente observados con algún grupo establecido. En general, los visitantes fueron perseguidos vigorosamente por el grupo residente; sin embargo en dos ocasiones un individuo ajeno al grupo pasó a formar parte del grupo residente sin ningún signo de rechazo. Los individuos en estos grupos transeuntes cuya edad se pudo estimar basada en peso y desgaste de dientes, fueron en su mayoría adultos maduros, y de ambos sexos.

El tamaño del 'home range' en tres grupos fue de 7.8, 7.8 y 10.0 hectáreas, respectivamente, con un amplitud de variabilidad correspondiente en el tamaño del grupo. Los mínimos y máximos dados a continuación fueron determinados durante el estudio: 13 a 5; 5 a 3; y 6 a 3 individuos. La sobreposición entre grupos contiguos fue de 20 a 30 por ciento. La densidad basada en estos tres grupos tuvo una variabilidad de 0.3 a 1.8 tamarinos por hectárea.

Los patrones de movimiento, actividad diaria, cohesión de grupo y sus relaciones con otras especies se describen brevemente. Las características de 25 sitios para dormir se resumen. Estos sitios fueron a menudo reutilizados. La mayoría de ellos eran esencialmente árboles con gruesas ramas laterales, unos pocos fueron densas masas de ramas. La cantidad de cobertura presente fue altamente variable. Con la excepción de dos casos, todos los miembros de un grupo dormían juntos.

La alimentación consistió basicamente en frutos e insectos. Los frutos provenían de árboles, enredaderas y algunas epífitas. Con menor frecuencia fueron ingeridas hojas nuevas o yemas, hojas, tallos, y en un caso, una rana. Material no identificado fue recogido de algunas flores, valvas de ciertas frutas, ramas o troncos y de algunas porciones aereas de ramas en descomposición.

La variación en el desgaste de las dientes, probablemente indica un espectro de edades entre los adultos de los grupos más grandes examinados (8 miembros). Sólo uno o un par de infantes o juveniles fue observado en algun grupo. De tal modo, los datos aquí presentados no son suficientes para decidir si los grupos pertenecen al tipo de familia extendida, hipotetizada por varios autores basados en trabajo de laboratorio (Epple, 1972b; Eisenberg, Muckenhirn and Rudran, 1972). No obstante, la hipótesis no predecería los substanciales cambios en el tamaño y la composición del grupo observados en este estudio y algunos recientes tratando de ostras especies de *Saguinus* (Dawson, 1976, 1977; Thorington, 1968; Durham and Durham, en imprenta; Castro and Soini, 1977; Izawa, 1976). Estudios a largo plazo en poblaciones marcadas son necesarias para completar el esquema parcial que tenemos ahora de la estructura social de estas especies.

El presente estado del hábitat de *Saguinus oedipus* es discutido en extension con énfasis en el hecho de que las cifras otorgadas por los exportadores llevan fácilmente a errores con respecto a la abundancia de esta especie. El substancial agotamiento del hábitat que está ocurriendo actualmente es la mayor amenza para la especie. Sólo a través del establecimiento inmediato de reservas protegidas puede asegurar su superviviencia. A pesar de los problemas la singular riqueza de la fauna colombiana hace que esta medida sea valiosa de tomar.

Introduction

Saguinus oedipus[2], the cotton-top tamarin, was one of the first callitrichids to become well known in the United States. The proximity of its range to Barranquilla, Colombia's second major animal export center, ensured that this species early became one of the cheapest and most easily available New World primates for the North American pet and biomedical market. However, despite the large numbers that have been held in captivity, there have been relatively few studies of its basic biology. Some information is available on the following: reproductive behavior and physiology (J. K. Hampton et al., 1966, 1971; Epple, 1967, 1970); general behavior (J. K. Hampton et al., 1966; Wendt, 1964); vocal repertoire (Epple, 1968; Muckenhirn, 1967); scent marking behavior (Epple, 1972a); skin and scent gland morphology (Wislocki, 1930; Perkins, 1969) and cytotaxonomy (DeBoer, 1974). Some of these topics have been treated for three other species of *Saguinus* in captivity: *S. geoffroyi* (Moynihan, 1970; Muckenhirn, 1967); *S. fuscicollis* (Epple, 1970, 1971, 1972a, 1972b, 1977); and *S. midas* (Mallinson, 1971). Substantial information has accumulated on the maintenance and propagation of these species in captivity (Epple, 1970; Lorenz, 1972; J. K. Hampton et al., 1966; S. H. Hampton and J. K. Hampton, 1967; S. H. Hampton et al., 1972).

Field observations of callitrichid species have been short-term or lacking, except for a one-year study of *Saguinus geoffroyi* (Dawson, 1976, 1977). Reports based on one to several weeks of field observation are available for the following species: *S. geoffroyi* (Moynihan, 1970; Muckenhirn, 1967); *S. midas* (Thorington, 1968; Durham and Durham, in press); *S. nigricollis* (Mazur and Baldwin, 1968—semi-natural conditions); *S. fuscicollis* (Izawa, 1975, 1976; Castro and Soini, 1977); *S. mystax* (Castro and Soini, 1977); *Cebuella pygmaea* (Ramirez et al., 1977; Izawa, 1975, 1976) and *Leontopithecus rosalia* (Coimbra-Filho, 1977; Coimbra-Filho and Mittermeier, 1973). Studies of free-ranging populations are completely lacking for the remaining 23 callitrichid species (cf., Napier and Napier, 1967), and for *Saguinus oedipus.*

This paper restricts itself to group characteristics, general aspects of the use of space and resources, and relations with other species by cotton-top tamarins in a single study area, plus a review of problems concerning conservation of the species. Methods have been described in some detail because of their possible applicability to future field studies of callitrichids. Other results and conclusions from the study will be presented in subsequent publications.

[2]Although Hershkovitz, 1966, suggested that the rufous-naped tamarin (*S. geoffroyi*) and cotton-top tamarin (*S. oedipus*) be considered subspecies of *Saguinus oedipus*, they will be considered as separate species in this paper in accordance with Hernandez and Cooper (1976).

Study Area

The study area was located about 15 km to the east-northeast of Tolú, Sucre, on the Caribbean coast of Colombia, at about 9°34′N, 75°27′W (Figure 1). It forms part of an alluvial plain at about 100 m elevation at that point, formed from the San Jacinto hills lying just to the east (560 m maximum altitude). The area appears level, but after a heavy rain the standing water can be seen flowing along the surface of the ground and the area drains rapidly into a network of gullies (*arroyos*), 3 m to 7 m deep and 5 m to 15 m wide which carry the water seaward.

Using the Holdridge classification, Espinal and Montenegro (1963) described the region as one capable of supporting "very dry tropical forests," with relatively low rainfall (500 mm to 1000 mm annually) distributed in a highly seasonal pattern. Between December and April when there is practically no precipitation, an estimated 60 percent of trees lose their leaves. During the heavy rainfall months of August through November, large areas of the forest become flooded. About 30 percent of the study area was flooded to three feet or less at the worst of a very rainy year (1974). In the preceding very dry year, the forest floor remained nearly dry during the same period. Even during the wet season three to four days without rain—a not uncommon happening—has a marked effect in drying out the forest floor. During the dry season, the water found in *arroyos* is the sole water supply for the entire area.

The study area is in one of the larger remnant forests still existing in that part of Colombia (Figure 1), totaling about 600 hectares, and surrounded by pasture. It contains an essentially isolated population of *S. oedipus,* although some dispersal is possible via fence lines which connect to *arroyos* since both often are lined with trees. The study groups frequented the southeastern corner of the forest block (about 45 hectares) plus an adjoining strip of forest along a fence line and *arroyo* (about 7 hectares) (Figure 2).

Human exploitation of the forest probably dates

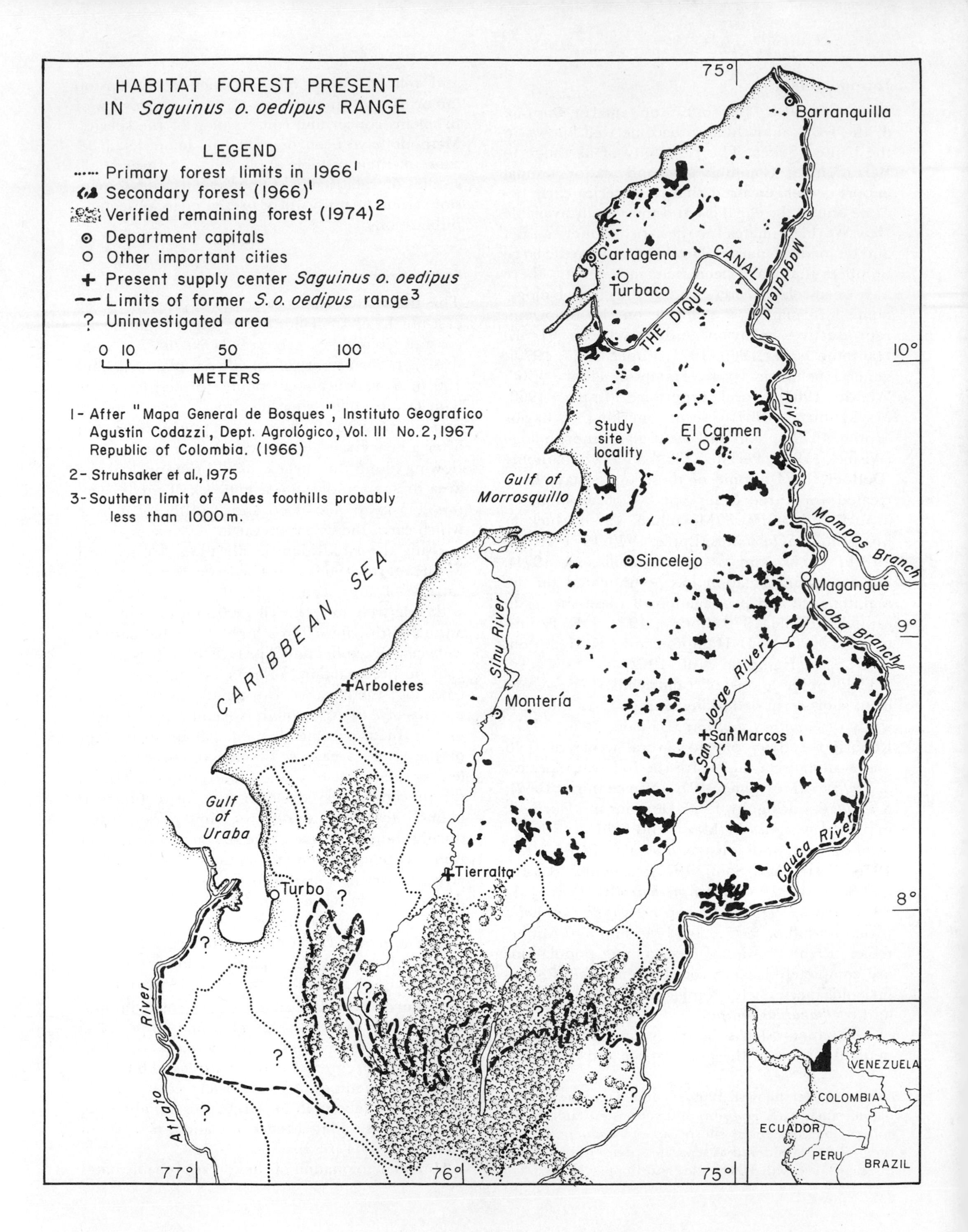
HABITAT FOREST PRESENT
IN Saguinus o. oedipus RANGE
LEGEND
Primary forest limits in 1966[1]
Secondary forest (1966)[1]
Verified remaining forest (1974)[2]
Department capitals
Other important cities
Present supply center Saguinus o. oedipus
Limits of former S. o. oedipus range[3]
? Uninvestigated area
0 10 50 100
METERS
1- After "Mapa General de Bosques", Instituto Geografico Agustin Codazzi, Dept. Agrológico, Vol. III No. 2, 1967 Republic of Colombia. (1966)
2- Struhsaker et al., 1975
3- Southern limit of Andes foothills probably less than 1000 m.
75°
Barranquilla
Cartagena
Turbaco
THE DIQUE CANAL
Magdalena
River
10°
Study site locality
El Carmen
Gulf of Morrosquillo
Mompos Branch
Sincelejo
Magangué
CARIBBEAN SEA
Loba Branch
9°
Sinu River
San Jorge River
Arboletes
Montería
San Marcos
Gulf of Uraba
Cauca River
Tierralta
Turbo
8°
Atrato River
VENEZUELA
COLOMBIA
ECUADOR
PERU
BRAZIL
77°
76°
75°

Figure 1. Location of study site and distribution of forest remaining in the original range of *Saguinus oedipus* (as given by Hernandez and Cooper, 1976). This is a composite showing 1966 forest limits together with 1974 aerial reconnaissance estimates of remaining forest in the southernmost section (Struhsaker et al., 1975). Part of the latter may be primary forest, but extensive clearcutting was already occurring there in 1966. Forest designated in 1966 as secondary is shown as black patches. Many of these may no longer exist.

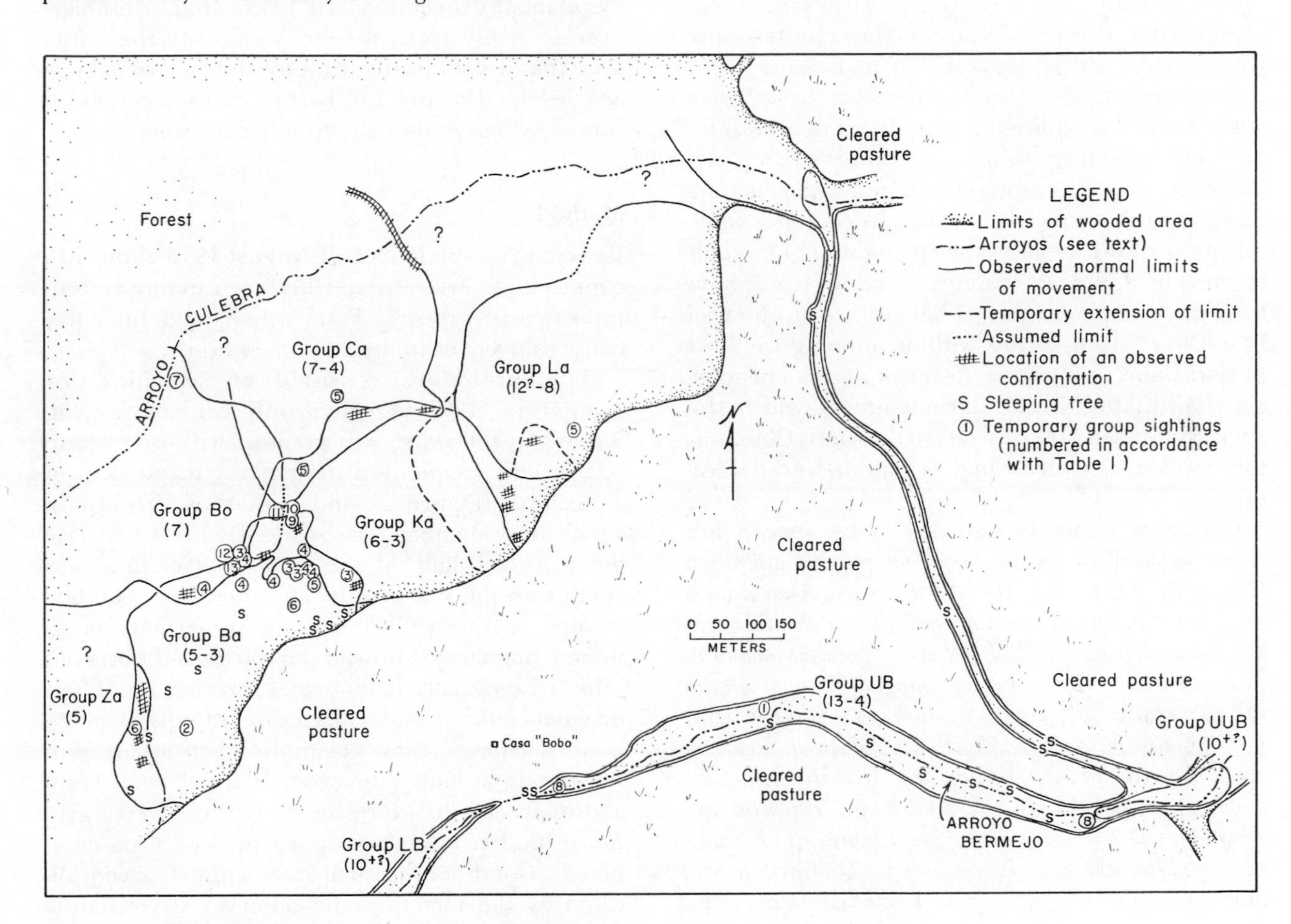

Figure 2. Map of main study area (see text). Surrounding areas are pasture, as indicated. Partial or complete limits of eight "established" groups in and adjacent to the study area are shown, together with observation locations of "transient" or temporary groups. The latter are numbered in accordance with Table 3. Sleeping trees for the Ba and UB groups are indicated. Maximum and minimum size of groups shown in parentheses. Hatched areas show points of inter-group encounters.

back to at least precolonial times, as Indian remains are common in nearby areas. The pastures are fairly recent in origin; those immediately adjoining the study area were cleared less than fifteen years ago. Exploitation of the forest is continuing. Middle-stratum trees are utilized for fence posts, corrals, houses, and bridges. Also some selective cutting is occurring, chiefly for *Cedrela, Ceiba, Bombacopsis,* and *Lecythis.* Few usable large specimens of the first three species remain. Hunters also frequent the study area. As far as is known, the resident primates (*Alouatta seniculus, Cebus capucinus,* and *Aotus trivirgatus* in addition to *S. oedipus*) are not presently subject to hunting pressure; however, they are commonly chased when encountered in vulnerable places, such as on fence lines, on the ground, or in low vegetation. The notable shyness of *A. seniculus* suggests that they may have been molested in the past. They react to an observer by hiding, often sitting without moving for long periods, and attempting to sneak away. This contrasts with the typical alarm patterns which this species shows in other forests in northern Colombia (branch shaking, grunting, urination, and defecation).

Common easily recognizable tree species are those typical of secondary forests in northern Colombia: *Luehea* sp., *Bursera simaruba, Anacardium excelsum, Cavanillesia platanifolia, Pseudobombax septenatum, Cecropia* sp., *Inga* sp., *Spondias mombin, Pithecellobium saman, Lecythis magdalenica, Gustavia* sp., *Triplaris* sp., *Calycophyllum* sp., *Guazuma ulmifolia, Brosimum* sp., *Muntingia calabura, Swartzia* spp., *Garcia nutans, Sapium* sp., *Ormosia* sp., *Casearia* spp., *Mayna* sp., *Nectandra* sp., *Trichilia* sp., *Urera* sp., *Hybanthus prunifolius, Clavija* sp., *Picramnia* sp., *Randia* sp., *Panopsis* sp., *Quararibea* sp., and others including various species of palms (e.g., *Bactris* spp., *Astrocaryum* spp., and *Sabal* sp.) Much less common, but also present are: *Ceiba* spp., *Bombacopsis septenatum, Cedrela* sp., *Enterolobium cyclocarpum, Prioria copaifera, Fagara* sp., *Sterculia apetala, Cochlospermum* sp., and *Cordia aliodora.* In many places the forest floor is dominated by a very spiny palm which branches from ground level (tentatively identified as *Astrocaryum* sp.). In other areas, a mixture of saplings and low palms (cf., *Cryosophila* and *Heliconia* spp.) predominates. In places more recently cleared and therefore receiving more sun, the *Heliconia* may form dense stands to over 3 m in height. All areas can be penetrated with the aid of a machete, though not always at the speed necessary to maintain contact with a *Saguinus* group.

The relative scarcity of densely vined areas and the openness of the floor in most places suggest forest of some maturity. Nevertheless, the brokenness of the canopy is marked—only infrequently is travel possible for more than 50 m in any one direction at a height of over 15 m. Breaks in the canopy are due to both selective logging and tree falls occurring during seasonal high winds (September–November). In 1973 a high wind blew over so many trees in one section of the study area that a continuous canopy did not remain at any level. Also, the fall of one large tree always causes extensive damage to adjacent trees.

Methods

Between August 1973 and August 1975 about 750 contact hours were spent with free-ranging cotton-top tamarin groups. Total active field time was conservatively estimated at 2,500 hours.

Basic procedure consisted of searching for groups and following them until contact was lost. Success in following was greatest with one group (UB) which frequented the narrow forest strip of 7 hectares (Figure 2) and with those groups in which one member was marked with a bell (see below). Over half of the total contact time was spent with the UB group. The three most-studied groups (UB, Ka, Ba) became somewhat conditioned, but always showed nervousness if observed intently, especially if binoculars were used. Their disquiet could often be allayed by rapidly glancing away whenever they began to stare at me, but frequently attempts to observe them even from a distance of 20 m resulted in their moving to more hidden branches or leaving the area completely. Conditioning such small animals is complicated by the fact that the observer is constantly disappearing and reappearing in their visual field as they forage.

Minimal data were obtained on unmarked groups, as it was not possible to distinguish individuals (except in one case), or to be certain whether the same group was encountered on different days in a given location. Age and sex composition of groups, often used by field workers to identify troops, was not helpful here. Tamarins cannot be sexed at a distance, and the rapid growth of juveniles renders them indistinguishable from adults after the age of about ten months if seen from a distance (unless adjacent). Group size was not a useful criterion either, as difficulties in detecting all members of moving groups often rendered group counts questionable. Marked animals were indispensable therefore for group and individual identification as well as for making complete group counts. Some individuals were always bolder than others and were seen repeatedly while others, particularly juveniles and adults carrying young, tended to hide.

The route followed by a group was crudely marked with a machete in order to follow or map its route later. An attempt was made to keep the noise to a minimum. The tamarins did not emit alarm vocalizations, make avoidance movements, or show other apparent signs of disturbance during this activity. Alarm was given only when I actually came into view again.

The Trapping Program

The trapping program was initiated in January 1974 and the first animals marked in May 1974. National Live Traps similar to those successfully used by Dawson (1977) in the capture of *S. geoffroyi* and using the same bait (*Musa* sp.) proved ineffective. This was probably due to the greater variety of strata available for movement and the lack of predictable crossing routes reachable by climbing in this study area. The National traps also required more labor to set and maintain than the decoy-type traps finally used. The latter were fabricated locally from wire and measured 1 m × 1 m × 0.8 m

Plate 1a. Trap unit containing captured cotton-top tamarin (nearest) and decoy. Door is still set open in compartment opposite the captured individual. Protective wire mesh covering the decoy's half can be clearly seen.

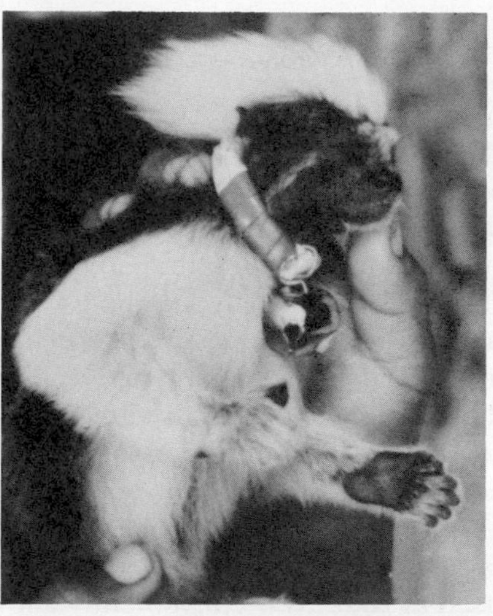

Plate 1b. Cotton-top tamarin wearing collar and bell. This "jingle" type bell was not very satisfactory for localization purposes, despite its size (see text).

Plate 1c. Sleeping tree, *Lecythis magdalenica*, used by *S. oedipus* group. Sleeping site and animals are circled.

Plate 1d. Sleeping tree, *Pseudobombax septenatum*, with the site and animals circled. The cotton-tops are clearly visible.

(Plate 1a). The trap units were placed on poles tied between trees at a height of about 1 to 1.2 m, in an area with visibility from above (yet shaded from the midday sun) and where vines or trees provided easy access for approaching tamarins. A live decoy tamarin was maintained continually in the trap as an attractant, and nearby tamarins would respond to its calls and approach. There was usually only a few days' delay before catching the first animals if the trap was placed in an area where the resident group had recently been seen.

With these traps, 118 captures were made in a total of 1,016 trap-days (12 percent "success"). Fifty-three individuals were caught, 21 of which were recaptured one or more times (one enthusiastic tamarin was retrapped eight times and various others were recaptured four or five times).

Trap Design Problems

Visiting tamarins were often aggressive towards the decoy. In order to prevent injury, double walls or an extra layer of wire mesh were added to the decoy's section. A protected corner was insufficient because rather than retreat to it when surrounded, the decoy jumped back and forth in panic, thus exposing itself to being bitten by one individual while trying to escape from another. The double wall design eliminated actual injuries to the decoy, but did not of course reduce the considerable stress involved in the experience.

A further design problem was the protection of the decoy's tail. An open wire mesh floor was used at first, as it conveniently allowed excrement and food residues to drop through, but the decoy's tail often hung outside the cage and was thus vulnerable to mauling by visitors. A solid cage floor was also unsatisfactory, since the tail was then dragged repeatedly through food and excrement. These sticky substances soon were transferred to the rest of the fur. A compromise solution was a solid partition or tray placed a few inches below the wire floor; this reduced the contact of the tail with waste to a tolerable level and still protected it from outside attacks. Openings in the sides of the cage large enough for the visitors to reach in had to be eliminated; otherwise they could grab the decoy and draw it to within biting reach. Food and water were placed so that visitors could not reach or disturb it and fecal matter would not fall into it. One corner was provided with a shelf for sleeping and covered with plastic to provide protection from rain.

It was most convenient to manufacture the trap section of the unit separately from the decoy section to facilitate transportation to or through the forest (by mule or foot). The trap section was subdivided (as in the complete unit) into two compartments (Plate 1a). On arrival at the trapping site a trap and decoy section were wired together. By adding a second and third trap section on top of the two original sections, the trap's capacity could be increased to a maximum of six compartments. Additional trap sections without an adjacent decoy were sometimes placed a few feet away.

Tamarins were captured in the latter despite being farther from a decoy, indicating that the bait (*Musa* sp., "platano") had an attractive effect. On the other hand, many captures were made in unbaited compartments adjacent to a decoy, indicating that the trap was entered during the course of interactions with the decoy.

Decoy Maintenance

Maintaining live decoy tamarins was difficult in a remote area with an irregular and scarce supply of fruit and protein. Obtaining food for them required frequent time-consuming trips to the nearest reliable supply of fruit and eggs, in this case a city some three hours travel one way. The lack of refrigeration limited the quantity of supplies that could be bought in any one trip.

Marking Methods

By drawing them into reach from outside the trap using the tail, captured animals were administered the anaesthetic "Ketalar" in the tail or thigh. They were fitted either with a leather collar about 0.8 cm wide wrapped with colored fiber-backed plastic tape in different color combinations and designs (Plate 1b), or with a light neck chain bearing a numbered tag (18 mm diameter). The tag was attached to the chain by a wire ring, which assured that the tag faced forward. The tamarins were measured, weighed, and examined for dental wear, signs of reproductive state, and external parasites. In some cases, a drop of blood was taken by a puncture in the heel of the forefoot. (The toes yielded no blood.) Most animals were released as soon as they recovered from anaesthesia but some were kept overnight or several nights and used as decoys to attract the remainder of their group, a tactic that was often successful.

Animals recaptured many months after marking were found to be in good condition. They showed no hair loss below their collars and generally the collars were undamaged. One captive, however, developed hair loss and lesions under a too-snug collar, although it exhibited no signs of distress such as pulling at the collar. Another managed to get the collar in its mouth but could not free its jaw, illustrating a possible danger of a collar that is too loose.

Chains and tags are probably preferable to collars since they appear more comfortable to the animal. Light-weight ordinary ball-link chain is suitable, providing the linking piece is crimped with pliers (uncrimped chains were opened in at least two cases.) The tags, however, can only be read when the wearer is stationary and facing the observer. The identity of an animal overhead could often not be ascertained because the face of the tag was not visible. On the other hand collar colors were sometimes difficult to distinguish. Yellow and white, red and orange, and green, blue and black may be confused in the deep shade frequented by the tamarins. For example, an orange and yellow collar appeared red and white until it was possible (after observing the animal for some time) to see it in the sunlight.

A third promising marking method (R. Cooper, pers. comm.) utilizes colored plastic beads (about 3/8-inch diameter) on a chain. These should be more comfortable for the animals than the collars while retaining the desired readability, providing nonconfusable colors are used. Also they may prove more durable, since leather and plastic tape could succumb to tropical heat and high humidity over a long period. Movement of the beads around the chain is simply prevented by placing a chain-link connector on either side of the group of beads.

One or two animals in each group were marked with a bell as an aid in following the group. The bells of lowest tone were the most easily localizable and carried furthest—the sound of the "jingle bell' type shown in Plate 1b did not carry far. The bell with the lowest tone doubled the range of detection (normally a maximum of 15 to 25 meters), and facilitated tracking without visual contact, as when a group entered a dense viny area, or when I fell behind. Nearly all bells were lost almost immediately, probably through continual twisting of the wire with which they were attached. Bells and attachment should be cast in one piece—an inserted or soldered wire is easily worked loose.

Animals wearing tags, collars, and bells seemed to be accepted normally by group members since they participated in social interactions such as grooming and continued to travel with the same group with which they were observed when captured.

Age Estimation

Three measures were taken as possible indices of age: weight, head and body length, and dental state. Written descriptions of the teeth of captured individuals were fitted into two juvenile and four adult "dental age" categories, defined as follows:

- J1 Upper permanent canines not yet erupted
- J2 Upper permanent canines partly erupted
- A1 Fully grown permanent canines, very new white teeth with no visible wear
- A2 Slightly worn canines and incisors, teeth quite white
- A3 Canines and incisors moderately worn, teeth somewhat to quite discolored, tooth decay sometimes evident
- A4 Very worn canines and incisors, teeth very discolored, one or more canines frequently broken

Ease of examination and measurement dictated the focus on incisors and canines. These are rooted teeth and undoubtedly of great importance in feeding; there is no reason to suspect they would not wear according to age.

Molar eruption was not examined since all but one of the juveniles exhibited partly erupted permanent canines. Permanent canines in marmosets are too strikingly different from deciduous ones to be confused, and are probably the last permanent teeth to erupt, as in *Callithrix jacchus* and *S. nigricollis* (Johnston et al., 1970; Chase and Cooper, 1969). Tooth eruption sequences have never been published for *Saguinus oedipus.*

The juveniles captured were about ten to fourteen months old (Figures 3 and 4). They were still distinguishable from the adults by various nondental characters. Overall size and weight was smaller (Figures 3 and 4). Also they still retained the typically juvenile extra facial hair and had a shorter topknot than the adults (see photographs in S. H. Hampton and J. K. Hampton, 1967). Lastly, in the male, testicles measured only about 3 to 6 mm (as compared to 13 to 15 mm in the adults), and the circumgenital gland in the females was not developed. All A1 adults, however were indistinguishable from the other adult dental-age classes in these characters.

In Table 1 the "dental age structure" of various groups is presented. The interpretation of these findings is hindered by the lack of a known-age series of teeth with which to estimate actual ages. Apparently A1 and A2 embrace a growth phase, since seven of eight recaptured individuals from these categories showed a weight gain seemingly unseasonally related (Figure 3), and the two classes also were nonoverlapping with respect to weight in the expected direction. The A3 and A4 catego-

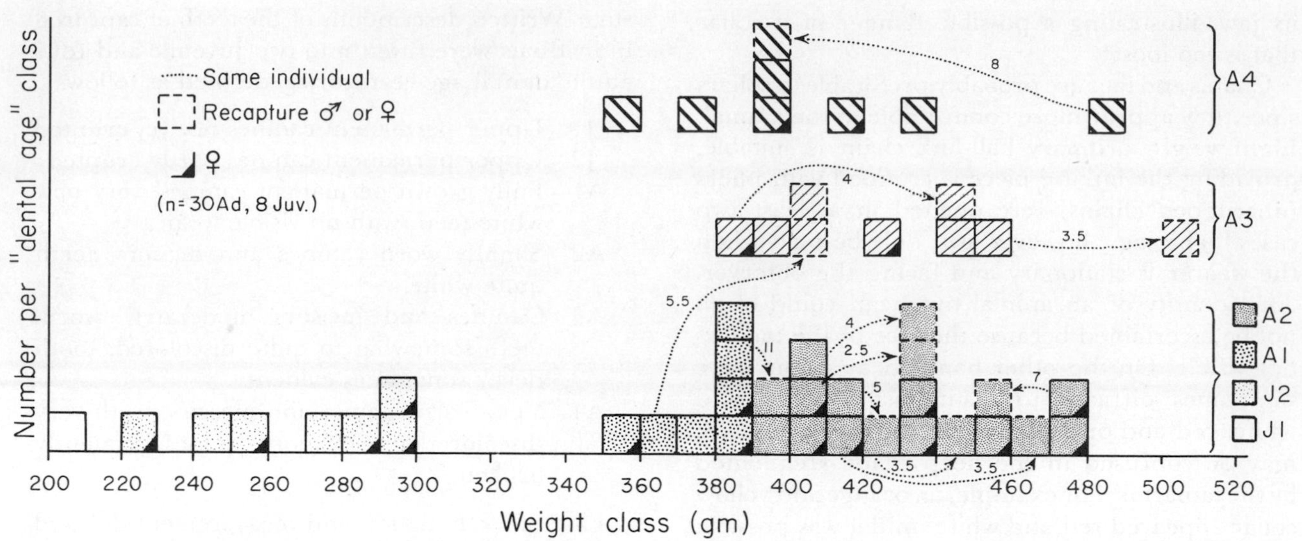

Figure 3. Weight distribution of individual *S. oedipus* in each "dental age" category (see text). Dotted lines connect different weights for a given individual; the arrow shows direction of change. Figures beside arrows are approximate number of months between captures.

Table 1. Age-sex composition of six *S. oedipus* groups during the month indicated (1975).

"Dental Age" Class	Definition	La (Mar) M	La (Mar) F	Ca (Mar) M	Ca (Mar) F	Za (July) M	Za (July) F	Bo (Feb) M	Bo (Feb) F	Ba (Jan) M	Ba (Jan) F	Ka (Jan) M	Ka (Jan) F
J1	Permanent canines not yet erupted	1											
J2	Permanent canines partially grown		1^{i}		**1**^{i}		2^{i}	1				1	1^{i}
A1	Teeth very white, sharp, no wear	2	1^{i}	1	**1**^{i}	1							
A2	Teeth white, barely noticeable wear		1^{d}				1^{d}			1	1^{i}		
A3	Definite wear, and discoloration			1	**1**^{m}	1		**2**	1*	1			
A4	Very discolored— canines very worn/broken, incisors worn down	1	1^{m}	1								1	
A?	Teeth not examined							2	1		**1**^{i}	1	
	Juveniles	1	1		1		2	1				1	1
Subtotals:	Adults	3	3	3	2	2	1	4	2	2	2	2	
	Adults not captured			1									
Group totals		8		7		5		7		4		4	

Nipple-length designation
- n—nipples undeveloped (nulliparous) (½–1 mm long)
- i—intermediate development (1-½–2 mm)
- d—well developed, possibly parous (3-4 mm)
- m—maximal development, probably parous (5 mm)
- (not all females were examined for this character)

Boldface individuals left group before August 1975.

*This female (Lo) had increased weight by 60 gm since 4 months previously, nipples had changed from i to m, and she "seemed fat." Possibly pregnant—twins seen in this group late in April.

ries overlapped both with each other and with the A1 and A2 categories. Longitudinal growth probably ceases or becomes too slight to detect at an earlier age than weight gain: there was no difference between the adult "dental-age" classes in head and body length distributions. A more repeatable length measure depending on fewer growing elements, such as knee-heel length, might be a more sensitive measure of growth.

Males and females were combined in Figure 3, as within dental-age class means showed no significant difference, due to the wide distribution and low number of values. The lack of sexual dimorphism in weight and body length (Figure 4) is evident at all ages. Nor was a sexual difference in canine length or breadth evident, as reported for *Callithrix jacchus* (Johnston et al., 1970).

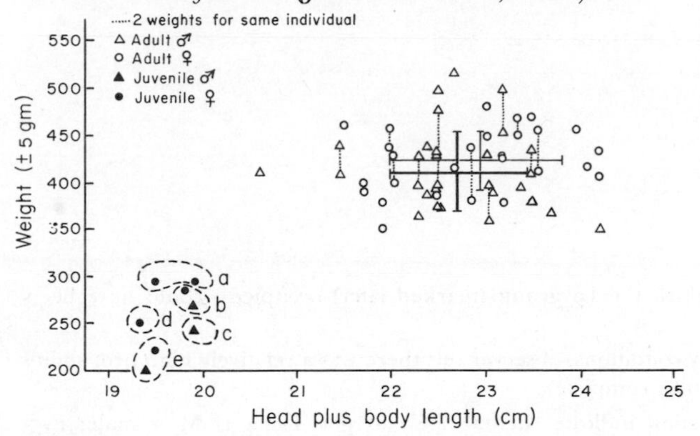

Figure 4. Relationship between weight and head and body length in *S. oedipus*. Cross arms are one standard deviation in length (heavy line = males, light line = females). Juveniles from differing groups are circled, "a-e." The birth date of pair "b" (Ka group) could be placed within an 11-day period: at the date of capture, they were nearly a year old (352-363 days). Pair "e" was known to be around 10 months old when captured (group La). The latter had considerable facial hair (see text), while the year-old ones had very little remaining. Pair "a" was about 13-14 months old (estimated by weight) and had extremely little, if any, extra facial hair, and their upper canines were almost the adult length (about 4½ mm—adults range 5 to 6 mm maximum).

Nipple length was taken on many females (Tables 1 and 3) as a possible means of distinguishing between parous and nulliparous adults. The nipples of parous females become more elongate (R. W. Cooper, pers. comm.). Whether some elongation may also occur at sexual maturity or in parous females who lose their young and do not suckle is not known.

In this sample, length ranged from 1/2 to 5 mm. The measurements fell into four classes (see legend for Table 1). The 5 mm class contained only A3 and A4 females while the lowest (1/2 to 1 mm) group contained all the juveniles and A1 adults. Aside from this, there was no clear correlation with "dental age": A2 females occurred in all three lower classes and an A4 female occurred in the 3 to 4 mm class. One A3 female (Lo), moved from the 1-1/2 to 2 mm class to the 5 mm class in four months; on the latter occasion, she appeared pregnant (Table 1). Thus, some females can show seasonal (?) changes. There was a tendency for nipple color to darken with age (whitish or pinkish to dark); but a few had dark though small nipples (1/2 to 1 mm).

Results

Group Size and Composition

Groups frequenting established home ranges in or adjacent to the study area (hereinafter referred to as "established" groups) contained between three and thirteen individuals, including carried young (Table 2). Groups only temporarily present in the study area ("transient" groups) consisted of one to five individuals.

The age-sex composition of six groups is analyzed in Table 1. None of these (or any other group observed) contained more than two juveniles. Two to six adults were present, of both sexes, except in one group (and later also possibly a second) which lacked females. No more than three adults of a given "dental age" were present but not all ages were represented in every group. Five groups whose composition was known in early 1975 contained a total of 31 individuals: 6 juveniles about 6 to 10 months old (3♂♂, 3 ♀♀) and 25 adults (15 ♂♂, 9 ♀♀, 1 sex unknown). By the end of July the composition of a sixth group was known and the other groups had undergone changes. The six groups then contained 30 animals: 7 juveniles (3 ♂♂, 4 ♀♀) and 23 adults (15 ♂♂, 7 ♀♀, one sex unknown). The sex ratio among adults in these samples is insignificantly unbalanced in favor of males (1.8:1, omitting the one of unknown sex). The sex ratio among juveniles up to one year is about equal.

Various changes in composition were observed in "established" groups, despite the fact that they were followed in some cases for only a few months. These changes are summarized in Table 2 and Figure 5, but the Ka group's history may serve as an example. When first observed, this group consisted of three animals—two adult males and an adult female. It increased to five with the birth

Table 2. Size of "established" groups in the study area and adjacent areas, cause of changes therein, and known facts concerning individuals involved.

Group	*Dates*	*Group size* [1] *Begin*	*End*	*Cause for change in group size* [5] *Born*	*Appearance*	*Reappear*	*Disappear*	*Death*	*Leave*
UB[6]	8/72–8/75	13	5	4	?	?	11	1+?	?
Ka	3/73–8/75	3	4	2	1(MA4)[4]		2(FA,MA)	?	
Ba	7/74–8/75	5	3		1(FA2)		1(J)	?	2(FA2,FA)
Bo	7/74–8/75	7	7?[7]	2		2(2MA)			3(2MA)
Ca	2/75–8/75	7	4						3(FA3, FA1,FJ)
La	3/75–7/75	8	7?[8]						

The following only observed one or a few times, as shown by dates:

La?[2]	3/74–4/74	Min.[3] 12
Ua	5/73	Min. 8
	7/74	Min. 9
	7/75	4
Za	4/75	5
UUB	9/72	Min. 10
LB	10/72	10
	4/75	6
F	8/72	Min. 8

[1]Includes carried infants and juveniles;

[2]Unmarked group followed for over a month in the same area which the La group (marked later) occupied—could have been La group;

[3]Min. = minimum, the number actually verified; but due to visibility conditions, observer felt there was a relatively high probability that not all group members were seen. (Counts lacking "Min" felt to be complete);

[4]A = adult; J = juvenile, the numbers following A or J designation indicate "dental age" (as per Table 1) M = male; F = female;

[5]For category definitions, see text. Animal "disappearing" could have wandered into rest of woods (study area constituted only 52 hectares of a 600 hectare woods). This was suspected in cases of "disappearance" but was not differentiatable from death. If the individual was observed at least once apart from its group, it was considered to have "left" its group;

[6]Uncertainties in UB group are due to its containing very few or no (most of study) marked members;

[7]Only one infant seen of two possible (3 mos. old), but both could well have been present;

[8]Bad visibility, 8th may well have been present.

of twins, and then to six by the addition of another adult male (Oa). When the twins were about six months old the female disappeared, followed soon afterward by one of the original males. Eight months later, at the end of the study (August 1975) the group still consisted of the same two adult males and two juveniles, then 17 months old. One other group (Ca) lost three of its seven members (all females) in the month after it was marked. Some groups maintained their numbers, however, during this same period. Four of the six established groups showed an overall decrease in size during the period observed (5 to 22 months depending on the group). The group whose size decreased most (UB) may have had some members captured by local people, since its home range was a narrow strip of woods flanking a well traveled footpath. There was no reason, however, to suspect human interference of other groups.

Fluctuations in numbers prevented the calculation of an "average" group size, as any method of selecting data from the varying time periods that the different groups were under observations is arbitrary and does not give a sample from which a central tendency can be derived. Nor can data from study area groups be combined with that of unmarked infrequently observed peripheral groups (Table 2, lower half). Until more detailed data are available, it seems better to view cotton-top tamarin groups in terms of maximum and minimum size rather than average size.

Regular group splitting or coalescing such as reported for *Saguinus fuscicollis* (Castro and Soini, 1977), and *S. midas* (Thorington, 1968; Durham

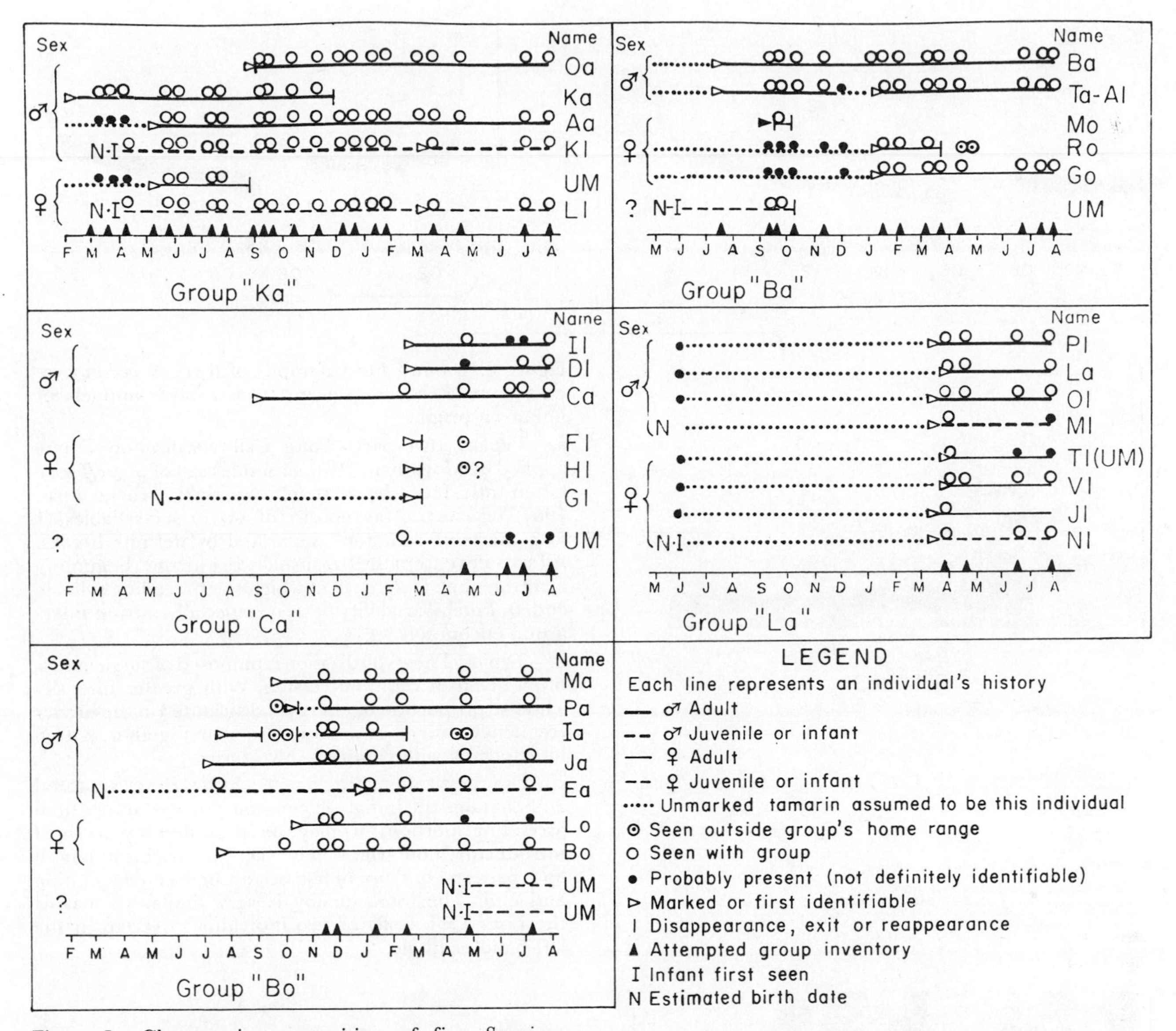

Figure 5. Changes in composition of five *Saguinus oedipus* groups over 5 to 18 months, ending in August 1975. Shaded triangles indicate those occasions on which a reasonably complete group count was obtained. Group size on those dates equals the total number of horizontal lines (solid, dotted, or dashed), each of which represents an individual's history in the group. Named individuals were all marked. UM = unmarked, (UM) = marking lost. In three groups (Ka, Ca, La) one adult remained unmarked. The Ka group UM individual must have been a female, since infants appeared and the other two group members were males. Infants and juveniles were assumed to remain with the same group, and so are figured as identifiable although unmarked.

and Durham, in press) did not occur in the established groups studied. In over 300 hours of contact with group UB, no instance of prolonged splitting or joining of subgroups was seen.[3] Two cases of temporary splitting were observed. In both cases the separated subgroups exchanged contact vocalizations until one moved to join the other. Individuals separated from their group typically ran back and forth through the trees calling loudly ("Long Calls," Figure 6). When other group members vocalized in response such isolated animal(s) moved toward the source of the sound, and the exchange of calls ceased as soon as they rejoined the group. Thus, a certain group cohesion was evident, apparently reinforced by the tendency of separated individuals to find and rejoin their groups. Nevertheless, individuals did at times lag behind the rest of the group by several minutes. More important, some individuals were observed to leave their home groups. The most striking case was that in which an adult male previously seen with the Bo group

[3]The UB group numbered 13 when observations began in August 1972 and 7 in July 1973.

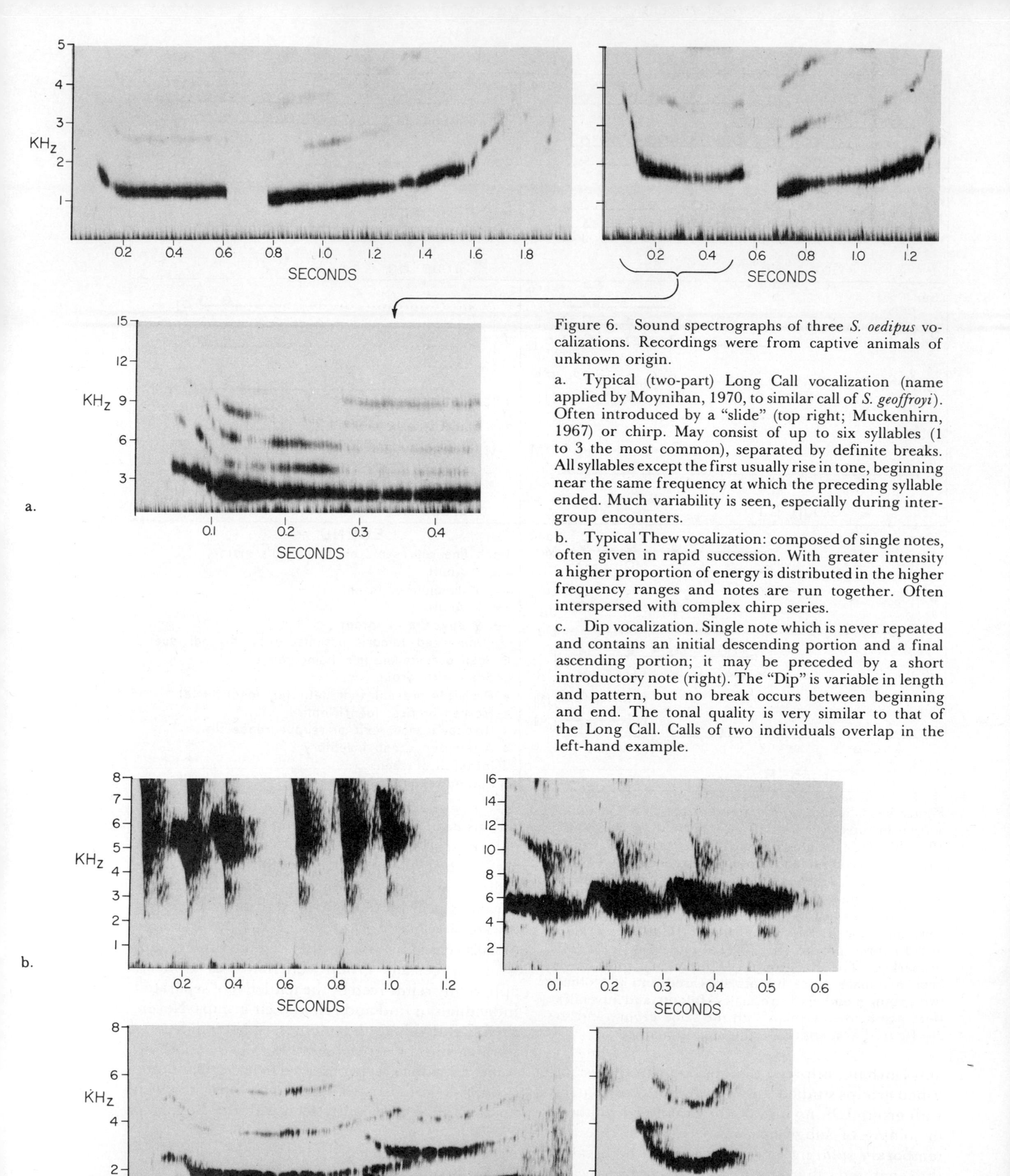

Figure 6. Sound spectrographs of three *S. oedipus* vocalizations. Recordings were from captive animals of unknown origin.

a. Typical (two-part) Long Call vocalization (name applied by Moynihan, 1970, to similar call of *S. geoffroyi*). Often introduced by a "slide" (top right; Muckenhirn, 1967) or chirp. May consist of up to six syllables (1 to 3 the most common), separated by definite breaks. All syllables except the first usually rise in tone, beginning near the same frequency at which the preceding syllable ended. Much variability is seen, especially during intergroup encounters.

b. Typical Thew vocalization: composed of single notes, often given in rapid succession. With greater intensity a higher proportion of energy is distributed in the higher frequency ranges and notes are run together. Often interspersed with complex chirp series.

c. Dip vocalization. Single note which is never repeated and contains an initial descending portion and a final ascending portion; it may be preceded by a short introductory note (right). The "Dip" is variable in length and pattern, but no break occurs between beginning and end. The tonal quality is very similar to that of the Long Call. Calls of two individuals overlap in the left-hand example.

(Ia—Figure 5) along with a second male (Pa) formed a group with two females (Ho, Jo) which was seen at various times during a three-week period, usually near the edge of the neighboring (Ba) group's home range (Figure 2). The Ba group was twice observed chasing them vigorously during encounters. The two males joined the Bo group after three weeks, and the two females were later observed or trapped together several times elsewhere in the study area. The females were not seen during the last five months of the study. Some six months after the return of the two males to the Bo group one of them (Ia) was again observed outside of his group's area, this time with a female from the Ba group (Ro, Figure 2). These two were observed twice within a few days in the Ba group's area, but not again during the remaining four months of the study. The second of the two males (Pa) was sighted during this period carrying a new infant in the Bo group.

The individuals involved in these cases were not "lost"; they were in or near the area frequented by their original group and probably could have rejoined it. This suggests that at least some "disappearances" (Table 2) of marked animals resulted from individuals choosing to leave the home group, possibly voluntarily.

For convenience, animals observed within "established" groups' home ranges but not part of the group have been designated as "transient" groups (Table 3, Figure 2). They were observed only once or a few times and contrasted with

Table 3: Summary of observed "transient" groups and characteristics of captured members.

Group Number[1]	*Number of Individuals*	*Marked Ind. Present*	*Date*	*No. Times Seen*	*Sex*	*Weight (gm)*[3,7]	*Dental Age*[2]
1	1		7/73	1			A
2	4	Mo	9/18/74	1	F	470	A2, 3A
3	1	Mo	9/16,18/74	3	(above)		
4	4	Ia,	9,10/74	many	M	430	A
		Pa,			M	400	A2
		Ho,			F	430	A
		Jo			F	450n	A2
5	2	So[4]	2,3/75	4	F	400n	A2
		Jo			(above)		
6	2	Ro,	4/75	2	F	420i	A
		Ia			M	430i	A
7	2(3?)[5]	F1,	4/75	1	F	420m	A3
		H1,			F	380n	A1
		(G1)			F	250n	J2
8	5	Au,	7/75	various	F	455	A2
		Bu,			M	350	A4
		Cu,			M	395	A3
		Eu			F	435i	A2
9	2		7/13/74	1	M[6]		
10	2	Jo	7/24/74	1	(above)		
11	2	Yo	7/18/74	1	F	430m	A
		To			F	480n	A
12	2	Yo	9/20/74-	3	(above)		
		No	9/24/74		F	390n	A
13	1	Wa	9/11/74	1	M	430	A4

[1]Keyed to Figure 2.
[2]A, adult, J, juvenile, dental age categories as in Table 1.
[3]Weight to nearest 10 gm;
[4]Probably same individual as Ho;
[5]The third individual not seen, but disappeared at same time from Group Ca;
[6]Escaped before marked;
[7]Letters in this column indicate nipple length (Table 1).

"established" groups in lack of attachment to a specific area. One of the females (Jo) which formed the group with Ia and Pa, for example (above), was captured or observed in the territories of several groups (Table 3, Figure 2). The aggression toward transient groups typically shown by "established" groups has already been mentioned. One "transient" group of five (Table 3, number 8) first was seen near the end of the study at the tip of the UB group's home range. Three of four captured individuals from that group had fresh slash wounds which probably resulted from fighting. These were the only animals ever caught with wounds. The injuries could have resulted from the aggression of resident animals. The limited extent of the woods in that area and its conformation—a strip—(Figure 2) would have prevented the intruding group from escaping or leaving the residents' home range as normally must occur.

By contrast, there was no apparent aggression shown toward extra-group individuals on two other occasions. One female (Mo) was observed to join the Ba group temporarily. One hour before joining with them, Mo (who had been trapped and marked in another part of the study area two days previously) had been moving with three unmarked individuals which were not seen again after she approached the Ba group. Mo's approach to the Ba group occurred on an open branch and was clearly observed. A grooming session lasting over an hour was immediately initiated between her and three (two males, one female) of the four members of the Ba group. During the grooming bout, the various partners alternated. Mo traveled with the group for some days, but her association with the group ended shortly thereafter, and she was later observed alone in the Ba group's area. On two of three subsequent sightings, she was observed to approach the Ba group, but not to stay with them except perhaps for a short period. During the last observation of Mo, an adult male from the Ba group appeared to chase her for a short distance. She was not seen again. The other case, that of Oa and the Ka group has already been discussed in the summary of the Ka group's history. I did not observe Oa joining the group, but he remained with it until the end of the study nearly a year later, while two of the original three adults present left.[4] The ability of the above-described animals to join "established" groups without apparent aggression contrasts with the reception normally accorded to extra-group individuals both in this study and in captivity. That immigration and emigration may be normal in wild populations is supported by Dawson's observations of *S. geoffroyi* in Panama (Dawson, 1976, 1977). Only the long term study of marked animals can ascertain whether individuals which enter "established" groups without aggression are related to members of those groups.

The summary of "transient" groups (Table 3) includes all extra-group individuals observed subsequent to the initiation of marking, together with weight, "dental age," and nipple length where known. Of a total of 23 individuals observed, 22 were adults, 11 of which were examined. There was a strong skew in the sex ratio, 4 males: 7 females. The majority of females (4 of the 7) fell into the A2 category, with one additional individual in each of the other three adult categories. Judging by nipple condition, two parous females were included. The males were distributed only in the A2 (1), A3 (1), and A4 (2) categories. It appears from this small sample that older males are more prone to become transients, while females of any age may become transients, including parous and so presumably dominant females. These data include adults which "left" but not those which "disappeared" from their groups (Table 2) among which one other parous female is included.

[4]Such turnover makes the use of the term "established" in reference to these groups dubious; it may well become more dubious as further investigation reveals more complications. For the present, however, the term serves to emphasize the cohesion of these groups and their attachment to a specific area despite the turnover observed.

Home Range and Density

Data concerning home-range size, overlap, density, and daily travel for the three best known groups are shown in Table 4. The observed movements of groups (Figure 7) leave no doubt as to the site attachments of "established" groups. Boundaries remained predictable despite changes in group composition, but extensions of both Ka group and possibly the La group were seen into neighboring areas.

Due to changes, group size and density are best expressed in terms of maxima and minima. For the three best known groups, sizes were: UB (13 to 4), Ka (6 to 3) and Ba (5 to 3), with corresponding home ranges of 7.8, 7.8, and 10.0 hectares. Including only half of the overlap area (20 to 30 percent of the home range), the density for these groups was between 0.3 and 1.8 individuals per hectare.

Relations Between "Established" Groups

Nothing resembling the morning calling between groups of *S. midas* (Thorington, 1968) was noted

Table 4. Group-size maxima and minima, estimated home range, shared area, density, and daily travel distance of three *Saguinus oedipus* groups.

Group	*Group Size*	*Home Range*[1] *(Hectares)*	*Shared (Ha)*	*Portion %*	*Density*[2] *Individ/Ha Max*	*Min*	*Path length*[3] *Km/day*	*M/hr*
UB	13-5	7.8	1.6	20	1.8	0.7	1.5–1.9	120–140
Ka	6-3	10.0	2.5	25	0.7	0.3		
Ba	5-3	7.8	2.1	27	0.7	0.4		

Average of maximum and minimum density estimations for these three groups: 0.78 tamarins per hectare.

[1]As shown in Figure 2.
[2]Density calculations include only half of the area shared with neighboring groups.
[3]Refers to the actual distance over which the group moved.

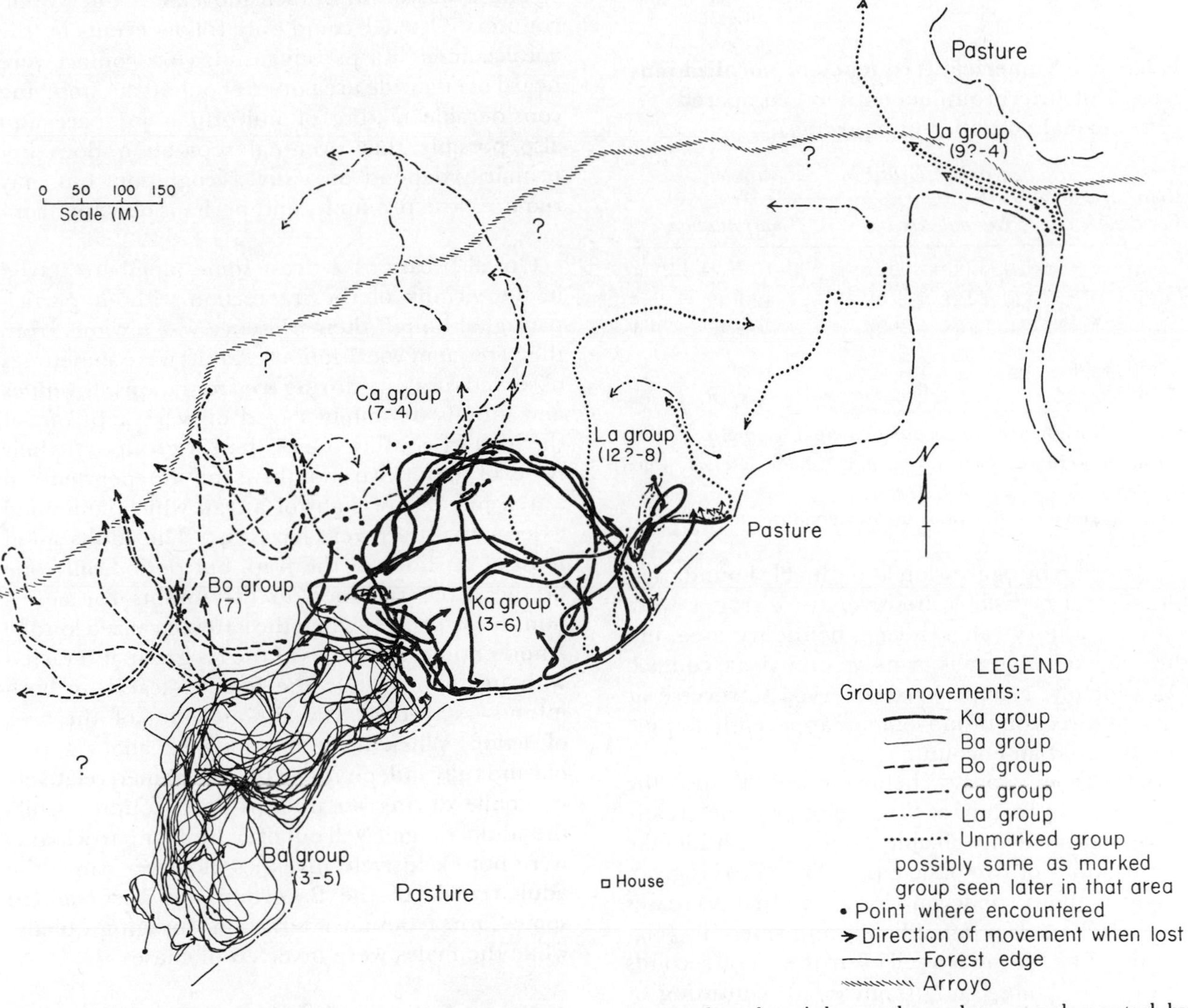

Figure 7. Map of major study area, showing routes taken by followed groups. Home-range outlines in Figure 2 were based mainly on these data, supplemented by point localizations of marked groups.

in this species. Contact between neighboring groups occurred at irregular intervals about once every few days. Chasing, occasional grappling, and approach and staring at the other group from a distance were common during encounters, as was the "Rasp" (Table 5), a vocalization often associated with chasing or grappling. Encounters usually occurred near home-range boundaries in the overlap area and terminated gradually with the two groups drifting apart toward their own area. On one occasion, the two groups moved simultaneously along the presumed boundary for about 230 m before separating (Figure 2). Once when the Ka group had intruded over 45 m farther than the norm for the preceding months into the La group's area, they retreated precipitously on appearance of that group to within the more usual limits of their movements, suggesting the existence of a mutually recognized boundary. In addition, repeated observations of agonistic interactions between neighboring groups at boundary areas suggest that *Saguinus oedipus* is territorial in the sense of exclusion by aggression at definable boundaries. On several occasions, however, two groups were within auditory range near a boundary area, but did not exchange calls or move into visual contact. Occasionally, groups were observed to reverse or change direction of movement, apparently to purposely avoid an encounter.

Table 5. Numerical frequency of vocalizations typical of intergroup encounters, compared with normal activities in *S. oedipus*.

Name of Vocalization	*Number Heard in Five Minutes*	
	Normal Activity	*Confrontation*
Rasp[1]	0–6 ($\bar{x}$ = .082)	0–10 ($\bar{x}$ = 4.2)
Long Call[1]	0–22 ($\bar{x}$ = 1.5)	7–42+ ($\bar{x}$ = 25.4)
Dip[3]	0–32 ($\bar{x}$ = 2.2)	7–70+ ($\bar{x}$ = 29.4)
Number of samples[2]	98	5

[1]As per Moynihan, 1970 (*Saguinus (oedipus) geoffroyi*).

[2]Samples were routinely taken at half-hour intervals when possible.

[3]Not mentioned or described by Moynihan (see text).

Aside from agonistic behavior and "Rasps," the most striking feature of an encounter is the greatly increased incidence (Table 5) of clear whistle-like vocalizations of two basic types. The "Dip" is given singly, is about one second long, and first decreases in pitch then rises immediately and smoothly (Figure 6c). The second is a call of up to several seconds duration ("Long Call") (Figure 6a), consisting of one to six syllables separated by definite breaks. Two and three syllables are most common in the "Long Call." Often the first indication I had that a group encounter was beginning was the alternation of "Long Calls" from the two groups often in not quite synchronous "choruses" as though one member's beginning a "Long Call" stimulated the others to vocalize. During encounters, some individuals are engaged in chases while others are immobile, emitting frequent "Dips" and "Long Calls." These same calls also are used to maintain or regain group contact in situations not involving intergroup encounters (see following section).

In the few encounters which involved marked animals, the most obvious participants were males. They were seen chasing, feinting, and grappling. Females, however, were difficult to localize. Either they did not attract attention because they remained relatively motionless or their movements more often carried them out of my sight.

There was always much movement during encounters. Often I could only follow events by the vocalizations. "Rasps" suggesting close contact were heard over a wide area up to about 40 m^2, implying considerable mixing of individuals in space and also possibly that eventual separation does not primarily depend on visual recognition, but may require close proximity and perhaps olfactory contact.

On all occasions at least some members stayed in the vicinity of the interaction without participating, although their presence was obvious from their frequent vocalizations. Adults were sometimes observed foraging during confrontations. Juveniles and infants definitely stayed or were kept out of the area of conflict. Once the Ka group left their twins of about two months (semi-independent) in a tree just out of sight of a trap which contained a decoy that they were harassing. The adults spent most of an hour at the trap, but periodically one of the three returned to the infants for a few minutes, especially when the latter vocalized loudly. Another time, the males of the Ka group interacted with neighbors while the female—carrying both infants—stayed nearby although out of the area of action. When these infants were about a year old and fully independent, they remained relatively immobile during an intergroup encounter while the adults ranged well out of sight. Nonparticipants were not exclusively juveniles, however, since the adult female of the Ba group was once sighted some 25 m from an interaction, foraging quietly, while the males were involved in chases.

Movement Patterns and Daily Routine

Cotton-top tamarins move through the trees using mainly a quadrupedal gait, at times jumping from one vertical support to another. Bridging of gaps, typical of howlers and spider monkeys, was not observed. Instead, gaps are traversed by jumping. Landing surfaces may be branch-end masses of foliage and all sizes of branches or vines. The tamarins often jumped onto single branches no more than 1 cm in diameter and crossed open spaces on vines of similar dimension. Travel routes occasionally incorporated palms (although not in proportion to their abundance in this forest), including the spineless top surface of the very spiny *Astrocaryum*, but only when other movement routes were restricted. They never foraged in palms.

Animals were observed to fall eight times. Twice intruders fell while being chased by residents, but the actions were so abrupt and in such unlikely places as to seem deliberately evasive rather than accidental. Both animals moved along the ground after dropping, and were not followed by their pursuers. In a third case, one of two grappling animals fell and caught itself on vegetation below. Twice adults fell to the ground while traveling during foraging (from 3 and 27 m, respectively), and once an adult fell from 13 meters but caught itself on vegetation below. All immediately ascended again. Twice juveniles fell to the ground (ages one and one-half and two months, from heights of 12 and 17 m) and were retrieved by adults. In both cases, the juvenile moved along the ground vocalizing until picked up. One of these may have died as a result of the fall since a dead juvenile was observed being carried by a group member the following day; however, there were two young in the group at the time, so identification could not be certain. The juvenile died during the night and was carried for about two hours in the morning, then was left in a tree when the group moved on.

Like *S. geoffroyi* (Moynihan, 1970), the cotton-top tamarin begins moving and feeding relatively late in the morning compared to most primates—up to an hour and twenty minutes after dawn (0550 to 0650). Observations of animals lifting heads and stretching indicate that the tamarins are awake before movement begins, at which time they all immediately leave the tree in quick succession.

A foraging group may spread out over an area up to 35 m or more in diameter. Certainly, they frequently lose visual contact. "Dips" and "Long Calls" (or parts thereof) (Figure 6) were sometimes emitted during foraging, but their frequency and the incidence of "answering" varied. While individuals followed devious paths, the group as a whole traveled in a seemingly definite direction and tended to repeat previous routes.[5] There was no obvious leader in either of the two best studied marked groups (Ka and Ba). Routes were apparently dictated by custom or location of currently available fruits. Trees in fruit were sometimes visited daily or more than once a day. Nevertheless, the home range was fairly well covered in the course of a few days (Figure 7 shows typical patterns of movement). The rate of movement (UB and Ba groups) varied between 0.12 and 0.24 km per hour, and total daily path length was between 1.5 and 1.9 km.

After about 0730 hours, the monkeys began taking short rests, often of only a few minutes, during which they lay along a wide branch with the legs hanging down or tucked under, or engaged in allogrooming. Occasionally, they rested in dense vine entanglements. Individuals lay alone or in pairs or trios within a few feet of each other or groomed in changing combinations. Some group members might continue foraging nearby or interrupt their rest with a short foraging bout. "Dips" and "Long Calls" were emitted occasionally during rest periods. Inactive periods of 30 to 60 minutes occurred as early as 0900 (in the sun) while the longest rests, occasionally up to two hours, occurred around midday (in the shade).

Around 1630 hours, the behavior of the tamarins acquired a characteristic pattern: the group became more cohesive and often traveled quickly and quietly, though foraging still sometimes occurred. Vocalizations were noticeably subdued. The group often appeared to be trying to get away from me. If they saw me near a sleeping tree as they approached it, they would not enter but continued traveling (see Figure 8). This behavior was effective in "losing" me and may function similarly with predators. When the animals were apparently unaware of me, they would gradually approach the sleeping site and forage sporadically near it for about an hour, often sitting quietly and looking around (a normal foraging component) for long periods between progressions. Then, one by one, they would enter the tree.

Most social interaction occurred around concentrated food sources (fruiting trees, scarce favorite foods), during rest periods, and probably in the sleeping tree, which usually was entered well before dark. Foraging, however, was individualistic. There was no food announcement vocalization, or other means of advertizing a favorable food source, except in one possible case. On this occasion, a juvenile spotted a favored and scarce food (*Her-*

[5]Routes were to some extent influenced by the irregular nature of the middle and upper strata (see section on Study Area).

members had left the vicinity. The juvenile then uttered a long series of distinct loud chirps upon which the group immediately returned and approached the food. The frequency of "Rasp" vocalizations was noticeably greater around favored foods, especially in trees where fruits were few in number, large, and sparsely distributed.

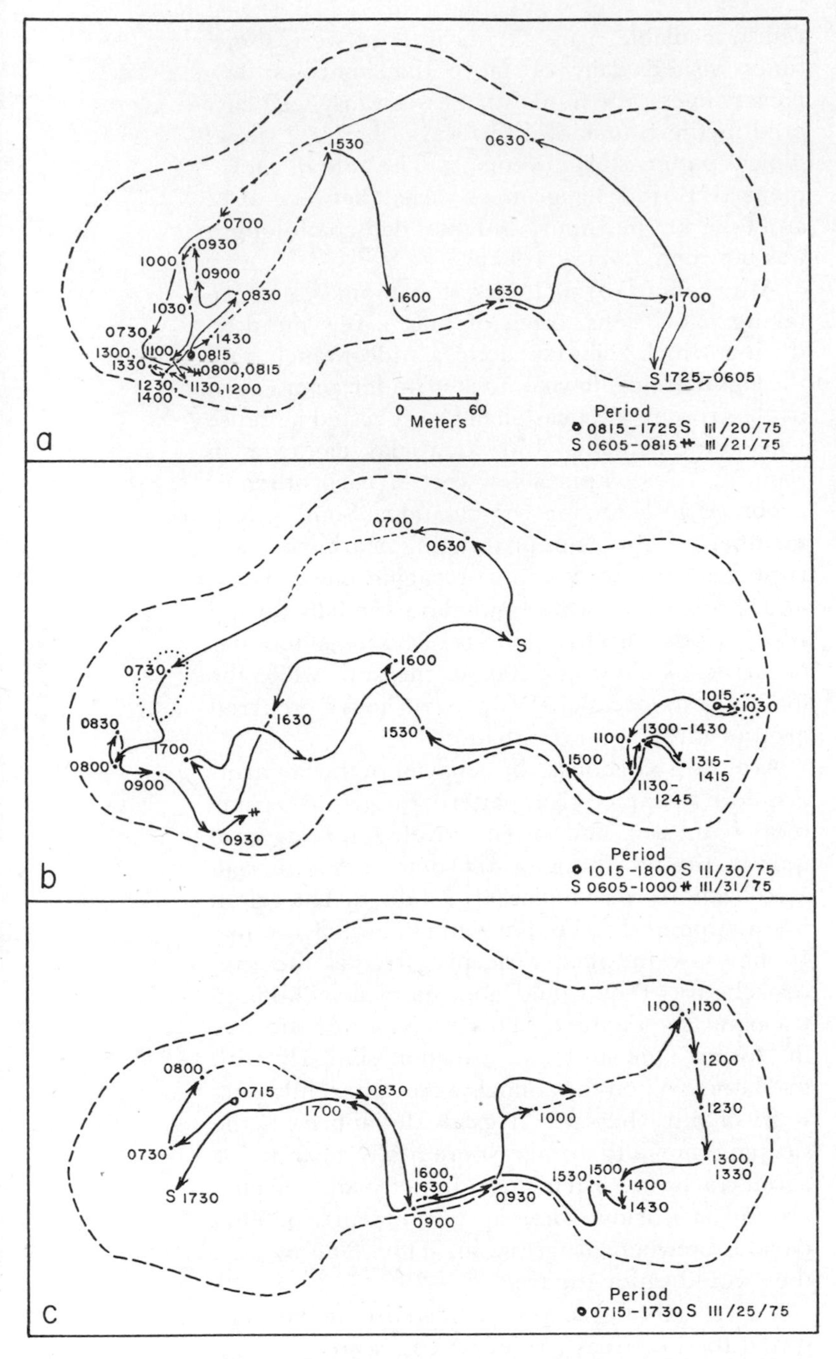

Figure 8. Typical movement patterns of the Ba group within its home range. In a and b the group was picked up during the morning (o), followed to its sleeping tree (S), and picked up the next morning as it left the tree. Dotted circles enclose the approximate area of an encounter with a neighboring group. At 1700 hours on b, the group was headed toward a sleeping tree, but they immediately changed direction when they saw me near it and passed two other previously used trees before settling at dusk in a thickly vine-covered tree.

rania sp.) but hesitated to approach due to my presence, particularly as the height of thre tree was only about 1.5 m. Meanwhile, the other group

Sleeping Habits

Between 1630 hours and 1830 hours a tamarin group entered one of various sleeping trees in its home range (Figure 2). Over a period of about 18 months, the UB group was traced 32 times to a sleeping tree, accounting for 14 different trees. The Ba group was traced to a sleeping site 13 times over a 5-month period, and a total of 9 trees were used. The 25 different sleeping spots used in these and other trees identified over the course of the study were of four types:

	Group		
	Ba	UB	Other
1. Broad tree branch near trunk or wide main fork (at 10-22 m) (Plate 1c, 1d)	3	13	2
2. Among bases of grown-out leafy branches of broken-off tree (at 13-20 m)	4	—	—
3. Low trees with dense crown (at 3-7 m)	1	2	—
4. End branch among dense viny mass (at 17 m)	—	1	—

Group location was verified by observations on the following morning, except for the latter two categories. The greater proportion of category 1 sleeping sites used by the UB troop is probably due to the greater relative frequency of tall broad-branched trees in their home range.

In some sleeping sites, very little cover was present and the animals were visible from the ground (Plate 1c, 1d), while in others they were largely or completely hidden by leafy branches or vines. In a few cases, I was not able to determine exactly where in the tree they were.

Certain species of trees were favored as sleeping sites: *Lecythis magdalenica* (9 sites), *Ceiba* sp. (1 often used site), and *Pseudobombax septenatum* (3 sites). Also used were *Prioria copaifera*, *Spondias mombin* and *Samana samanea.*

In all but two cases, the entire group slept together. Splitting was seen twice in the UB group when it had 13 and 10 members and thus may have been related to limited space, rather than social factors. In one instance, the tamarins bedded down in the same tree with a group of howlers

(*Alouatta seniculus*) which had arrived beforehand. The howlers occupied forks well out on the branches (*Lecythis magdalenica*) while the tamarins settled in two crotches close to the trunk.

The use of a hole in a tree for sleeping was not observed, in contrast to the habits of *Leontopithecus rosalia* (Coimbra-Filho, 1977).

Food Habits

Cotton-top tamarins were observed eating fruits of trees, vines and epiphytes, insects, newly sprouting leaves or buds, leaves, leaf stems, and in one case a frog. They may also lick nectar or gather pollen or insects from certain flowers or fruits. They extracted unidentifiable material from the surfaces of the branches and trunk of some trees by pressing the mouth to the surface and possibly pulling. In some trees the same sites were visited by a succession of individuals. Such spots were frequently but not always decaying. It was not possible to ascertain whether insects or perhaps sap or resin was taken. *Cebuella pygmaea* and *S. fuscicollis* have been observed to utilize sap and resin (respectively) as a food resource (Ramirez et al., 1977; Izawa, 1975, 1976), so the latter possibility cannot be dismissed. In one case, caterpillars were found on the trunk surface at about 2 m suggesting that these, if present on higher parts of the tree also, could have been the prey.

Known genera and species of trees, vines, and epiphytes which were eaten are summarized in Table 6. It was not possible to determine what proportion of the diet each type of food represents, since both insects or small fruits may be taken during long periods when the monkeys are foraging in dense vegetation. Also, some flowers and fruits are not eaten, but may be visited for their pollen or nectar, or for the insects they attract (Table 6). This was suspected when the monkeys visited different fruits or flowers of a species, contacting them with their mouth but apparently removing no part of them (for example the fruits of *Pithecellobium saman* or *Sterculia apetala* which seem too woody to be edible). Vegetative parts usually constituted a minor proportion of the diet. They are most important during the December to April/May dry season when fruit is less available and trees are flowering and leafing out.

The tamarins were observed foraging and feeding on the ground (fallen guava of *Psidium guajava* fruits) and in all strata, although the majority of food trees species identified were middle-canopy (5 m to 15 m) species (see Table 7).

A group was observed licking water from leaf surfaces after a rain. Tamarins were not seen to come to the ground for water or other resources, such as minerals, as Izawa observed (1975) in *Ateles belzebuth* and *Alouatta seniculus* in the Amazon.

Interspecific Relations

Cotton-top tamarins clearly compete in this forest with squirrels (*Sciurus granatensis*), other diurnal primate species (*Cebus capucinus* and *Alouatta seniculus*) and various birds, as evidenced by repeated observations of overlap in food habits. Many nocturnal species might also be competing including *Aotus trivirgatus, Didelphis marsupialis, Marmosa cinerea, Caluromys* sp. and other fruit- and insect-eating species, including bats.

Only three cases of aggressive interspecific interactions were noted:

1. A *Cebus* was seen to chase a tamarin for a few meters when it entered a fruiting *Ficus* where the *Cebus* were feeding. The tamarin left the tree immediately.

2. A toucan (*Pteroglossus torquatus*) was observed to fly away twice when a cotton-top moved as though to approach it.

3. On one occasion a tamarin chased a squirrel from a tree. It is worth noting that *Cebus* clearly dominated the larger *Alouatta*—howlers twice were observed rapidly leaving trees where they were feeding when a *Cebus* group approached. Several times, I had the clear impression that tamarins also avoided *Cebus* when they were heard approaching.

Other interspecific contacts observed were neutral. In particular, both squirrels and howler monkeys fed in the same tree with *Saguinus* without interacting. No polyspecific feeding associations between the tamarins and either *Alouatta* or *Cebus* were observed as described for various other South American primate species (Klein and Klein, 1973; Castro and Soini, 1977). In fact, two primate species were rarely seen within 100 feet of each other. There were also no associations between any of the diurnal primates and birds, such as those described for *Saimiri* (Klein and Klein, 1973).

Important predators could be arboreal or aerial, and the tamarins used different alarm vocalizations for the two. *Eira barbara*, a mustelid with marked arboreal tendencies reported to prey on *Saguinus geoffroyi* (Dawson, 1976), was sighted seven times near tamarin groups, usually at a distance greater than 35 m, either on the ground or in the trees. In one instance, an *Eira* began feeding in the same

Table 6. Some plant food sources utilized by *Saguinus oedipus*.

Family	*Species*	*Type*		*Local Name*	*List Number*
Anacardiaceae	*Anacardium excelsum*	Th	F	Caracolí	27
	Spondias mombin	Tmh	F	Hobo	18
Anonaceae		Tm	F	Yaya	1
Apocynaceae		Tlm	F	Tomate del Monte	46
Araceae	*Monstera pertusa*	E	F		25
Bignoniceae	*Bignonia* sp.	V	(F1)		21
Bombacaceae	*Cavanillesia platanifolia*	Th	(F1)	Volandero	21
	Quararibea sp.	Th	F	Palo de Leon	3
Boraginaceae	*Tournifortia* sp.	V	F		36
Capparidaceae	*Capparis* sp.	Tl	F		29
Elaeocarpaceae	*Muntingia calabura*	Tlm	F	Niguito	2
Flacourticaceae	*Cassearia* sp.	Tm	F		48
	Cassearia sp.	Tm	F		56
	Mayna sp.	Tm	F		52
	Hasseltia floribunda	Tm	F		51
		Tm	F		59
Guttiferae		Tm	F		12
Leguminosae	*Inga punctata*	Tm	F	Guamo	16
	Pithecellobium saman	Th	(F)	Campano	37
Malpighiaceae		Tl	F		30
Marantaceae		V	F		43
Meliaceae	*Trichilia* sp.	Tm	F	Mangle	47
Moraceae	*Brosimum* sp.	Tm	F,Fl	Caucho	8
	Cecropia sp.	Tm	F	Guarumo	34
	Ficus palmicida	V	F	Abrazopalo	10
	Ficus sp.	Th	F		13
	Ficus sp.	Th	F		13c
	Ficus sp.	Th	F		13d
	Ficus sp.	Tm	F		31
Myrsinaceae	*Ardisia* sp.	Tm	F	Corosita	39
	Stylogyne turbacensis	Tlm	F	Pie Paloma	45
Myrtaceae	*Psidium guajava*	Tm	F	Guayabo del Monte	35
Phytolacaceae	*Trichostigma octandrum*	V	F		17
Piperaceae	*Piper* sp.	Tl	F		44
Rosaceae	*Hirtella* sp.	Tm	F		23
Rubiaceae		Tm	F	Loma de Caiman	4
		Tm	F		57
Sapindaceae	*Melicoccus bijugus*	Tm	F	Mamón	*
	Talisia oliviformis	Tm	F	Mamón del Monte	49
	Paullina sp.	V	F		85
	Serjania sp.	V	F		50
Sapotaceae	*Chrysophyllum* sp.	Tm	F	Caimito Morado	R
Sterculiaceae	*Guazuma ulmifolia*	Tm	F		28
	Herrania sp.	T1	F	Cacao	42
	Sterculia apetala	Th	(F)	Camajón	60
Ulmaceae	*Zizyphus* sp.	V	F		7
Urticaceae		Tm	F		32
Verbenaceae	*Cytanexilum* sp.	Tm	F		26
Unidentified					
		Tm	F		9
		Tl	F		14
		Th	F	Manao	24
		Tm	F	Mangle	53
		Tm	F		54
		V	F		58
		V	F		38

Key: T, tree; V, vine; E, epiphyte; l, low (0–5m); m, medium (5–15m); h, high (over 15m); F, fruit eaten; Fl, flowers eaten; (Fl)(F), visited flowers or fruit but could not verify consumption; R, reported to me by a reliable source; *, at Los Borrachos site. Identifications supplied by Dr. Jesus Idrobo B., Herbario Nacional, Universidad Nacional, Bogotá, Colombia.

Table 7. Heights of food-resource tree species shown in Table 6.

Food Item Type and stratum(a)	*Number of Species*	*Proportion (of tree species used)*
Low (height 5 m and below) tree	5	.11
Low and medium tree	3	.07
Medium (5–15 m) tree	27	.60
Medium and high tree	1	.02
High (over 15 m) tree	9	.20
Epiphyte	1	
Vine	10	

large fig tree as the tamarins. The tamarins always reacted with extended series of piercing elongate calls falling in tone mnemonically described by the word "Thew" (Figure 6b) (possibly equivalent to the "Loud Sharp Notes" of *S. geoffroyi* described by Moynihan, 1970) mixed with loud chirps in varying combinations. The *Eira* appeared to ignore them completely. The same vocalization was elicited by two other potential terrestrial predators, humans and dogs. Vocalizing individuals would stop frequently and look around at or for other group members. As in *S. geoffroyi* (Moynihan, 1970), this type of alarm response tended to be contagious; however, only group members in the immediate vicinity responded while those out of sight usually continued foraging.[6]

Birds flying overhead elicit a series of 5 to 10 loud short chirps, "Chirp Burst," possibly similar to the components of the "Short Whine" of *S. geoffroyi* (Moynihan, 1970), but never repeated. The reaction is commonly although not invariably given to hawks (Table 8). Once a hawk has alighted, even if relatively close to the group (e.g., in one instance, 3 m away) no further alarm is given until it flies again, when the alarm is usually repeated, even if the hawk is only changing perches. An unsuccessful predation attempt on *Saguinus* by a hawk was seen.

[6]During alarm calling there is a strong impression that all group members are visible, especially since the alarm response often continues for 15 minutes or more. From personal experience, I would rarely count more than 5 to 7 individuals in a group of 10 or 12 during such an alarm (or even after hours of following) except under unusual conditions where there was very open vegetation. With smaller groups invariably one or two individuals were missed. Since a group often renews the alarm response when the observer begins to follow, repeated low group counts may be obtained. I believe that some published estimates of callitrichid group size may be too low, because group counts are often made under such conditions.

Table 8. Stimuli evoking the aerial predator call ("Chirp Burst"), over a period of several months (UB group).

Stimulus	*Number of instances*
Hawks (various species)	10
Buzzards (2 species)	6
Toucan	2
Parrot (Macaw)	2
Ibis	1
Hawks flying overhead without eliciting the response	3

Discussion

J. K. Hampton et al. (1966), Epple (1972b, 1975), Eisenberg et al. (1972), and Moynihan (1976) have hypothesized that the basic social unit of callitrichids in the wild is probably the extended family group. The term "family group" here means a nuclear family, an adult pair with one or two sets of offspring, as exemplified by *Hylobates lar* and *Callicebus moloch* (Ellefson, 1968; Mason, 1968). In these species, the subadults leave or are forced to leave the family group in order to reproduce. Moynihan (1976) argues that, despite considerable variability in group size, the social organizations of New World monkeys clearly are of two types, families and troops: ". . . there seems to be a significant hiatus between families, no matter how extended, and troops, no matter how reduced. Even when groups are of the same size, the types may be distinguished by the nature of the internal sexual and parental relations within them." (p. 116). (Callitrichids, of course, are included in the first type). Unfortunately, there exist practically no data on the genetic relationships between group members in any New World primate species. The callitrichid "extended family" may actually include members not related by birth as suggested by data from this study and Dawson's (1976, 1977).

Six types of observations on callitrichids have been used to support the extended nuclear family group hypothesis: (1) the small size usually reported for wild groups (Table 9); (2) the formation

Table 9. Reported group size in various free-ranging callitrichid species.

Species	*Duration of study*[4]	*Size range*	*No. units observed*	*Splitting coalescing observed*	*Location*	*Reference*
S. midas	1 wk(i)	1–7	8	yes[1]	Brazil	Thorington, 1968
S. midas		5–20			Surinam	Geijesko, in Husson, 1957
S. midas	9 wks(W)	2–17	11	C, S[2]	Surinam, French Guiana	Durham and Durham in press
S. mystax	months	2–6[5]	12		Peruvian Amazon	Castro and Soini, 1977
S. fuscicollis	months	2–10[6]	25	yes	Peruvian Amazon	Castro and Soini, 1977
S. fuscicollis	months(W)	20–40+	46		Colombian Amazon	Izawa, 1976
S. fuscicollis	months(D)	to 12			Colombian Amazon	Izawa, 1976
S. geoffroyi	Intermittent	1–9	28	C	Panama	Moynihan, 1970
S. geoffroyi	1 year(m)	1–19	71[8]	C[3]	Panama	Dawson, 1977
S. oedipus	years(m)	1–13	25[8]	no	Colombia	Neyman, this study
Cebuella pygmaea		1–10	9		Peruvian Amazon	Izawa, 1976; Castro and Soini, 1977
Leontopithecus r. rosalia		2–8[7]		poss.[7]	Brazil	Coimbra-Filho and Mittermeier, 1973

[1]Coalescing for travel and defense.
[2]Coalescing during travel.
[3]Two largest groups (14 and 19) around concentrated food sources.
[4]W = "wet" season; D = "dry" season; m = marked animals; i = animals individually distinguishable by natural markings.
[5]Often traveling in company with *S. fuscicollis.*
[6]A "supertroop" of 15–26 reported around concentrated food source.
[7]Reports of "up to 15" in the literature thought by authors to be attributable to temporary congregations around favorable food sources.
[8]Including as one unit the various sightings of same troop/group.

of long lasting preferential bonds between pairs in mixed groups of adults in captivity; (3) the universal participation by the male in carrying and caring for offspring, often exceeding that of the female in captive groups; (4) the fact that strangers introduced to an established group are usually attacked, often by all adults; (5) the fact that only one female reproduces in captive groups; and (6) the markedly greater stability of family groups in captivity compared to groups of unrelated adults. Instability in the latter is due to fighting between members of the same sex, especially females. By contrast, aggression serious enough to require separation rarely occurs in even large family groups where the eldest offspring are several years old. Also in contrast to observation 4, family members can often be reunited successfully after a long period of separation. In both cases, fighting is most likely to involve the like-sexed member of the dominant pair. The literature on these topics has been reviewed recently by Epple (1975).

The nuclear family hypothesis implies a relatively stable pair bond, lasting over at least a number of years. It has been suggested that in the wild the pair bond functions to ensure the female of the help of the male which is necessary for successful reproduction (Epple, 1975; Eisenberg, 1977). Since male or female *Callithrix j. jacchus, Saguinus oedipus, Saguinus midas* and *Saguinus fuscicollis* often attempt to mate with members of other groups if given the opportunity (Epple, 1975; Rothe, 1975; J. K. Hampton et al., 1966), Epple has suggested that the aggression often evident between like-sexed individuals may function in the wild to maintain the exclusiveness of the pair bond (Epple, 1975). The limitation of reproduction to one female, in captive *Callithrix jacchus* at least, is effected by hormonal changes in the subordinates (Hearn, 1977). Field data also indicate the restriction of reproduction to one female. No *Saguinus oedipus* group observed contained more than two juveniles or infants. Dawson (1976, 1977) saw a maximum of three young (one case) in *Saguinus geoffroyi* groups and cites indirect observations to indicate the presence of only one reproductively active female per group.

Nevertheless, the data presented here indicate that *Saguinus oedipus* social organization does not fit the nuclear family model. The mobile sector of the population did not correspond in age structure to that expected in a social system based on stable nuclear pairs where offspring leave the group in order to breed. Although females were mainly in the younger tooth-wear and nipple-condition brackets (see Methods: Age Estimation and Table 2), at least two and possibly three parous individuals were included. The origin of one (Yo) is unknown. The second (F1) left the Ca group, probably together with the subadult and older juvenile females from the group. This left four in the group, of which three were known to be males. The sex of the fourth is unfortunately unknown. In the third case, the female's exit (as opposed to possible death) was not verified, but here the three remaining adults were definitely all males (Ka group).

Males also left their groups, and some transients appeared to be old animals (e.g., Oa). According to laboratory observations and the nuclear family theory, parous females would not be expected to be transients, nor to leave their groups. Thus the data are suggestive, although certainly not conclusive, that pair bonds may be of a different nature than has heretofore been surmised, perhaps centered only around the rearing of young, or varying between pairs due to other factors. *Saguinus geoffroyi* groups at Dawson's (1976, 1977) study site showed greater instability (and less area defense) in drier upland areas where food supplies were inadequate during the dry season. No such correlation was obvious in my study area, which is equally as seasonal, but this point needs further investigation. The markedly stable group size that was maintained despite composition changes in Dawson's study contrasts to the fluctuations seen in the present one.

The size of some groups of *Saguinus oedipus* exceeded that which the nuclear family model would predict. Four groups of ten or more were observed; the UB group with 13 members was followed for several months and was stable and cohesive during that period,[7] not merely a temporary feeding aggregate or uniting of two groups. What size would groups be expected to reach? Let us assume that reproductive maturity occurs at two years and that two sets of twins are produced annually, both reasonable given laboratory findings. Although triplets occur, they are relatively infrequent even under presumably optimal laboratory conditions (J. K. Hampton et al., 1966). An extended nuclear family could be expected to contain a maximum of eight individuals, if young were ejected from the group at reproductive maturity. If only one set of twins (or less) was born annually, as was the case in this study area, the maximum group size which did not include reproductively mature offspring would be six. A group of 13, if it was an extended nuclear family, might contain offspring up to five years old, assuming 100 percent survival. Alternatively, larger groups may indicate a higher rate of reproduction than is so far apparent (perhaps during a series of unusually favorable years), or the frequent inclusion of unrelated adults.

The prevalent concept of very small group size in wild tamarins, a keystone to the nuclear family hypothesis, may be misleading. Moynihan (1976) suggests that the small mean group size in Panamanian tamarins may be a result of human interference. Perhaps the same may be true in other studies done in disturbed areas with high human density. Complications in making accurate group counts might in many cases have also resulted in misleadingly low figures. It should be noted, however, that the majority of groups in the present study area did have less than eight members most of the time. The forest they inhabited could be said to be semidisturbed and certainly the population was almost completely isolated.

By contrast, in the Peruvian and Colombian Amazon, recent observers have reported *Saguinus fuscicollis* groups numbering over 20 individuals (Castro and Soini, 1977; Izawa, 1976). In the latter

[7]The group did however gradually become reduced in size until by August 1975 it numbered four members. See footnote 3 and discussion on group size in text.

case, apparently seasonally correlated differences in group size were evident. In one area, five groups whose size was estimated at 40 or more were present, although groups of about 10 were also frequently seen. The "large" groups were observable from November 1971 to February 1972 when no dry season was discernable, and again during similar weather in September to October 1973. But in the same and another location during the following dry period (November 1973 to February 1974), only groups of 12 or less could be found (Izawa, 1976). Group splitting and coalescing is implied, and has been reported by others for *S. geoffroyi*, *S. midas*, and *S. fuscicollis* (Dawson, 1977; Thorington, 1968; Durham and Durham, in press; Castro and Soini, 1977). Dawson (1977) observed his largest groups (14 and 19) of *S. geoffroyi* in the vicinity of concentrated food sources.

Lastly, the exclusiveness of group membership and expulsion of extra-group individuals predicted from laboratory observations, although usually evident in wild groups, was not observed consistently. As Epple has already suggested (1975), under natural conditions, interactions may be affected by factors not present in captivity; for example, season or previous relations to the intruding individuals. Aggression toward strangers may also be exaggerated in captivity by restricted space, constant contact (visual or olfactory) and food competition (Rowell, 1972).

Further study should aim at following the life history of marked individuals, their relationships, and the longitudinal changes that occur in groups. Reliable aging criteria for individuals at close range and at a distance need to be developed. These will necessitate samples of individuals of known ages, obtainable only in conjunction with long-term studies. Growth and tooth-wear studies on captives could provide potentially useful supplemental data, although it appears that infant and juvenile growth rates in captivity are considerably higher than in the wild (compare Figure 4 with Chase and Cooper, 1969). Finally, insight into the immediate and ultimate causes for changes in group size will require analysis of relevant environmental factors, particularly seasonal and regional differences in the resource base on which different tamarin species rely.

Conservation Status

The cotton-top tamarin, endemic to northwestern Colombia (Hershkovitz, 1949; Hernandez and Cooper, 1976), occupies an area that supported an extensive indigenous pre-Colombian population, and is today a densely inhabited region. By 1966 at least 70 percent of the original forest cover in the original range of the cotton-top tamarin[8] had been replaced with pasture and farmland (Figure 1). The more densely settled northern three quarters of the area accounted for only about 5 percent of remaining forest, which was scattered in over 270 isolated tiny secondary patches. Some of these are known to lack tamarins even though they appear to be suitable habitat (Struhsaker et al., 1975; pers. obs.). The future of these forest patches is at best uncertain, not only because wood and wildlife are constantly being extracted, but because in Colombia forested land not yielding cuttable timber is considered to be "unexploited." By law and custom such land may be colonized, a not uncommon event, which discourages private owners from maintaining naturally forested areas.

As of 1966 (the date of the most recently available vegetation map), the less accessible southern portions of the cotton-top tamarin's range (Andean foothills, Figure 1) contained fairly extensive tracts of primary forest. Satellite photographs in 1973 (NASA) and 1974 low-altitude flights (Struhsaker, 1974; Struhsaker et al., 1975, Figure 1) over the area indicated extensive deforestation in the intervening years, including disappearance of much of the primary forest. Since cotton-top tamarins survive well in secondary forests, a scattered discontinuous population probably remains there. Indeed, where secondary forest has recently replaced primary, some increase in numbers might even be temporarily expected. Moynihan (1970) cites evidence that the closely related rufous-naped tamarin of Panama (*S. geoffroyi*) does better in secondary than in primary growth. Nevertheless, the prognosis for the long-term future of even those populations is bleak. Habitat destruction will continue at an even more rapid pace than in the past, as the density of colonizers increases in this

[8]While Hershkovitz (1949) and Hernandez and Cooper (1976) postulate the Magdalena River as the eastern boundary of the original range in the north, Struhsaker et al. (1975) obtained reports of its previous possible occurrence in areas to the east of the Magdalena. They suggest that "the apparent control of distributions (of *S. o. oedipus* and *S. leucopus*) by major rivers may be only an artifact of agricultural patterns that obscure the true former distributions. *Saguinus oedipus* may be characteristic of the drier forests of all of northernmost Colombia while *S. leucopus* appears in the wetter southern forests" (p. 61).

newly opened-up area. At best, any remaining forest will be reduced to tiny patches such as remain in the northern and central parts of the cotton-top tamarin's range.

The present population situation is, however, even more serious than this brief review of habitat changes would indicate. Animal dealers in Colombia maintain that the cotton-top tamarin is still easy to obtain and therefore not endangered. This index of abundance is, however, misleading. The number of tamarins that exporters can obtain is probably less a reflection of actual population levels, than of the level of habitat disturbance in the southern part of its range, containing the majority of the surviving populations. These same areas until recently[9] represented the primary supply source for commercial exploitation (S. Daza, pers. comm.).

It is important to realize that the animal trade in Colombia is organized only loosely through established buying and selling contacts, originating in the most rural areas where local traders buy animals brought to them by local inhabitants. They are then passed from buyer to seller until they reach the major exporters (pers. obs.; Green, 1976). In some cases, exporters contract with buyers for certain species to be shipped at regular intervals. Although some trapping of tamarins was occurring, my observations (1972–1974) indicated that an exporter's stock was drawn from a wide area and that most tamarins were captured by local inhabitants who encountered them by chance in low vegetation or in some other similarly vulnerable location created by clearing and settling activities. At such times, country dwellers will frequently chase the tamarins, hoping to run them down or cause them to fall. They then capture an animal by hand (or by throwing something at it). Thus, the rate of capture is not primarily related to tamarin density, but to two others factors: the frequency with which tamarins enter vulnerable places, and the frequency with which they are encountered by people, in situations allowing capture by hand. These factors indisputably increase as the human population density increases. They also increase according to the degree of habitat disturbance, as clearing activities may force tamarin groups to cross exposed areas in order to traverse their accustomed home range. They may cross on fences (jumping from one fence post to another) or on the ground (C. A. Leon, pers. comm.). The relatively high price that a tamarin brings on the local market (equivalent in 1974 to two or three days' wages) also is some incentive to their capture, although they are also valued as pets and may be sold only after being held for some time, when economic necessity prevails.

Since the situation just described is prevalent in the very areas containing the majority of presently surviving tamarin populations, the number available for export will tend not to respond to decreased tamarin numbers until they have already become dangerously low. Continued availability does not reflect sustained cropping from the total species' range, but the progressive depletion of populations as new portions of its habitat are colonized. Therefore, I believe that the cotton-top tamarin should be considered at least a threatened if not actually endangered species (in terms of remaining numbers).

It is not possible to provide an estimate of numbers of remaining cotton-top tamarins since we have limited knowledge of the state and extent of the remaining forest, much less the numbers contained in the various isolated forest remnants. It is unlikely that there are many forests currently large enough to maintain sufficient tamarins for a viable long-term breeding population.

At present, Colombia does not have even one park or reserve with an established population of cotton-top tamarins.[10] Ideally, a series of areas should be set aside which includes representative habitat and fauna within the cotton-top tamarin range from the very dry deciduous forests in the north to the humid tropical forests in the Andean foothills. Within INDERENA (Instituto de Desarallo de los Recursos Naturales Renovables), there is considerable interest in the establishment of these reserves. International support for these measures would go a long way toward circumventing the financial and political problems involved. Also, as in so many countries, conservation interests often lose when balanced against immediate needs of an expanding population.

An even more basic need then, as in all countries, is to convince the public that their natural resources are uniquely important and worthy of preservation in all their diversity. Conservation of diverse natu-

[9]INDERENA resolution No. 392 (1973) stopped all legal export of primates from Colombia; exceptions were temporarily made for scientific use until 1974 when all legal export was halted.

[10]While this paper was in press, two reserves were established in areas where *S. oedipus* could occur.

ral systems is in fact not a luxury, but a basic need (Watt, 1972). Each species constitutes a reservoir of genetic information which is unique and not replaceable once destroyed; each region possesses complex relationships which are still inadequately understood and which could serve as models for solutions to man-made problems arising from manipulations of the environments. Yet these resources are being destroyed daily because no "economic value" can be demonstrated for them. This as true for Colombia as for many countries, including the United States. Colombia is, however, a country of an unusual faunal diversity. It has more primate species (22) and genera (13) than any other country in South America, except Brazil, and more bird species than any other country. Thus, efforts at establishing reserves in Colombia are extremely important despite the problems involved.

Although habitat destruction is clearly the major threat to the cotton-top tamarin's future, exportation has taken its toll; and demand for tamarins abroad provides impetus for their capture. Between 1968 and 1970, about 2,700 to 3,800 imports into the United States were registered yearly (USDI, cited in Green 1976). In 1972, the United States Department of Interior reports of imports totaled about 2,400, while increases in demand for 1974 were indicated in an Institute for Laboratory Animal Resources (ILAR) survey (Muckenhirn, 1975, p. 30). The actual number extracted from the wild is of course greater than these figures, as considerable attrition (3 to 33 percent in marmosets; Thorington, 1972) must occur between capture and export, particularly as tamarins are delicate and difficult to maintain in captivity. The numbers are still relatively small compared to many other primates, but they are large for a primate of such restricted range, and their impact on the total remaining population could be considerable.

Halting export alone will not assure the survival of the cotton-top tamarin, but, since it is not a species relied on for food, the removal of the monetary incentive would reduce the number removed from the wild. In recognition of this, and the inadequate knowledge of the state of natural populations, Colombia in 1973 prohibited the export of cotton-top tamarins. The United States Endangered Species Act of 1970 provides for United States collaboration in preventing the import of any species whose export has been declared illegal by the government of the country involved, including the cotton-top tamarin. Nevertheless, there have been recent (1975) cases of cotton-top tamarins being offered for sale in the United States. They are imported "legally" by securing the required papers in an intermediate country with no regulations regarding wildlife exportation, such as Bolivia or Panama. Clearly, the law requires changes. Imports should require permits from the country of origin of the species, not the country of shipment. This could do much to protect a species with localized distribution whose point of origin is unquestionable. Of course, import officials must have reference sources enabling them to distinguish different species to avoid name falsification.

Since the biomedical community is the major user of imported cotton-top tamarins, it has a particular responsibility to adjust its research strategy with regard to conservation needs. Future usage should take into account the range, size, and abundance of natural populations, and where the need is greater than the sustained yield of natural populations, researchers should establish breeding colonies. Breeding programs, investigations of basic husbandry problems, and basic biological studies are long overdue. Interest in these subjects has recently increased in the biomedical community since various countries have restricted primate exports. Nevertheless, such projects are still not being funded in proportion to their urgency and relative importance. Furthermore, field research on basic biology is being ignored almost entirely. Only a few species of New World monkeys are presently serving as models for biomedical investigations. Although this approach is understandable, given the need for research continuity, the biomedical community must take a long-range view. Callitrichids have barely been tapped as a resource. Only a minority of species have either been utilized in the laboratory or are known from field studies, and others could perhaps serve as better models for certain studies than species being presently utilized. The cost of the necessary basic field research is cheap compared to the cost of laboratory studies, and given the rate at which habitats are disappearing, such studies cannot be postponed indefinitely.

Acknowledgments

I am greatly indebted to Mr. Miguel N. and his family for permission to conduct this study on their property and for their hospitality during my stay. Among his employees, the Narvais family and Sr. Marcos Fuentes have my special appreciation for teaching and helping me in many ways. I also thank

Dr. Gabriel Oliver E. for offering me the hospitality of his property for a short time.

Dr. Jorge Hernandez Camacho, Dr. Hernando-Chirivi, Dr. Simon Max Franky and Dr. Arturo Leon of INDERENA (Instituto de Desarallo de los Recursos Naturales Renovables), and Dr. Ernesto Barriga, Instituto de Ciencias Naturales, Universidad Nacional, in Bogotá provided help and encouragement without which the study would not have been initiated. The late Dr. Carlos Lehmann deserves special mention as the person who convinced me that the cotton-top tamarin was urgently in need of study. I also owe thanks to Dr. Manual Mercado and Dr. Alberto Berrio of INDERENA in Sincelejo, who helped me in many ways. Dr. Mercado was instrumental in locating the study area and arranging for me to work there, and later helped me obtain much needed decoy animals. Dr. Ricardo A. Tinoco and Carmen Comas B. of the Barranquilla Zoological Park also helped me obtain decoys and traps, for which I thank them. Many other INDERENA personnel in the Bogotá, Barranquilla, and Sincelejo offices, too numerous to mention, also assisted me in many ways.

I thank Dr. Jesús Idrobo of the Instituto de Ciencias Naturales, Universidad Nacional, Bogotá, and also Drs. Joe Kirkbride and J. Cuatrecasas of the United States National Museum Herbarium, and Dr. H. E. Moore, Jr., of Cornell for plant identification. The illustrations were done by Mrs. Emily Reid.

Lastly, I especially thank Ariel Powers for his preparation of the Spanish summary, Drs. N. Muckenhirn and C. Snowdon for their help with the sound spectrographs, John Cassidy for his help and encouragement, and Drs. Devra Kleiman, C. Snowdon, John Eisenberg, Thelma Rowell, and Jorge Hernandez C. for their critical reading of the manuscript.

The study was carried out under the auspices of INDERENA, the Colombian governmental arm charged with all affairs relating to renewable resources, while the author was a member of the United States Peace Corps. Support for the writing phase was provided by the Office of Zoological Research, United States National Zoological Park, and through grant NSF GZ3770 (C. P. Snowdon).

Literature Cited

Castro, R., and P. Soini
1977. Field studies on *Saguinus mystax* and other callitrichids in Amazonian Peru. In *The Biology and Conservation of the Callitrichidae,* edited by Devra G. Kleiman, pp. 73–78. Washington, D.C.: Smithsonian Institution Press.

Chase, J. E., and R. W. Cooper
1969. *Saguinus nigricollis*—physical growth and dental eruption in a small population of captive-born individuals. *Am. J. Phys. Anthrop.*, 30:111–116.

Coimbra-Filho, A. F.
1977. Natural shelters of *Leontopithecus rosalia* (Linnaeus, 1766) and some ecological implications (Callitrichidae, Primate). In *The Biology and Conservation of the Callitrichidae,* edited by Devra G. Kleiman, pp. 79–89. Washington, D.C.: Smithsonian Institution Press.

Coimbra-Filho, A. F., and R. A. Mittermeier
1973. Distribution and ecology of the genus *Leontopithecus* Lesson, 1840 in Brazil. *Primates*, 14:47–66.

Dawson, G. A.
1976. Some aspects of the population ecology of the Panamanian marmoset, *Saguinus oedipus geoffroyi*. Ph. D. dissertation, Michigan State University, East Lansing, Michigan.
1977. Composition and stability of social groups of the tamarin, *Saguinus oedipus geoffroyi* in Panama: Ecological and behavioral implications. In *The Biology and Conservation of the Callitrichidae,* edited by Devra G. Kleiman, pp. 23–37. Washington, D.C.: Smithsonian Institution Press.

DeBoer, L. E. M.
1974. Cytotaxonomy of the Platyrrhini (Primates). *Gene Phaenen.*, 17:1–115.

Durham, N. M., and L. H. Durham
In press. Observations on the social groupings of the red-handed tamarin, *Saguinus midas*, of the Guianas.

Eisenberg, J. F.
1977. Comparative ecology and reproduction of New World monkeys. In *The Biology and Conservation of the Callitrichidae,* edited by Devra G. Kleiman, pp. 13–22. Washington, D.C.: Smithsonian Institution Press.

Eisenberg, J. F.; N. A. Muckenhirn; and R. Rudran
1972. The relation between ecology and social structure in primates. *Science*, 176:863–874.

Ellefson, J. O.
1968. Territorial behavior in the common white-handed gibbon, *Hylobates lar*. In *Primates*, edited by P. Jay, pp. 180–199. New York: Holt, Rinehart and Winston.

Epple, G.
1967. Vergleichende Untersuchungen über Sexual und Sozialverhalten der Krallenaffen (Hapalidae). *Folia Primat.*, 7:37–65.

1968. Comparative studies on vocalization in marmoset monkeys (Hapalidae) *Folia Primat.*, 8:1–40.
1970. Maintenance, breeding, and development of marmoset monkeys (Callithricidae) in captivity. *Folia Primat.*, 12:56–76.
1971. Discrimination of the odor of males and females by the marmoset *Saguinus fuscicollis* ssp. *Proc. 3rd Int. Congr. Primat. (Zurich)*, 3:166–171.
1972a. Social communication by olfactory signals in marmosets. *Internat. Zoo Yearb.*, 12:36–42.
1972b. Social behavior of laboratory groups of *Saguinus fuscicollis*. In *Saving the Lion Marmoset*, edited by D. D. Bridgwater, pp. 50–58. Wheeling, West Virginia: Wild Animal Propagation Trust.
1975. The behavior of marmoset monkeys (Callithricidae). In *Primate Behavior*, edited by L. A. Rosenblum, pp. 195–239. New York: Academic Press.
1977. Notes on the establishment and maintenance of the pair bond in *Saguinus fuscicollis*. In *The Biology and Conservation of the Callitrichidae*, edited by Devra G. Kleiman, pp. 231–237. Washington, D.C.: Smithsonian Institution Press.

Espinal, T., and E. Montenegro-M.
1963. *Formaciones Vegetales de Colombia*. Republica de Colombia, Instituto Geográfico "Augustin Codazzi," Departamento Agrológico.

Green, K.
1976. The non-human primate trade in Colombia. In *Neotropical Primates: Field Studies and Conservation*, edited by R. W. Thorington and P. Heltne, pp. 85–98. Washington, D. C.: ILAR, National Academy of Sciences.

Hampton, J. K., Jr.; S. H. Hampton; and B. T. Landwehr
1966. Observations on a successful breeding colony of the marmoset, *Oedipomidas oedipus*. *Folia Primat.*, 4:265–287.

Hampton, J. K., Jr.; S. H. Hampton; and B. M. Levy
1971. Reproductive physiology and pregnancy in marmosets. In *Medical Primatology, 1970*, edited by E. I. Goldsmith and J. Moor-Jankowski, pp. 527-535. Basel: Karger.

Hampton, S. H., and J. K. Hampton, Jr.
1967. Rearing marmosets from birth by artificial laboratory techniques. *Lab. Anim. Care,* 17:1-10.

Hampton, S. H.; J. K. Hampton, Jr.; and B. M. Levy
1972. Husbandry of rare marmoset species. In *Saving the Lion Marmoset,* edited by D. D. Bridgwater, pp. 70-85. Wheeling, West Virginia: Wild Animal Propagation Trust.

Hearn, J. P.
1977. The endocrinology of reproduction in the common marmoset, *Callithrix jacchus*. In *The Biology and Conservation of the Callitrichidae*, edited by Devra G. Kleiman, pp. 163–171. Washington, D.C.: Smithsonian Institution Press.

Hernandez-Camacho, J., and R. W. Cooper
1976. The non-human primates of Colombia. In *Neotropical Primates: Field Studies and Conservation,* edited by R. W. Thorington and P. Heltne, pp. 35–69. Washington, D.C.: ILAR, National Academy of Sciences.

Hershkovitz, P.
1949. Mammals of northern Colombia. Preliminary report no. 4: Monkeys (Primates) with taxonomic revisions of some forms. *Proc. U. S. Nat. Mus.*, 98:323-427.
1966. Taxonomic notes on tamarins, genus *Saguinus* (Callithricidae, Primates) with descriptions of four new forms. *Folia Primat.*, 4:381-395.

Husson, A. M.
1957. Notes on the primates of Suriname. *Studies on the Fauna of Suriname and Other Guyanas*, 2:13-40.

Izawa, K.
1975. Foods and feeding behavior of monkeys in the upper Amazon basin. *Primates,* 16:295-316.
1976. Group sizes and their compositions of monkeys in the upper Amazon basin. *Primates,* 17: 367-400.

Johnston, G. W.; S. Dreizen; and B. M. Levy.
1970. Dental development in the cotton ear marmoset (*Callithrix jacchus*). *Am. J. Phys. Anthro.*, 33:41-48.

Klein, L. L., and D. J. Klein
1973. Observations on two types of neotropical inter-taxa association. *Am. J. Phys. Anthro.*, 38:649-654.

Lorenz, R.
1972. Management and reproduction of the Goeldi's monkey, *Callimico goeldii* (Thomas 1904) Callimiconidae, Primates. In *Saving the Lion Marmoset*, edited by D. D. Bridgwater, pp. 92-109. Wheeling, West Virginia: Wild Animal Propagation Trust.

Mallinson, J. J. C.
1971. Observations on the breeding of red-handed tamarin, *Saguinus* (= *Tamarin*) *midas* (Linnaeus 1758) with comparative notes on other species of Callithricidae (= Hapalidae) breeding in captivity. *Ann. Rep. Jersey Wildlife Pres. Trust*, 8:19-31.

Mason, W. A.
1968. Use of space by *Callicebus* groups. In *Primates*, edited by P. C. Jay, pp. 200-216. New York: Holt, Rinehart and Winston.

Mazur, A., and J. Baldwin
1968. Social behavior of semi-free ranging white-lipped tamarins. *Psych. Rep.*, 22:441-442.

Moynihan, M.
1970. Some behavior patterns of platyrrhine monkeys. II. *Saguinus geoffroyi* and some other tamarins. *Smithson. Contribs. Zool.* 28:1-77.

1976. *The New World Primates*. Princeton: Princeton University Press.

Muckenhirn, N. A.
1967. The behavior and vocal repertoire of *Saguinus oedipus* (Hershkovitz, 1966) (Callithricidae, Primates). M. S. Thesis, University of Maryland, College Park.
1975. Supporting data. In *Non-human Primates, Usage and Availability for Biomedical Programs*, pp. 11-90. Washington, D. C.: ILAR, National Academy of Sciences.

Napier, J. R., and P. H. Napier
1967. *A Handbook of Living Primates*. New York: Academic Press.

Perkins, E. M.
1969. The skin of primates. XL. The skin of the cottontop pinche *Saguinus* (= *Oedipomidas*) *oedipus*. *Am. J. Phys. Anthro.*, 30:13-28.

Ramirez, M. F.; C. H. Freese; and J. Revilla C.
1977. Feeding ecology of the pygmy marmoset, *Cebuella pygmaea*, in northeastern Peru. In *The Biology and Conservation of the Callitrichidae*, edited by Devra G. Kleiman, pp. 91–104. Washington, D. C.: Smithsonian Institution Press.

Rothe, H.
1975. Some aspects of sexuality and reproduction in groups of captive marmosets (*Callithrix jacchus*). *Z. Tierpsychol.*, 37:255-273.

Rowell, T.
1972. *Social Behavior of Monkeys*. London: Penguin.

Struhsaker, T. T.
1974. The dim future of La Macarena. *Oryx*, 13:298-302.

Struhsaker, T. T.; K. Glander; H. Chiriví; and N. J. Scott
1975. A survey of primates and their habitats in northern Colombia (May-August, 1974). In *Primate Censusing Studies in Peru and Colombia*, pp. 43-78. Washington, D.C.: Pan American Health Organization.

Thorington, R. W., Jr.
1968. Observations of the tamarin, *Saguinus midas*. *Folia primat.*, 9:85-98.
1972. Importation, breeding, and mortality of New World primates. *Internat. Zoo Yearb.*, 12:18-23.

Watt, K. E. F.
1972. Man's efficient rush toward deadly dullness. *Natural Hist.*, 81:74-92.

Wendt, H.
1964. Erfolgreiche Zucht des Baumwollköpfechens oder Pincheäffchens, *Leontocebus* (*Oedipomidas*) *oedipus* (Linnè, 1758) in Gefangenschaft. *Säugetierk. Mitt.*, 12:49-52.

Wislocki, G. B.
1930. A study of scent glands in the marmosets, especially *Oedipomidas geoffroyi*. *J. Mammal.*, 11:475-483.

ROGERIO CASTRO
Proyecto Primates, P.O. Box 621, Iquitos, Peru, and
Universidad Nacional Mayor de San Marcos—IVITA
and
PEKKA SOINI
Proyecto Primates, P.O. Box 621, Iquitos, Peru, and
Dirección General Forestal y de Fauna,
Ministerio de Agricultura

Field Studies on *Saguinus mystax* and Other Callitrichids in Amazonian Peru

ABSTRACT

Preliminary results of field studies under way on Peruvian callitrichids are presented. *Saguinus mystax, Saguinus fuscicollis,* and *Cebuella pygmaea* were studied in Río Aucayo and Río Tahuayo basins from May through July 1975. Additional data on *S. fuscicollis* were gathered at our field station in the Río Nanay basin since September 1973.

Saguinus mystax lives in parental family troops of two to six animals. The largest troops are composed of one adult pair, two subadults, and one pair of dorsally carried infants. The daily path lengths for one troop on two consecutive days were 1209 m and 1100 m with corresponding travel distances of 580 m and 480 m. Intraspecific territorial defense was seen once. Locomotion consists of quadrupedal walking and running and horizontal or diagonal leaping. Postural behavior includes vertical clinging to large tree trunks and hanging from hindfeet. The majority of *S. mystax* troops traveled in mixed groups with *S. fuscicollis* in apparently stable associations. The interspecific cohesion varied largely, but the speed and direction of travel was always set by *S. mystax*, who also tended to forage at a slightly higher level in the trees than *S. fuscicollis.*

Troop size counts of *Saguinus fuscicollis*

ranged from two to ten animals. Recurrent merging of at least two adjacent troops was observed at the Nanay field station. A possible twice-a-year reproduction rate in natural populations is discussed.

Four *Cebuella pygmaea* troops studied consisted of five to ten animals, including a pair of twins born into the largest troop in May 1975. Three troops inhabited edge vegetation of the narrow and sinuous Aucayo river and had small, contiguous non-overlapping home ranges. The overall area encompassing the three home ranges measures 3.4 hectares, which gives a density of 0.88 troops or 5.6 individuals per hectare in this apparently optimum habitat. The fourth troop occupied one large "home tree" in which they foraged and slept. Their principal sap source consisted of an epiphytic shrub growing on the trunk of their home tree.

Introduction

Field studies of Peruvian callitrichids were initiated in 1973 as part of the Pan American Health Organization's (PAHO) two-year primate study project, AMRO-0719. These studies focused primarily on estimating primate population sizes and densities in both undisturbed areas and areas highly affected by human activities (see Neville et al., 1976; Freese, 1975; Freese et al., 1977) and included the following callitrichids: *Cebuella pygmaea, Saguinus fuscicollis, S. imperator* and *S. nigricollis.* Only *Cebuella pygmaea* was studied in more detail, principally by Marleni Ramirez, at Mishana in the Río Nanay basin (see Ramirez, 1975, and Ramirez et al., 1977). The remaining Peruvian callitrichid, *Saguinus mystax,* was not seen in the two-year study period.

After termination of the PAHO project at the end of 1974, primate studies have continued under the aegis of the Peruvian Ministry of Agriculture and with participation of the Ministry of Health and the Instituto Veterinario de Investigaciónes Tropicales y de Ahora (IVITA) of the University of San Marcos. A new agreement was recently signed between the Peruvian government and PAHO for participation of the latter in forthcoming field studies and a captive breeding project of callitrichids and other primates in Amazonian Peru.

We are currently studying geographic distribution, population densities, ecology, and behavior of free-living callitrichids, and this paper reports our preliminary findings.

Methods

Several field trips were made to Río Aucayo and Río Tahuayo basins (Figure 1) to gather data on abundance, ecology, and behavior of *Cebuella, S. fuscicollis nigrifrons,* and particularly *S. mystax.* Additional observations on *S. fuscicollis lagonotus* were made sporadically at and near our permanent field station in the Río Nanay basin over the course of the last two years. We recorded data principally on troop size and composition, troop movement, inter- and intra-troop behavior, feeding, and locomotion.

In *Cebuella,* complete counts of individuals in a troop are difficult to obtain; they were best achieved by locating sleeping trees and counting the animals as they entered and left them, which they usually do in single file. We also climbed up to one sleeping tree at night to check the troop's exact sleeping site. We outlined the home ranges

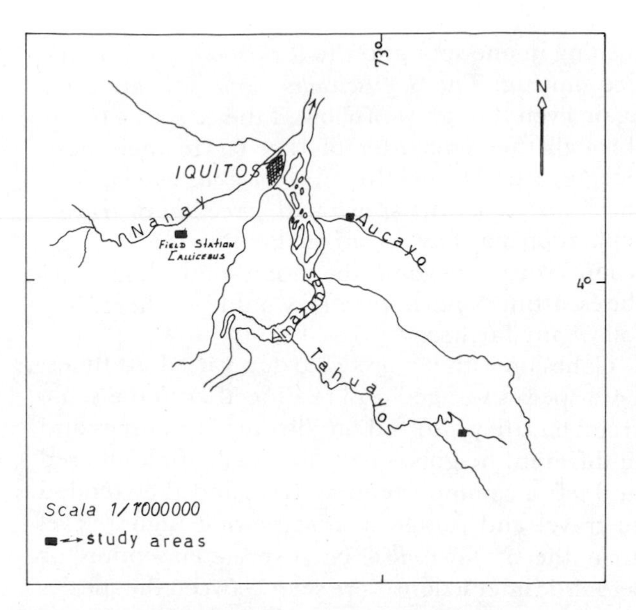

Figure 1. Map showing location of the three study sites in Department of Loreto, Peru.

of three *Cebuella* troops on the basis of their daily movements, location of their sap-source trees, including those no longer in use, and extant ecological boundaries.

At the Río Tahuayo study site, we measured and marked a base trail of 1500 m in primary forest and made a preliminary survey of the primate fauna along this and several accessory trails. The location and movements of the *Saguinus* troops we encountered were plotted in relation to the base trail. One *S. mystax* troop was followed uninterruptedly for two complete days and its daily range, travel path, and sleeping trees recorded. This and other *S. mystax* troops were followed for shorter periods of time in both the Aucayo and Tahuayo study areas. Also *S. fuscicollis* troops were followed, particularly when they were found associated with *S. mystax* troops. In the Nanay field station, we have watched *S. fuscicollis* troops sporadically since September 1973 and recorded data on their troop size, movements, and feeding.

Results and Discussion

Saguinus mystax

Saguinus mystax was usually found in a fairly undisturbed primary forest habitat, but a few troops were also observed in old secondary forest. They are also known to make sporadic incursions into unguarded fruit tree orchards, particularly when the domesticated "uvilla" tree (*Pourouma* sp.) is in fruit.

At the Aucayo study site, where *S. mystax* at least occasionally is hunted for food, and where a number of them were live-trapped in October and November 1974, they have become quite wary and difficult to approach. At the Tahuayo study site, where they are not hunted nor commercially trapped, most of the troops could be followed at close range. A preliminary survey of a rectangular area of approximately 1 km^2 of primary forest at Tahuayo indicated that at least five to seven *S. mystax* troops had their home ranges entirely or partly within that area, suggesting a fairly good population density.

The number of animals in troops of which complete counts could be obtained ranged from two to six. Once, we observed a congregation of eleven individuals moving along together, but after we had followed them for 1.25 hours, they separated into two groups of five and six animals. Each group moved in a different direction, and it became obvious that actually two independent troops were involved. It may have been a temporary merging of two adjacent troops, as seen in *S. fuscicollis* (see below), but since our sudden appearance had caused visible alarm in these monkeys, it seems as likely that a circumstantial congregation of two troops foraging near each other occurred as they were moving away from us.

The smallest troops consisted invariably of one adult male and one adult female. Two troops of six animals each were composed of two full-size adults, two self-locomoting juveniles or young adults, and one pair of very young infants that were carried dorsally either by one adult or singly by two larger troop members. The parental family social unit is clearly indicated in this species. In our principal study troop of six animals, it was the female rather than the male that carried the infants, though one of the infants was carried part of the time by another troop member. Most of the time the progression of the troop was led by the adult male, and the rear was brought up by the infant-carrying and suckling female.

The path lengths for two complete days were plotted and measured for this troop and amounted to 1209 m and 1100 m with corresponding travel distances of 580 m and 480 m (Figure 2). A line connecting the peripheral points in which the troop was observed encloses an area of 0.5 hectares. The actual home range of the troop will undoubtedly prove to be larger.

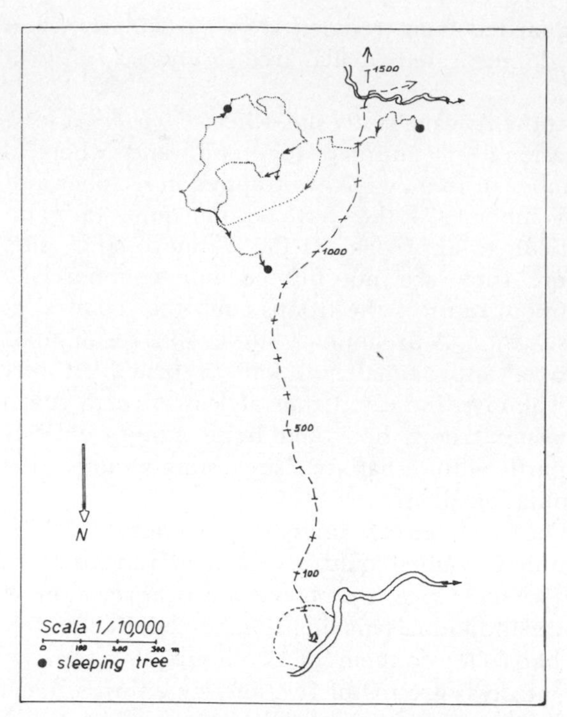

Figure 2. Daily paths and sleeping trees of the principal study troop of *S. mystax* on two consecutive days.

We observed once what seemed to be an attempt to defend a territory, when a troop of five *S. mystax* approached an area where another troop of two animals was foraging and feeding. One of the two occupants of the tree advanced toward the approaching troop and perched, visibly excited, on a horizontal branch about 10 m in front of the newly arrived troop while producing sharp, clearly hostile vocalizations. After a brief vocal battle the defender made a sudden, rapid retreat and was followed by the intruding troop, which, however, soon resumed travel and left the area.

In seven out of twelve encounters with *S. mystax* troops, they were found in mixed groups with *S. fuscicollis* troops. This interspecific association is apparently quite stable; on all three days that we followed our principal study troop of *S. mystax,* it was found traveling together with apparently the same troop of *S. fuscicollis,* which consisted of one adult pair and two self-locomoting juveniles. The first time the troop was followed for an entire day, this association continued all day, and both troops spent the night together among the fronds of a tall *ungurahui* palm (*Jessenia batana*). On the following morning, the troops continued foraging together for four hours and separated at about 1030 hours when the *S. fuscicollis* troop remained feeding in one spot and the *S. mystax* troop continued moving. The *S. fuscicollis* troop was not seen again even though we followed the *S. mystax* troop through the remainder of the day to their next sleeping tree. During this mixed species association, the *S. mystax* set the speed and direction of travel; separation may have occurred because the *S. mystax* troop crossed beyond the home-range limits of the escorting *S. fuscicollis* who would not, therefore, follow any farther.

Cohesion within a mixed group varied. At times, both species foraged and fed together in the same trees, but they also fed on different resources and at different heights. *S. mystax* usually first entered and left a common feeding tree, and they tended to travel and forage at a somewhat higher level than the *S. fuscicollis.* No overt competition or agonistic interactions were seen between the species members.

Progression in the trees consisted of quadrupedal walking and running along branches and tree trunks, combined with horizontal and diagonal leaps between terminal branches. Postural behavior included hanging by the hindfeet to reach a particular fruit and upright vertical clinging to large tree trunks while searching for food items in tree holes and knots or licking exudates from the tree trunks. Leaping from and/or to vertical supports, as described for *Cebuella* by Kinzey et al., (1975) and observed by us in *S. fuscicollis,* was not seen in *S. mystax.* The quadrupedal locomotion of *S. mystax* is distinctly less jerky and more feline-like than *S. fuscicollis.*

Saguinus fuscicollis

Saguinus fuscicollis inhabits both primary forest and secondary growth. Like *S. mystax,* it occasionally raids fruit trees in unguarded native orchards. It occurs in fair numbers in all our study areas and Freese (1975; Freese et al., 1977) found no significant differences in population levels between unaltered forest areas and areas of heavy human exploitation. At the Tahuayo study site, we found four or five troops within the 1 km^2 survey area, of which three were associated with *S. mystax* troops. At the more exploited Aucayo site, however, they greatly outnumbered *S. mystax.* At our field station in the Nanay basin, we and other field workers have recorded at least four or five troops foraging by a forest trail of 1,200 m that connects our two base camps, and a transect census conducted by Curtis Freese there in 1974 gave a relative density of three to four troops per km^2 in non-inundatable primary forest habitat (for the census method employed, see Freese et al., 1977).

In 25 observed troops, the number of animals ranged from two to ten; however, most troops had two to six animals, and we suspect that some of the largest troops recorded may actually have consisted of two closely feeding or temporarily merged troops (see below). A large, "supertroop" of 15 to 26 *S. fuscicollis* was observed repeatedly by one of us and by other field workers in the same area in a span of 15 months. In June 1975, the "supertroop" was followed for almost an hour and after about 90 m of travel, it split into two separate troops, each of eight or more monkeys, that moved away in different directions (John McDermott and Dawn Starin, pers. comm.). On at least one occasion, the troops obviously had been attracted together by a common food source, a large, fruiting *leche caspi* tree (*Couma* sp.). Such temporary mergings or congregations of adjacent troops into large foraging and feeding units, presumably in overlapping areas of their home ranges, may be characteristic of *Saguinus,* as they were also observed in *S. midas* by Thorington (1968) and, apparently, in *S. (oedipus) geoffroyi* by Moynihan (1970).

One semi-tame pair of *S. fuscicollis illigeri* living unconfined in a large backyard garden in Iquitos had five sets of twins over the course of the last three years, at approximately six-month intervals. The two last births took place in late November or early December of 1974 and mid-May 1975 (Sydney McDaniels, pers. comm.). None of the previous offspring had survived long, presumably due to predation by people and cats in the neighborhood. In laboratory breeding colonies, interbirth intervals of six months and less have been reported for *S. fuscicollis* whose offspring were separated early from their parents (Wolfe et al., 1972; Wolfe et al., 1975); however, when the parents were allowed to rear the offspring, the intervals were considerably longer. Whether a semi-annual reproduction rate is maintained in wild populations, with stable, normal infant loss rates, remains to be investigated.

Cebuella pygmaea

At the Aucayo study site, we located five troops of *Cebuella* living within and near the native community of Centro Union (Figure 1), and found evidence of the presence of several other troops along the banks of the river. Three of these inhabited vegetation of the inundatable edges of the Aucayo river, a fourth lived 180 m inland at the edge of the village clearing, and a fifth at the edge of an old banana field, 0.5 km farther inland. Their presence was hardly noticed by the villagers, and since the animals were quite habituated to human proximity, they could be observed at close range.

Complete counts of individuals in all but one of the troops ranged from five to ten. The only troop of ten animals consisted of eight self-locomoting individuals and a pair of infants, born in May 1975. The three river-edge troops occupied small, contiguous, non-overlapping home ranges, all bounded on one side by the village clearing.

Although the home ranges of at least two of the troops included a portion of edge vegetation on either side of the narrow (about 7 m wide) and sinuous water course, connected by one or two established branch bridges, the river curves formed clear-cut boundaries between adjacent home ranges. The core area of each troop included one or several sap-source trees and at least one sleeping tree (Figure 3), and the linear distance between the adjacent troops' core areas was only a little over 100 m. The troop of ten had a home range of approximately 1.3 hectares and the others about 0.8 hectares each, including the water surface within the home ranges. These figures compare well with Ramirez' findings in the Río Nanay basin (Ramirez, 1975; Ramirez et al., 1977), indicating that *Cebuella* as a rule maintains small home ranges. The total overall area encompassing the three home ranges, with 19 self-locomoting animals, amounts to 3.4 hectares, of which roughly 15 percent is non-utilized land and river surface. This would give a density of 0.88 troops or 5.6 individuals per hectare, for the river-edge study site that apparently represents optimum habitat for *Cebuella* (see Ramirez et al., 1977; and Freese et al., 1977).

The fourth troop, consisting of five animals, was isolated from the other troops by the village clearing. Its core area consisted of one large emergent tree that could properly be termed a "home tree," surrounded by rather low, mixed secondary vegetation. The monkeys were seen foraging all day long in that tree and also spent the night there. Their principal sap source consisted of the aerial root, stem, and branches of a large epiphytic shrub growing on the trunk of the home tree 12 m above ground. The only other sap-source plant in use was a slender *quillo-sisa* tree that grew 5 m off the home tree. Judging from the very few holes on it, it was used infrequently.

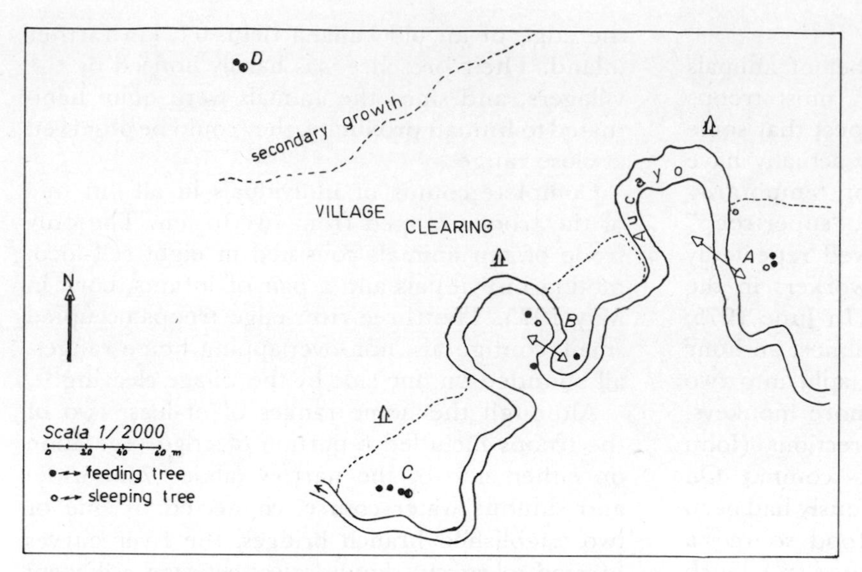

Figure 3. Location of sleeping trees and sap-source trees of four *Cebuella* troops at Centro Unión, Río Aucayo.

Acknowledgments

We thank Melvin Nelville and Curtis Freese, consultants for the PAHO study project in Peru, under whose direction the primate field studies were initiated and conducted in 1973 and 1974. We are indebted to Warren Kinzey, John McDermott, Dawn Starin, and Sybil Vosler for sharing their information on monkey sightings at the Nanay field station. Sybil Vosler also participated in the field work at the Aucayo and Tahuayo study sites. We are grateful to Mary Pearl for reviewing critically the manuscript; to Marleni Ramirez for introducing us to the fascinating world of *Cebuella*. We would also like to thank Antonio Rojas, our indispensable right-hand man on the river and in the bush.

Literature Cited

Freese, C.
1975. Final report: A census of non-human primates in Peru. Unpublished manuscript, PAHO Project AMRO-0719.

Freese, C. H.; M. A. Freese; and N. Castro R.
1977. The status of callitrichids in Peru. In *The Biology and Conservation of the Callitrichidae,* edited by Devra G. Kleiman, pp. 121–130. Washington, D.C.: Smithsonian Institution Press.

Kinzey, W.: A. Rosenberger; and M. Ramirez.
1975. Vertical clinging and leaping in a neotropical anthropoid. *Nature,* 255:327–328.

Moynihan, M.
1970. Some behavior patterns of platyrrhine monkeys. II. *Saguinus geoffroyi* and some other tamarins. *Smithson. Contribs. Zool.*, 28:1–77.

Neville, M.; N. Castro; A. Marmol; and J. Revilla C.
1976. Censusing primate populations in the reserved area of the Pacaya and Samiria Rivers, Department Loreto, Peru. *Primates,* 17:151–181.

Ramirez, M.
1975. Observaciónes sobre la ecología y comportamiento de *Cebuella pygmaea* Spix 1823 (Primates, Callithricidae) en la Amazonia Peruana. Unpublished thesis, Univ. San Marcos, Lima.

Ramirez, M. F.; C. H. Freese; and J. Revilla C.
1977. Feeding ecology of the pygmy marmoset, *Cebuella pygmaea,* in northeastern Peru. In *The Biology and Conservation of the Callitrichidae,* edited by Devra G. Kleiman, pp. 91–104. Washington, D.C.: Smithsonian Institution Press.

Thorington, R.
1968. Observations of the tamarin, *Saguinus midas. Folia primat.*, 2:95–98.

Wolfe, L.; F. Deinhardt; J. Ogden; M. Adams; and L. Fisher
1975. Reproduction of wild-caught and laboratory-born marmoset species used in biomedical research (*Saguinus* sp., *Callithrix jacchus*). *Lab. Anim. Sci.*, 25:802–813.

Wolfe, L.; J. Ogden; J. Deinhardt; L. Fisher; and F. Deinhardt
1972. Breeding and hand-rearing marmosets for viral oncogenesis studies. In *Breeding Primates,* edited by W. I. B. Beveridge, pp. 145–157. Basel: Karger.

ADELMAR F. COIMBRA-FILHO
Departamento de Conservação Ambiental,
FEEMA, Rio de Janeiro, Brazil

Natural Shelters of *Leontopithecus rosalia* and Some Ecological Implications (Callitrichidae: Primates)

ABSTRACT

Recent visits to the regions where *Leontopithecus r. rosalia* and *L. r. chrysopygus* occur produced new information on the present state of their habitats as well as certain aspects of their ecology. The adoption of a special method to capture specimens of these lion tamarins for the *Leontopithecus* Biological Bank in Rio de Janeiro, Brazil, was not only successful in yielding the needed simians, but also led to the discovery of six of their natural shelters. Four of these refuges belonged to *L. r. rosalia* and two to *L. r. chrysopygus.* Worthy of mention is the wide variation in the height of these shelters; entrances were found at distances ranging from 1.5 m to 15.0 m above the forest floor. Originally, the lion tamarins inhabited primary forests. Today, with the alteration and destruction suffered by these habitats, they are forced to live in lesser forests or secondary formations, many of which present limited possibilities for finding the natural refuges so indispensable to these animals.

Introduction

Very little is known about the shelters and nocturnal behavior of neotropical primate species. Researchers have chosen Old World species for their studies, leaving the scientific investigation of the New World primates undeveloped. Work such as that of Gautier-Hion (1971) on *Cercopithecus* (*Miopithecus*) *talapoin* in Gabon, Africa, for example, is indispensable for a broader and more precise ecological concept of the neotropical forms, which are practically unknown bionomically.

Of all the Catarrhini and Platyrrhini, *Aotus trivirgatus* (Cebidae) is the only species with nocturnal habits (Moynihan, 1964). It is much in demand for scientific research and consequently has been captured in large numbers by various techniques, including catching the animals by day in the tree holes used for shelter and sleep.

The shelter-seeking behavior adopted by *A. trivirgatus* seems to be a normal occurrence among species of Callitrichidae and most likely *Callimico goeldii* (Callimiconidae) also. Species of the former family, however, and probably the monotypic form of the latter, are the only anthropoid primates which seek refuge in hollow tree trunks and branches at night (the Prosimii are not considered here). It is my opinion that marmosets and tamarins seek cover among the components of the epiphytic community only when they cannot possibly occupy these cavities.*

Projects IUCN No. 16-2 and WWF No. 793[1] carried out by A. Magnanini and myself are aimed at establishing a nucleus for the reproduction in captivity of the three *Leontopithecus rosalia* subspecies for conservation purposes. In the course of capturing several animals for the Biological Bank certain information on the nocturnal shelters of the lion tamarins was obtained. The various shelters used by *L. r. rosalia* and *L. r. chrysopygus* are described here, as well as aspects of the present state of their habitats (see also Magnanini, 1977). Certain ecological implications of the use of shelters are also suggested.

Methods

In a narrow remnant of forest located in Guapi, Araruama County, Rio de Janeiro (Coimbra-Filho, 1969), I first sighted a natural shelter belonging to *L. r. rosalia* in September 1967. At that time, two individuals (possibly mates) retreated into a slit in the trunk of a leguminous tree around 1700 hours. The golden lion tamarins did not leave the hole again, at least during the observations which continued until dusk. The two animals had been followed for almost one hour before they took shelter. Though skittish due to the disturbance caused by constant tree felling, the lion tamarins seemed very calm as they entered their shelter.

From what was observed then, it seemed possible to accompany these animals until they sought nightly cover in holes. After having spotted the shelters, the tamarins could be captured in these same holes since they do not leave after nightfall.

This procedure was adopted for the discovery of shelters, as well as for the capture of 17 specimens of the nominal subspecies and seven of *L. r. chrysopygus* from 1973 to 1975.

Instructions for locating shelters and capturing animals were given to an experienced person who first found people living in the area with information on localities occupied by lion tamarins. Once these areas had been pinpointed, small paths were cut, penetrating the forest in such a way as to facilitate pursuit of the animals.

After discovery of the animals' shelters, the location of the tree was noted. Since all observed cavity entrances had small diameters, they were easily closed with pieces of wood or cloth. A small section of plastic tubing was placed in each cavity so the animals could breathe. With the entrances blocked, the lion tamarins could be captured at sunrise. Shelters high up in the trees were closed during the night by climbing up to the spot; the next day the tree was felled. To force the animals

Plate 1a. Rio de Janeiro State, Silva Jardim County. The beginning of forest destruction where a group of six individuals of *L. r. rosalia* were captured in May 1975. The diameter of the trees indicates that they are of quite recent secondary formation.

Plate 1b. An area, now completely devastated, where three specimens of *L. r. rosalia* were captured as late as July 1974. All the large trees were converted to charcoal. During the burning of these woods, several golden lion tamarins must surely have been killed.

**Editor's note:* Neyman (this volume) observed *S. oedipus* in sleeping trees on numerous occasions; they were never observed to use holes, but it is not clear whether such refuges were available in her study area.

[1]The Tijuca Bank of Lion Marmosets (IUCN No. 16-2, WWF Project No. 793) is a project, jointly sponsored by the International Union for the Conservation of Nature, World Wildlife Fund, and the Brazilian Government, that is devoted to the conservation and captive breeding of the genus *Leontopithecus.*

1a.

1b.

out, a small opening was made with an ax near the basal part of the cavity (Plate 2 d, Plate 3a).

The seven specimens of *L. r. chrysopygus* were taken in reasonably protected forests of the State of São Paulo, Morro do Diabo State Reserve, with proper authorization. Capture of the 17 *L. r. rosalia* specimens, also authorized, took place in privately owned forests located in Silva Jardim County (Rio de Janeiro). Unfortunately, more than 70 percent of the privately owned land where the 17 golden lion tamarins were captured no longer exists. Trees and saplings are being converted to charcoal and the rest of the biota is suffering destruction by fire with the owner's consent (Plate 1a, 1b).

Moving around is very difficult in the woods which represent the last refuges of *L. r. rosalia* in nature. Efforts are made very tiring by the existence of swampy areas, the frequent occurrence of spiny palms (*Astrocaryum ayri*) as well as the rarity of the animal. Moreover, the lion tamarins can only be successfully followed during rainy periods when the forest floor is humid and the easily frightened animals are not alarmed by footsteps crunching on dry leaves and twigs.

At no time was it possible to obtain sufficient botanical material for precise identification of the trees containing the shelters; either the trunks were dead or the tree was not in flower. For this reason determination of tree species had to be made by microscopic wood anatomy, employing commonly used techniques. Once prepared, the test material was cut in a special microtome, then stained with Delafield hematoxylin and safranin. Interpretation of the sections permitted identification to the generic level.

Results

Present Habitat of L. r. rosalia

Two authors analyzed and classified in different ways the forests where *L. r. rosalia* occurs at the present time. Rizzini (1963) designates them as "Lower Montane Rain Forest," a subdivision of the "Atlantic Forest" which covers the lower altitudes. On the other hand, Veloso (1966) considers these communities as edaphic forms of the "Tropical Seasonal Rain Forest," which also includes the arboreal forms of coastal plain vegetation and the mangroves.

The forest communities which originally covered the lower altitudes of the State of Rio de Janeiro were noteworthy for their richness of botanical and zoological species. Currently, the remnants of a once magnificent and diverse biota reflect a wan shadow of what once was. There no longer exists in this part of the state an entirely primitive forest formation. The remaining sections of better quality woods, reduced to mere fragments, suffer uninterrupted ecological degradation and are gradually being transformed into secondary formations in various stages of succession. In these extremely altered communities, the commercially valuable tree species have disappeared as have those of great ecological significance such as the larger Sapotaceae, for example. The once extremely abundant *Euterpe edulis*, an important food element for the regional fauna, has practically disappeared.

The lowland forests of today and those of small hills in the State of Rio de Janeiro are all secondary and are usually humid with some swampy areas. These forests are similar structurally but species distribution may vary, especially in relation to soil type.

No ecological, botanical or zoological study of any scope has yet been undertaken to examine these disappearing communities. Thus, the regional biota remain practically unknown to scientific research.

In the best existing forests, noteworthy species are *Plathymenia foliolosa, Apuleia leiocarpa, Schizolobium parahyba, Piptadenia* spp., *Machaerium* spp., several *Ficus*, notably *F. cattapaefolia, F. clusiaefolia, F. enormis* and *F. insipida, Nectandra* sp., *Astrocaryum ayri*, all of which are still relatively abundant at higher, better drained elevations; in the wet lowlands are found *Tapirira guianensis, Tabebuia cassinoides, T. obtusifolia, Ficus* sp., *Inga* sp. In wet places, epiphytes are common, especially various species of Bromeliaceae, Orchidaceae, etc. The beautiful orchid, *Cattleya intermedia*, was quite common a few years ago and typified the zone of occurrence of the golden lion tamarins.

With the degradation or total destruction of the primary forests originally forming this subspecies' habitat, *L. r. rosalia* has been compelled to live in very poor secondary communities. Habitat destruction has advanced at such an unbroken pace, temporally and spatially, that many species of the biota have been exterminated. As a consequence of the marked ecological instability in these relict forests, the ecological niche of the golden lion tamarin has been impaired substantially. Recently captured adult specimens of this subspecies have been relatively small in comparison to previously obtained specimens, and with inferior pelage in terms of color and shine. These wild caught specimens even looked inferior to individuals born and raised in captivity in the *Leontopithecus* Biological Bank.

It seems that populations of *L. r. rosalia* may have also suffered a serious alteration of family structure, for a group consisting solely of five females was captured in one cavity, indicating a strong abnormality in population dynamics.

The descriptions of Wied (1940, translated edition), examination of old engravings from the priceless personal collection of Gilberto Ferrez, in Rio de Janeiro, and my familiarity with the region for three decades, confirm the radical ecological changes in the habitat of the golden lion tamarin.

Description of L. r. rosalia *Shelters*

The first shelter was found on March 26, 1974, in a forest north of Juturnaiba lagoon, and a group of five lion tamarins was captured, all of which were females. The males may have been caught by predators, but more likely were shot down.

The tree containing the refuge was identified as being a species of *Piptadenia* (Leguminosae, Mimosoideae), approximately 20 m high and 0.5 m in diameter. The shelter's entrance was a narrow, vertical slit, 0.5 m long and 2.5 m above the ground. The tamarins could only enter and leave through a small, slightly wider section (about 0.06 m) located approximately in the center of this slit. The hollow interior was relatively roomy as shown in Plate 2a.

The golden lion tamarins had been frequenting this refuge for some time as evidenced by the thick layer of hair which lined the bottom, most likely deposited during moulting which normally takes place in the hottest period of the year. Pregnant females may moult a little later, some time after the birth of the young. This refuge was well protected by a leaf of the spiny palm, *Astrocaryum ayri.*

The second refuge was found on July 26, 1974, in the same forest, approximately 800 m from the previously described shelter. The group included three individuals, an adult pair and a subadult female (about one year old).

The dead tree, a species of *Nectandra* (Lauraceae) with a diameter of 0.4 m, was peppered with many interconnecting cavities. The height of the shelter's entrance was about 11 m and its width 0.08 m (Plate 2b).

On August 15, 1974, the third shelter was discovered in a forest northeast of Juturnaiba lagoon in the same region as the others. Four lion tamarins were captured: two males, one female, and an individual of undetermined sex. During the process of forcing the animals from the hole, one escaped, thus leaving three individuals.

The live tree was a *Piptadenia* sp. (Leguminosae, Mimosoideae) about 16 m high; at the height of the shelter the trunk measured 0.4 m in diameter. At the base, the tree diameter was greater due to prop roots. The two entrances, with diameters of 0.06 m and 0.08 m, respectively, were located 1.5 m from the ground. Internally, the cavity extended to a distance of only 0.5 m from the forest floor (Plate 2c, d).

The fourth refuge was found on May 8, 1975, in the same region as the previous ones. Six individuals were captured: three adults (one male and two females), two subadults (one male and one female, possibly twins born in October 1974), and one juvenile (probably born in February 1975). Ages of the immature animals were estimated by comparison with individuals born in captivity.

This refuge was located in a *Nectandra* sp. (Lauraceae) with a stump diameter of 0.4 m. The circular cavity entrance (0.08 m in diameter) was located 13 m above the ground. Observable scarring at the margins gave this shelter the appearance of having been caused by an initial insect attack. In the bottom of the refuge was a layer of hairs from the moulting period of the gold lion tamarins (Plate 3a, b, c).

Present Habitat of L. r. chrysopygus

São Paulo State covers an area of 248,600 km^2. Cavalli et al. (in Victor, 1975) reconstructed the primitive arboreal vegetation and calculated that 81.8 percent of this total area had been covered by forest. They estimate that in 1854, or 35 years after Natterer had discovered *L. r. chrysopygus* (Pelzeln, 1883), forest vegetation still covered about 79.7 percent of São Paulo. During this period, the forests in the western half of the state where all known specimens of the golden-rumped lion tamarin have been captured, was practically unaltered. In the last five decades, however, destruction of São Paulo's native forests has reached unequalled proportions, and the remnants are about 8 percent of the total original coverage (Serra-Filho, et al., 1974).

The geographic distribution of *L. r. chrysopygus,* as well as some aspects of the habitat, were treated in a previous paper (Coimbra-Filho and Mittermeier, 1973). Distribution of the race in question was restricted to the Pontal do Paranapanema region. At the present time, the geographic range of *L. r. chrysopygus* is limited to less than half of

2a.
2b.
2c.
2d.

Plate 2a. A *L. r. rosalia* shelter in a *Piptadenia* sp. (Leg. Mimosoideae). The animals can enter only through the widest part of the opening, which is at a height of 2.5 m above the forest floor.

Plate 2b. A *L. r. rosalia* shelter in the dry trunk of a *Nectandra* sp. (Lauraceae). The shelter was located at a height of 11 m and presented external openings, internally interconnecting. The two holes shown on the branch at the right were formerly the nest of a medium-size species of woodpecker.

Plate 2c. A *L. r. rosalia* shelter in a *Piptadenia* sp. (Leg. Mimosoideae), similar to the shelter found in a tree of the same genus seen in Plate 2a. The entrance to this shelter was nearer the ground, about 1.50 m in height.

Plate 2d. Detail of the base of the same tree showing the holes hacked open to facilitate capture of the golden lion tamarins.

Plate 3a. Section of the trunk of *Nectandra* sp. (Lauraceae) showing the natural entrance to the shelter and the holes which were opened with an ax. Observe the scarring of the cavity entrance.

Plate 3b. The trunk section shown in Plate 3a belongs to the distal part of the lighter trunk which appears in the foreground. The darker trunk, lying to the right, served to soften the fall of the tree where the six golden lion tamarins were sheltered.

Plate 3c. The base of the same tree as in Plate 3b. Observe the spot from which test material for microscopic wood anatomy was obtained.

the 37,147 hectares of Morro do Diabo Reserve (MDSR).[2] The heterogeneous vegetation of this forest preserve has some primary features; a goodly portion, however, consists of degraded woods or secondary formations. Many sections of the original forest have degenerated greatly, with many clearings caused by the felling of large trees some years ago.

In MDSR, the lion tamarins most frequently occupy the type I woods of Campos and Heindijk (1970), mainly near the Paranapanema River where the first live specimens were captured.

The habitat of *L. r. chrysopygus* is, biotically, superior to that of the nominal subspecies in Rio de Janeiro State, in spite of certain negative aspects. Total destruction of the forests of western São Paulo has clearly provoked ecological instability in this area. Also the MDSR is presently isolated and surrounded by some highly prejudicial human activities, such as the recent use of the defoliants 2,4,5T and 2,4D. Finally, a railroad line already cuts through the park, and construction of a wide highway has divided MDSR in two. Be that as it may, the golden-rumped lion tamarin has a much greater chance of finding adequate hollow shelters in trees than does *L. r. rosalia* in its present habitat.

Some of the larger forms of Picidae and Psittacidae are still relatively abundant in the MDSR forests. The presence of these birds is of great ecological importance to the Callitrichidae for their abandoned shelters may be taken over by the tamarins.

Description of L. r. chrysopygus *Shelters*

On November 11, 1973, the first shelter belonging to *L. r. chrysopygus* was discovered in Morro do Diabo State Reserve. The forest was rich in fructiferous species, principally Myrtaceae and some *Ficus.* Initially, the group was made up of four, apparently adult animals, but at night they separated into two groups of two individuals each. An adult pair was captured in the abandoned nest of a large woodpecker, built 15 m above the forest floor in the trunk of a dead, branchless *Cabralea* sp. (Meliaceae) (Plate 4a). The entrance to the

[2]A second population of *L. r. chrysopygus* was discovered in August 1976 in an area of remnant forest in Galia, central São Paulo State. The population is tiny, but indicates a new and important genetic potential.

Plate 4a. A *L. r. chrysopygus* shelter in the dry trunk of a *Cabralea* sp. (Meliaceae). The golden rumped lion tamarins were sheltered in the abandoned nest of a large species of woodpecker. The hollow interior is outlined on the photograph.

Plate 4b. A *L. r. chrysopygus* shelter in the fissure of a branch of *Holocalyx* sp. (Leg., Caesalpinoideae), about 14 m above the ground.

4a.

4b.

shelter measured 0.06 m in diameter.

The second *L. r. chrysopygus* shelter was discovered on November 18, 1973, in a tree located about 500 m from where the first pair was captured. On this date, five adult individuals were taken, two males and three females. Two well-developed individuals, one male and one female, were assumed to be the founding pair of the group.

The tree containing the shelter was a tall, aging individual of the genus *Holocalyx* (Leguminosae, Caesalpinoideae). The cavity opening measured 0.07 m at its widest part and was located in a branch 14 m from the ground. Since this shelter was not photographed, a sketch is included (Plate 4b).

Discussion and Conclusions

The constant state of alert of the lion tamarins, their swift flight when frightened, as well as their early evening retirement, around 1700 to 1730 hours under normal conditions, are indicative of the intense predation of which they were victims in the past, when their habitats were primary communities. I believe that any study of predation on the lion tamarin would be impaired by the fact that natural predation must be substantially diminished at present, at least in the case of *L. r. rosalia.* The principal diurnal predators of this subspecies, especially *Spizaetus t. tyrannus* and other great birds of prey of the family Accipitridae, have practically disappeared from the area. The black hawk-eagle is thought to be the most efficient predator of *L. r. chrysomelas* in the forests of Bahia, whereas *L. r. chrysopygus* would potentially be preyed upon by the ornate hawk-eagle (*Spizaetus o. ornatus*). In the region where the latter lion tamarin occurs, *S. o. ornatus* has always been more frequently encountered than *S. t. tyrannus.*

Although we have never observed these rare hawk-eagles attack primates, they do eat various simian species. The captive lion tamarins exhibit intense fear upon seeing stuffed, mounted specimens of these birds through the wire screen of their cages and all three subspecies manifest great agitation and retreat when any large bird flies over their cages. Even the low, abrupt, noisy flights of icterid *Cacicus haemorrhous affinis,* a relatively small bird, may startle the animals under certain circumstances, as was recently observed in the natural habitat of *L. r. rosalia,* Silva Jardim County. Perhaps this behavior supports Epple's suggestion (1968) that the fear evoked in these animals by the flight of any bird may be an innate character since it is obvious that *C. h. affinis* is not a predator of *L. r. rosalia.*

Another aspect of the behavior of *L. r. rosalia* and *L. r. chrysopygus* worth mentioning is that, especially upon sensing imminent danger, they often descend to the lower strata of the forest. This may function to evade an aerial attack as well as to capture certain prey. Moynihan (1970) refers to an observation of C. M. Hladik that *Saguinus (oedipus) geoffroyi* also descends to the forest floor in search of food.

The small diameter of all observed shelter openings suggests that they may function to bar entrance to potential predators larger than the lion tamarins, such as *Bubo virginianus* (Strigidae) and certain mammals, especially the smaller felines (*Felis pardalis, F. yagouaroundi, F. wiedii* and *F. tigrina*). If the lion tamarins are hindered from seeking the safety of the shelters, principally at night, they may be easy prey to the aforementioned predators. A recent occurrence (March 1975) at the *Leontopithecus* Biological Bank enhances this supposition. One afternoon a horde of *correição* ants (*Eciton* sp.) ransacked several cages, expelling a pair of *L. r. chrysopygus* with two grown young from one of the box shelters. Frightened by the bites of these aggressive ants, the golden-rumped lion tamarins remained outside the shelter until nightfall when they were discovered by the night watchman on his rounds. The four animals were huddled in a ball at the far end of one of the perches. Even though the ants had already left, the animals would enter only after the caretaker had carefully examined the nest. In a natural setting, the lion tamarins would be extremely vulnerable to nocturnal predation for they have no means of defending themselves in the dark.

Competition may exist between *Leontopithecus r. chrysopygus* and *Cebus apella vellerosus.* In the forests of Morro do Diabo Reserve, *C. apella vellerosus* is quite common, while *L. r. chrysopygus* is rare. Even though numerous individuals of the black-capped capuchin have been spotted in this preserve, no opportunity was presented to observe them interacting with the lion tamarins.

Klein and Klein (1973) refer to the meeting of *Cebus apella* and *Saimiri sciureus,* the latter of about the same size as *L. r. chrysopygus.* According to these authors, *C. apella* and *S. sciureus* gained mutual advantage from the association, most conspicuously during foraging. They verified that these two species may stay together the whole day. Thorington (1968), who also observed relations between these same species, considers such meetings accidental and remarks that individuals of one species sometimes isolate themselves in groups of the other species, but they are sometimes persecuted.

By contrast, F. M. de Oliveira, who was responsible for the capture of *L. r. chrysopygus* specimens in São Paulo State, noted that even when foraging in a favorite fructiferous species, the golden-rumped lion tamarins left immediately upon sighting individuals of *C. apella.* The fear on the part of the lion tamarins may signify aggressive actions by *C. apella* against this subspecies.

Nothing is known about the possible interaction of *L. r. rosalia* and *C. apella* in the forests where they are, or at least were up to a few years ago, sympatric. In the region where the last specimens of the golden lion tamarin survived, *C. a. nigrittus,* as well as *Alouatta f. fusca,* are found in extremely reduced numbers, bordering on extinction. Four individuals of *A. f. fusca* were recently seen in the same place where four specimens of *L. r. rosalia* were captured. Since these two species occupy quite diverse niches, they most likely do not compete directly.

In the State of Bahia, *L. r. chrysomelas* is sympatric with a race of *Cebus apella,* but nothing is known as of now about their interactions.

With the discovery of shelters used by *Leontopithecus,* it is apparent that certain types of tree holes are indispensable for the nightly protection and consequent survival of these individuals. It also seems evident that the cavities are chosen very carefully and occupied, when adequate, for quite some time. In the bottom of two *L. r. rosalia* shelters, hair deposits were found, forming a layer about 1 cm deep, which demonstrates a certain fealty to the refuges. Estimates of a one-year occupancy, or more, were made from the quantity of hairs found.

The number of sufficiently large trees, old enough to provide adequate shelter was originally much greater in the habitats of *L. rosalia,* especially in that of the subspecies *L. r. rosalia.* Considering these ecological peculiarities of *L. rosalia,* the distribution of shelter-boxes which would be hung from the trees in areas specifically designated for the preservation of these small primates is of importance. This measure is recommended especially in young secondary forest where the trees are of slight diameter and adequate shelters do not exist. Although such action might cause a reduction in the areas occupied by each family group, especially if the favorite fructiferous species were placed under agronomical management, it would be admissible in degraded or secondary forest areas, specifically established to preserve forms threatened with extinction. This procedure would also improve such forests from an ecological aspect, aiding the economic utilization of marmosets, tamarins, and even *Aotus trivirgatus,* by the ranching method. Of course, these measures should never be recommended for biological preserves established in primary ecosystems in the climax stage which have as their main aim the safeguarding of the regional biotic equilibrium.

Acknowledgments

I would like to express my sincere thanks to Mr. Francisco M. de Oliveira for the remarkable dedication shown in performing all the tasks assigned him. Much information concerning the shelters of lion tamarins was only possible due to his willingness and tireless field work.

I am indebted to Dr. A. Magnanini and Dr. A. D. Aldrighi, our collaborators in the enterprise of administering the *Leontopithecus* Biological Bank for their help in some aspects of the work. Special thanks are due to Dr. A. D. Aldrighi, director of the Tijuca National Park, for having lent some rare stuffed mounted specimens from the Fauna Museum (IBDF-MA) for my research, and to Dr. A. Mattos-Filho, for his help in microscopic wood anatomy.

I am greatly indebted to Dr. Devra Kleiman for her invitation to participate in the "Conference" and for her efforts in obtaining from the Smithsonian Institution the resources guaranteeing my presence, as well as to Dr. Benjamin D. Blood for completing coverage of expenses through the National Institutes of Health.

I would finally like to thank Dr. Dorothy Dunn de Araujo for her generous collaboration in translating my paper into English and offering comments.

The author is a fellow of the Brazilian National Council for Scientific and Technological Development (CNPq).

Literature Cited

Campos, J. C. C. and D. Heinsdijk.

1970. A floresta do Morro do Diabo. *Silvicult., São Paulo,* 7:43–58.

Coimbra-Filho, A. F.

1969. Mico-leão, *Leontideus rosalia* (Linnaeus 1766), situação atual da espécie no Brasil. (Callithricidae-Primates). *An. Acad. Brasil. Ciênc.*, 41 (Supl.) :29–52.

Coimbra-Filho, A. F., and R. A. Mittermeier.
1973. Distribution and ecology of the genus *Leontopithecus* Lesson 1840 in Brazil. *Primates*, 14(1) :47–66.

Epple, G.
1968. Comparative studies on vocalization in marmoset monkeys (Hapalidae). *Folia Primat.*, 8:1–40.

Gautier-Hion, A.
1971. L'ecologie du talapoin du Gabon. *La Terre et la Vie*, 25(4) :227–490.

Klein, L. L. and D. J. Klein.
1973. Observation on two types of neotropical primate intertaxa associations. *Am. J. Phys. Anthrop.*, 22:233–238.

Magnanini, A.
1977. Progress in the development of Poço das Antas Biological Reserve for *Leontopithecus rosalia rosalia* in Brazil. In *The Biology and Conservation of the Callitrichidae*, edited by Devra G. Kleiman, pp. 131–136. Washington, D.C.: Smithsonian Institution Press.

Moynihan, M.
1964. Some behavior patterns of Platyrrhine monkeys. I. The night monkey (*Aotus trivirgatus*). *Smithsonian Misc. Coll.*, 146(5) :1–84.
1970. Some behavior patterns of Platyrrhine monkeys. II. *Saguinus geoffroyi* and some other tamarins. *Smithson. Contrbs. Zool.*, 28:1–77.

Pelzeln, A. von.
1883. Brasilische Säugethiere. Resultate von Johan Natterer's Reisen in den Jahren 1817 bis 1835. Wien.

Rizzini, C. T.
1963. Nota prévia sobre a divisão fitogeográfica do Brasil. *Rev. Bras. Geogr.*, 25(1) :1–64.

Serra-Filho, Ra., A. C. Cavalli, J. R. Guillaumon, J. V. Chiarini, C. M. de A. Ivancko, J. L. Barbieri, P. L. Douzeli, A. G. de S. Coelho, and J. Bittencourt.
1974. Levantamento de cobertura vegetal natural e do reflorestamento no Estado do São Paulo. *I. F., São Paulo, Bol. Tec.*, 11:1–53.

Thorington, R. W., Jr.
1968. Observations of squirrel monkeys in a Colombian forest. In *The Squirrel Monkey* edited by L. A. Rosenblum and R. A. Cooper, pp. 69–85. New York: Academic Press.

Veloso, H. P.
1966. *Atlas Florestal do Brasil.* Rio de Janeiro: Serv. Inform. Agric., Min. Agric.

Victor, M. A. M.
1975. *A Devastação Florestal.* São Paulo: Ed. Soc. Bras. Silv. 48 pp.

Wied-Neuwied, M. von.
1940. *Viagem ao Brasil.* Translation by E. S. de Mendonça and F. P. de Figueiredo. 551 pp. Cia. Ed. Nac., Brasil.

MARLENI FLORES RAMIREZ
Puerto Rico Nuclear Center,
Tropical Agro-Sciences Division,
Mayaguez, Puerto Rico, 00708

CURTIS H. FREESE
Department of Pathobiology,
Johns Hopkins University
615 N. Wolfe Street,
Baltimore, Maryland, 21205
and

JUAN REVILLA C.
Universidad Nacional Mayor
de San Marcos,
Lima, Peru

Feeding Ecology of the Pygmy Marmoset, *Cebuella pygmaea,* in Northeastern Peru

ABSTRACT

From July to November 1974, a troop of *Cebuella pygmaea* was studied in a seasonally inundated forest along the Rio Nanay of northeast Peru. Data were also obtained from other areas on *Cebuella* ecology. Number of individuals in the main study troop varied from seven to nine with twins born in July. Most of the feeding-associated time of the troop was spent securing exudate from five tree species. Searching for and eating insects, and infrequent fruit eating, constituted the rest of their feeding time. The home range measured approximately 75 m by 40 m.

Findings indicate that within a given river basin the local distribution and density of *Cebuella* are largely a function of the distribution and density of the two or three most preferred species of exudate-source trees. Thus, in a well-surveyed forest along approximately 1.5 km of a small tributary stream of the Nanay only two *Cebuella* troops were found, and these two troops were located in the only two sites where the exudate-source trees occurred in distinctly higher densities. Also, species used as exudate sources by *Cebuella* changed between the Nanay and the Ampiyacu rivers. This was associated with a major difference in *Cebuella* distribution: *Cebuella* appear to inhabit only inundatable forests along the Nanay whereas they inhabit both inundatable and non-inundatable forests along the Ampiyacu.

Data are presented on more specific aspects of *Cebuella* resource utilization, particularly with respect to their exudate-eating habit.

Introduction

Information on the ecology and behavior of wild troops of the pygmy marmoset, *Cebuella pygmaea*, has begun to appear only recently from observations in Colombia (Izawa, 1975; Hernandez-Camacho and Cooper, 1976; Moynihan, 1976) and Peru (Freese, 1975; Kinzey et al., 1975; Castro and Soini, 1977), but detailed ecological data on this species are still scarce. One of the most important and interesting findings reported was that *Cebuella* obtain tree "sap" for food by digging holes in the trunks of certain tree species. Whether the tree exudates eaten by *Cebuella* are actually sap, gum, or both is unclear and awaits further study. Thus, in this paper we will refer to the substances eaten as exudates.

With the objective of learning more about the ecology of *Cebuella* and the factors that affect their population levels and distribution, a study of one troop of this species was done from July to November 1974 along the Rio Nanay in northeastern Peru. Additional information on *Cebuella* ecology was obtained during expeditions along the Nanay and Ampiyacu rivers in Peru.

Description of Study Areas

The meanderous Nanay and Ampiyacu rivers are northern tributaries of the Amazon, the Nanay discharging its waters just below Iquitos and the Ampiyacu about 170 km downstream from Iquitos (Figure 1). The Nanay is a large, black-water river, at least 200 m wide along much of its lower section. The forests along both sides of the river are flooded extensively each year from about February to July. Occasional small hamlets occur along the Nanay, with residents practicing slash-and-burn agriculture. Selective cutting in the inundatable forests of the Nanay has been intense although the intensity decreases as one gets farther from Iquitos. Except for the small areas in agriculture, the non-inundatable forests have received little cutting. The locals depend greatly on hunting and fishing for protein. The larger monkeys are intensively hunted, but *Cebuella* is caught only infrequently for the local pet trade.

The Ampiyacu is less than half as wide as the Nanay, superficially appears to be intermediate between a white- and black-water river, and has less extensive areas of inundated forest than the Nanay. Human disturbance has been similar to that of the Nanay.

The main study troop was located on a 40 m to 50 m wide peninsula formed at the junction of the Nanay River and one of its small (approximately 8 m wide) tributaries, San Pedro stream, a very meanderous stream entering from the south (right side, facing down river) (Figure 2). This study site is 0.5 km below the small hamlet of

Figure 1. Locations of the Nanay River, Ampiyacu River, and San Pedro Stream.

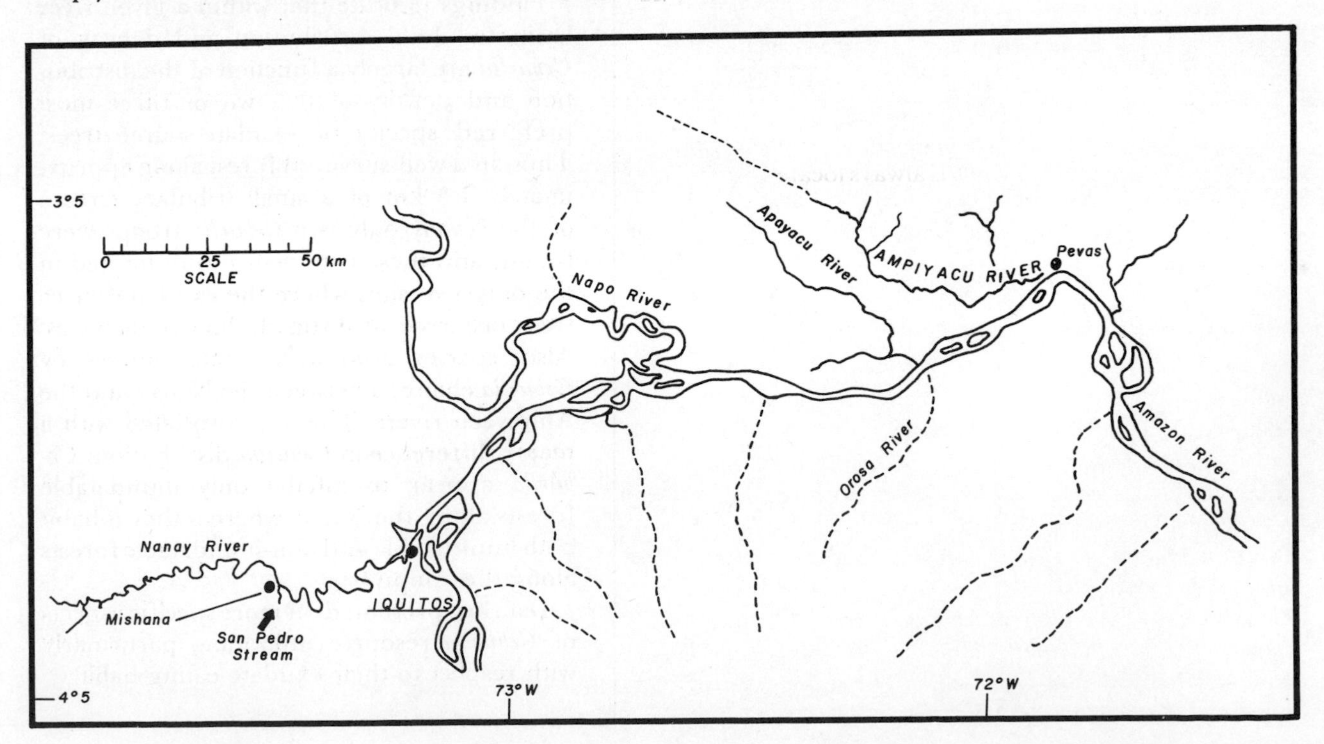

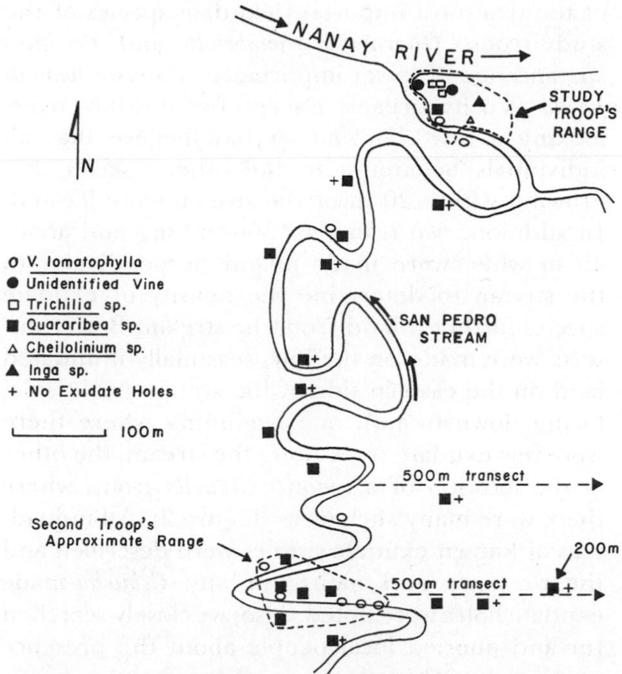

Figure 2. Location of the ranges of the main study troop and another troop of *Cebuella*, and of individuals of exudate species. (Tree species have been identified as *Quararibea rhombifolia* and *Cheiloclinium cognatum.*)

Mishana, and approximately 40 km southwest of Iquitos (Figure 1). A family lived directly across San Pedro stream from the study troop until shortly before the study began; apparently they never bothered the troop.

The soils of the main study area are inundated annually for four to five months. The forest, which has received heavy selective cutting, can be classified into two major types in the range of the main study troop. The *marañal alto* is always located along the edge of rivers, lakes, or large streams, and is characterized by dense growth comprised of many vines and densely branched trees averaging 20 m tall. The *marañal alto* forms a narrow belt no more than 10 m wide which surrounds the other forest type in the study troop's range, the *tahuampal alto.* A virgin *tahuampal alto* is characterized by a closed canopy with trees averaging 25 m tall, and by scarce undergrowth. Substantial secondary forest, however, occurs within the *tahuampal alto* habitat in the study area, and the forest averages around 20 m tall. The most common tree species in the principle study area are *Campsiandra laurifolia* (Caesalpiniaceae), *Eschweilera* sp. (Lecythidaceae), *Trichilia* sp. (Meliaceae), and *sachacaimito.* Figure 3 shows the approximate location of all trees (with the possible exception

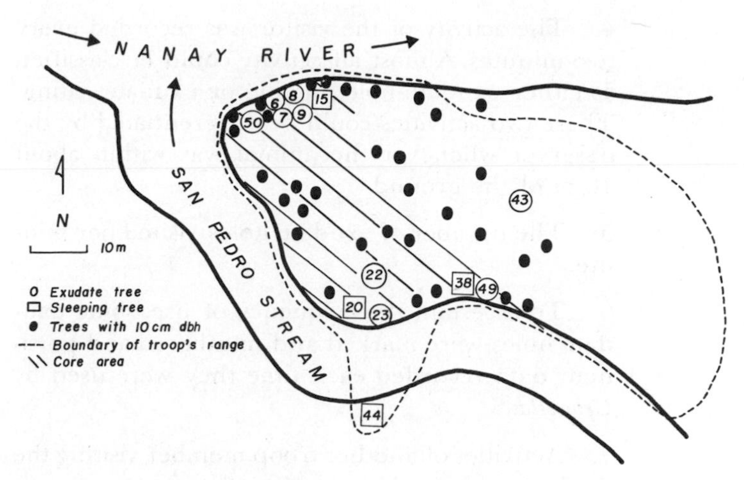

Figure 3. The range and core area of the study troop showing locations of sleeping trees, exudate trees, and trees from which *Cebuella* ate fruit. In the western two-thirds of the range, all trees (except possibly one or two) with greater than 10 cm dbh are mapped. See text for identification of numbered trees.

of one or two) with greater than 10 cm diameter breast height (dbh) in the western two-thirds of the troop's range.

Methods

Two hundred forty hours of observation of the main study troop, plus 20 hours of work in their range without contact, were made by M. Ramirez intermittently from July 14 to November 20, 1974. The troop also was observed for several hours by the other two authors.

The animals were watched with the naked eye and with 7x50 and 7x35 binoculars. We tried to minimize disturbing troop members and the animals almost immediately habituated to observers; individuals would approach to within two meters of us while feeding.

Two data recording methods were employed for the study troop. One method, termed here the "focal tree" method, consisted of continuously watching a tree that *Cebuella* used as an exudate source (referred to henceforth as exudate trees), and recording all *Cebuella* activity in that tree. The following data were recorded with this method.

1. Age, and when possible, sex of the visitor.

2. Time when the visitor entered and left the tree.

3. The visitor's height in the tree, recorded every minute,with categories of 0 to 5 m, 5 to 10 m, 10 to 15 m, etc.

4. The activity of the visitor was recorded every two minutes. Almost all activity could be classified as either exudate-hole digging or exudate eating. These two activities could be differentiated by the observer whenever the animal was within about 10 m of the ground.

5. The number of exudate holes visited per minute.

6. To ascertain the frequency of use, some exudate holes were marked and numbered and pertinent data recorded each time they were used by *Cebuella*.

7. Activities of another troop member visiting the tree simultaneously.

The most important species of exudate trees (one, *Cheiloclinium cognatum* (Miers) A. C. Smith was a tree-like vine) were watched a total of 172.5 hours with this method, with the following distribution for each tree (see Figure 3 for the location of each tree): *Quararibea* aff. *rhombifolia* (Cuatr.) Macbr. (Bombacaceae) (No. 22), 69 hrs; *Trichilia* sp. (No. 7), 34.5 hrs; *Trichilia* sp. (No. 8), 23 hrs; *Trichilia* sp. (No. 6), 23 hrs; *Cheiloclinium cognatum* (Hyppocrataceae) (No. 49), 23 hrs. Of this total, 23 hours of observation during two consecutive days were obtained for each tree. Other focal-tree observations were in periods of 6 to 7 hours. When the focal tree was not occupied by *Cebuella*, other data could be recorded opportunistically since animals often were within view elsewhere in the range.

For daily activity patterns of individuals, the "focal animal" method was employed. The adult male *Cebuella* was followed as closely as possible for three days, totaling 34.5 hours, and the adult female for one day, 11.5 hours. The focal animal was in view of the observer approximately 80 percent of the time, during which its activity was continuously recorded to the nearest minute. Periods when they could not be seen by the observer are not included in activity calculations.

Data obtained from the above two methods were analyzed separately. We will refer to the method used for each section of the results.

Within the western two-thirds of the range of the main study troop, every tree of more than 10 cm dbh, except possibly one or two, was described, identified if possible, and mapped. All exudate trees within the entire range of the troop were mapped (Figure 3).

Along the lower 1.5 km of San Pedro stream, we attempted to locate and map all individuals of the two most important exudate species of the study troop, *Quararibea rhombifolia* and *Trichilia* sp., and one of lesser importance, *Vochysia lomatophylla* (Vochysiaceae). Except for possibly overlooking a few *Trichilia* sp., we believe that all individuals belonging to the other species and growing within 20 m of the stream were located. In addition, two transects, 500 m long and about 40 m wide, were made inland perpendicular to the stream to determine the density of exudate species farther inland from the stream. Both transects were made on the low, seasonally inundated land on the eastern side of the stream (right side, facing downstream), one beginning where there were few exudate trees along the stream, the other at the location of a second *Cebuella* troop where there were many such trees (Figure 2). All individuals of known exudate species were described and the presence and nature of any *Cebuella*-made exudate holes were noted. Also, we closely searched for and queried local people about the presence of any other *Cebuella* troops along the stream.

Monkey censusing expeditions along the entire lengths of the Nanay and Ampiyacu rivers by C. Freese and J. Revilla yielded additional data on *Cebuella* ecology. During an expedition to the headwaters of the Nanay in May 1974, people were questioned about the presence and habits of *Cebuella*. In two expeditions to the Ampiyacu river in August and November 1974, general notes were taken on the distribution and abundance of tree species whose exudates were known to be eaten by *Cebuella*. Two *Cebuella* troops were briefly observed and notes were taken on four other areas known to be inhabited by *Cebuella*. Especially during the second Ampiyacu trip, we carefully watched for *Cebuella* exudate holes in trees, both from the motorboat and during extensive foot and canoe work through the forest. People along the Ampiyacu also were questioned about the habits of *Cebuella*.

The chemical composition of the exudates of *Q. rhombifolia* and *V. lomatophylla* are being analyzed by Claude-M. Hladik.

Results

The Main Study Troop

The study troop was initially composed of seven individuals: one adult male, one adult female, two sub-adult males, two juveniles, and one independently locomoting infant. In July, soon after the study began, twins were born, bringing the total to nine, but one of these disappeared in late August.

The troop occupied a range measuring roughly 75 m by 40 m (3000 sq m). As shown in Figure 3, this entire area fell within the peninsula except for a site across the stream where the troop had an unidentified (taxonomically) sleeping tree (No. 44). Castro and Soini (1977) reported *Cebuella* troops of five to ten animals, but the home ranges of these troops were three to four times larger than the home range of our study troop. Troops of three to six (Moynihan, 1976) and of ten to fifteen (Hernandez-Camacho and Cooper, 1976) have been reported in Colombia.

The focal animals spent 32 percent of observed daily activity eating exudate and excavating exudate holes, 16 percent foraging for insects, 11 percent traveling, 32 percent inactive, and 9 percent in other activities, such as grooming and play. Thus, 67 percent of the focal animals' total feeding time was dedicated to obtaining tree exudate and 33 percent to insect foraging. Although focal animals did not eat fruits, the fruit of a *Ficus* sp. (Moraceae) and an unidentified Sapotaceae were occasionally eaten by *Cebuella*.

The relative intensity of utilization by *Cebuella* (eating exudate or digging holes) of the five focal trees was determined from the data from two consecutive days of focal tree observation. The percent of time that each of the five focal trees was in use out of the total utilization time for the five trees was calculated. Utilization time was not weighted to account for more than one *Cebuella* in a tree. *Quararibea rhombifolia* (No. 22) accounted for 44 percent of the total utilization time, the three trees of *Trichilia* sp. represented 10 percent (No. 6), 22 percent (No. 7), and 4 percent (No. 8) of the total, and a vine, *Cheiloclinium cognatum* (No. 49) represented 20 percent of the total. Two other species, *V. lomatophylla* (No. 23) and *Inga* sp. (Leguminosae) (No. 43), were used very infrequently during the study. Two dead, unidentified vines (No. 9 and No. 50) also were found with abandoned exudate holes made by *Cebuella*. The single *Q. rhombifolia* was clearly used more than any other plant for exudate feeding by the troop; the percentage for *Q. rhombifolia* would be even higher if times were weighted to account for simultaneous utilization by two or more *Cebuella*. The *Q. rhombifolia* also had the greatest absolute number, and probably density, of exudate holes. Table 1 gives the heights, dbh, and the relative density of exudate holes for each of the exudate species. Some characteristics of the exudate holes and exudate of these species are listed in Table 2.

Cebuella excavate their exudate holes almost exclusively on the trunk and sometimes major branches of the trees. We found exudate holes always densest within a few meters of the base of the trunk, with some holes usually within .5 m of the ground. Izawa (1975) also found that "sap stains" made by *Cebuella* were conspicuous from near the ground to 6 m high. Digging is done with the semi-procumbent lower incisors while the head is moved up and down. During periods of hole digging and sap feeding, *Cebuella* adopt a variety of orientations varying from a vertical, headup position, to, less commonly, the head being lower than the body. They move up and down the tree head first. The positions and movements of the body while exudate feeding and notes on their locomotion, especially concerning vertical clinging and leaping, are described by Kinzey et al. (1975).

In a single exudate tree four intergrading types of exudate holes usually occur: 1) apparently young, extremely shallow holes that do not exude; 2) small holes, but sufficiently deep to exude a little; 3) large holes that exude a lot (Figure 4a); 4) abandoned holes, of varying sizes but usually large, showing varying degrees of cicatrization (Figure 4b). In general, we had the impression that there were more abandoned holes than young holes low on the trunk, but that the trend tended to reverse higher on the trunk. The same pattern was reported by Moynihan (1976). These observations suggest that *Cebuella* first dig holes low on the trunk and subsequently dig higher.

The shape and size of holes varies greatly between tree species (Table 2). Exudate-producing holes of the *Q. rhombifolia* (No. 22) were usually round and 12–15 mm in diameter, although there were a few oblong ones greater than 25 mm long. Most were from 5–10 mm deep, averaging 8–9 mm. In the *Q. rhombifolia*, it appeared to be common practice for *Cebuella* to excavate three to four round holes in a tight clump so their edges overlapped. Holes in the *Trichilia* sp. were oblong, usually measuring 8–12 mm wide by 30–60 mm long, but reaching lengths of over 100 mm. These holes averaged 4–5 mm deep (Figure 4a). The round exudate holes of a giant *Parkia* sp. (Leguminosae) along the Ampiyacu were very large, some more than 20 mm in diameter and approximately 15 mm deep.

A vertical cross-section through these various kinds of *Cebuella* excavations would show that the upper edge of the hole is relatively perpendicular to the trunk's face, while the lower edge enters

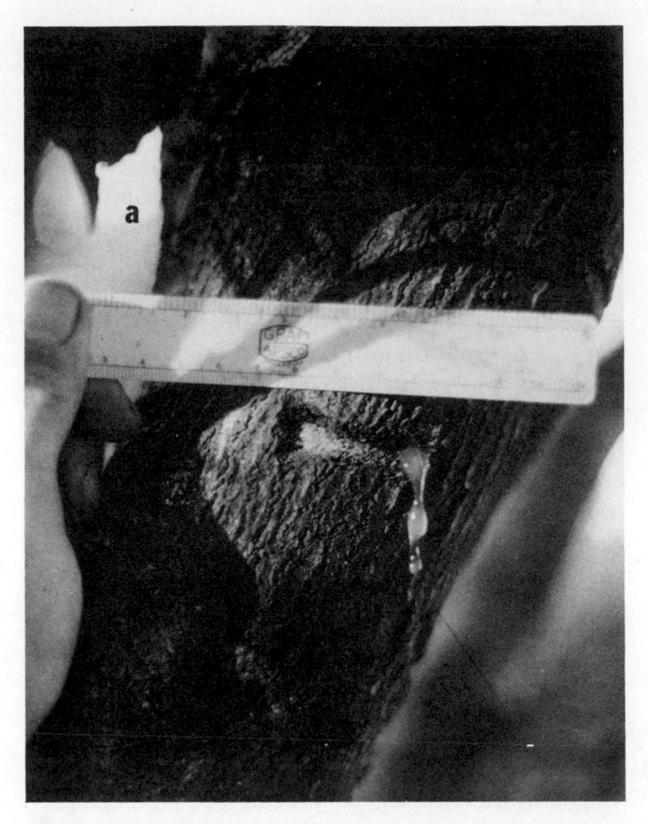

Figure 4. Exudate holes in a *Trichilia* sp. in the range of the study troop: a, a highly productive exudate hole and the exudate; b, an abandoned hole showing advanced cicatrization.

the bark at a gentler slope. This shape would be the expected result of a head-up animal digging with its lower incisors.

Rates of exudation from the holes vary greatly. A hole in the *Q. rhombifolia* may exude a glob the size of a marble in less than one-half hour, whereas the exudation of a *Trichilia* sp. in one-half hour would be less than raindrop size.

As noted above, the density of holes decreases as one proceeds up the tree trunk. On the *Q. rhombifolia* in the range of the study troop, the density at about chest to head height was around 480 holes per square meter. Less than 10 percent of these holes appeared to be exuding at any one time. The percent of holes with exudations on the *Q. rhombifolia* appeared to decrease higher up the trunk, despite the fact that there was probably a lower proportion of abandoned holes high on the trunk.

Exudate holes often occur in distinct clumps or in horizontal bands of several hundred (Figure 5a, b). One *Quararibea* cf. *spatulata* Ducke (different species from the one in the study troop's range) along the Ampiyacu was found with an incredibly high density of approximately 1,300 holes per square meter at head height, and with an unusual variation in the overall distribution of its holes (Figure 5a). As with other exudate species, the holes on the main trunk were concentrated near the base (the lowest hole was 5 cm from the ground), the density dropping off sharply above 3.5 m with essentially no holes on the top few meters of the main trunk. However, at about 13 m the trunk divided into three branches, and the base of each branch had a high density of holes. The same pattern was seen on another tree of the same species in this area. On a giant *Parkia* sp. along the Ampiyacu, exudate holes were concentrated along both sides of a pair of pencil-thin vines clinging to much of the trunk's length.

Using the "focal tree" technique, we determined the pattern of use of the exudate trees by the *Cebuella*. The exudate trees usually were first entered shortly after 0600 hours, the troop's first activity of the day. As shown in Figure 6, the first two hours or more were primarily dedicated to eating exudate, especially the first hour. After 0900 hours, and throughout the rest of the day, however, most (approximately 70 percent) time in exudate trees was spent excavating holes, with the very last visits almost entirely for excavation. Overall, on the lower 10 m of the trunk, 57 percent of their active time in exudate trees was dedicated to excavating holes, and 43 percent to eating

Table 1. Height and dbh (diameter at breast height) of the exudate trees in the range of the study troop and the relative density of exudate holes in them.

Tree	*Height (m)*	*dbh (cm)*	*Relative Density of Exudate Holes*
Quararibea rhombifolia (No.22)	25	45	High
Trichilia (No.6)	8	12	Low
Trichilia (No.7)	15	15	High
Trichilia (No.8)	10	8	High
Cheiloclinium cognatum (No.49)	13	10	Low
V. lomatophylla (No.23)	23	25	Low
Inga (No.43)	15		Low
Vine (No.9)	12	5	Low
Vine (No.50)	20	8	Low

Table 2. Some characteristics of the exudates and exudate holes of the exudate tree species in the range of the study troop.

	Trees					Vines
Exudate characteristics	*Quararibea rhombifolia (No. 22)*	*Trichilia (Nos. 6, 7, 8)*	*Cheiloclinium cognatum (No. 49)*	*V. lomatophylla (No. 23)*	*Inga (No. 43)*	*(Nos. 9, 50)*
Consistency	viscous	viscous	liquid	viscous		
Color	clear	clear	clear, to milky in air	viscous		
Taste to human	no taste	no taste	sweet-sour	no taste		
Odor to human	no odor	no odor	aromatic	no odor		
Exudate Holes						
Shape	usually round	usually oblong	oblong	round	round	round
Average diameter	12–15 mm	30–60 mm × 8–12 mm	20 mm × ?			
Average depth	8–9 mm	4–5 mm	4–5 mm			
Vertical distribution on trunk	.5–18 m	.5–10 m	0–9 m	.3–4 m	1–2.5 m	2–3 m
% producing exudate[a]	<10%	No. 6–8% No. 7–2% No. 8–5%	50%	0%[b]	0%	0%
% young and shallow, nonproducing[a]		No. 6–40% No. 7–60% No. 8–30%		10%[b]	0%	0%
% abandoned[a]	>90% (young and abandoned)	No. 6–52% No. 7–38% No. 8–65%	50% (young and abandoned)	90%[b]	100%	100%

[a]Approximation [b]At beginning of study

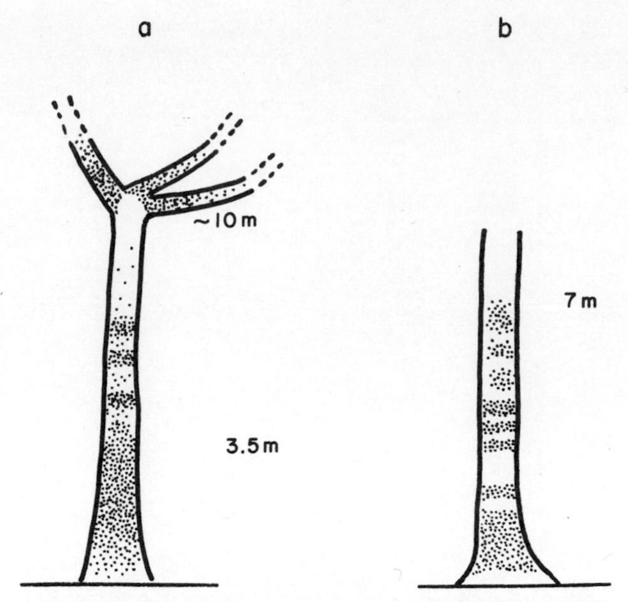

Figure 5. A sketch showing the distribution of exudate holes on two trees observed along the Ampiyacu River: a, *Quararibea spatulata;* b, Leguminoseae.

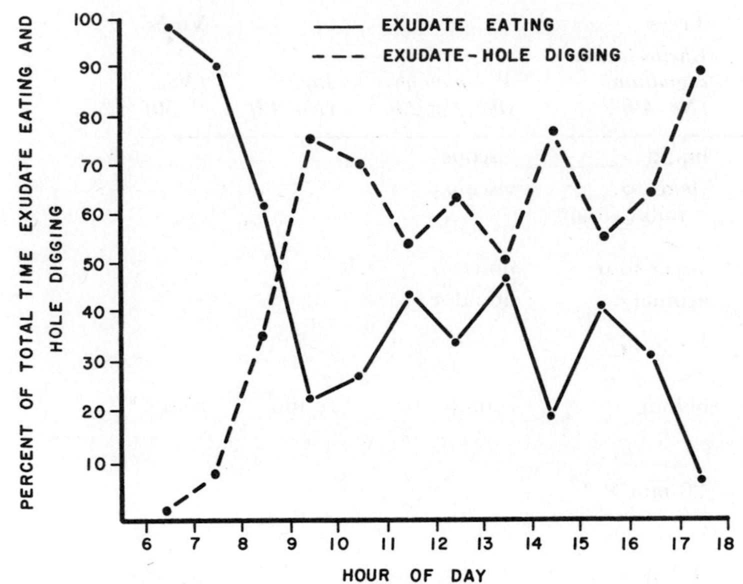

Figure 6. Daily pattern of exudate eating and exudate-hole digging as determined from focal-tree observations.

exudate. This contrasts with Moynihan's (1976) observation that digging was infrequent. Sometimes on the *Q. rhombifolia*, the animals would become inactive for a short period and just sit, usually high on the trunk.

As the vertical distribution of exudate holes suggests, *Cebuella* concentrated their activities on the lower part of the trunks. Little time was spent on the trunk above 10 m. Within the 0–10 m zone on the *Q. rhombifolia*, *Cebuella* were in the 0–5 m level 64 percent of the time and in the 5–10 m level 36 percent of the time (from eight visits of 15 minutes duration or more). More time was spent digging than eating in both levels, but proportionately more digging was done in the 5–10 m level. This latter trend was even stronger on the *Trichilia* sp. and *C. cognatum*.

The duration of a visit by a single animal to the *Q. rhombifolia* averaged 20 minutes, ranging from 6 to 37 minutes. (One visit of 63 minutes, of which 59 minutes were sitting on the trunk, is excluded.) In the largest *Trichilia* sp. (No. 7) visits ranged from 3 to 11 minutes in duration, with an average of 7 minutes. The two smaller *Trichilia* sp. had visits lasting from 2 to 8 minutes and averaging 4 minutes. Visits to the *C. cognatum* lasted 2 to 9 minutes and averaged 6 minutes. During the two consecutive days of observation of the *Q. rhombifolia*, the adult male visited the tree four and five times, and the adult female four times both days. The total number of visits by individuals of the troop to the *Q. rhombifolia* was 21 the first day and 24 the second.

Indicative of how heavily *Cebuella* utilized the *Q. rhombifolia* is the fact that from 0600 hours to 1800 hours during the two consecutive days of observation, the tree had at least one visitor more than 65 percent of the time. Of the total time on the *Q. rhombifolia*, 53 percent of the time there was one individual, 41 percent two individuals, and 5 percent three individuals. A few times in the late afternoon as many as four or five *Cebuella* were on the *Q. rhombifolia* trunk, mostly digging holes.

Generally, a *Cebuella* entered an exudate tree at the upper part of the trunk, descended to near the base, and ate exudate and excavated on its way up the trunk, leaving it from the upper part again. This pattern agrees with Moynihan's (1976) observations. Usually, five to eight holes were visited per minute on the *Q. rhombifolia.* Time spent eating from a hole averaged 3.3 seconds, and digging 4.7 seconds (different types of holes included). An individual usually did not consume all of the exudate in a hole, but left a little and continued on to other holes. One high-producing exudate hole may receive visits from different individuals in a short time interval, the first visits for exudate eating and the latter often for excavation. In a tree that exudes rapidly, such as *Q. rhombifolia,* holes that have been exhausted of exudate and/or recently excavated can be fed upon again within a few minutes, while holes in *Trichilia* sp. may require much longer between feeding visits. We do not know how long it takes *Cebuella* to develop a hole to an exuding condition, probably several weeks. The exuding life-span of holes or the duration of their use by *Cebuella* must be at least 3 or 4 months for the species used by our study troop. We believe a hole must eventually be abandoned because continuous excavation, necessary to prevent cicatrization, eventually makes the hole too big to be dug in and economically utilized by *Cebuella,* or because of some other overriding defense mechanism of the tree.

Some long term, temporal variation in hole digging by *Cebuella* was observed in two tree species in the range. In July, the lone individual of *V. lomatophylla* had what appeared to be only abandoned holes up to 2.5 m on the trunk. In September, the *Cebuella* dug some new holes and reopened some abandoned ones. Exudate holes in this tree became more numerous, mostly on the lower 3 m of the trunk, during the study. In the last week of November, we observed that the only individual of *Inga* sp. in the range was possibly just beginning to be used as an exudate source by *Cebuella,* for it had only a few holes low on the trunk, none of which were abandoned; however, we had not noticed the tree before November.

Small arthropods constitute an important part of the *Cebuella* diet. Ants, orthopterans, and spiders were frequently eaten. While foraging for insects, the focal animals spent 75 percent of their time between 5 m and 15 m above the ground, 24 percent below 5 m, and only 1 percent above 15 m. Twice, a *Cebuella* was seen pursuing a grasshopper to the ground and catching it, a foraging behavior also reported by Moynihan (1976). While insect foraging they look under limbs and leaves, turning over the latter with their hands while moving close to and slowly along a branch. Capture frequently is made with a quick jump.

Besides the fruits of the *Ficus* sp. and Sapotaceae that were eaten, local people stated that the fruits of *Inga* sp. and *Pourouma* sp. are consumed by *Cebuella* living near cultivated fields.

We once saw what was probably drinking of water by *Cebuella.* In July, around midday, the adult male with two newborn infants on his back descended to the ground and crossed about 5 m of ground to a position partially out of view but next to a water puddle. He remained there for 15 to 20 seconds, and then returned to a tree and ascended.

Activities of individuals in the troop often showed little coordination. While some individuals were eating exudate, others would hunt insects and

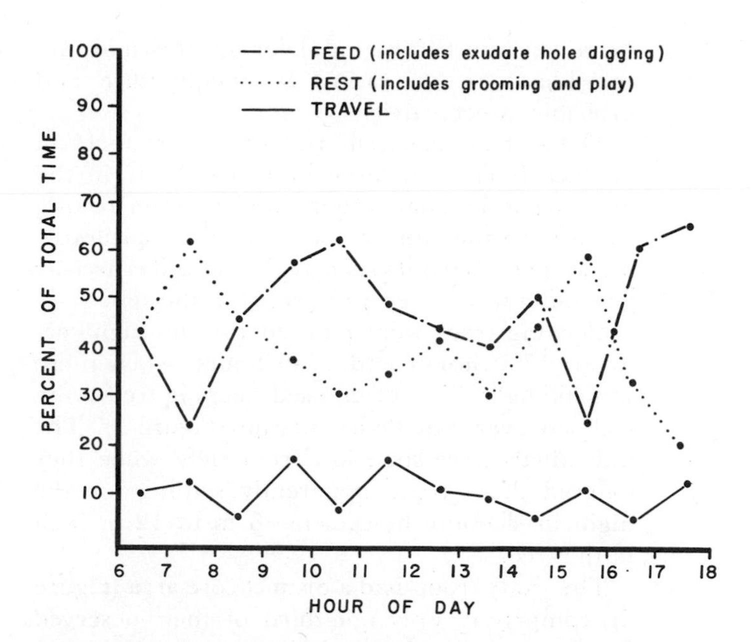

Figure 7. Daily activity pattern of the two focal animals (both adults) based on four days of observation.

others would rest in the crown of a tree, so that at any one time the troop members were dispersed, often over a large portion of their range.

Except during insect foraging, *Cebuella* movements occur in fast spurts, i.e., they move rapidly a short distance and stop. Such movements must make them difficult to spot or follow for a predator.

Figure 7 shows the daily pattern of feeding, travel, and resting (includes grooming and play) as determined from 4 days of observation of the two adult, focal animals. Observations of the juveniles indicated that they had similar activity profiles, although, as noted above, members often did not synchronize their activities during any given minute. Peaks in resting, including many grooming bouts, occurred from 0700 hrs to 0800 hrs and from 1500 hrs to 1600 hrs, and these coincided with lulls in feeding (includes hole digging). A peak in absolute inactivity—no grooming or play—was evident at midday. The proportion of insect-foraging time to exudate-eating/hole-excavation time shifted during the day. The early morning feeding peak was almost entirely exudate eating, the peak around 1000 hrs to 1100 hrs was primarily insect foraging, in midafternoon they were in approximately equal proportions, and the peak at the end of the day was primarily hole excavation. By excavating many holes just before nightfall, they presumably maximize total exudate flow during the next 12 to 13 hours, which they then harvest immediately in the morning. The surge in insect foraging later in the morning coincides with increasing air temperature and probably insect activity.

The animals generally rested at 10 m to 15 m in thick foliage on large branches. Early in the morning individuals often could be seen resting exposed to the sun, at times quite conspicuously in the open. Activities such as playing and especially grooming were often performed in the sun.

Sleeping trees were entered for the night between 1700 hours and 1745 hours, when there was still light. Two often-used sleeping trees were inclined over San Pedro stream (Figure 3). The individuals were soon lost from sight when they entered these trees, apparently settling for the night in sleeping holes some 8 m to 12 m high in the tree.

The study troop had a distinct core area (Figure 3), comprising only one-third of their observed 3,000 sq m range, but including all of the exudate trees and principle sleeping trees. An estimated 80 percent of their daylight time was spent in the core area.

Only two possible competitors for exudate were observed. Ants were seen visiting the exudate holes of *Q. rhombifolia* and *V. lomatophylla* although never in great numbers. Some ants are thus probably incidentally ingested by *Cebuella* via the exudate. Once, a small lizard was seen moving from hole to hole on the *Q. rhombifolia*, but we could not determine if it was ingesting exudate or possibly ants.

On several consecutive days a troop of five to seven *Saguinus fuscicollis*, and on one occasion a troop of *Saimiri sciureus*, entered the study troop's range to feed on some *Ficus* sp. fruits. During these intrusions two or three of the older *Cebuella* entered the *Ficus* and closely approached the intruders, but no further interactions occurred. David Pearson (pers. comm.) informed us that *Cebuella* and *Saguinus* sp. (*fuscicollis* or *nigricollis*) traveled in mixed troops at his study site along the Napo River in Ecuador. After the Indians pointed the *Cebuella* out to him, he saw up to five or six individuals with every *Saguinus* troop, usually on the outskirts of the *Saguinus* troop. The *Cebuella* would feed close to the tree trunks (he believes mainly for insects), then move with sudden spurts to the next foraging area, barely keeping up with the *Saguinus*. He occasionally saw *Cebuella* troops alone.

The most obvious interaction with another vertebrate occurred when a toucan (*Ramphastos* sp.) landed near one of the troop members. The *Cebuella* slowly approached the toucan several times, but the toucan retreated on the branch until it finally flew away. A similar interaction occurred when a toucan landed in a *Ficus*, possibly to eat fruits. When a hawk landed near a troop member, all individuals in the immediate vicinity would flee.

Distribution of Cebuella *Along the San Pedro Stream and the Nanay River*

Along the 1.5 km of the San Pedro stream that we surveyed, we found only two *Cebuella* troops, the study troop and another troop 1 km upstream. These two troops were located in the two areas where exudate species (those used by the study troop) occurred in distinctly higher densities, with *V. lomatophylla* and *Q. rhombifolia* abundant in the area of the second troop (Figure 2). All of the nine individuals of exudate species located in the range of the study troop and the ten to eleven exudate trees in the approximated range of the second troop had exudate holes, with one possible exception. (The inclusion of this possible exception in the range of the second troop is questionable.) By contrast, over one-half of the sixteen to seventeen individuals of exudate species outside of the two ranges had no holes. Furthermore, exudate holes were in use and relatively numerous in most trees in the ranges of the two troops, whereas there were few exudate holes in trees outside the troops' ranges and all holes were abandoned. These data strongly indicate that the density and distribution of certain exudate species can be a primary determinant of the density and local distribution of *Cebuella.*

All individuals of exudate species recorded in the two, 500 m transects were within 200 m of the San Pedro stream (Figure 2). It appears that these species are restricted to frequently inundated soils along waterways. We found none of these exudate species in the non-inundatable forest of the nearby biological station, Campamento *Callicebus,* and with the possible exception of *Trichilia* sp. and *Inga* sp., we feel sure that they do not occur there. Also, no *Cebuella* inhabit the extensive non-inundatable forest of the biological station. Residents along the Nanay River unanimously concurred that *Cebuella* inhabit only seasonally inundated forest along the Nanay, usually on the edge of the river, a stream, or lake. Locals reported that at least two other *Cebuella* troops inhabited seasonally inundated forests near Mishana. We know of no reports of *Cebuella* inhabiting non-inundatable forest along the Nanay River. People of the Samiria River basin in Peru, where almost all forests are inundatable, claim that *Cebuella* live primarily in forests on the edge of streams or lakes there (Freese, 1975).

Cebuella *Along the Ampicayu River*

Cebuella inhabit both inundatable and non-inundatable forests along the Ampiyacu River. Troops were observed by us, and reported by locals to occur, in both forest types, and numerous exudate holes were found on several trees at two sites in non-inundatable forests. We did not locate trees with exudate holes in inundatable forests, but the four species we located in the non-inundatable forests were different from the exudate species along San Pedro stream. The four were: *Parkia* sp. (Leguminosae), *Parkia oppositifolia,* another unidentified leguminous tree (*Inga*?), and *Quararibea* cf. *spatulata* Ducke. Of these four, we have found only the two *Parkia* species along the Nanay, both on high ground. *Parkia* sp. has been reported to be an important exudate tree for *Cebuella* in Colombia (Hernandez-Camacho and Cooper, 1976), and Izawa (1975) saw *Cebuella* eat the exudate of *Parkia oppositifolia* in Colombia. People along the Ampiyacu reported that *Cebuella* also feed on the exudate of *Spondias* sp. (Anacardiaceae).

We found two of the exudate species of our Nanay study site along the Ampiyacu during our brief stay. (The others could have been easily overlooked if they occur at low densities.) *V. lomatophylla* was a common tree on inundatable ground in the basin, growing in quite dense stands in some areas. *Trichilia* sp. was also encountered. We did not see these species with exudate holes, however, even though one stand of *V. lomatophylla* was near an area occupied, or recently occupied, by *Cebuella.*

Discussion

Descriptions of saps and gums in the literature do not allow one to distinguish easily between the two. Howes (1949, p. vii) stated that the terms gum and resin "have been applied to almost any form of plant exudation," and Smith and Montgomery (1959, p. 1) noted that there is disagreement as to the origin of gum. According to Howes (1949, p. 3), however, it is generally agreed that gum production is due to infection by microorganisms, and Esau (1965, pp. 46, 260-261) stated that "pathological or physiological disturbances that induce a breakdown of walls and cell contents" result in the formation of gums. Speaking of the gums eaten by some prosimians, Charles-Dominique (1977) noted that gums often are found at the base of fissures in the bark or where the tree has been damaged, and that they are sometimes produced as a defense mechanism in response to damage by certain insects, a point also noted by Esau (1965, p. 261). It thus appears that gums are produced locally in or under the bark.

The term sap generally refers to the liquids transported up the xylem tissue and down the phloem tissue, the latter carrying food materials produced in the leaves (Salisbury and Ross, 1969, p. 168). Sap flow from the phloem can be triggered in certain trees by cutting into the inner bark or phloem tissue (Zimmerman, 1960; Salisbury and Ross, 1969, p. 179), but callose formation seals the injury and stops the flow quickly in many trees (Zimmerman, 1961). Saps and gums apparently differ chemically (see Smith and Montgomery, 1959; Salisbury and Ross, 1969; Esau, 1965).

It is apparently gums that constitute an important part of the diet of some lemurs and lorises (Charles-Dominique, 1977; Claude-M. Hladik, pers. comm.; and see Coimbra-Filho and Mittermeier, 1977). All of these prosimians, except possibly *Phaner furcifer* (Petter et al., 1971, cited by Coimbra-Filho and Mittermeier, 1977), do not induce gum production, but harvest it where it forms due to other causes. This contrasts with the hole digging of *Cebuella* and at least some *Callithrix* species (Coimbra-Filho, 1972; Coimbra-Filho and Mittermeier, 1977) that directly results in the release of exudates. The tree exudates tapped and eaten by such animals as aphids and yellow-bellied sapsuckers (*Sphyrapicus varius*) have been recognized as phloem exudates (sap) by both botanists and zoologists (e.g., Crafts and Crisp, 1971, p. 65; Kilham, 1964). It seems likely that whether gum or sap is exuded after hole digging by *Cebuella* (and probably *Callithrix*) could depend largely on the trees species attacked. Indeed, we think it is conceivable that *Cebuella* eat both gum and sap. Both can have high nutrient concentrations, especially sugars (e.g., Machado and Leite, 1957, cited by Coimbra-Filho and Mittermeier, 1977; Zimmerman, 1961) and thus are potentially important food sources. The viscous nature of the exudate of three of the species utilized by *Cebuella* suggests that these are gums, but the liquid exudate of the fourth suggests a sap. Chemical analyses of these exudates should help resolve this question.

Cebuella has evolved some obvious behavioral and morphological adaptations for feeding on exudates. As pointed out by Moynihan (1976) and

Kinzey et al. (1975), their locomotor skills, which resemble the vertical clinging and leaping category of Napier and Walker (1967), are a specialization for feeding from vertical tree trunks. The "short-tusked" relationship of the lower canines to the lower incisors, found only in *Cebuella* and *Callithrix* spp. (Napier and Napier, 1967), is probably an adaptation for digging exudate holes (Coimbra-Filho and Mittermeier, 1977).

As mentioned above, at least some *Callithrix* species also have developed the hole-digging and exudate-feeding habit. *Callithrix jacchus* has been observed in the wild and in captivity to excavate numerous holes in certain tree species and eat the exudate, and four other *Callithrix* species have demonstrated similar habits in captivity (Coimbra-Filho, 1972; Coimbra-Filho and Mittermeier, 1977). *Callithrix* also urinate in these holes, and Coimbra-Filho (1972) suspects that the holes therefore may play an important role in territorial and hierarchical behavior. We observed *Cebuella* rub their ano-genital region and urinate on sap holes also, but this and other behavior will be discussed in a later paper.

The intensity of exudate use by *Cebuella* probably differs between habitats. Also, observations over a 12 month period are needed to ascertain if *Cebuella*'s utilization of exudate fluctuates seasonally in our study area. Nevertheless, it appears that tree exudate is essential in the diet of *Cebuella* in our study area, and except for proteins, *Cebuella* probably can obtain most nutrients from exudates. The diversity and abundance of exudate sources necessary to maintain a *Cebuella* troop in a particular area must be partly a function of the nutritional content of exudate from each species utilized. Whether or not the exudate of a plant is utilized must primarily depend on whether the exudate's nutritional value multiplied by its rate of exudation outweighs the energy required to travel to the tree, dig the holes, and digest and detoxify, if necessary, the exudate. Certainly, the energy expenditure in digging the holes must be tremendous. Competition from other animal species which utilize the exudate or some other food resource of a particular species may also be important in some cases. For example, Moynihan (1976) found squirrels, *Microsciurus* sp., eating bark and *Cebuella* eating exudate from the same tree, but from different branches. High predation risk could deter the use of an otherwise good exudate tree; feeding on a tree with an extremely light-colored bark, making the otherwise cryptic-colored *Cebuella* conspicuous, would be such a risk.

The peculiarities of the exudate holes pose some interesting problems. Why are the holes concentrated near the trunk's base, and also near the base of major branches in the case of the *Q. spatulata* along the Ampiyacu? If, for the moment, we assume that the exudate is phloem sap, some possible explanations can be generated. As noted by both Moynihan (1976) and us, *Cebuella* appear to dig the first holes, and thus create a higher concentration of them, low on the trunk, a pattern similar to that shown by yellow-bellied sapsuckers (Tate, 1973). This pattern might assure that sap flow to newly made holes is not impeded by injuries to the phloem tissue inflicted when older holes were made, which would occur if holes were made progressively lower on the trunk. Perhaps there is greater sap pressure near the base of the trunk and major branches resulting in a higher rate of exudation. It seems doubtful that nutrient concentration of phloem sap is greater lower in the trunk; in fact, there is evidence to the contrary (Zimmerman, 1961). From our observation that high on a trunk the ratio of hole digging to exudate eating was greater than low on the trunk, the cost of digging holes must eventually become greater than the return (in exudate), and hole digging therefore should cease above this height.

Cebuella's habit of digging holes in bands or clumps also is exhibited by the yellow-bellied sapsucker. Kilham (1964) speculated that sapsuckers do this because nutrient flow might be increased to areas of repeated injury, thereby resulting in a greater nutrient supply for the sapsucker.

The life span of a hole, or period during which it produces exudate, is an important factor in analyzing the resource utilization and population ecology of *Cebuella*. In view of the numerous abandoned holes and constant opening of new holes on most exudate trees, a steady turnover of holes must occur. One would predict then, that a *Cebuella* troop eventually, over many years, may exhaust the capacity of a tree to exude substantial quantities for them (assuming that old cicatrized holes cannot be reopened). The exudate-eating activities of *Cebuella* also might result directly in the death of a tree because of nutrient depletion, or indirectly through increased susceptibility to fungal or bacterial infection. The *Q. spatulata* individual along the Ampiyacu (Figure 5a) certainly may have reached one of these points; it was both saturated with holes and appeared unhealthy. Depending on how quickly one of these processes, or death from old age, occurs in the exudate trees of a troop's range, and on the species of new trees growing in, the *Cebuella* troop (or possibly only some individuals) occasionally may be forced to search for another area with an adequate diversity and density of exudate trees. Tate (1973) found that the sap-eating activities of yellow-bellied sapsuckers killed many trees, thereby forcing them to shift to other feeding areas. Our Nanay study area was still occupied by a *Cebuella* troop (presumably the same one) in September 1975, and the *Q. rhombifolia* and three *Trichilia* sp. appeared to be receiving heavy feeding use with no obvious increase in the density of holes on their trunks.

The feeding niche occupied by *Cebuella* is relatively unique within its geographic range, and therefore competition with other animals for its major food source is probably much less intense than that experienced by other New World monkeys. In addition, as pointed out by Coimbra-Filho and Mittermeier (1977), exudate eaters probably have to contend with less seasonal fluctuation in their major food source. By specializing in exudate eating, however, it appears that *Cebuella* must be more selective than leaf and fruit eaters in their choice of food-source species. As our findings indicate, this means that the local distribution and numbers of *Cebuella* are largely determined by the distribution and abundance of relatively few tree species.

Acknowledgments

Pekka Soini and Rogerio Castro offered many stimulating discussions during the field research and development of this paper. We also thank Paul Heltne, Charles Southwick, Marge Freese, Russell Mittermeier, and Alfred Rosenberger for their helpful comments on the manuscript. Support for this study was provided by the National Academy of Sciences through Contracts PH 43-64-44 (T.O.12) with the National Institutes of Health and DADA 17-71-C-1117 with the United States Army Medical Research and Development Command under terms of the subcontract BA 22/23-72-30 with the Pan American Health Organization.

Literature Cited

Castro, R., and P. Soini.
1977. Field studies on *Saguinus mystax* and other callitrichids in Amazonian Peru. In *The Biology and Conservation of the Callitrichidae,* edited by Devra G. Kleiman, pp. 73-78. Washington, D.C.: Smithsonian Institution Press.

Charles-Dominique, P.
1977. *Biology of Nocturnal Primates.* London: Duckworth.

Coimbra-Filho, A. F.
1972. Aspectos ineditos do comportamento de sagüis do genero *Callithrix* (Callithricidae, Primates). *Revista Brasileira de Biologia,* 32(4):505-512.

Coimbra-Filho, A. F., and R. A. Mittermeier.
1977. Tree-gouging, exudate-eating and the "short-tusked" condition in *Callithrix* and *Cebuella.* In *The Biology and Conservation of the Callitrichidae,* edited by Devra G. Kleiman, pp. 105–115. Washington, D.C.: Smithsonian InstitutionPress.

Crafts, A. S., and C. E. Crisp.
1971. *Phloem Transport in Plants.* San Francisco: W. H. Freeman and Company.

Esau, K.
1965. *Plant Anatomy.* New York: John Wiley and Sons, Inc.

Freese, C. F.
1975. A census of non-human primates in Peru. In *Primate Censusing Studies in Peru and Colombia,* pp. 17-41. Washington, D.C.: Pan American Health Organization.

Hernandez-Camacho, J., and R. W. Cooper
1976. The non-human primates of Colombia. In *Neotropical Primates—Field Studies and Conservation,* edited by R. W. Thorington, Jr. and P. G. Heltne, pp. 35-69. Washington, D.C.: National Academy of Sciences Press.

Howes, F. N.
1949. *Vegetable Gums and Resins.* Waltham, Massachusetts: Chronica Botanica Company.

Izawa, K.
1975. Foods and feeding behavior of monkeys in the Upper Amazon Basin. *Primates,* 16(3):295-316.

Kilham, L.
1964. The relations of breeding yellow-bellied sapsuckers to wounded birches and other trees. *Auk,* 81(4):520-527.

Kinzey, W. G., A. L. Rosenberger, and M. Ramirez.
1975. Vertical clinging and leaping in a neotropical anthropoid. *Nature,* 255:327-328.

Machado, A., and O. C. Leite.
1957. Goma de Cajueiro. *Bol. Inst. Quim. Agric.,* 50:7-15.

Moynihan, M.
1976. Notes on the ecology and behavior of the pygmy marmoset, *Cebuella pygmaea,* in Amazonian Colombia. In *Neotropical Primates—Field Studies and Conservation,* edited by R. W. Thorington, Jr., and P. G. Heltne, pp. 79-84. Washington, D.C.: National Academy of Sciences Press.

Napier, J. R., and P. H. Napier.
1967. *A Handbook of Living Primates.* New York: Academic Press.

Napier, J. R., and A. C. Walker.
1967. Vertical clinging and leaping, a newly recognised category of locomotor behaviour among Primates. *Folia Primat.,* 6:204–219.

Petter, J. J.; A. Schilling; and G. Pariente.
1971. Observations Eco-Ethologiques sur deux Lemuriens Malagaches Nocturnes: *Phaner furcifer* et *Microcebus coquereli. La Terre et la Vie,* 118(3):287-327.

Salisbury, F. B., and C. Ross.
1969. *Plant Physiology.* Belmont, California: Wadsworth Publishing Company, Inc.

Smith, F., and R. Montgomery.
1959. *The Chemistry of Plant Gums and Mucilages.* New York: Reinhold Publishing Corporation.

Tate, J., Jr.
1973. Methods and annual sequence of foraging by the sapsucker. *Auk,* 90(4):840-856.

Zimmerman, M. H.
1960. Transport in the phloem. In *Annual Review of Plant Physiology, Vol. II.,* edited by L. Machlis and W. R. Briggs, pp. 167-190. Palo Alto, California: Annual Reviews, Inc.

1961. Movement of organic substances in trees. *Science,* 133(3446):73-79.

ADELMAR F. COIMBRA-FILHO
Department of Environmental Conservation,
FEEMA, Rio de Janeiro, Brazil
and
RUSSELL A. MITTERMEIER
Department of Anthropology and
Museum of Comparative Zoology,
Harvard University,
Cambridge, Massachusetts 02138

Tree-gouging, Exudate-eating and the "Short-tusked" Condition in *Callithrix* and *Cebuella*

ABSTRACT

The "short-tusked" lower anterior dentition in the callitrichid genera *Callithrix* and *Cebuella* is used to perforate tree bark and induce the flow of exudates (gum and sap), which are important food sources. *Callithrix* also uses holes made by the lower anterior dentition as sites for marking behavior. Tree-gouging is conspicuously absent in the two "long-tusked" callitrichid genera, *Saguinus* and *Leontopithecus.* Occasional exudate-eating and occasional tree-gouging are fairly common behavior patterns among primates; however, the use of teeth to specifically elicit exudate flow is rare and, on the basis of available data, limited to the two "short-tusked" callitrichids and *Phaner furcifer,* a Madagascan cheirogaleine lemur.

Introduction

The four genera comprising the New World primate family Callitrichidae can be divided into two major groups on the basis of lower canine to lower incisor relationships. Napier and Napier (1967) coin the terms "short-tusked" and "long-tusked" to describe the differences. The "long-tusked" genera, *Saguinus* and *Leontopithecus,* have a "normal" incisor-canine relationship, with canines much longer than incisors (Figure 1). On the other hand,

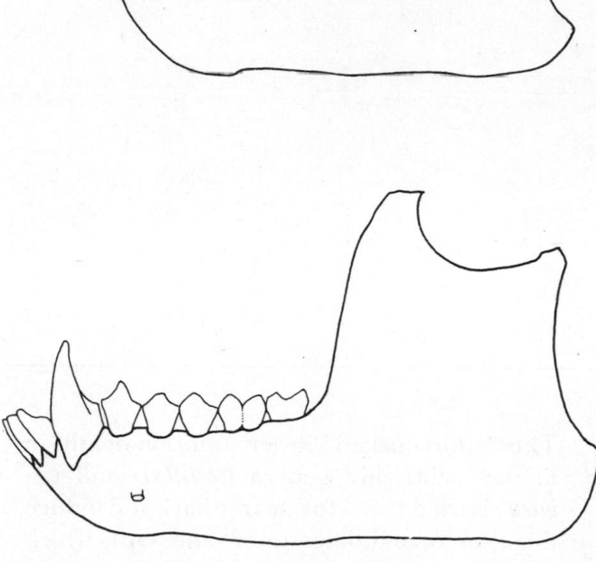

Figure 1. Mandibles of *Saguinus midas* (above) and *Leontophithecus rosalia* (below) showing the "long-tusked" condition, with canines longer than incisors.

Figure 2. Mandibles of *Cebuella pygmaea* (above) and *Callithrix geoffroyi* (below) showing the "short-tusked" condition, with incisiform canines and long incisors.

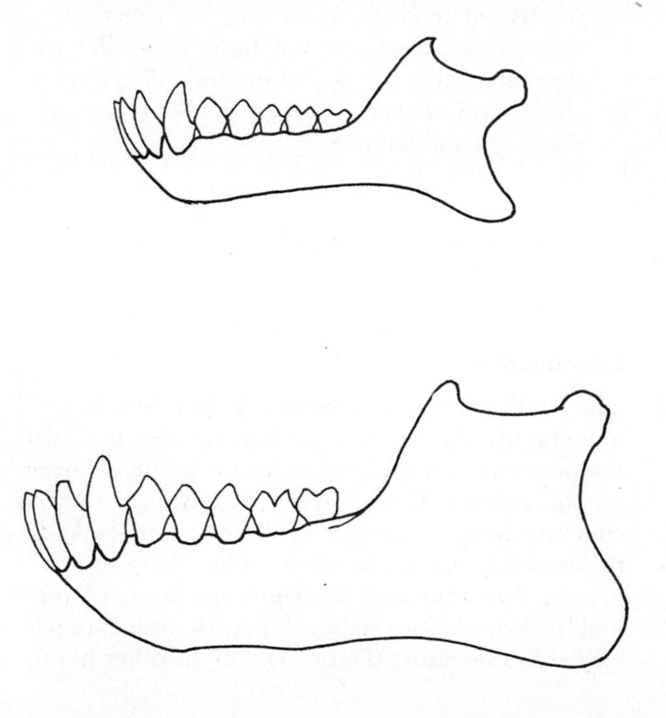

the "short-tusked" group, which consists of *Callithrix* and *Cebuella,* has long incisors that are roughly equal in length to the canines which thus makes the canines appear short (Figure 2).

Although the distinction between the two groups has long been recognized, the function of the unusual lower anterior dentition in *Callithrix* and *Cebuella* has remained unclear. Several recent field studies and short-term observations of *Cebuella pygmaea* in the Peruvian and Colombian Amazon regions (Izawa, 1972, 1975; Kinzey et al., 1975; Hernandez-Camacho and Cooper, 1976; Moynihan, 1976; Ramirez et al., 1977) have shown that this small primate uses its lower anterior dentition to gouge holes in certain trees and induce the flow of exudates[1] (gum and sap), which are important food sources.

During captive studies and short-term field observations in southern Brazil, we have noted similar "tree-gouging" behavior in *Callithrix* (Coimbra-Filho, 1971, 1972). In contrast, we have not seen this behavior during captive and field studies of *Saguinus* and *Leontopithecus.* In this paper, we discuss tree-gouging as a feeding adaptation in *Callithrix* and *Cebuella* and its role in *Callithrix* marking behavior. In addition, we briefly compare these two genera with other primates that exhibit similar behavior.

Methods

Short-term field observations of *Callithrix* and *Leontopithecus* were made by Coimbra-Filho in the Brazilian states of Alagoas, Bahia, São Paulo, Rio de Janeiro, and the former state of Guanabara.[2]

[1]In the above-mentioned *Cebuella* studies, the term sap has been used to refer to the substances that flow from trees attacked by these monkeys. However, Freese (pers. comm.) and Ramirez et al. (1977) point out that *Cebuella* may be eating both gums and sap. The *Callithrix* discussed in this paper ate what we consider to be gums, but may have been eating small quantities of sap as well. Part of the confusion results from imprecise definitions of gum and sap and the fact that so little is known of these substances in most South American trees. The more general term *exudate-eating* (as suggested by Freese, pers. comm.) will be used to refer to the feeding patterns discussed in this paper.

[2]The state of Guanabara, in which the *city* of Rio de Janeiro was located, united with the larger *state* of Rio de Janeiro in 1975. The resulting state is known as Rio de Janeiro.

1a.

1b.

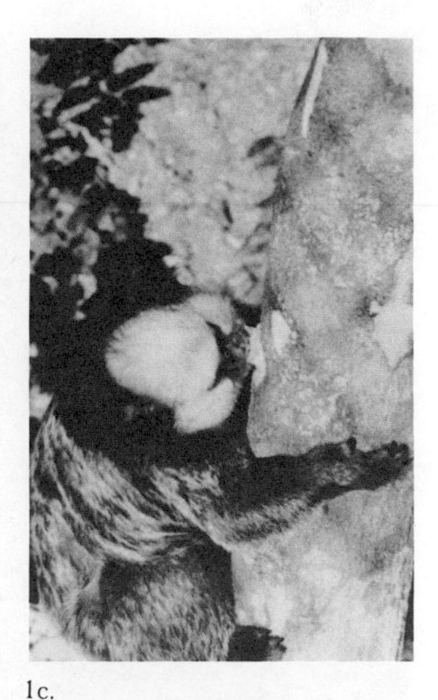

1c.

Plate 1a. Close-up of trunk of *Tapirira guianensis* showing holes gouged by *Callithrix jacchus.* Photo taken in Maceio, state of Alagoas.

Plate 1b. Captive *Callithrix jacchus* gouging hole in trunk of *Chrysalidocarpus lutescens,* a Madagascan palm common in Brazilian gardens.

Plate 1c. Captive *Callithrix* hybrid (*C. p. penicillata* × *C. geoffroyi*) gouging hole in branch in its cage.

Captive observations of these two genera (including the forms *Leontopithecus rosalia rosalia, L. r. chrysomelas, L. r. chrysopygus, Callithrix jacchus, C. penicillata penicillata, C. p. kuhlii, C. flaviceps,* and *C. geoffroyi*) were made by both of us at the Department of Environmental Conservation and the Tijuca Bank of Lion Marmosets, both in Rio de Janeiro. Brief field observations of *Saguinus midas midas* were made by Mittermeier in Surinam and French Guiana (Mittermeier et al., 1977).

Captive animals were provided with freshly cut branches of *Anacardium occidentale* (Anacardiaceae), one of the species eaten by wild marmosets. Their reactions to these branches and to the wooden fixtures in their cages were noted.

Observations

Field Observations

Field observations of exudate-eating and tree-gouging are limited to *Callithrix jacchus.* In the state of Alagoas, local people report that *C. jacchus* frequently feeds on exudates of several tree species, among them *Tapirira guianensis* (Anacardiaceae), *Anacardium occidentale* (Anacardiaceae), *Terminalia catappa* (Combretaceae—an exotic species now well established in Brazil). These trees are often riddled with holes which penetrate as deep as the superficial layer of the cambium (Plate 1a). Coimbra observed a pair of *C. jacchus* feeding on the trunk of an *Anacardium occidentale.* One animal was in the process of gouging a hole in the tree, while the other was eating gum from a hole already made.

In the city of Rio de Janeiro, introduced *C. jacchus* (established at least since 1919—Coimbra-Filho et al., 1973) have also been seen feeding on gums of *Piptadenia colubrina* (Leguminosae, Mimosaceae) and using their teeth to gouge holes in the peduncle of unripe fruits of *Artocarpus heterophyllus* (Moraceae). When attacked, the peduncle of *Artocarpus* fruit exudes a latex which *C. jacchus* also eats.

Captive Observations

All captive *Callithrix* observed quickly gouged holes into the freshly cut branches of *A. occidentale* and ate any gums that exuded from them. A single privately owned *C. jacchus* did the same to the trunk of a *Chrysalidocarpus lutescens* (Palmae—a common Brazilian garden plant of Madagascan origin) (Plate 1b). Captive *Leontopithecus,* on the other hand, did not perform such behavior.

In addition to gouging holes in fresh branches, captive *Callithrix* riddle the wooden fixtures of their cages with holes (Plates 1c, 2a). In contrast, identical wooden fixtures in the cages of captive *Leontopithecus* show no such holes.

Plate 2a. Close-up of holes gouged by captive *Callithrix* in dry wooden-cage fixtures.

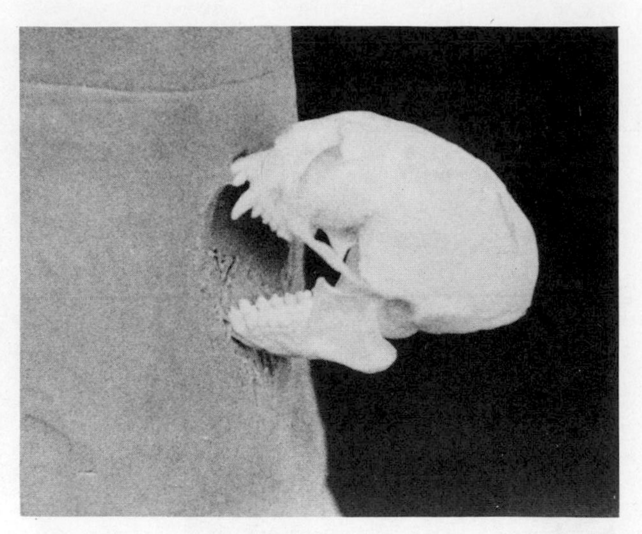

Plate 2b. *Callithrix* skull placed in gouging position. The lower anterior teeth scrape and gouge, while the upper anterior teeth remain in a relatively fixed position.

In captive *Callithrix,* gouged holes also serve as sites for marking behavior. Marmosets, especially high-ranking males and females, frequently urinate in these holes and perform genital rubbing along wood in the vicinity of them. Thus, holes made by the lower anterior dentition seem to serve as foci for marking and retain a dark coating from glandular secretions and urine as well as the scent itself.

The gouging process is usually accomplished by anchoring the upper anterior teeth in a fixed position and then scraping with the lower teeth (Plates 1b, 1c, 2b). Gums are either licked up or scraped with the teeth.

Discussion

Tree-gouging as a Feeding Adaptation in Callithrix *and* Cebuella

All callitrichids feed on insects, fruit and small vertebrates (Hill, 1957; Hladik and Hladik, 1969; Moynihan, 1970, 1976; Coimbra-Filho and Mittermeier, 1973; Izawa, 1975; Hernandez-Camacho and Cooper, 1976). However, in contrast to *Saguinus* and *Leontopithecus, Cebuella* and at least some *Callithrix* apparently eat a high proportion of exudates and regularly use their lower anterior dentition to induce exudate flow.

Hernandez-Camacho and Cooper (1976) discuss the stomach contents of a number of *Cebuella* from the vicinity of Puerto Leguizamo in the Colombian Amazon. The stomachs examined contained "jelly-like, dirty-whitish-colored *guarango* sap (*Parkia* sp.) in addition to some finely crushed insects (mostly Coleoptera) and evidence of fruit pulp." Moynihan (1976) noted that sap was a major food source for *Cebuella* in the region of Mocoa and Puerto Asís in the Colombian Amazon. The *Cebuella* group studied by Kinzey et al. (1975) spent 77 percent of its feeding time clinging to vertical supports and feeding on sap (see also Ramirez et al., 1977). Our field observations and information from local people indicate that the gums are an important part of the diet for *C. jacchus.* Judging from the avidity with which other captive *Callithrix* species attacked the branches of *A. occidentale,* gums are probably included in their diets as well.

On the other hand, there are very few observations of exudate-eating in the two "long-tusked" genera. *Leontopithecus* has never been seen eating exudates and only two species of *Saguinus* have been reported eating them (*Saguinus* (*oedipus*) *geoffroyi,* Hladik and Hladik, 1969; *Saguinus fuscicollis,* Izawa, 1975). A number of other field studies and short-term observations of *Saguinus* (Moynihan, 1970, 1976; Hernandez-Camacho and Cooper, 1976; Dawson, 1977; Neyman, 1977; Mittermeier, unpublished observations) have failed to note exudate-eating. On the basis of evidence currently available, the "long-tusked" callitrichids eat far fewer exudates than some of their "short-tusked" relatives and, most important, lack the ability to elicit exudate flow.

Captive *Callithrix* gouge into empty screw holes or other areas where a hole has already been started. This may be because they find it easier to get purchase with their teeth or because they are searching for insect larvae in such holes. Gouging in dry wood may be a means of procuring insect larvae in the wild.

Exudates as Food Sources

Although little is known about the exudates of most South American trees, the gum of *A. occidentale* has been fairly well studied and consists of 7.80 percent substances insoluble in water, 76.00 percent substances soluble in water, 1.05 percent minerals and 16.20 percent moisture and with an acidity of 1.90 (Rosenthal, 1951). Ghiggino (in Rosenthal, 1951) analyzed the ash content by spectrography and found that it contained iron, aluminum, calcium, magnesium, silicon, potassium, and traces of manganese and sodium.

Machado and Leite (1957) obtained the following results for *A. occidentale*:

Moisture	14.67%
Soluble in water:	
Arabin	66.48%
Reducing sugars	1.63%
Mineral salts, undetermined substances	1.25%
Total	69.36%
Insoluble in water:	
Cerasin (soluble in sodium carbonate)	2.98%
Bassorin (insoluble in sodium carbonate)	12.99%
Total	15.97%

By hydrolysis, arabin produces the sugars arabinose and galactose; whereas cerasin produces arabinose and arabin and bassorin produces galactose, xilose, and methylpentose.

The gum of *A. occidentale,* therefore, appears to be rich in both sugars and minerals and may serve as a source of energy and of trace elements. In addition, gums and other exudates are probably a more dependable food source than seasonal fruits.

Significance of Exudate-eating for Callithrix *and* Cebuella

Exudate-eating is an important adaptation for small primates like *Callithrix* and *Cebuella,* providing them with a high energy, carbohydrate food source not utilized to any great extent by larger primate species and other mammals and birds. By eating gums, they may avoid competition for fruit, the major high-energy food source for many arboreal mammals and birds in the tropics. In West Africa, for example, the galago, *Euoticus elegantulus,* avoids competition with two other sympatric galagos (*Galago alleni* and *Galago demidovii*), two sympatric lorisines (*Perodicticus potto* and *Arctocebus calabarensis*) and other larger primates because its diet consists primarily of gums throughout most of the year (Charles-Dominique, 1971; Table 1). *Cebuella* occurs in the Upper Amazon region where primate diversity is higher than anywhere else in the world and where competition with other primates and fruit-eating mammals and birds should be high.

Exudate-eating may help *Callithrix jacchus* survive in the comparatively dry habitat of northeastern Brazil where other high-energy food sources are probably in short supply. Exudate-eating may also account for the great adaptability of *C. jacchus,* which has become firmly established where it has been introduced by man (e.g., the city of Rio de Janeiro).

Exudate-eating in Other Primates

Exudate-eating has been observed in a wide variety of other primates, ranging in size from mouse lemurs to baboons and chimpanzees (e.g., *Microcebus murinus,* Martin, 1972, 1973; *Microcebus coquereli, Phaner furcifer,* Petter et al., 1971; *Lemur catta,* Sussman, 1973; *Lemur fulvus,* Sussman, pers. comm.; *Propithecus verreauxi,* Petter et al., 1971; *Perodicticus potto, Galago demidovii, Euoticus elegantulus,* Charles-Dominique, 1971; *Galago senegalensis,* Sauer and Sauer, 1963; Doyle, 1974; *Galago crassicaudatus,* Allen and Loveridge, 1942; Bearder and Doyle, 1974; *Saguinus fuscicollis,* Izawa, 1975; *Saguinus (oedipus) geoffroyi, Aotus trivirgatus, Alouatta palliata, Cebus capucinus,* Hladik and Hladik, 1969; *Cercopithecus aethiops,* Gartlan and Brain, 1968; *Macaca radiata,* Rahaman and Parthasarathy, 1969; *Papio cynocephalus,* Altmann and Altmann, 1970; J. Fleagle, pers. comm.; *Presbytis entellus,* Ripley, 1970; S. B. Hrdy, pers. comm.; N. Bishop, pers. comm.; and *Pan troglodytes,* Goodall, 1963). It is only in several of the smaller species, however, that these substances appear to play a major role

Table 1. Primates that depend on tree exudates as an important food source.

Species (Average Weight)	*Mode of Procuring Exudates*	*Percent Exudate in Diet*	*Plant Species Providing Exudates*	*References*
Cebuella pygmaea (150 g—males; 136 g—females (1))	Gouges holes in trees with "short-tusked" lower anterior dentition.	Sap would appear to be a major food source (3); 77% of feeding time spent clinging to vertical supports feeding on sap (4).	*Parkia* sp. (2) *Parkia oppositifolia* (5) *Inga* sp. (3) *Matisia cordata* (3) *Cedrela odorata* (3) *Palicourea macrobotrys* (4) *Vallea stipularis* (5) *Vochyzia lomatophylla* (6) *Trichilia* sp. (6) *Cheiloclinium* sp. (6)	1. Christen, 1974 2. Hernandez-Camacho and Cooper, 1976 3. Moynihan, 1976 4. Kinzey et al., 1975 5. Izawa, 1975 6. Ramirez et al., 1977
Callithrix jacchus (206 g—males; 236 g—females (1), (2))	Gouges holes in trees with "short-tusked" lower anterior dentition.	Apparently high	*Anacardium occidentale* (3) *Tapirira guianensis* (3) *Terminalia catappa* (3) *Piptadenia colubrina* (4) *Chrysalidocarpus lutescens* (3) —captivity only	1. Hill, 1957 2. Wettstein, 1963 3. Coimbra-Filho, 1972 4. Coimbra-Filho et al., 1973
Galago crassicaudatus (1241 g—males; 1034 g—females (1))	Eats gums (2)	Not specified	Not specified	1. Napier and Napier, 1967 2. Bearder and Doyle, 1974
Galago senegalensis (300 g—males; 229 g—females (1))	Bark may be chewed to expose gum; gum is licked or eaten in small pieces (2). The tooth comb is sometimes used as a scoop in gum collection (3).	Insects and *Acacia* gum from several species seem to be the main food sources in S.W. Africa (4). *Acacia* gum available throughout year; more time spent eating gum in summer than winter (2).	*Acacia* spp. (2), (4)	1. Napier and Napier, 1967 2. Doyle, 1974 3. R. D. Martin, pers. comm. 4. Sauer and Sauer, 1963
Galago demidovii (61 g—(1))	Eats gums from areas where insect attacks or mechanical damage result in gum flow (1).	10%—eats gums mainly in rainy season when they are abundant (1).	Not specified	1. Charles-Dominique, 1971
Euoticus elegantulus (300 g—(1))	Eats gums from areas where insect attacks or mechanical damage result in gum flow; gums scraped or scooped from branches with tooth comb (1).	75%—eats gums all year round (1).	Not specified	1. Charles-Dominique, 1971

Table 1. Primates that depend on tree exudates as an important food source (Continued).

Species (Average Weight)	*Mode of Procuring Exudates*	*Percent Exudate in Diet*	*Plant Species Providing Exudates*	*References*
Perodicticus potto (1100 g—(1))	Eats gums from areas where insect attacks or mechanical damage result in gum flow (1).	21%—eats gums mainly in rainy season when they are abundant (1).	Not specified	1. Charles-Dominique, 1971
Microcebus murinus (60—70 g—(1))	Frequently "scores" and licks tree trunks and lianas (2). May "score" fine branches by using upper canine and upper anterior premolar in rotary action around branch (3). Sap scraped or scooped from branches with tooth comb or licked up with tongue (4).	Not specified	Not specified	1. Charles-Dominique and Martin, 1972 2. Martin, 1972 3. Martin, pers. comm. 4. Martin, 1973
Microcebus coquereli (385 g—(1))	Licks exudates from areas where they flow in response to insect attack (1).	Supplements diet with exudates in October and November when fruit and insects are scarce (1).	*Hippocratea* sp. (1) Vine wrapped around *Anacardium occidentale* (1)	1. Petter et al., 1971
Phaner furcifer (440 g—(1))	Licks gums from areas where they exude in response to insect attack; scrapes bark with highly procumbent upper incisors, upper canines and mandibular tooth comb and licks exudates that appear in response to these attacks; also uses "pincer" formed by occlusion of upper and lower incisors to extract gums from irregularities in bark (1).*	High (1)	2 *Terminalia* spp. (1)	1. Petter et al., 1971

*One other lemur species, *Allocebus trichotis*, has large upper incisors similar to those of *Phaner* and may perform similar exudate-eating behavior. Nothing, however, is known of the behavior of this extremely rare animal (Petter et al., 1971).

in the diet. Table 1 lists those species that depend on exudates as an important food source at least part of the year. Of the ten species listed, seven (*Microcebus murinus, Microcebus coquereli, Galago demidovii, Galago crassicaudatus, Galago senegalensis, Euoticus elegantulus,* and *Perodicticus potto*) eat exudates only from sites where insect infestation or mechanical damage has resulted in exudate flow. At least three of these seven (*Microcebus murinus, Galago senegalensis,* and *Euoticus elegantulus*) sometimes use specialized "tooth combs" or "tooth scrapers" formed by the lower anterior dentition as "scoops" to "collect" exudates (R. D. Martin, pers. comm.). However, the regular use of teeth as gouges to dig holes in trees and elicit exudate flow has thus far only been observed in *Callithrix, Cebuella,* and *Phaner furcifer,* the fork-marked mouse lemur. *Phaner* apparently uses its highly procumbent upper incisors, its upper canines and the mandibular "tooth comb" to scrape the bark of *Terminalia* sp. and induce exudate flow (Petter et al., 1971). (It is interesting to note that another *Terminalia* introduced into Brazil is exploited by *Callithrix jacchus* in the state of Alagoas.)

Wear patterns on the occlusal surfaces of the upper incisors of Cebuella and Callithrix indicate that they are used to hone the tips of the lower incisors, making them efficient gouging implements. On the other hand, the only consistent wear patterns on the teeth of small-bodied, exudate-eating prosimians are the whip-marks produced by the hair in grooming (R. D. Martin, pers. comm.).

Tree-gouging and Related Behavior in Other Primates

A number of other primates use their teeth to gouge, pry loose, scrape or break off woody parts of trees, usually because they feed on bark and sometimes cambium (*Lemur catta, Lemur fulvus,* Sussman, 1974; *Indri indri, Propithecus diadema, Avahi laniger,* Petter, 1962a, 1962b; *Propithecus verreauxi,* Petter, 1962a, 1962b; Richard, 1973; *Arctocebus calabarensis, Galago alleni,* Charles-Dominique, 1971; *Leontopithecus rosalia rosalia,* Coimbra-Filho and Mittermeier, 1973; *Saimiri sciureus,* R. Bailey, pers. comm.; *Cebus apella,* Izawa, 1975; *Alouatta seniculus,* Neville, 1973; *Alouatta palliata,* Hladik and Hladik, 1969; *Ateles belzebuth,* Klein, 1972; *Lagothrix lagotricha,* J. Cassidy, pers. comm.; *Cercopithecus aethiops,* Gartlan and Brain, 1968; *Cercopithecus nictitans martini,* Gartlan and Struhsaker, 1972; *Macaca mulatta,* Lindburg, 1971; *Macaca fuscata,* Suzuki, 1965; *Papio* spp., Hall, 1963; Altmann and Altmann, 1970; *Colobus badius tephrosceles,* Clutton-Brock, 1975; *Presbytis entellus,* N. Bishop, pers. comm.; *Presbytis cristatus,* Bernstein, 1968; *Pongo pygmaeus,* Mackinnon, 1971; *Pan troglodytes,* Goodall, 1963; Reynolds and Reynolds, 1965; *Gorilla gorilla,* Schaller, 1965; Jones and Sabater-Pi, 1971).

Most of these animals only occasionally eat bark or other woody tree parts (although during the winter *Macaca fuscata* lives primarily from bark in some parts of its range, Suzuki, 1965), and none of the Anthropoidea mentioned above have any noteworthy dental specializations related to wood-eating behavior. The closest thing to the *Cebuella-Callithrix* pattern can be seen in *Propithecus verreauxi,* which uses its mandibular tooth comb or tooth scraper to gouge out chunks of bark and cambium from living trees and both its tooth comb and premolars to tear splinters from dead wood and strips of bark from living trees (Richard, 1973). *Indri indri, Microcebus murinus* and several other prosimians use the tooth comb to gouge out pieces of soft fruit, but not wood (R. D. Martin, pers. comm.).[3]

Although it has not been observed to eat wood or exudates, the aye-aye (*Daubentonia madaga-*

[3]Data on tree-gouging in *Callithrix* and *Cebuella* may add to our understanding of the evolution of the prosimian tooth comb or tooth scraper. This unusual structure, formed by the lower anterior dentition, is found in all prosimians except *Daubentonia* and *Tarsius* (if one considers the latter a prosimian). Several authors (Roberts, 1941; Buettner-Janusch and Andrew, 1962) have, on the basis of captive observations, suggested that the primary function of the tooth comb is grooming the fur and that it is not used for eating. Stein (1936) and Avis (1961), on the other hand, maintained that it was primarily a feeding adaptation. Martin (1972) favors the argument that the comb developed primarily as a feeding adaptation, and that the grooming function was largely secondary. There is no question that the tooth comb evolved at least in part as a grooming device. The structure is very similar in morphology and is used in much the same way in all the Lorisidae, Lemuridae, and Indriidae. It would be difficult to account for this uniformity if it had evolved for different reasons in different animals; however, in light of the facts that recent field observations of a number of lemurids, indriids and lorisids indicate that the tooth comb is important in feeding behavior ranging from gum-collecting to bark-prying and gouging out pieces of soft fruit and that a similar (but not as specialized) complex has evolved primarily as a feeding adaptation in two New World genera, it seems highly probably that dietary considerations also played a role in the evolution of the tooth comb.

scariensis) has the most specialized tree-gouging dentition among the primates. It uses its large, rodent-like, continuously growing incisors to gouge or gnaw holes into trees and expose wood-boring insect larvae, which are then extracted with the elongate middle digit on the manus (Cartmill, 1974). A similar adaptation is also found in several species of New Guinean and northern Australian phalangerids of the genera *Dactylopsila* and *Dactylonax* (Cartmill, 1974).

This comparison with other primates shows that neither sporadic exudate-eating nor tree-gouging is an unusual behavior pattern; however, the regular use of tree-gouging to specifically *elicit* exudate flow has thus far been found in only three genera, *Cebuella, Callithrix* and *Phaner.* The combination appears to be rare not only among primates, but among vertebrates in general.[4]

Tree-gouging and Marking Behavior in Other Primates

The brachial marking of branches by male *Lemur catta* using a forearm spur is the only other behavior pattern comparable to *Callithrix* urinating in and marking in the vicinity of holes it gouges. The spur is used to cut comma-shaped slits into branches. These cuts retain the scent of the antebrachial gland located near the spur and also of the brachial gland with which the spur is annointed prior to marking (Schilling, 1974). As in the case of *Callithrix* urine, the secretions left by male *L. catta* in the comma-shaped cuts probably remain longer than if simply placed on the surface of the branch.

Acknowledgments

We would like to thank R. C. Bailey, J. G. Fleagle, C. Freese, R. D. Martin, and R. W. Sussman for helpful comments on early drafts of this paper. R. D. Martin and R. W. Sussman also provided unpublished data on prosimians. The senior author is a fellow of the Brazilian National Research Council (Conselho Nacional de Pesquisas). Mittermeier's research in South America was partly funded by a National Science Foundation Graduate Fellowship and partly by the New York Zoological Society. Illustrations were prepared by G. Carter and K. Kamata.

[4]Aside from the three primate genera mentioned, the combination of tree-gouging and exudate-eating has been mentioned for only one other mammal, the neotropical didelphid *Philander opossum* (DuBost, in Charles-Dominique, 1971). No details are available, however, on the tree-gouging/exudate-eating complex in this animal. Another neotropical didelphid, *Marmosa,* occasionally bites and frays bark and may lick the bark or perform marking behavior on the bitten surfaces (J. F. Eisenberg, pers. comm.).

Literature Cited

Allen, G. M., and A. Loveridge
1942. Scientific results of a fourth expedition to forested areas in East and Central Africa, Part I. Mammals. *Bull. Mus. Comp. Zool., Harvard,* 89:145-214.

Altmann, S. A., and J. Altmann
1970. *Baboon Ecology. Bibliotheca Primatologia,* 12. Basel: S. Karger.

Avis, V.
1961. The significance of the angle of the mandible: an experimental and comparative study. *Amer. J. Phys. Anthro.,* 19:55-61.

Bearder, S. K., and G. A. Doyle
1974. Ecology of bushbabies, *Galago senegalensis* and *Galago crassicaudatus,* with some notes on their behaviour in the field. In *Prosimian Biology,* edited by R. D. Martin, G. A. Doyle, and A. C. Walker, pp. 109-130. Pittsburgh: University of Pittsburgh Press.

Bernstein, I. S.
1968. The Lutong of Kuala Selangor. *Behaviour,* 32: 1-16.

Buettner-Janusch, J., and R. J. Andrew.
1962. The use of the incisors by Primates in grooming. *Amer. J. Phys. Anthro.,* 20:127-129.

Cartmill, M.
1974. *Daubentonia, Dactylopsila,* woodpeckers and klinorhynchy. In *Prosimian Biology,* edited by R. D. Martin, G. A. Doyle, and A. C. Walker, pp. 655-670. Pittsburgh: University of Pittsburgh Press.

Charles-Dominique, P.
1971. Eco-ethologie des prosimiens du Gabon. *Biol. Gabon.,* 7:121-228.

Charles-Dominique, P., and R. D. Martin.
1972. Behaviour and ecology of nocturnal Prosimians. *Fortschr. der Verhaltensforsch.,* 9:1-91. Berlin: Verlag Paul Parey.

Christen, A.
1974. Fortpflanzungsbiologie und Verhalten bei *Cebuella pygmaea* und *Tamarin tamarin. Fortschr. der Verhaltensforsch.,* 14:1-79. Berlin: Verlag Paul Parey.

Clutton-Brock, T. H.
1975. Feeding behaviour of red colobus and black and white colobus in East Africa. *Folia primat.*, 23:165-207.

Coimbra-Filho, A. F.
1971. Os sagüis do genero *Callithrix* da região oriental brasileira e um caso de duplo-hibridismo entre tres de suas formas (Callithricidae, Primates). *Rev. Brasil. Biol.*, 31(3):377-388.

1972. Aspectos inéditos do comportamento de sagüis do gênero *Callithrix* (Callithricidae, Primates). *Rev. Brasil. Biol.*, 32(4):505-512.

Coimbra-Filho, A. F.; A. D. Aldrighi: and H. F. Martins
1973. Nova contribuição ao restabelecimento da fauna do Parque Nacional da Tijuca. *Brasil Florestal*, 4(16):7-25.

Coimbra-Filho, A. F., and R. A. Mittermeier
1973. Distribution and ecology of the genus *Leontopithecus* Lesson, 1840 in Brazil. *Primates*, 14(1):47-66.

Dawson, G.
1977. Composition and stability of social groups in the tamarin, *Saguinus oedipus geoffroyi*, in Panama: ecological and behavioral implications. In *The Biology and Conservation of the Callitrichidae*, edited by Devra G. Kleiman, pp. 23–37. Washington, D. C.: Smithsonian Institution Press, 1977.

Doyle, G. A.
1974. The behavior of the lesser bushbaby (*Galago senegalensis moholi*). In *Prosimian Biology*, edited by R. D. Martin, G. A. Doyle, and A. C. Walker, pp. 213–231. Pittsburgh: University of Pittsburgh Press.

Gartlan, J. S., and C. K. Brain
1968. Ecology and social variability in *Cercopithecus aethiops* and *C. mitis*. In *Primates: Studies in Adaptation and Variability*, edited by P. C. Jay, pp. 253-292. New York: Holt, Rinehart and Winston.

Gartlan, J. S., and T. T. Struhsaker
1972. Polyspecific associations and niche separation of rainforest anthropoids in Cameroon, West Africa. *J. Zool., Lond.*, 168:221-266.

Goodall, J.
1963. Feeding behaviour of wild chimpanzees. A preliminary report. *Symp. Zool. Soc. Lond.*, 10:39-47.

Hall, K. R. L.
1963. Variations in the ecology of the Chacma baboon, *Papio ursinus. Symp. Zool. Soc. Lond.*, 10:1-28.

Hernandez-Camacho, J., and R. W. Cooper
1976. The nonhuman primates of Colombia. In *Neotropical Primates: Field Studies and Conservation*, edited by R. W. Thorington, Jr. and P. G. Heltne, pp. 35-69. Washington, D. C.: U. S. National Academy of Sciences Press.

Hill, W. C. O.
1957. *Primates. Comparative Anatomy and Taxonomy. Vol. III. Hapalidae.* Edinburgh: University Press.

Hladik, A., and C. M. Hladik
1969. Rapports trophiques entre végétation et primates dans la forêt de Barro Colorado (Panama). *La Terre et la Vie*, 1:25-117.

Izawa, K.
1972. Monkeys in the Upper Basin of Amazon. *Monkey*, 16(2):4-18. (Japanese).
1975. Foods and feeding behavior of monkeys in the Upper Amazon Basin. *Primates*, 16(3):295-316.

Jones, C., and J. Sabater-Pi
1971. *Comparative Ecology of Gorilla gorilla (Savage and Wyman) and Pan troglodytes (Blumenbach) in Rio Muni, West Africa. Bibliotheca Primatologica*, 13. Basel: S. Karger.

Kinzey, W. G.; A. L. Rosenberger; and M. Ramirez
1975. Vertical clinging and leaping in a neotropical anthropoid. *Nature*, 255:327-328.

Klein, L. L.
1972. The Ecology and Social Organization of the Spider Monkey, *Ateles belzebuth.* Unpublished Ph. D. dissertation, University of California, Berkeley.

Lindburg, D. G.
1971. The rhesus monkey in North India: an ecological and behavioral study. In *Primate Behavior, Vol. 2*, edited by L. A. Rosenblum, pp. 2-106. New York: Academic Press.

Machado, A., and O. C. Leite
1957. Goma de cajueiro. *Boletim do Instituto de Química Agrícola*, 50:7-15.

MacKinnon, J.
1971. The orang-utan in Sabah today. *Oryx*, 11(2-3):141-191.

Martin, R. D.
1972. Adaptive radiation and behaviour of the Malagasy lemurs. *Philosophical Trans. Roy. Soc. Lond.*, 264:295-352.
1973. A review of the behavior and ecology of the lesser mouse lemur (*Microcebus murinus* J. F. Miller 1777). In *Comparative Ecology and Behavior of Primates*, edited by R. P. Michael and J. H. Crook, pp. 1-68. New York: Academic Press.

Mittermeier, R. A.; R. C. Bailey; and A. F. Coimbra-Filho.
1977. Conservation status of the Callitrichidae in Brazilian Amazonia, Surinam, and French Guiana. In *The Biology and Conservation of the Callitrichidae*, edited by Devra G. Kleiman, pp. 137–146. Washington, D. C.: Smithsonian Institution Press, 1977.

Moynihan, M.
1970. Some behavior patterns of Platyrrhine monkeys. I. *Saguinus geoffroyi* and some other tamarins. *Smithson. Contrbs. Zool.*, 28:1-77.
1976. Notes on the ecology and behavior of the pygmy marmoset, *Cebuella pygmaea,* in Amazonian Colombia. In *Neotropical Primates: Field Studies and Conservation,* edited by R. W. Thorington, Jr. and P. G. Heltne, pp. 79-84. Washington, D. C.: U.S. National Academy of Sciences Press.

Napier, J. R., and P. H. Napier
1967. *A Handbook of Living Primates.* New York: Academic Press.

Neville, M. K.
1972. The population structure of red howler monkeys (*Alouatta seniculus*) in Trinidad and Venezuela. *Folia primat.*, 17:56-86.

Neyman, P.
1977. Some aspects of the ecology and social organization of free-ranging cotton-top tamarins (*Saguinus oedipus*) and the conservation status of the species. In *The Biology and Conservation of the Callitrichidae,* edited by Devra G. Kleiman, pp. 39–71. Washington, D.C.: Smithsonian Institution Press, 1977.

Petter, J. J.
1962a. Recherches sur l'écologie et l'éthologie des lémuriens Malagaches. *Memoires Musee d'Histoire Naturelle, Paris,* Sér. A, 27(1):1-146.
1962b. Écologie et éthologie comparée des lémuriens Malagaches. *La Terre et la Vie,* 109:394-416.

Petter, J. J., A. Schilling, and G. Pariente
1971. Observations eco-éthologiques sur deux lémuriens Malagaches nocturnes: *Phaner furcifer* et *Microcebus coquereli. La Terre et la Vie,* 118(3):287-327.

Rahaman, H., and M. D. Parthasarathy
1969. Studies on the social behaviour of bonnet monkeys. *Primates,* 10:149-162.

Ramirez, M.; C. Freese; and J. Revilla
1977. Feeding ecology of the pygmy marmoset, *Cebuella pygmaea,* in northeastern Peru. In *The Biology and Conservation of the Callitrichidae,* edited by Devra G. Kleiman, pp. 91–104. Washington, D.C.: Smithsonian Institution Press, 1977.

Reynolds, V., and F. Reynolds
1965. Chimpanzees in the Budongo Forest. In *Primate Behavior: Field Studies of Monkeys and Apes,* edited by I. DeVore, pp. 369-424. New York: Holt, Rinehart and Winston.

Richard, A.
1973. Social Organization and Ecology of *Propithecus verreauxi* Grandidier 1867. Ph.D. dissertation, London University, London, England.

Ripley, S.
1970. Leaves and leaf-monkeys: the social organization of foraging in gray langurs (*Presbytis entellus thersites*). In *Old World Monkeys: Evolution, Systematics, and Behavior,* edited by J. R. Napier and P. Napier, pp 481–509. New York: Academic Press.

Roberts, D.
1941. The dental comb of lemurs. *J. Anat.*, 75:236-238.

Rosenthal, F. R. T.
1951. *Goma de Cajueiro.* Rio de Janeiro: Instituto Nacional de Tecnologia.

Sauer, E. G. F., and E. M. Sauer
1963. The southwest African bush-baby of the *Galago senegalensis* group. *J. Southwest Afr. Sci. Soc.*, 16:5-35.

Schaller, G.
1965. The behavior of the mountain gorilla. In *Primate Behavior: Field Studies of Monkeys and Apes,* edited by I. DeVore, pp. 324-367. New York: Holt, Rinehart and Winston.

Schilling, A.
1974. A study of marking behaviour in *Lemur catta.* In *Prosimian Biology,* edited by R. D. Martin, G. A. Doyle and A. C. Walker, pp. 347-362. Pittsburgh: University of Pittsburgh Press.

Stein, R. M.
1936. The myth of the lemur's comb. *Amer. Nat.*, 70:19-28.

Sussman, R. W.
1972. An ecological study of two Madagascan primates: *Lemur fulvus* Audebert and *Lemur catta* Linnaeus. Ph. D. dissertation, Duke University, Durham, North Carolina.
1974. Ecological distinctions in sympatric species of *Lemur.* In *Prosimian Biology,* edited by R. D. Martin, G. A. Doyle and A. C. Walker, pp. 75-108. Pittsburgh: University of Pittsburgh Press.

Suzuki, A.
1965. An ecological study of wild Japanese monkeys in snowy areas focused on their food habits. *Primates,* 9:31-72.

Thorington, R. W., Jr.
1968. Observations of the tamarin (*Saguinus midas*). *Folia primat.*, 9:95-98.

Wettstein, E. B.
1963. Variabilität, Geschlechtunterschiede und Alterveränderungen bei *Callithrix jacchus* L. *Morphol. Jahrb.*, 104, 2:185-271.

SECTION II:

THE STATUS AND CONSERVATION OF CALLITRICHIDS

Introduction

The reports in this section suggest that most callitrichids are not now severely threatened, but will only remain so if habitat destruction is abated. Callitrichids are not hunted for food nor do they compete with local populations for food. Their preservation is primarily dependent upon the continued existence of uninterrupted forest. The precarious status of the three subspecies of *Leontopithecus rosalia* and *Saguinus oedipus* has resulted primarily from the near complete eradication of forest within their range in response to increasing population pressure. As several people point out, the more inaccessible and inhospitable an area, the safer are the species therein.

Even if existing habitats are preserved, there will be pressure on marmoset and tamarin populations by the biomedical community for research purposes. To supply sufficient animals for research from the wild, however, it is essential that censusing techniques be improved so that harvesting programs can be initiated which will not deplete natural populations. In the Panel Discussion, these problems are discussed in detail. To date, there has not been a single instance where a population of any primate species has been monitored for long enough to determine how many individuals could be safely removed on an annual basis.

The question of where and how to distribute resources for conservation is an enduring one covered at length in the Panel Discussion. The golden lion tamarin, *L. r. rosalia*, is nearly extinct and the proposed reserve (see Magnanini chapter) is probably not large enough to maintain a genetically viable population. But the golden lion tamarin is photogenic and the international symbol of a primate on the verge of extinction. Do we use energy that cannot be wasted for a hopeless cause because of its symbolic value? The consensus was affirmative, but mainly since it is inconceivable to declare a cause hopeless until the species is extinct; however, the need for preserving large tracts of undisturbed habitat with a diversity of species was recognized as a major priority for conservation efforts.

One problem brought up by Cooper and discussed by others in the Panel Discussion is the effect of cultural preconceptions on the usefulness of conservation efforts. Where overpopulation and poverty are the rule in a country, there may be little economic or political incentive to conserve the natural fauna and flora. Obviously, every effort must be made to promote conservation in a manner that is compatible with the culture of the country. One alternative is to develop methods of creating preserves which are economically viable through regular harvesting of natural primate populations or, of course, through tourism.

The Status of Callitrichids in Peru

CURTIS H. FREESE
Department of Pathobiology,
Johns Hopkins University,
615 N. Wolfe Street,
Baltimore, Maryland, 21205

MARGARET A. FREESE
Department of Medicine,
Johns Hopkins University,
Baltimore, Maryland, 21205

and

NAPOLEON CASTRO R.
Ministerio de Agricultura,
Lima, Peru

ABSTRACT

An eight-month survey, including systematic censusing, of primate populations in Amazonian Peru was conducted in 1974. Primary objectives included an assessment of the status of natural populations, and of factors affecting population sizes. *Saguinus nigricollis, S. fuscicollis, S. imperator,* and *Cebuella pygmaea* were observed during field surveys. Populations of *Saguinus* species appear not to be significantly affected in heavily hunted areas, while populations of the larger primate species are usually greatly depressed in these areas. Estimated densities of callitrichids show much less fluctuation between areas than densities of cebids. Although *S. mystax* was not observed in the wild, other evidence suggests that they are common in the region south of the Amazon River. *Saguinus* populations are relatively unaffected in hunted areas partly because they, in contrast to larger monkeys, are not commonly shot for food. However, heavy trapping for some future export demand of any one species, particularly those other than *S. fuscicollis,* could seriously reduce populations.

Recent reports indicated that *Cebuella* inhabits a wider range in Amazonian Peru than was previously supposed. *Cebuella* is the only Peruvian monkey generally considered too small to be hunted for food; however, *Cebuella* is probably only locally abundant even in undisturbed areas. Thus, while *Cebuella's* survival is not currently threatened in Peru, its populations are susceptible to and could be severely reduced by any future heavy trapping.

Factors affecting Peru's callitrichid populations are discussed and recommendations are given concerning their conservation.

Introduction

Five, and possibly six, species of Callitrichidae occur in Peru: *Cebuella pygmaea*, the pygmy marmoset; *Saguinus fuscicollis,* the saddle-back tamarin; *Saguinus nigricollis* (including *S. n. graellsi*), the black-mantled tamarin; *Saguinus mystax*, the moustached tamarin; and *Saguinus imperator*, the emperor tamarin. *Callimico goeldii* is also found in Peru and will be discussed in this report. According to a recent survey in Bolivia (Heltne et al., 1975), *S. labiatus,* the red-bellied tamarin, occurs in extreme southeastern Peru, but this species will not be discussed in this paper. Exportation of callitrichids, particularly *S. fuscicollis,*from the Iquitos area of Peru for biomedical research and the pet trade was heavy through 1973 (Soini, 1972; Ministry of Agriculture statistics for Iquitos). Excessive exportation, habitat destruction, and the hunting of monkeys for meat (Neville, 1974; Castro et al., in press) have created concern for the status of wild populations, particularly in the Iquitos area.

This paper presents results of field surveys of nonhuman primates in Peru conducted from April to December 1974. Systematic field censuses of monkeys were performed in seven geographical areas of Amazonian Peru to assess the population status of monkey species there. Population data on individual callitrichid species are compared to cebid populations as a whole, and population densities of both families are discussed in relation to levels of human exploitation. Some notes on distribution are presented, and conservation needs for Peru's callitrichids are discussed.

Study Areas

The seven geographical areas surveyed are shown in Figure 1. A brief description of each area is given below.

1. Nanay River–Iquitos area. Censuses were conducted between the Itaya and Amazon Rivers from 2 to 4 km south of Iquitos, and at several locations along the Nanay River, a tributary of the Amazon which, near its mouth, roughly forms the northern and western limits of Iquitos. Censusing was done along the Nanay from Iquitos to nearly 400 km upriver, including about 16 km along each of the headwater rivers, Aguas Negras and Aguas Blancas. Scattered hamlets occur along the Nanay, diminishing in size and frequency until the last permanently inhabited house we encountered at the 307 km mark. Several camps used intermittent-

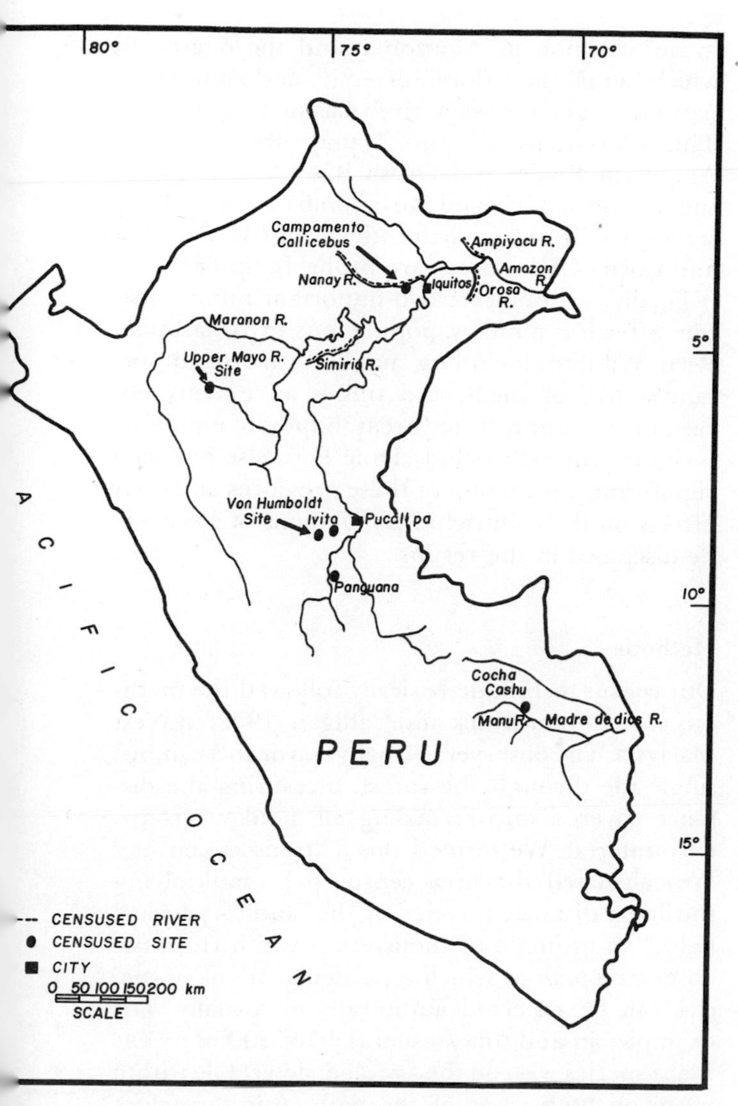

Figure 1. Location of study areas during censusing in Peru.

ly by hunters and loggers exist along the entire river. Hamlets are only along the river or its ox-bow lakes, never inland. This is generally true for all lowland Amazonia.

The Nanay is a black-water river. (For the possible significance of this characteristic for biological productivity, see Janzen, 1974). Selective cutting in the inundatable forests has occurred all the way up the Nanay although the amount of cutting diminishes as one gets farther from Iquitos. Non-inundatable forests suffer little or no cutting, except in the immediate areas around villages where slash-and-burn agriculture is practiced. These trends in forest cutting are the same along the Orosa and Ampiyacu rivers discussed below.

The sub-area between the Itaya and Amazon has been subjected to more human pressure than any other area censused. The forest has received heavy selective and clear cutting although primary growth may still dominate. The area is inundated annually by the Amazon and Itaya.

2. Ampiyacu River basin. The Ampiyacu River is a northern tributary of the Amazon, discharging its waters 170 km downstream (east) from Iquitos. The Ampiyacu basin, in many respects, is a miniature of the Nanay basin, but some notable differences do exist. During the survey, the Ampiyacu was fairly turbid and, from appearance alone, seemed to be somewhere between a white-water and black-water river. The water level was low during the survey. Inundatable forests occur almost exclusively along the lower section (roughly the lower 50 km) of the river.

Almost the entire human population along the river lives within 35 km of its mouth; above 35 km, where almost all censusing was done, there are only two or three lone habitations.

3. Orosa River basin. The Orosa River enters the Amazon from the south side approximately 120 km downstream from Iquitos. The Orosa is very similar to the Ampiyacu, including its appearance as an almost white-water river and the distribution of its human population.

4. Samiria River basin. The black-water Samiria River lies in the fork between the Marañon and Ucayali Rivers and empties into the Marañon near the confluence of the two great rivers. The features of the Samiria basin are quite distinct from those of the previous river systems. Foot and canoe work, aerial photographs and flyovers, and accounts of local people all indicate that almost the entire Samiria basin, as well as the basin of its sister river, the Pacaya, and the forest between the two, is seasonally or permanently inundated.

Also, the Samiria basin lies within the Empresa Publica de Servicios Pesqueros (EPSEP) fishing reserve in which commercial hunting and lumbering have been prohibited since 1944. Several guard stations are located along the river, and, consequently, violations probably are infrequent even though subsistence hunting and logging are allowed. The hamlet populations have dwindled in recent years, but recent ingress and exploration by hundreds of petroleum workers, who are permitted to hunt for food, must be detrimental to large animals in the area.

5. Pucallpa area. Three localities were censused near Pucallpa. One is a biological field station, Panguana, located along the white-water Llullapichis River, a small affluent of the Pachitea River entering it from the east. Panguana is approximately 140 km south-southwest of Pucallpa near 9.6° S latitude and 74.9° W longitude. The other two localities censused are 59 and 90 km west-southwest of Pucallpa near the Lima-Pucallpa road. The 2 sq km of Panguana have been an official preserve since 1972 and it probably has received only light hunting since then. But compared to the Samiria and Cocha Cashu areas, it has been subjected to heavy hunting, especially before 1972, and we include it as an unprotected area for the comparison of census results. Human settlements are common near all three localities, and the two along the Lima-Pucallpa road are subject to intensive hunting. No extensive inundatable forests occur in the survey localities, and logging has varied from selective to none at all.

6. Cocha Cashu Biological Station. The Cocha Cashu Biological Station is located on the northeast side of the white-water Manu River in Manu National Park at about 11.8° S latitude and 71.4° W longitude. Manu National Park lies mostly within the department of Madre de Dios, the most sparsely populated department in Peru. The park was established in 1969 and since then has received rigorous protection. Even before 1969 the inaccessability of the park area made human disturbance a very limited factor.

The river rarely inundates the forest of the field station, and the virgin forest resembles forests near Pucallpa more than forests near Iquitos.

7. Moyobamba area. A short investigation was made in the rugged highland (approximately 1000 m) forest along the upper Mayo River 80 km northwest of Moyobamba, around 5.7° S latitude and 77.5° W longitude.

Differences in biological productivity and other ecological factors between the study areas almost certainly affect primate abundance and diversity. These differences, however, have received little to no attention in Amazonia, and the degree to which faunal (and floral) diversity and abundance naturally vary between river basins is unknown. The ecosystems of the Nanay River–Iquitos, Ampiyacu River, and Orosa River areas appear similar, the Samiria and Moyobamba areas are each somewhat distinct from the others, and the Pucallpa and Cocha Cashu areas are probably similar.

Finally, we reiterate two important human factors affecting monkey populations in Amazonian Peru. Wildlife, including monkeys, are an important source of meat, and this is an everpresent factor in the unprotected areas. Export of monkeys, including the callitrichids, from Peru also has been significant. The extent of these pressures and their effects on the callitrichid populations of Peru will be discussed in the results.

Methods

Our census technique basically followed the methods used by Southwick and Cadigan (1972) in West Malaysia. The observers, usually two or three, move single file through the forest, measuring the distance covered and recording all monkey groups encountered. We termed this a "transect census." We calculated the area censused by multiplying the linear distance traveled by the "auditory / visual field," an estimate of the average width from the observer's path at which a particular monkey species can be detected auditorally or visually. For example, an auditory / visual field of 100 m means that a species was, on the average, detectable within 50 m on both sides of the path. Auditory clues do not include the dawn or troop spacing choruses of *Alouatta* and *Callicebus*. The auditory / visual field ranged from 40 m for the inconspicuous *Callicebus* and *Alouatta* to 60 m for *Saguinus* to 100 m for *Saimiri*. This method should provide reliable comparative data on intraspecific differences in population densities. Also, depending on the accuracy of the auditory/ visual field estimate for each species, population sizes can be compared interspecifically and estimates of absolute densities are obtained. The accuracy of this method needs to be tested in areas of known population densities.

Because *Cebuella* is so small and inconspicuous, the transect-census method was inadequate for determining numbers of this monkey, and we had to rely on different methods of assessing *Cebuella* populations (see Ramirez et al., 1977). Similarly, *Aotus* densities could not be estimated with transect censuses. Therefore, in the results, terms such as

Table 1. Months when the study areas were censused and a breakdown of total distances (unique and repeated) censused. The numbers in parentheses show the distance that was repeat census.

Area	*Months*	*Distance by foot (km)*	*Distance by canoe (km)*	*Total (km)*
Nanay-Iquitos	April-May	56.09 (6.20)	42.72 (0.25)	98.81 (6.44)
Orosa	June-July	2.35	20.12	22.47
Ampiyacu	Oct-Nov	19.27	9.73	29.00
Samiria	Oct	16.36 (3.50)	39.87	56.23 (3.50)
Pucallpa	June-July	60.44 (18.34)	0.00	60.44 (18.34)
Cocha Cashu	Aug-Sept	30.24 (19.00)	0.00	30.24 (19.00)
Moyobamba	Nov	5.50	0.00	5.50
Total		190.25 (47.04)	112.44 (0.25)	302.69 (47.28)

"total monkey abundance" and "Cebidae population densities" do not include these two species. Total monkey abundance is in units of monkeys of all species (excluding *Aotus* and *Cebuella*) per sq. km.

Transect censusing was done on foot and by canoe. A total of 302.69 km, during 267 hours of censusing, was covered by transect censuses—112.44 km by canoe and 190.25 km by foot. Table 1 gives the breakdown of distances for each area. Also, 620 km of riverside forest were censused by motorboat, but *Saguinus* were seen only seven times during these censuses and the results are not presented here. Many more hours and kilometers of nontransect-census field work were also performed.

During surveys of each area, we intensively questioned local inhabitants about monkeys. We usually found their information reliable.

Results

Group Sizes

Four different *Cebuella* groups were observed during field work, at least 35 sightings were made of *S. fuscicollis* groups, probably six sightings of *S. nigricollis* groups, and four sightings of *S. imperator* groups. *S. mystax* and *Callimico* were never observed in the wild. Good group-size counts of the small, fast-moving *Saguinus* species are difficult to obtain, and usually one can never be sure of a complete count unless a group is followed for several hours or even days. Precise data on group sizes and composition are much needed. A summary of group-size counts for each species is given below. Our estimated average group size for each *Saguinus* species is used for the density estimates presented in Table 2. We then give the results obtained from each survey area.

1. *Cebuella*: One complete count of a *Cebuella* group along the Nanay was obtained during the surveys (see the paper by Ramirez et al., 1977). This group varied from 7 to 9 individuals. A local guide reported that another *Cebuella* group in the area contained 7 individuals. In Colombia, Moynihan (1976) observed 6 groups with 3 to 6 individuals and Hernandez and Cooper (1976) report groups as large as 10 to 15.

2. *S. fuscicollis*: Our best counts of *S. fuscicollis* groups ranged from 1 to 9, with a large majority registering 4 to 8. No discernable differences in group sizes between areas were detected. Rogerio Castro (pers. comm.) once counted a group of 24 at Campamento *Callicebus* along the Nanay River, and Hernandez and Cooper (1976) report that *S. fuscicollis* groups in Colombia have 5 to 20 members. Neville et al. (1976) observed groups of 4 to around 11 along the Samiria and Pacaya rivers, and used 4 as the average group size. We estimated an average of 6 for *S. fuscicollis* groups in our study areas.

3. *S. nigricollis*: The few observations of this species, seen only along the Ampiyacu River, suggest that its group sizes are similar to *S. fuscicollis*. Our best count registered 6 individuals, which we use for the average group size. Hernandez and Cooper (1976) believed the average size for this species was 5 to 20 in Colombia.

4. *S. imperator*: Four sightings of *S. imperator* groups yielded counts of 1, 3, 3, and 3 individuals. We feel that these were good counts and consider 3 the average group size. Table 2 shows individual and group density estimates of each *Saguinus* species and of the Cebidae collectively for each study area, except Moyobamba where no monkeys were observed. These data were obtained from transect censuses.

Monkey Abundances in the Study Areas

1. Nanay River–Iquitos. *Cebuella* probably are locally abundant along the Nanay, but restricted to areas with sufficiently high concentrations of exudate-source trees. It appears that along the Nanay these suitable areas are only in inundatable forests (for details, see Ramirez et al., 1977). Two years of work by scientists in the high-ground forest of Camp *Callicebus* have revealed no signs of *Cebuella*. Locals, likewise, claim that *Cebuella* are found only in the annually inundated forests along the Nanay. As elsewhere, we found it occupying the unique position of being the only monkey too small to hunt for food, and it is generally unmolested in this area.

S. fuscicollis is the second most common monkey species in this area according to transect census estimates and comprises 29 percent of total monkey abundance (Table 2). The ability of *S. fuscicollis* to thrive in highly disturbed, mixed primary and secondary forest is illustrated by census results from inundated forest between the Itaya and Amazon within 4 km of Iquitos. From 13.67 km of censusing the estimated density of *S. fuscicollis* was 5.0 groups/sq km, or 30 individuals/sq km.

There is a slight possibility that one or two poorly seen groups that we assumed to be *S. fuscicollis* were *S. nigricollis graellsi.* We know of no sighting of *S. nigricollis* along the Nanay, however, and locals never mentioned the existence of a second kind of tamarin in the basin. Soini (1972) stated that natives report *S. nigricollis graellsi* to be rare along the Nanay. Only a few people along the Nanay recognized our pet *Callimico* and claimed that the species occurs there in small groups. Soini (1972) collected one (and observed another) near Mishana on the south side of the Nanay River, where in 1974 he observed another lone individual (Pekka Soini, pers. comm.).

2. Ampiyacu River basin. *Cebuella* appear to be common along the Ampiyacu River. Locals often reported *Cebuella* groups living nearby, and we saw several *Cebuella* signs, i.e., sap holes dug in trees by *Cebuella*, along the Ampiyacu. As noted by Ramirez et al. (1977), *Cebuella* inhabits both inundatable and non-inundatable forests along the Ampiyacu. *Cebuella* groups can also be found in highly disturbed areas; for example, *Cebuella* were observed in a large tree (with *Cebuella*-dug exudate holes) standing alone in a clear-cut field with only short secondary forest nearby. At one hamlet, the people said they recently had captured two *Cebuella* to sell (unsuccessfully) as pets in Pevas.

Our observations indicate that *S. nigricollis* is the most numerous monkey in the Ampiyacu basin, comprising 50 percent of total monkey abundance, but that *S. fuscicollis* is not found there or is very rare. Although we could not identify positively two *Saguinus* groups as being *nigricollis*, for density calculations we assumed they were. Four other groups observed definitely were *nigricollis*, and locals claimed that only one kind of *Saguinus* inhabits the area. (That local hunters distinguish between *S. fuscicollis* and *S. nigricollis* was demonstrated by our guide having a different name for each.) Hershkovitz (1968) has a question mark concerning the presence of *S. fuscicollis tripartitus* in this region, and Hernandez and Cooper (1976) note that this species is absent in the immediate area of Leticia, Colombia. Our observations are in line with the lack of reports or collections of this species in this region.

No one in the area was familiar with a monkey fitting our description of *Callimico*.

3. Orosa River basin. *Cebuella* were reported present in the Orosa River basin by locals, but we have no idea how common they are.

Density estimates of *S. fuscicollis* were somewhat lower in this area than for *S. fuscicollis* and *S. nigricollis* in the previous two areas (Table 2). Our estimate shows it to be the fourth most abundant monkey species in the area, accounting for only 11 percent of total monkey abundance. No *S. mystax* were observed, although the area is in the reported range of this species. Unfortunately, locals were not questioned about this species.

One of our permanent guides who was born in this area informed us that the Yagua Indians of the Ampiyacu kill many *Saguinus* with blowguns. This, in part, may explain their lower density here. Fortunately, *Cebuella* is considered too small to hunt.

The Yagua hunters and other locals did not recognize our *Callimico* mascot and claimed that it did not inhabit the area. Indeed, the Yaguas thought that our mascot was a very strange monkey.

4. Samiria River basin. According to local guards, *Cebuella* inhabits only the river and lake-side forests in the area, but they are seldom seen. Neville et al. (1976), like us, did not see *Cebuella* in the wild along the Samiria, but they found two in captivity and state that it probably occupies the area.

S. fuscicollis abundance in this area was similar to *Saguinus* numbers in the Nanay River-Iquitos and Ampiyacu areas (Table 2). Although it is fairly

Table 2. Estimated group densities (Gr/km²) and individual densities (Ind/km²) of three callitrichid species and of Cebidae (except *Aotus*) in the six major areas surveyed.

	Area Surveyed											
	Nanay-Iquitos		*Ampiyacu*		*Orosa*		*Samiria*		*Pucallpa*		*Cocha Cashu*	
Taxon	*Gr/km²*	*Ind/km²*	*Gr/km²*	*Ind/Km²*	*Gr/km²*	*Ind/km²*	*Gr/km²*	*Ind/km²*	*Gr/km²*	*Ind/km²*	*Gr/km²*	*Ind/km²*
S. fuscicollis	2.6	15.6			1.5	9.0	2.5	15.0	1.4	8.4	1.8	10.8
S. nigricollis			3.2	19.2								
S. imperator											1.8	5.4
Cebidae	5.9	38.8	4.4	19.3	8.0	71.0	12.8	143.5	2.2	25.2	16.8	192.5

abundant, our estimates place it as the fourth most abundant monkey species, representing only 9 percent of total monkey density. Neville et al. (1972) felt *S. fuscicollis* was moderately to heavily abundant in the Samiria and Pacaya river basins.

Hershkovitz's (1968) distribution map places *S. mystax* in this area, and although we saw none, some guards claimed that another "pichico" (usually refers to tamarins) inhabits the basin; however, their description was too sketchy for us to say with any confidence that they were referring to *S. mystax*. Neville et al. (1976) also saw none during their surveys along the Samiria and nearby Pacaya River.

Our *Callimico* was not recognized by any of the local guards, and Neville et al. (1976) did not find evidence of its existence there.

5. Pucallpa area. Our survey indicates that *S. fuscicollis* is the only callitrichid inhabiting the census sites or surrounding areas west of the Ucayali River. No one recognized our description of *Callimico*.

S. fuscicollis is found in comparatively low abundance in the Pucallpa area, approximating their estimated density in the Orosa basin. Overall monkey densities were low, however, and our censuses indicated that *S. fuscicollis* is the second most abundant monkey species, comprising 25 percent of total monkey density.

6. Cocha Cashu Biological Station. Our observations and questioning of locals provided essentially no evidence that *Cebuella* inhabit this area (except for some suspicious looking "exudate holes" in trees), but Charles Janson (pers. comm.) saw a single *Cebuella* once at the field station in 10 weeks of forest work. To our knowledge, this is the farthest south that this species has ever been reported. Philip Hershkovitz (pers. comm.) has no confirmed records of *Cebuella* in Manu, but includes the region within the probable range of the species. *Cebuella* must exist at very low numbers in the area.

S. fuscicollis and *S. imperator* were observed in Manu during our censuses. Their separate estimated densities at Cocha Cashu are not high compared to other areas, but their combined density is (Table 2). Even when combined, however, they are estimated to be only the sixth most abundant monkey and constitute just 8 percent of total monkey abundance. A fascinating relationship between *S. fuscicollis* and *S. imperator* was observed: the two were together in a closely knit group each of four times each species was seen. Hernandez and Cooper (1976) report apparently mixed troops of *S. nigricollis* and *S. fuscicollis* in Colombia.

The only evidence that we found of *Callimico* in the area was from a park guard. He stated that a monkey that looked like *S. fuscicollis*, except that it was entirely black, could be found near the mouth of the Manu River.

7. Moyobamba area. During our short survey in this area, we saw no callitrichids in the wild, but we did see a captive *S. fuscicollis* being sold in Moyobamba. *S. fuscicollis* is said to be common in secondary forest and around agricultural land near Moyobamba. This species ranges up to 500 m in the Andean piedmont of Colombia (Hernandez and Cooper, 1976).

A good description by our local guide of a monkey found in the area fit that of *Cebuella*. Its occurrence here would extend its known range farther west. Philip Hershkovitz (pers. comm.) notes that Moyobamba seems to be entirely outside the range of *Cebuella*. Grimwood (1969) reported that a specimen is known from as far west as Santa Cruz on the Huallaga River.

Discussion

There are, or have been, two major kinds of human exploitation of callitrichid populations: capture for the local pet trade or export and hunting for food. An average of 797 *Cebuella* were exported annually from Iquitos from 1962 through 1968, but they have received complete protection since 1970 (Soini, 1972). The number of *Saguinus* exported via Iquitos from 1962 through 1973 totaled 22,559, an average of 1880 per year (Soini, 1972; Ministry of Agriculture statistics for Iquitos). In October 1973, the government of Peru enacted a law (Supreme Decree 934-73-AG) that prohibits the hunting or commercialization of all monkeys as well as most other mammals, birds, reptiles, and amphibians. Since then no monkeys have been exported officially from Peru, but for the domestic pet trade and hunting the law is largely unenforceable.

The number of callitrichids involved in domestic trade is unknown, but *Cebuella* and *S. fuscicollis* frequently are pets.

In general, the larger the monkey, the more avidly it is hunted for food. For the large monkey species, Castro et al. (in press) conclude that in the Iquitos region human consumption has been responsible for a bigger drain on wild populations than has the export market. They did not find *Saguinus* being sold as meat in the markets of Iquitos, nor did we ever see it. Even though *Saguinus* are shot and consumed occasionally in many areas of Amazonian Peru, hunting probably has little effect on *Saguinus* populations in most areas.

Forest cutting is another human activity that must adversely affect monkey populations, particularly of the larger monkeys. Callitrichids, however, may benefit from the increased amount of secondary growth in cut areas. Hernandez and Cooper (1976) note that *S. fuscicollis* and *S. nigricollis* inhabit both primary and secondary forests in Colombia, and *S. (oedipus) geoffroyi* seems to thrive best in short, scrub, or secondary forest in Panama (Moynihan, 1970). Forest cutting is most serious in the inundatable forests where annual flooding facilitates logging, and in parts of the Andean foothills where land is being cleared for agriculture.

Whatever the extent of these forms of human depredation on monkeys and their habitats, our censuses show that Cebidae populations are seriously affected in areas subject to heavy human disturbance, but Callitrichidae populations are relatively unaffected. The data in Table 2 illustrate the point.

Cebidae population densities in unprotected areas are about one-fourth as dense as populations in the two protected study areas. The average cebid population densities in protected and unprotected areas are significantly different ($P < 0.001$, Student t test).

Saguinus population densities (and probably *Cebuella*) have almost identical averages between protected and unprotected areas, and compared to individual species of Cebidae, *Saguinus* densities fluctuate very little from one area to the next. Indeed, some cebids, e.g., *Ateles* and *Alouatta*, vary from abundant in protected areas to extinct in unprotected areas. Thus, according to transect census results in the four unprotected survey areas, *Saguinus* appear to be the second most abundant monkey, usually behind *Saimiri*, but in the two protected areas it is the fourth and sixth most abundant species. Similarly, *Saguinus* represents an estimated average of 29 percent of total monkey density in the four unprotected areas, but only 9 percent in the two protected areas.

The high densities of *S. fuscicollis* in the highly exploited inundatable forests adjacent to Iquitos exemplify the ability of *S. fuscicollis*, and apparently of *S. nigricollis*, to survive in areas of heavy human disturbance. Several factors may explain why these two species, and probably other callitrichids, fare so well compared to the Cebidae in highly disturbed areas: (1) callitrichid species are not hunted for food as intensively as the larger cebids; (2) callitrichids have a higher reproductive rate than any of the Cebidae and thus can tolerate a faster harvest rate; (3) callitrichids can probably exploit secondary forests better than most cebids, and may even thrive best in secondary forests, and (4) as populations of some cebid species are diminished or extirpated because of hunting or habitat destruction, reduced competition for food resources may favor higher callitrichid populations.

Unfortunately, we have little first-hand data on *S. mystax* populations in Peru. Although we failed to observe this species during expeditions to the Orosa and Samiria rivers, the observations by Castro and Soini (1977) and the recent capture of over 800 individuals near Iquitos in just two to three months suggest that substantial populations exist in some areas. Also, C. Freese has observed *S. mystax* in a tall forest mixed with extensive secondary growth and cultivated fields at Fonte Boa along the Brazilian Amazon. It is reported by locals to be common there.

S. imperator is found in one of the most undisturbed regions of Amazonian Peru. This species has the most restricted geographical range of any callitrichid species in Peru, and we found a low population density at Cocha Cashu. Thus, it may be the most susceptible to over-exploitation of any of Peru's callitrichids. It presently is well protected in the large area of Manu National Park as well as in the extensively uninhabited forests of southeastern Peru.

Because of its specialized feeding habits, *Cebuella* probably is abundant only locally and may have a quite disjunct distribution. If so, its populations, at least in areas such as the Nanay River basin where it is restricted to waterways, are highly vulnerable to over-exploitation if a market for them developed. *Cebuella* appear capable of thriving in areas of heavy human disturbance and degraded forest, a relationship noted by Hernandez and Cooper (1976) and Moynihan (1976) for *Cebuella* in Colombia. As Moynihan cautions, however, intense collecting pressures may endanger the species in some areas.

All evidence indicates that *Callimico* is rare in Peru and throughout its rather disjunct range, although there may be isolated pockets where this species occurs in good numbers. Field investigations of its range and population status are badly needed. *Callimico* certainly deserves full protection.

More field studies on the status and population ecology of many of Peru's monkey species are necessary before knowledgeable decisions concerning the establishment of reserves and export regulations can be made. Currently, no Peruvian reserve or park offers protection for large populations of *S. nigricollis*, most subspecies of *S. fuscicollis*, *S. mystax*, *C. goeldii*, possibly *C. pygmaea*, and four of Peru's cebid species. Areas where these and other fauna and flora can be protected and can live under natural conditions should be established soon. The development of primate breeding centers both in South America and in the United States can partly alleviate the pressure to export wild monkeys. In general, hunting and forest destruction probably have not significantly reduced populations of Peru's callitrichids, whereas cebid populations have been greatly depleted by these activities. Compared to some of the Cebidae, e.g., *Ateles* and *Lagothrix*, callitrichids enjoy a good status in Peru.

Acknowledgments

We thank Rogerio Castro, Marleni Ramirez, Juan Revilla, and Andrés Marmol for their participation in field expeditions and for their many helpful discussions. The logistical assistance provided by the Ministry of Health and the Instituto Veterinario de Investigaciones Tropicales y de Altura (IVITA) was greatly appreciated. The Ministry of Agriculture, Hans Koepke, and Empresa Publica de Servicios Pesqueros (EPSEP) gave us permission to work in Manu, Panguana, and the Samiria River Reserve. We especially thank Charles Southwick, Paul Heltne, Melvin Neville, and Pekka Soini for their encouragement and helpful suggestions throughout the surveys.

Literature Cited

Castro, N.; J. Revilla; and M. Neville.
In press. "Carne de Monte" como una fuente de proteinas en Iquitos, con referencia especial a monos. *Revista Forestal del Peru.*

Castro, R., and P. Soini.
1977. Field studies on *Saguinus mystax* and other callitrichids in Amazonian Peru. In *The Biology and Conservation of the Callitrichidae*, edited by Devra G. Kleiman, pp. 73-78. Washington, D.C.: Smithsonian Institution Press.

Grimwood, I. R.
1969. *Notes on the Distribution and Status of Some Peruvian Mammals, 1968.* Special publ., No. 21, American Committee for International Wild Life Protection and New York Zool. Soc.

Hernandez-Camacho, J., and R. W. Cooper.
1976. The non-human primates of Colombia. In *Neotropical Primates: Field Studies and Conservation,* edited by R. W. Thorington and P. G. Heltne, pp. 35-69. Washington, D.C.: U.S. National Academy of Sciences Press.

Hershkovitz, P.
1968. Metachromism or the principle of evolutionary change in mammalian tegumentary colors. *Evolution,* 22:556-575.

Janzen, D. H.
1974. Tropical blackwater rivers, animals, and mast fruiting by the Dipterocarpaceae. *Biotropica,* 6(2):69-103.

Moynihan, M.
1970. Some behavior patterns of Platyrrhine monkeys. II. *Saguinus geoffroyi* and some other tamarins. *Smithson. Contribs. Zool.*, 28:1-77.

1976. Notes on the ecology and behavior of the pygmy marmoset, *Cebuella pygmaea*, in Amazonian Colombia. In *Neotropical Primates: Field Studies and Conservation*, edited by R. W. Thorington and P. G. Heltne, pp. 79–84. Washington, D.C.: U.S. National Academy of Science Press.

Neville, M.
1974. "Carne de Monte" and its effect upon Simian populations in Peru. Paper presented at 73d Annual Meeting, American Anthropological Association, Mexico City.

Neville, M.; N. Castro; A. Marmol; and J. Revilla.
1976. The status of primate populations in the reserved area of the Pacaya and Samiria Rivers, Department Loreto, Peru. *Primates*, 17:151–182.

Ramirez, M. F.; C. H. Freese; and J. Revilla C.
1977. Feeding ecology of the pygmy marmoset, *Cebuella pygmaea*, in northeastern Peru. In *The Biology and Conservation of the Callitrichidae*, edited by Devra G. Kleiman, pp. 91–104. Washington, D.C.: Smithsonian Institution Press.

Soini, P.
1972. The capture and commerce of live monkeys in the Amazonian region of Peru. *Internat. Zoo. Yearb.*, 12:26–36.

Southwick, C. H., and F. Cadigan, Jr.
1972. Population studies of Malaysian primates. *Primates*, 13(1):1–18.

ALCEO MAGNANINI
Fundação Estadual de
Engenharia do Meio Ambiente
Departamento de Conservação Ambiental
Rio de Janeiro, Brazil

Progress in the Development of Poço Das Antas Biological Reserve for *Leontopithecus rosalia rosalia* in Brazil

In order to save a species, it is necessary to maintain viable populations in the wild. In Brazil, the establishment of Poço das Antas reserve would not only permit the existence of a sanctuary for *Leontopithecus rosalia* within the species' range, but also would be a pioneer effort in the development of conservation programs in all of South America.

Currently, the protection of existing natural families of captive *Leontopithecus rosalia* is a major priority, to be followed by the urgent action of capturing and relocating groups where habitat destruction is inevitable. In addition, conservation efforts are aimed at the establishment of an effective propagation program by the captive breeding of *L. r. rosalia, L. r. chrysopygus,* and *L. r. chrysomelas.* The ultimate aim is, of course, the release of specimens of *L. r. rosalia* into a protected reserve.

The history of the development of the Poço das Antas Biological Reserve from 1971 to 1975 will be reviewed here. The area was chosen in 1971 and many requests were made that the proposed area be designated as a reserve. At that time more than 70 percent of the land was densely forested and there were few human improvements, such as fences and roads.

After two and one-half years, in March 1974, the reserve was formally created (Decree No. 73791-11) and the expropriation of the land authorized (Decree No. 73792-11). The decrees, however, were not enforced during the following year despite the efforts and interest of numerous conservationists, such as G. Budowski, P. Scott, J. Perry, D. Bridgwater, P. Nogueira Neto, H. Jungius, C. R. Gutermuth, Governor F. Lima, J. L. Belart, and others. In April 1975 an aerial reconnaissance of the reserve showed that the

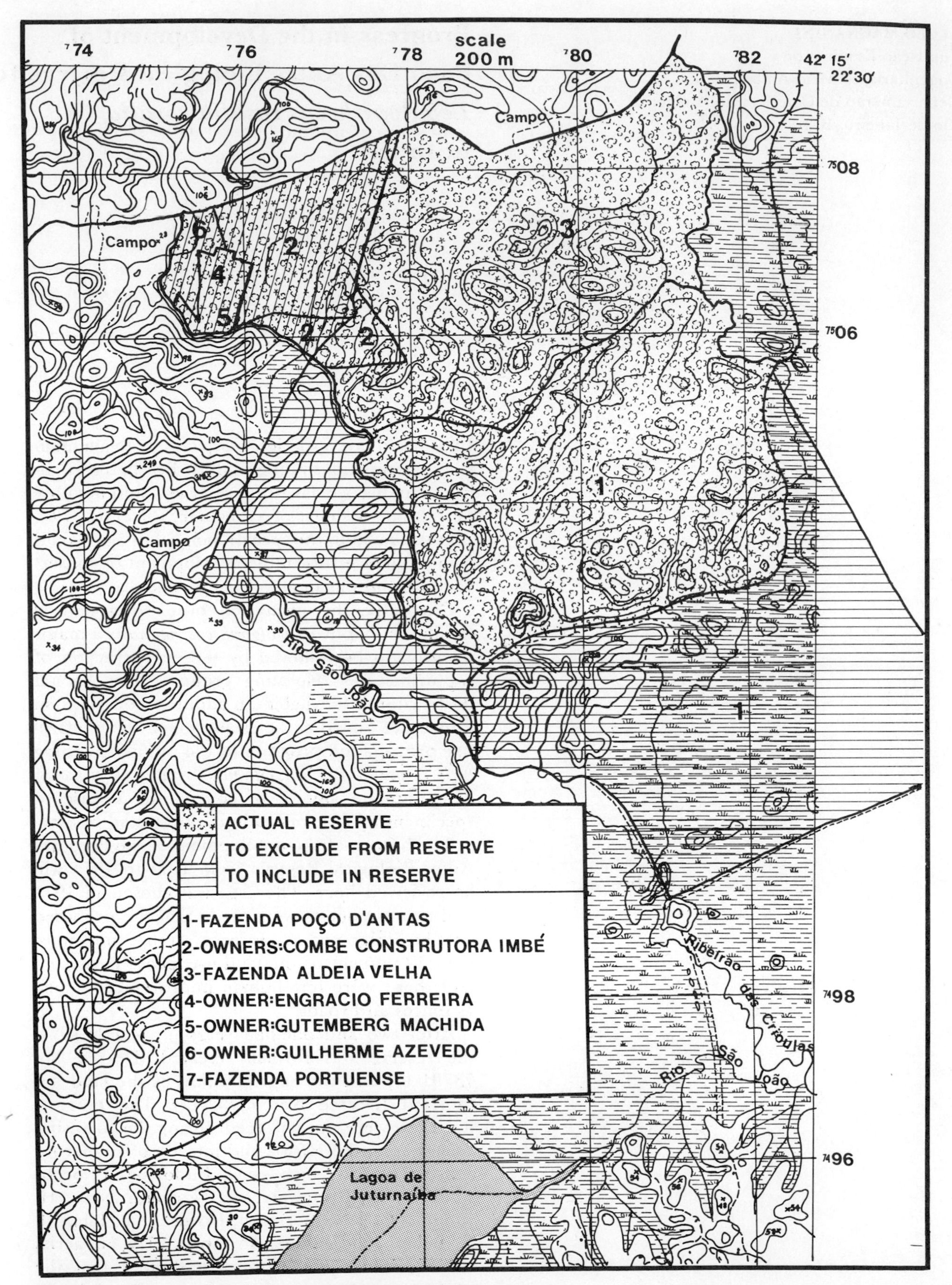

Figure 1. The actual Poço das Antas Reserve and the proposed changes in the reserve boundaries.

original area (about 3,000 ha) was less than 40 percent forested, with less than 10 percent of the area having dense forest. Many of the plots had been deforested by burning and planted in pasture or reforested with *Eucalyptus* trees. Roads, trails, drainage channels, houses, and fences have all been built since 1971.

These changes had two consequences. First, the usefulness of the area as a sanctuary for *L. r. rosalia* has decreased significantly; second, since it was required that improvements to the land be paid in cash, the budget allocated for the expropriation had been surpassed. Thus, the delayed expropriation had resulted in severe habitat alteration which both prevented the expropriation for financial reasons and made the area unfit for a reserve.

At this time, a third factor emerged, namely, that there were plans to alter the ecology of the São João River and its tributaries for economic reasons. The river is located south of the projected reserve (Figure 1).

With all of these changes, there was only one solution—to maintain the core area (2,700 ha) of the reserve, to exclude the altered areas (300 ha), and to include neighboring areas with few improvements (2,500 ha) which would be easier to expropriate (Figure 1). Thus, the total size of the reserve could be about 5,000 ha. In July 1975, a new decree was submitted to the president with these recommendations.

Two further steps were also taken. In May 1975, a Project (IUCN No. 24-1) was submitted to the president of IBDF (Instituto Brasileiro de Desenvolvimento Florestal) for the establishment of a Biological Reserve for golden lion tamarins in Brazil, for his official approval, and suggesting the remittance to IUCN-WWF. A copy of this outlined project was also sent to Dr. H. Jungius of the Survival Service Commission in IUCN. But Project IUCN No. 24-1 cannot be initiated unless the land is expropriated. Next, a request was made for a warden to enforce the reserve's boundaries.This request was made in July 1975 with the support of IBDF's agency in Rio de Janeiro State.

In November 1975, a new decree which excluded and added areas as suggested was finally signed (No. 76,584). Simultaneously, a decree was signed for the land expropriation (No. 76,583). As of June 1976, the expropriation process is in the final stage in the agrarian reform institute (INCRA) and IBDF is looking forward to taking possession of the land which totals about 5,000 ha. It is likely that the next steps, organizing and enforcing the reserve, will be carried out through the mutual collaboration of IBDF and FEEMA (Fundação Estadual de Engenharia do Meio Ambiente).

L. rosalia rosalia is vanishing, and there is little cause for optimism. In the past, its range was a continuous forested lowland of more than two million hectares with a population of thousands of individuals (Figure 2a). By 1968, the primitive forest had been reduced to less than 20,000 hectares with a population of approximately 600 animals. By 1971, primary primitive forest existed only in isolated plots and the population of golden lion tamarins was estimated to be less than 400 individuals. By 1975, it is probable that only between 100 and 200 animals survive in a few plots of altered forest with only some primitive trees (Figure 2b). The same habitat destruction has occurred in areas inhabited by *L. r. chrysomelas* and *L. r. chrysopygus* (Figures 3 and 4).

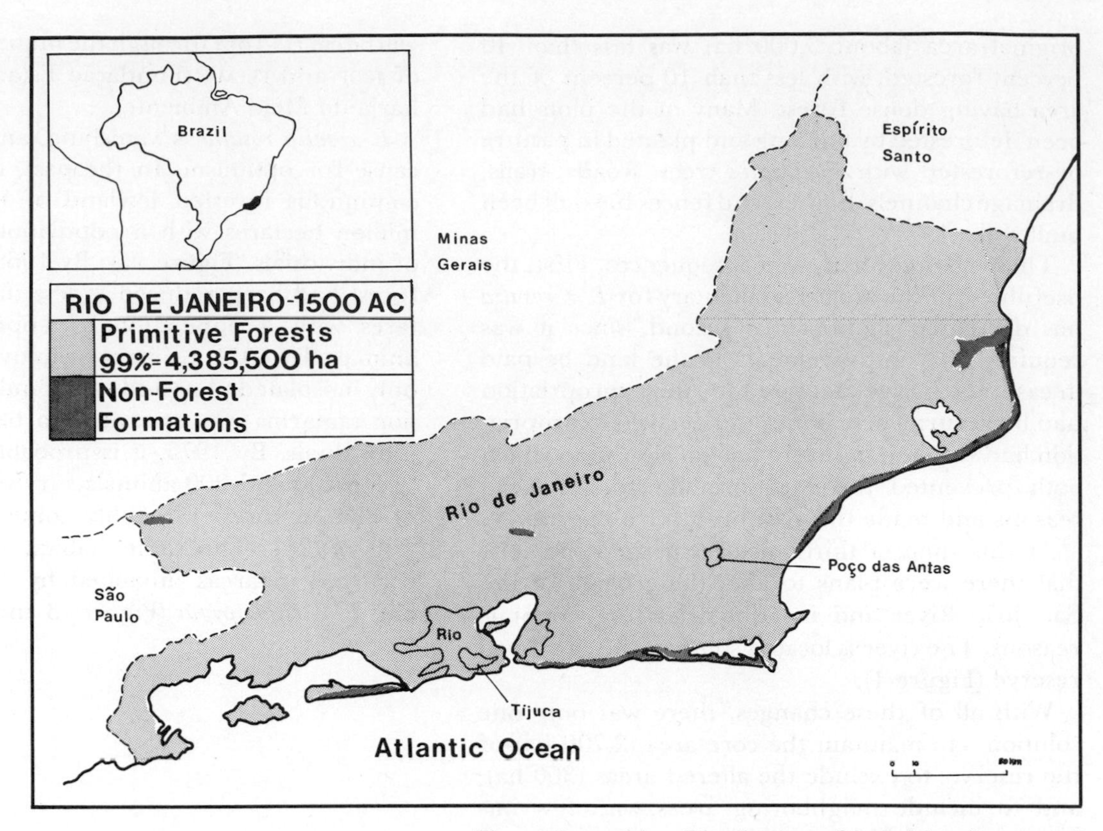

Figure 2a and b. Distribution of primitive forest and of *Leontopithecus rosalia rosalia* in Brazil in 1500 and 1975.

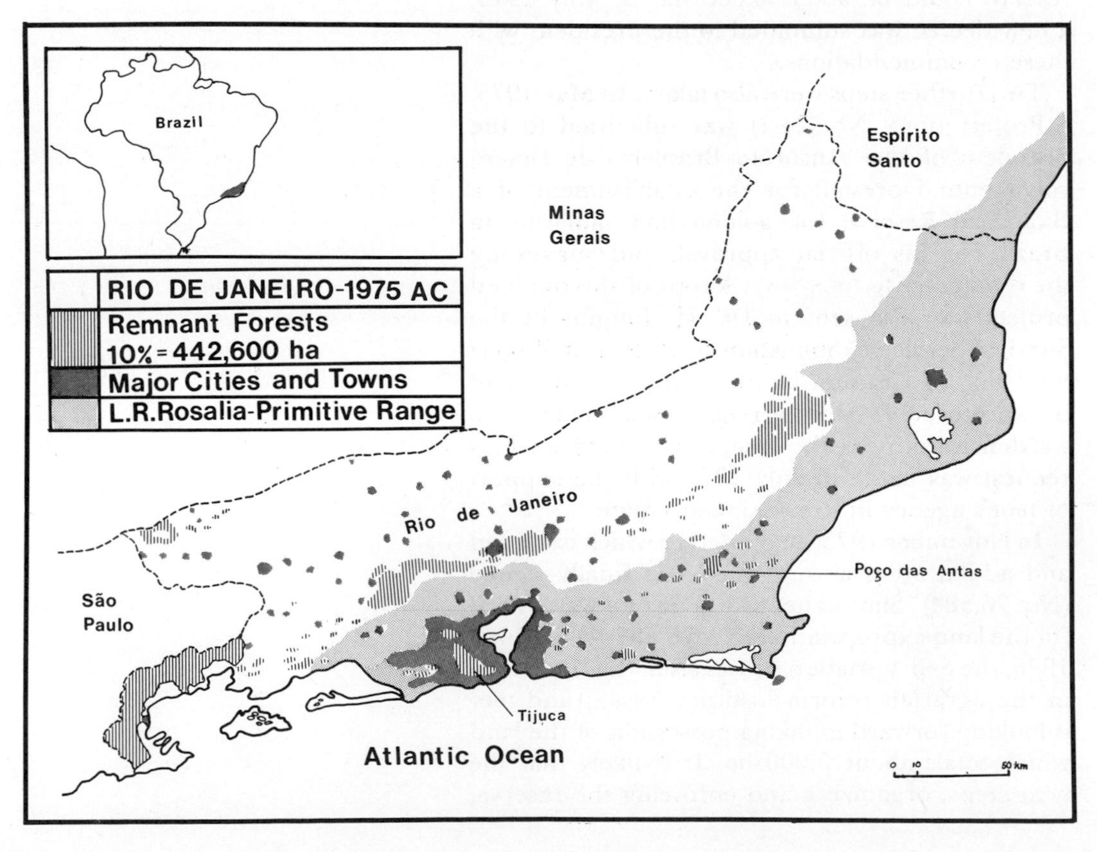

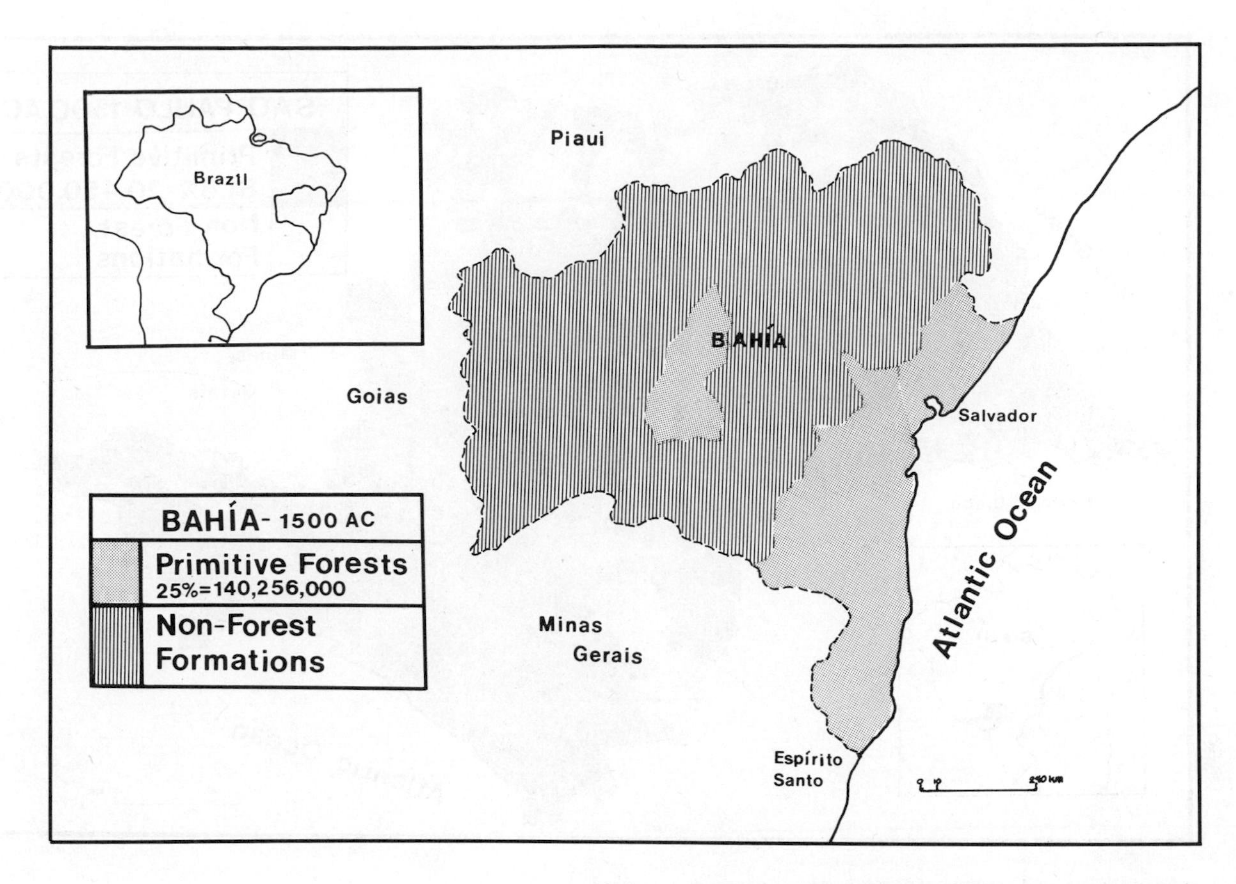

Figure 3a and b. Distribution of primitive forest and of *L. r. chrysomelas* in 1500 and 1975.

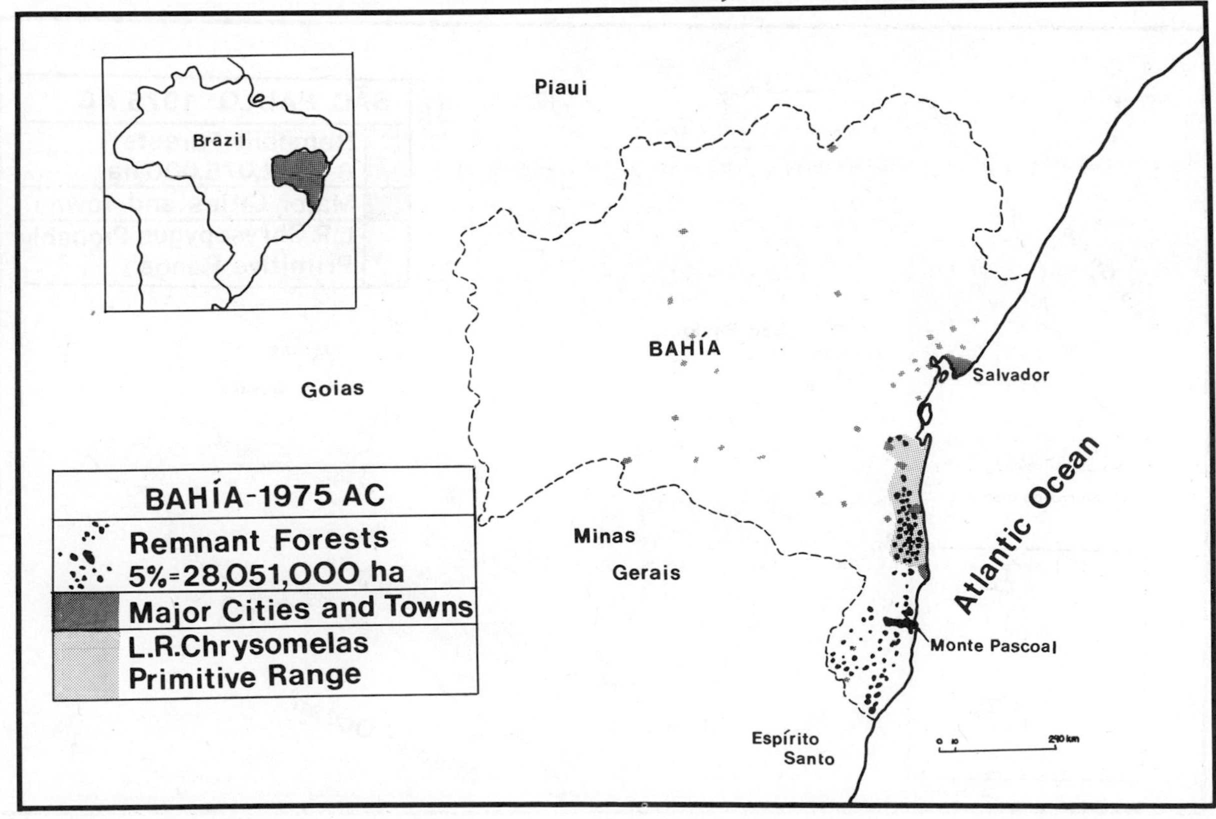

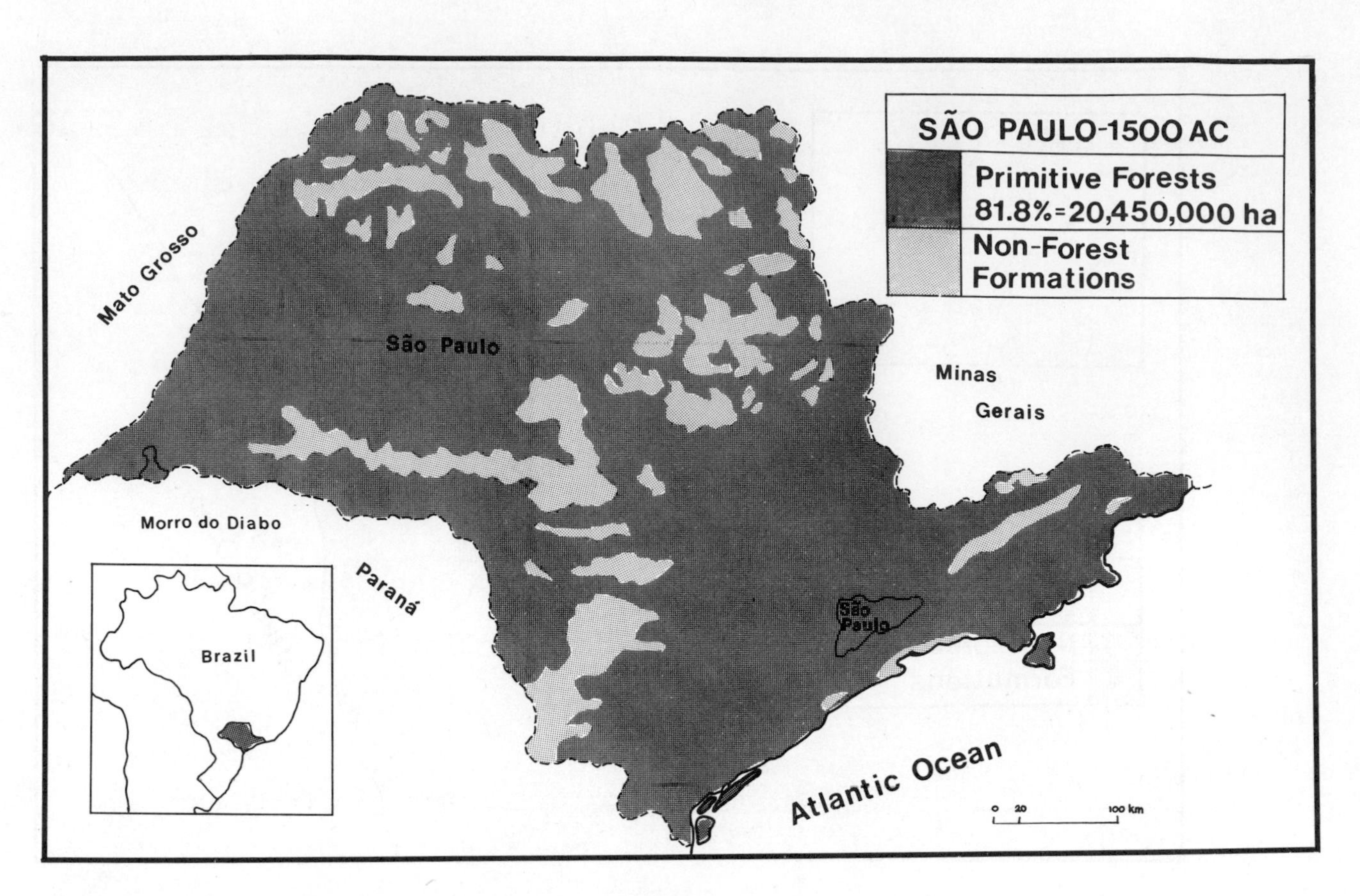

Figure 4a and b. Distribution of primitive forest and of *L. r. chrysopygus* in 1500 and 1975.

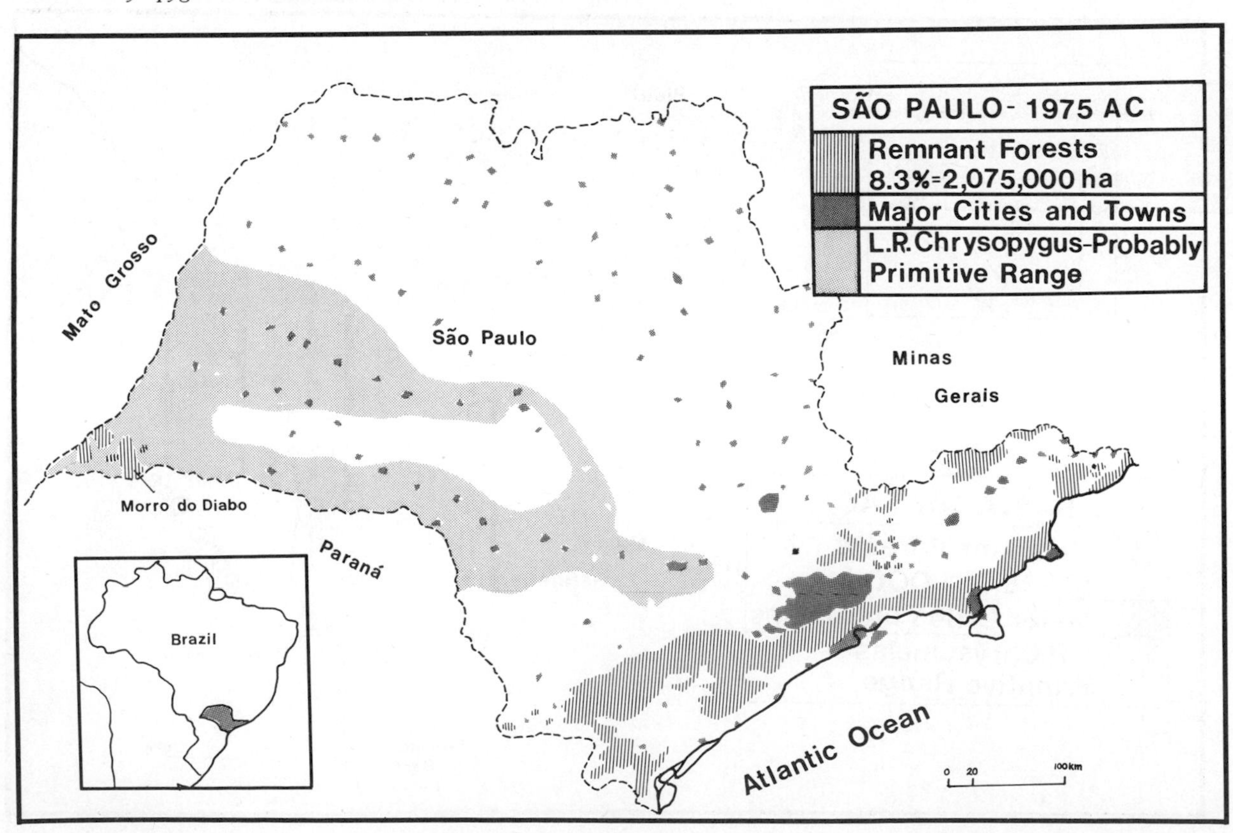

Conservation Status of the Callitrichidae in Brazilian Amazonia, Surinam, and French Guiana

RUSSELL A. MITTERMEIER
Department of Anthropology and
Museum of Comparative Zoology,
Harvard University,
Cambridge, Mass. 02138

ROBERT C. BAILEY
Matthews 34,
Harvard University,
Cambridge, Mass. 02138

and

ADELMAR F. COIMBRA-FILHO
Department of Environmental
Conservation,
FEEMA, Rio de Janeiro, Brazil

ABSTRACT

Eleven species of Callitrichidae occur in Brazilian Amazonia. They are rarely hunted by man and can in some cases exist in secondary forest situations created by man. There are, however, several species and subspecies in Brazilian Amazonia which are restricted in range, could be adversely affected by an increase in habitat destruction, and should be watched. *Saguinus midas midas* is the only callitrichid in Surinam and French Guiana. It is quite abundant, rarely hunted, and less affected by the presence of man than any other nonhuman primate in these two areas. None of the callitrichids discussed in this paper are considered to be in immediate danger of extinction.

Introduction

Few data are presently available on the behavior, ecology, and conservation status of nonhuman primates in Brazilian Amazonia and neighboring Surinam and French Guiana. Information on primates of Brazilian Amazonia still comes primarily from Bates (1863) and Cruz Lima (1944) and is supplemented by several shorter works (e.g., Miller, in Allen, 1916; Thorington, 1968). Data on Surinam primates is found mainly in Sanderson (1949, 1957), Geijskes (1954) and Husson (1957). Aside from locality data in museums, almost nothing is known of French Guianan primates. In this paper, we present some preliminary data gathered during several primate surveys conducted in Brazil, Surinam, and French Guiana, paying particular attention to the conservation status of callitrichids encountered during these surveys.

Methods

The data presented in this paper come primarily from four surveys, one conducted by R. A. M. in middle and upper Brazilian Amazonia from July to November 1973; another conducted by R. C. B. in upper Brazilian Amazonia near the Brazilian-Peruvian border in November 1973, and the third and fourth conducted by R. A. M. in Surinam and French Guiana from April to June 1975, and in Surinam from January to May 1976. A total of 22 localities was investigated during the 4-month Brazilian Amazonia survey (Figure 1). Since the primary purpose of this survey was to locate populations of the three uakaris (*Cacajao melanocephalus,*

Figure 1. Map of northern South America showing the localities covered during the two 1973 primate surveys in Brazilian Amazonia. Localities numbered 1-22 were investigated during the four-month survey conducted by R.A.M. They are listed below. The 20-day survey by R.C.B. was carried out in the Rio Quixito, a tributary of the Rio Itacuaí, which is itself a tributary of the Rio Javarí.

1. Taperinha
2. Monte Cristo, Rio Tapajós
3. Tapaiuna, Rio Tapajós
4. Oriximina, Rio Tapajós
5. Lower Rio Nhamundá
6. Boiaçú, Rio Branco
7. Rio Cuiuni, Rio Negro
8. Rio Araçá, Rio Negro
9. Rio Padauiri, Rio Negro
10. Rio Uneiuxi, Rio Negro
11. Lago Miuá, Rio Solimões
12. Costa da Batalha, Rio Solimões
13. São José (= Jacaré), Rio Solimões
14. Lago Marimari, Rio Auatí-Paraná
15. Santo Antonio do Içá, Rio Solimões
16. Rio Jacurapá, Rio Içá
17. Rio Panauá, Rio Auatí-Paraná
18. Bom Futuro, Rio Japurá
19. Lago Maraá, Rio Japurá
20. Rio Cuieras, Rio Negro
21. Itapiranga, Rio Solimões
22. Vicinity of *Podocnemis* turtle *taboleiro* (= nesting beach), Rio Trombetas

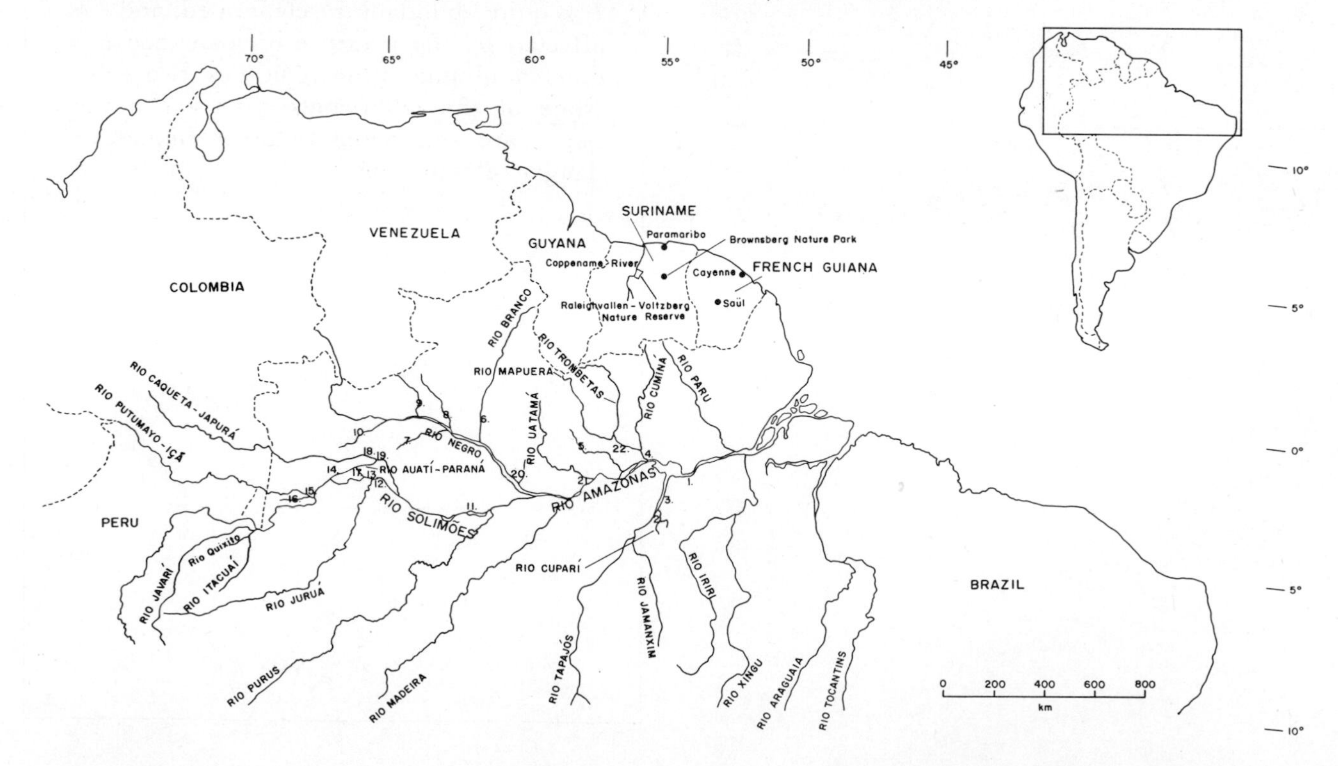

C. calvus calvus, C. c. rubicundus) and the white-nosed saki (*Chiropotes albinasus*), the areas investigated were mainly within the known ranges of these animals. In the November 1973 survey, 20 days were spent in the Rio Quixito, a small, white-water tributary of the Rio Itacuaí, which is itself a tributary of the Rio Javarí. None of the localities in Brazil were in reserves or parks and hunting pressure varied from very light to heavy.

The rain-forest habitats investigated in Brazil were all low altitude. We divided them into a number of categories according to type of river they were associated with and whether or not they were periodically (or permanently) flooded by seasonal changes in river level. *Terra firme* forests are elevated slightly above the high-water mark of the rivers and are not seasonally inundated. They make up some 98 percent (3,303,000 km^2) of all Amazonian forest (Meggers, 1971; Pires, 1974). For the purposes of this investigation, we subdivided the *terra firme* habitats studied into "primary" forests, which either had never been cut by man or had been cut so long ago that they had again reached a climax stage, and secondary forest or *capoeira*, which had been recently cut by man (usually for slash and burn agriculture) and was in some stage of regeneration.

Flooded forests in Amazonia were subdivided according to type of river. The seasonally inundated forests found on the flood plain of the nutrient-rich, silt-laden "white" water rivers like the mainstream of the Amazonas and the Madeira are referred to as *várzea* forests.They support high densities of animal life, but account for less than 2 percent of all Amazonian forests (55,000 km^2). The seasonally and sometimes permanently flooded forests of the nutrient poor "black" and "clear" water rivers are called *igapó* forest. The *igapó* forests of the black-water rivers like the Rio Negro are frequently quite extensive because the low banks of these rivers allow considerable flooding. The clear-water rivers like the Rio Tapajós usually have higher banks and flooding is generally limited to a narrow margin along the rivers. *Igapó* forests account for less than 0.5 percent of total Amazonian forest (15,000 km^2).

Additional, short-term observations relevant to the conservation of callitrichids of Brazilian Amazonia were made in the vicinity of Leticia, Colombia (on the Brazilian-Colombian border) in July 1971 (R.A.M.), June-July 1972 (R.A.M.) and from May 1972 to December 1973 (R.C.B.); in Belem, Brazil, in June 1966 (A.F.C.-F.); and in Manaus,Brazil, in November and December 1969 (A.F.C.-F.). Observations made in Leticia and vicinity were aimed at estimating the effects of the Leticia-centered primate trade on primates in nearby Brazil.

Three localities were investigated in the 1975 Surinam-French Guiana survey, and the two Surinam localities were reinvestigated in 1976. The two Surinam localities, Raleighvallen-Voltzberg Nature Reserve and Brownsberg Nature Park, are protected areas in which all hunting is prohibited. In contrast, the vicinity of Saül, the single locality investigated in French Guiana, is subject to heavy hunting pressure.

Brownsberg Nature Park, consisting of 8,400 hectares, includes the steep, forest-clad slopes and part of the 500 m plateau of the Brownsberg, northwest of the Brokopondo storage lake that was formed by the 1964 damming of the Surinam River (Schulz, 1971). All areas investigated in Brownsberg Nature Park were *terra firme* forest.

Raleighvallen-Voltzberg Nature Reserve covers some 56,000 hectares of low altitude tropical rain forest on the east bank of the Coppename River. The reserve includes Raleighvallen, the boulder-strewn rapids of the Coppename, and several granite *inselbergs*, most notably the Voltzberg (250 m) and the Van Stockumberg (360 m) (Schulz, 1971). Most of the area is *terra firme* forest, but there is some seasonal flooding along the river margins. During the survey in this area, we covered forest on the east side of the river in the reserve itself and on the west side of the river directly across from the reserve.

Saül is a small village located in the interior of French Guiana, about 185 km southwest of Cayenne. Most of the 75 inhabitants live by hunting, subsistence agriculture, and gold-digging. A trail system has been established for botanical studies by ORSTOM, the French Overseas Research Institute, and this facilitated surveywork in the area. Forests around Saül are mainly *terra firme* primary forest, except in the immediate vicinity of the village, where slash and burn agriculture is being conducted. The village itself is at 206 m, the small hills surrounding it average 300 m and the highest point in the area is 670 m (M. Condamin, pers. comm.).

Census work in the four-month Brazilian Amazonia survey was conducted on foot, in paddled canoe, or in a canoe equipped with a 4 h.p. outboard motor. During the course of travel from one survey locality to another within Amazonia, sporadic observations of primates were also made from the deck of the two expedition boats, which were too large and noisy for ordinary censusing. The 20-day survey in the Rio Quixito was carried out on foot and in paddled canoe. The survey work in Surinam and French Guiana discussed in this paper was conducted on foot. Primate densities/km^2 were estimated in the Surinam and French Guiana studies and will be presented in a later paper.

In all areas investigated, we supplemented field observations with information from local people. This proved especially helpful in determining the regional importance of primate hunting and the effects of hunting on different species. In Surinam and French Guiana, we also obtained anecdotal information on primate distribution and status from other researchers associated with STINASU (Surinam Nature Conservation Foundation) and ORSTOM.

Results

Brazilian Amazonia

Figure 2 is a species density map showing the number of species and genera of callitrichids per 500 km × 500 km quadrant in South America. The callitrichids reach their greatest diversity in Peru, and Colombia. The number of species decreases rapidly to the north, east, and south, although there is a second, smaller radiation in the coastal forests of southern Brazil. Eleven species have been recorded from Brazilian Amazonia (Hill, 1957; Hershkovitz, 1966, 1968, 1969, 1972). Of these, we have first-hand observations on seven (*Cebuella pygmaea, Callithrix argentata, Saguinus bicolor, S. fuscicollis, S. midas, S. mystax, S. nigricollis*), and no information on four (*Callithrix humeralifer, Saguinus imperator, S. inustus, S. labiatus*).

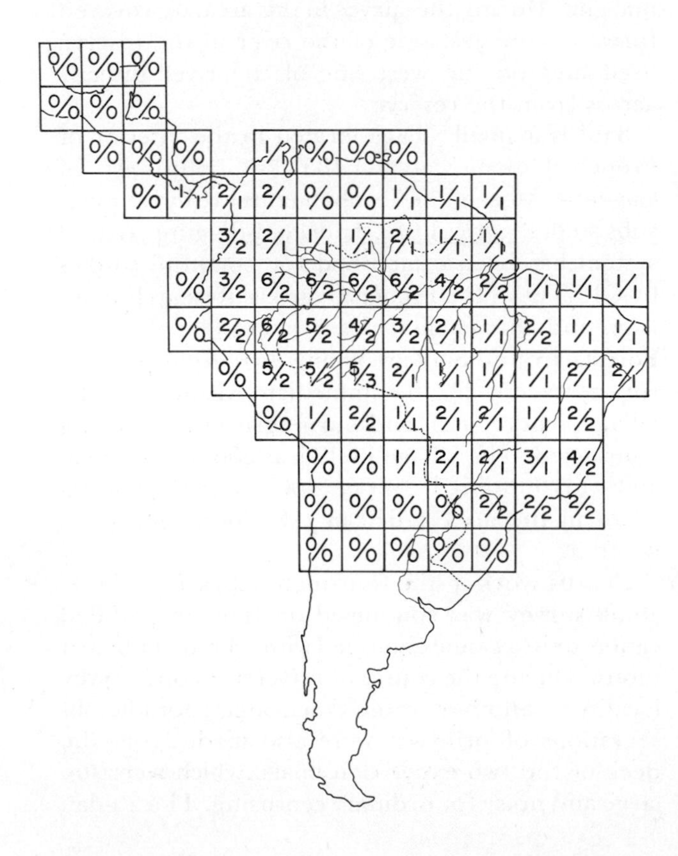

Figure 2. Map of South America showing the number of species/genera of callitrichids per 500 km × 500 km quadrant. (Taxonomic arrangement used in making this map follows Hershkovitz,1972, except for members of the southern Brazilian *Callithrix jacchus* group. We recognize five species in this group, whereas Hershkovitz recognizes only one.)

During the four-month survey in 1973, R.A.M. walked a total of 76.3 km of trail in *terra firme* and *várzea* forest, surveyed 315.9 km of seasonally flooded *várzea* and *igapó* forest by canoe and made sporadic observations of primates along 436 km of seasonally flooded forest in the large expedition boats. During the 20-day survey in the Rio Quixito, R.C.B. covered 110 km of seasonally flooded *várzea* forest in canoe and 17.5 km of trail in *terra firme* and *várzea* forest. Two species of callitrichids (*C. argentata, S. midas*) accounted for 7 of 95 primate groups seen in the four-month survey, whereas in the Rio Quixito two other species (*S. fuscicollis, S. mystax*) made up 6 of 49 groups located. Table 1 gives the total distances covered and the number of primate groups located in each of the major habitat types during the two Brazilian surveys.

Nowhere in the areas investigated did it appear that callitrichids were hunted for food and, of 36 primates being kept as pets by local people, only one was a callitrichid, *Callithrix argentata.* Of primate pets, 83 percent were either *Saimiri sciureus, Cebus albifrons, C. apella,* or *Lagothrix lagotricha.*

In the period from 1971 to 1973, we noted that three species of callitrichid from upper Amazonian Brazil (*S. fuscicollis, S. mystax, S. nigricollis*) were commonly brought across the border to animal dealers in Leticia, Colombia. Although we do not know how many Brazilian animals were shipped out of Leticia each year, local informants indicated that many of the primates exported from Leticia were of Brazilian origin.

Surinam and French Guiana

Only one species of callitrichid, *Saguinus midas*, has been recorded from Surinam and French Guiana. It was the most frequently encountered primate in the 106.3 km of *terra firme* forest covered

Table 1. Number of primate groups found in each of the major habitat subdivisions during the two Brazilian Amazon surveys. The distances covered in each habitat type are given under the respective heading. Distances in parentheses are those covered during sporadic observations made from the deck of the expedition boats. Groups in parentheses are those located during the sporadic observations.

Primate Species	*Terra Firme*		*Black Water igapó*		*Clear Water igapó*	*White Water várzea*		*Total*
	Secondary 12 km	*Primary* 51.1 km	*Small Rivers* 237 km	*Large Rivers* (276 km)	7.2 km	*Secondary* 30 km	*Primary* 182.4 km (160 km)	519.7 km (+436 km)
Callithrix a. argentata							1	1
Callithrix a. leucippe	1	1						2
Saguinus f. fuscicollis						1	1	2
Saguinus midas midas	3	1						4
Saguinus mystax mystax							4	4
Aotus trivirgatus			1					1
Callicebus moloch		1						1
Callicebus torquatus torquatus			1			1	18	20
Cacajao calvus calvus							3	3
Cacajao calvus rubicundus			2					2
Cacajao melanocephalus			7					7
Pithecia pithecia	1							1
Pithecia monachus		1					2	3
Chiropotes albinasus					1			1
Cebus albifrons			1				1	2
Cebus apella		4	1	(+1)	1		8	14 (+1)
Saimiri sciureus		4	7	(+3)	2	3	13 (+2)	29 (+5)
Alouatta belzebul		1						1
Alouatta seniculus		7	7		2		23 (+1)	39 (+1)
Total	5	20	27	(+4)	6	5	74 (+3)	137 (+7)

Table 2. Number of primate groups located in three localities surveyed in Surinam and French Guiana.

Primate Species	*Raleighvallen Voltzberg Nature Reserve, Surinam (ca. 9.3 km)*	*Brownsberg Nature Park, Surinam (ca. 38 km)*	*Saül, French Guiana (ca. 44 km)*
(Hunting pressure)	(None)	(None)	(Heavy, 5.7 shells per km trail)
Saguinus midas midas	7	10	8
Aotus trivirgatus	Not recorded from Surinam	Not recorded from Surinam	Reported by local people
Pithecia pithecia	2	4	Reported by local people
Chiropotes satanas chiropotes	2	1	Not known in vicinity of Saül
Cebus apella apella	4	3	Reported by local people
Cebus nigrivittatus	2	2	Reported by local people
Saimiri sciureus	4	Not known from Brownsberg	Occurrence in vicinity of Saül uncertain
Alouatta seniculus	7	6	1
Ateles paniscus paniscus	4	2	Reported by local people
Total	32	28	9

in Surinam and French Guiana, but was not seen along the 60 km of seasonally flooded riverside forest. Table 2 gives the total number of primate groups in each of the three localities surveyed.

S. midas is apparently rarely hunted for food and only occasionally kept as a pet by local people in Surinam. In French Guiana, it is sometimes hunted for food (M. Condamin, pers. comm.; pers. obs.), but is a low priority target because of its small size and rapid movements.

Discussion

Brazilian Amazonia

The number of callitrichid groups encountered during the two Brazilian surveys was surprisingly low, especially since local people usually reported that the animals were common. Only 13 of 137 groups (9.4 percent) seen in both surveys were callitrichids. In the Rio Quixito, *Saguinus mystax* and *Saguinus fuscicollis* accounted for 6 of 47 (12.8 percent) of the primate groups seen in *várzea* forest; however, in all the other areas covered in Brazil, callitrichids were very rare in flooded forest. None were seen in 237 km of black-water *igapó* or in 7.2 km of of clear-water *igapó* forest, and only one was seen in 88.9 km of *várzea* forest. Callitrichids were relatively more abundant in *terra firme* forest, where they accounted for 6 of 25 (24 percent) groups encountered, and especially in *terra firme* secondary forest.

The absence of callitrichids from flooded forest in most of the areas investigated is noteworthy. Certain callitrichid species (e.g., *S. mystax mystax*, Dawson, 1975; *S. midas midas*, pers. obs.) may prefer *terra firme* forest over flooded forest. It is also possible that the small primate niche in some flooded forest areas is largely filled by *Saimiri sciureus*, a characteristic species of the flooded forests and river margins. Thirty of 34 groups (88 percent) of *Saimiri* were found in *igapó* or *várzea* forest; they were the most abundant species encountered during the two Brazilian surveys (Table 1). This is clearly not the case in the Rio Quixito, where callitrichids (6 groups) were encountered almost as frequently as *Saimiri* (9 groups), but may be what is happening in insect-poor areas like the black-water *igapó* forests.[1]

If it turns out that some Amazonian callitrichids actually do prefer *terra firme* forest, it would bode well for their continued survival. *Terra firme* forests make up 98 percent of Amazonian rain forest and are usually far less accessible to man than riverside areas.

In any case, our observations and reports from local people clearly indicate that callitrichids are the only primates commonly found in *terra firme* secondary forest. *S. midas* (pers. obs.) and *S. mystax mystax* (Dawson, 1975), and perhaps other species as well, may even prefer second growth near human habitations. They also probably benefit since most potential predators are usually shot or scared away and larger competing primate species are often hunted out by man.

Callitrichids are very rarely or never hunted for food in the parts of Brazilian Amazonia investigated during the 1973 surveys. In addition, they are rarely kept as pets locally and little persecuted away from the major export centers. Eisenberg (1977) further points out that their reproductive rate is much faster than that of the larger New World monkeys, indicating that they are better able to withstand and recover from human exploitation. All of these facts suggest that the potential for the continued survival of callitrichids in Brazilian Amazonia is good. Nonetheless, several species and subspecies are restricted in range and should be carefully watched, especially in areas where forest clearance for pasture land is reducing the habitat available to them.

The following is a review of the available information on the conservation status of the eleven species of Callitrichidae found in Brazilian Amazonia.

Cebuella pygmaea

Not endangered at present. *Cebuella* occurs in the western part of Brazilian Amazonia as far as the Rio Japurá north of the mainstream and as far as the Rio Purús to the south (Avila-Pires, 1974). It is not hunted for food, but is occasionally captured by local people who keep it as a pet.

[1]Black-water rivers generally support a much lower biomass of animal life than white-water rivers (see Janzen, 1974, for a review), a point which has been noted for insects (Spruce, 1908; Stark, 1971), fish (Roberts, 1972; Marlier, 1973), mammals (Stark, 1971), and a variety of other organisms. The data gathered on all primate species during the two 1973 Brazilian surveys also indicate lower densities in *igapó* forests of the black-water rivers than in the *várzea* forests of the white-water rivers. In the Rio Negro and tributaries and several other black-water areas, a total of 27 primate groups were located in 237 km of canoe travel. This translates to 0.11 groups/km surveyed. In the *várzea* forests, on the other hand, we located 60 groups in 181.7 km of canoe travel or 0.33 groups/km surveyed—three times as many.

The traditional method of capturing *Cebuella*, used both in Colombia and Brazil (Hernandez-Camacho and Cooper, 1976), is to ring the trunk of a large tree frequented by the animals with a resin to which the animal sticks when it enters or leaves the tree. Nonetheless, such practices are of a small scale and do not represent a threat to the survival of *Cebuella* in Brazilian Amazonia. Brazilian *Cebuella* were sometimes brought across the border into Leticia, Colombia, but this species is not important in the primate trade. Moynihan (1976) reports that *Cebuella* in the Putumayo region of Colombia are very adaptable and sometimes live in hedgerows as commensals of man. In Peru, the species also occurs in flooded forest. It is quite possible that it also occurs in flooded forest in some of the areas investigated in the 1973 Brazilian survey. The small size and cryptic coloration of *Cebuella* make it very easy to overlook, even when the observer is only a few meters away.

Callithrix argentata

C. a. argentata: not endangered at present.

C. a. leucippe: restricted in range and vulnerable.

C. a. melanura: no data.

C. argentata occurs south of the Rio Amazonas, between the Rio Tapajós and the Rio Tocantins-Araguaía, south to the Rio Tacuari and the Rio Mamoré (Hershkovitz, 1968). *C. argentata argentata* is the northernmost race, occurring on the south bank of the Amazonas, between the lower Tapajós and the Tocantins and Irirí. Although it is occasionally captured as a pet and stuffed specimens sometimes appear in very small numbers for sale in Santarém, it is apparently common and little persecuted. *C. a. leucippe*, on the other hand, is very restricted in range, being found only between the Rio Jamanxim and the Rio Cuparí, east bank tributaries of the Tapajós. Its range is cut by the Trans-Amazonian highway and considerable clear cutting of forest is taking place in the region to make way for cattle pasture. If habitat destruction continues, the survival of this subspecies could be threatened. No data are available on *C. a. melanura*, the southern and most widespread representative of the species, but its large range and the remoteness of much of the territory it occupiés leads us to believe that it is not presently in any danger.

Callithrix humeralifer

No data. *C. humeralifer humeralifer* and *C. humeralifer chrysoleuca*, the two subspecies of this species, occur south of the Rio Solimões-Amazonas[2] between the Rio Madeira and the Rio Tapajós (Hershkovitz, 1968). The range of this species was not covered during the 1973 survey, so we have no information on its status. Hershkovitz (1972) considers it endangered because of its very small range.

Saguinus bicolor

S. b. bicolor: not endangered at present, but restricted in range and should be watched.

S. b. martinsi: no data.

S. b. ochraceous: no data.

The three subspecies of *S. bicolor* are found in a fairly restricted area between the lower Rio Negro and the lower Rio Paru de Oeste (or Cuminá), north of the Rio Amazonas. Since the range of the westernmost subspecies, *S. b. bicolor*, includes the vicinity of rapidly growing Manaus, the second largest Amazonian city, it could conceivably be endangered by habitat destruction. Hershkovitz (1972) mentions it in his list of endangered New World species, because of its small range and its proximity to Manaus. Nonetheless, *S. b. bicolor* appears to be quite adaptable and has been seen on several occasions in secondary forest within the city limits of Manaus (Coimbra-Filho, pers. obs.; D. Magor, pers. comm.) and is apparently not persecuted. No data are available on the other two subspecies.

Saguinus fuscicollis

S. f. fuscicollis: apparently common.

S. f. melanoleucus: reportedly common within its small range.

S. f. acrensis, S. f. avilapiresi, S. f. crandalli, S. f. cruzlimai, S. f. fuscus, S. f. weddelli: no data.

The range of this species extends into Brazilian territory as far east as the Rio Japurá north of the Solimões and the Rio Madeira south of the Solimões (Hershkovitz, 1968). Eight (and possibly nine, if *S. f. tripartitus* extends into Brazil) of the thirteen subspecies of *S. fuscicollis* occur entirely or partly in Brazil. Several of these (*S. f. acrensis, S. f. avilapiresi, S. f. crandalli, S. f. cruzlimai*) are fairly restricted in range, but nothing is known of their status. *S. f. fuscus* and *S. f. weddelli* have

[2]In Brazil, the mainstream of the Amazon is called Rio Amazonas only between the mouth of the Rio Negro and the mouth of the Amazon itself. The upper portion, between the mouth of the Rio Negro and the Colombian border, is referred to as the Rio Solimões.

larger ranges, but again nothing is known of their status. Freese (pers. comm.) saw a captive *S. f. melanoleucus* from the Rio Tarauacá in Fonte Boa and was told by its owners that the animal is common in the Tarauacá region; however, this subspecies also has a small range. *S. f. fuscicollis* was one of the three Brazilian species commonly brought across the border into Leticia, Colombia. It is apparently still common away from the major collecting areas.

Saguinus imperator

No data. *S. imperator* occurs in the southwestern part of Brazilian Amazonia in the vicinity of the upper Rio Purús and upper Rio Juruá, far from the areas investigated during this survey (Hershkovitz, 1968). We therefore know nothing of its status in Brazil.

Saguinus inustus

No data. *S. inustus* occurs north of the Solimões, in a restricted area bordered by the Rio Negro and Rio Japurá and extending into Colombia (Hill, 1957; Hershkovitz, 1969). We know nothing of its status in Brazil.

Saguinus labiatus

No data. This species is divided into two subspecies, *S. l. labiatus* and *S. l. thomasi* (Hershkovitz, 1968). *S. l. labiatus* is found south of the Solimões, between the Rio Madeira and the Rio Purús. *S. l. thomasi* is known only from the type locality, the Rio Tonantins on the north bank of the Solimões. We have no data on either subspecies.

Saguinus midas

S. midas midas: common.

S. midas niger: common.

S. midas has a large range in Brazil, being found from the Rio Negro east to the mouth of the Amazonas and north to the Guianas. East of the range of *S. bicolor*, it is the only callitrichid in a large area on the north bank of the Amazonas. South of the Amazonas, the subspecies *S. m. niger* is found between the Rio Xingu and the Rio Gurupí (Hershkovitz, 1969). *S. m. midas* was the most frequently encountered callitrichid during the four-month Brazilian survey and was reported to be common by local people. It was found only in *terra firme* forest and is apparently common in *terra firme* secondary forest. *S. m. niger* is also apparently still common, even though it occurs in the most developed part of Brazilian Amazonia. It can live in secondary forest and even occurs in close proximity to Belem, largest city in the Amazon region (Cruz Lima, 1944; Carvalho and Toccheton, 1969; Thorington, 1969).

Saguinus mystax

S. m. mystax: apparently common.

S. m. pileatus: no data.

S. m. pluto: no data.

S. mystax occurs in Brazil south of the Rio Solimões, as far east as the Rio Madeira (Hershkovitz, 1968). Brazilian specimens of the subspecies *S. m. mystax* were brought in considerable numbers across the border into Leticia, Colombia, and *S. mystax* was one of the three most commonly exported callitrichids. Nonetheless, it is apparently still common away from the major collecting areas and occurs in close proximity to towns and villages. Dawson (1975), for instance, found it abundant on the *terra firme* around Fonte Boa on the south bank of the Solimões. The subspecies *S. m. pileatus* and *S. m. pluto* occur further east in Brazil and were probably not affected by the Leticia trade. No information is available on them.

Saguinus nigricollis:

Indeterminate

S. nigricollis extends a short way into Brazil, between the Rio Solimões and the Rio Içá (Hershkovitz, 1966). Brazilian specimens of this species were also brought into Leticia for export. The comparatively small range of this species is centered around Leticia so it may have been more affected by the primate trade than other callitrichids.

Surinam and French Guiana

In Surinam and French Guiana, *S. midas* is one of the most abundant primates. In Raleighvallen-Voltzberg Nature Reserve, *S. midas* was one of the most common primate species. In Brownsberg Nature Park, *S. midas* was the most frequently encountered primate species, with *Alouatta seniculus* second (Table 2). In the vicinity of Saül, French Guiana, an area with heavy hunting pressure, *S. midas* was, with the exception of a single *A. seniculus* group, the only primate species encountered. It is occasionally hunted, but is not a high priority target because of its small size.

In addition to occurring in the *terra firme* primary rain forest areas investigated, *S. midas* is also reported to be common in "savanna forest" (a low xeromorphic forest always located close to open savanna) in northern Surinam and in various forest formations along the French Guiana coast (J. P. Schulz, pers. comm.; P. Teunissen, pers. comm.; M. Condamin, pers. comm.).

S. midas can be considered very common in Surinam and French Guiana and is less affected by hunting than any other primate species.[3] If commercial exploitation continues to be prohibited, it should remain abundant for a long time to come.

Acknowledgments

We would like to thank the New York Zoological Society for partly funding the four-month 1973 survey and the 1976 Surinam study. Other field work by R.A.M. in Amazonia was partly funded by an NSF Predoctoral Fellowship. R.C.B.'s work was supported by a grant from Merck, Sharp and Dohme. A.F.C.-F. is a fellow of the Brazilian National Research Council (Conselho Nacional de Pesquisas). Thanks also are due to the Fundação de Amparo à Pesquisa do Estado de São Paulo (FAPESP) for making available the University of São Paulo's Expedição Permanente de Amazonia (EPA) boats in which the survey was conducted. J. G. Fleagle read and commented on early drafts of this paper. Illustrations were prepared by G. Carter.

[3] We should note that we did not visit any areas inhabited by Indian tribes. Indians frequently hunt animals as small as hummingbirds, and at least some tribes hunt callitrichids as well (Lenselink,1972). In Brazilian Amazonia, Surinam, and French Guiana, however, Indians make up only a small part of the human population and their overall effect on primates like callitrichids is probably minimal.

Literature Cited

Allen, J. A.
1916. Mammals collected on the Roosevelt Brazilian Expedition, with field notes by Leo E. Miller. *Bull. Amer. Mus. Nat. Hist.*, 35:559-610.

Avila-Pires, F. D. de
1974. Caracterização zoogeografica da Provincia Amazonica. II. A familia Callithricidae e a zoogeografía Amazonica. *An. Acad. Brasil, Ciencias*, 46(1):159-185.

Bates, H. W.
1863. *The Naturalist on the River Amazons.* London: John Murray.

Carvalho, C. T. de and A. J. Toccheton.
1969. Mamiferos do nordeste do Para, Brasil. *Rev. Biol. Trop. San Jose*, 15(2):215-226.

Cruz Lima, E. da
1944. *Mamiferos da Amazonia. Introdução Geral e Primates.* Museu Paraense Emilio Goeldi de Historia Natural e Etnografia: Belem do Para.

Dawson, G. A.
1975. The Delta Regional Primate Research Center's *Saguinus mystax* trapping project: A field report. Unpubl. report, 5 pp.

Eisenberg, J. F.
1977. Comparative ecology and reproduction of New World monkeys. In *The Biology and Conservation of the Callitrichidae,* edited by Devra G. Kleiman, pp. 13–22. Washington, D.C.: Smithsonian Institution Press.

Geijskes, D. C.
1954. Het dierlijk voedsel van de Bosnegers aan de Marowijne. *Vox Guayanae*, 1(2):61-83.

Hernandez-Camacho, J., and R. W. Cooper.
1976. The nonhuman primates of Colombia. In *Neotropical Primates: Field Studies and Conservation*, edited by R. W. Thorington, Jr., and P. G. Heltne, pp. 35-69. Washington, D.C.: ILAR, National Academy of Sciences.

Hershkovitz, P.
1966 Taxonomic notes on tamarins, genus *Saguinus* (Callithricidae, Primates), with descriptions of four new forms. *Folia primat.*, 4:381-395.
1968. Metachromism or the principle of evolutionary change in mammalian tegumentary colors. *Evolution*, 22:556-575.
1969. The evolution of mammals on southern continents. VI. The recent mammals of the neotropical region: Zoogeographical and ecological review. *Quart. Rev. Biol.*, 44(1):1-70.
1972. Notes on New World monkeys. *Intern. Zoo Yearb.*, 12:3-12.

Hill, W. C. O.
1957. *Primates. Comparative Anatomy and Taxonomy, Vol. III. Hapalidae.* Edinburgh: University Press.

Husson, A. M.
1957. Notes on the primates of Suriname. *Studies on the Fauna of Suriname and Other Guyanas*, 1:13-40.

Janzen, D.
1974. Tropical black-water rivers, animals, and mast fruiting by the Dipterocarpaceae. *Biotropica*, 6(2):69-103.

Lenselink, J.
1972. De jachtopbrengst in een Surinaams trio-dorp. *De Surinaamse Landbouw*, 20(3):37-41.

Marlier, G.
1973. Limnology of the Congo and Amazon Rivers. In *Tropical Forest Ecosystems in Africa and South America: A Comparative Review*, edited by B. J. Meggers, E. S. Ayensu, and W. D. Duckworth,

pp. 223–238. Washington, D.C.: Smithsonian Institution Press.

Meggers, B. J.
1971. *Amazonia, Man and Culture in a Counterfeit Paradise.* Chicago: Aldine-Atherton.

Moynihan, M.
1976. Notes on the ecology and behavior of the pygmy marmoset, *Cebuella pygmaea* in Amazonian Colombia. In *Neotropical Primates: Field Studies and Conservation,* edited by R. W. Thorington, Jr. and P. G. Heltne, pp. 79–84.Washington, D.C.: ILAR, National Academy of Sciences.

Pires, J. M.
1974. Tipos de vegetação da Amazonia. *Brasil Florestal,* 5(17):48-58.

Roberts, T. R.
1972. Ecology of fishes in the Amazon and Congo Basins. *Bull. Mus. Comp. Zool., Harvard,* 143(2):117-147.

Sanderson, I. T.
1949. A brief review of the mammals of Suriname (Dutch Guiana), based upon a collection made in 1938. *Proc. Zool. Soc. Lond.,* 119:755-789.
1957. *The Monkey Kingdom.* Garden City: Doubleday.

Schulz, J. P.
1971. Nature preservation in Suriname *STINASU Verhandeling,* 2:1-24.

Spruce, R.
1908. *Notes of a Botanist on the Amazon and Andes* (2 vols.). London: A. R. Wallace.

Stark, N. M.
1971. Nutrient cycling. I. Nutrient distribution in some Amazonian soils. *Tropical Ecology,* 12:24-50.

Thorington, R. W., Jr.
1968. Observations of the tamarin, *Saguinus midas. Folia primat.* 9:95–98.
1969. The study and conservation of New World monkeys.*An. Acad. Brasil. Ciencias,* 41 (Suplemento):253-260.

Panelists:
R. W. Thorington (chairperson),
R. Castro, A. Coimbra, R. Cooper,
M. Freese, R. Mittermeier,
P. Neyman, P. Soini.

PANEL DISCUSSION:
AUGUST 18, 1975

Conservation of the Callitrichidae

THORINGTON: I would like to suggest that we start talking about censusing problems and population densities in different habitats, and disagreements that may come up with respect to censusing techniques and their consistency and their accuracy, remembering that consistency may not be the same thing as accuracy. From that, we can perhaps discuss the population densities and total numbers of animals that may be necessary to maintain viable populations of marmosets in the wild, followed by a discussion of what size areas need to be protected to maintain viable populations. We might also consider at the same time what forms of habitat management may be needed or desirable. Usually, habitat management means doing nothing and letting the regeneration of primary forest occur. There is some evidence with respect to marmosets that that would be a bad policy. Then, perhaps, we can discuss where reserves might be located and consider which marmosets seem to be the most endangered and which habitats are going most rapidly. Throughout all of this, the group as a whole should consider whether it wishes to present some resolutions or recommendations, or some other form of public utterance. Perhaps a group could draft something to be considered later on in conference, if this is of interest. Could I ask for some comments on the densities of marmosets that have been noted so far and the problems of survey techniques?

DAWSON: I had an area of about 15 hectares that I censused frequently from the roads at a standardized time of the morning. I found that both the time of day and the time of year had a lot to do with the density of the animals that you saw. In the dry season, the animals are moving more, and even if you correct for the increased visibility because of tree deciduousness, you still see more animals because they are more active. My data were standardized by seasons and by time of day and I found that, in going over the same census strip a number of times (the census strip was about 4-5 km long) to a total of about 1700 km, my standard deviation was actually greater than the mean estimate of population size per square kilometer that I had. It strikes me that strip census techniques for animals which exhibit large territories in habitats where visibility is quite limited, is really a very poor method of estimating density. I would invite some comments on this.

THORINGTON: What was your estimate of the density of marmosets in that area?

DAWSON: By the strip census method, it was about 23 plus or minus 25 per square kilometer. In my lowland group, by extrapolating from my density, accounting for home-range overlap and certain other things, I had a density of about 26 per square kilometer and, in the upland group, somewhere around 30 per square kilometer. So they are all in the same ball park, but that doesn't really say much about the precision or accuracy of any of the procedures.

EISENBERG: Well, it says something if they happen to be close.

NEYMAN: How many censuses did you have?

DAWSON: Well, whatever 1,700 divided by 5 is, we ran it that many times. I suppose that's about 340 censuses. The range of densities was from 0 to 56 animals.

FREESE: It seems to me that if you are in the same ball park, you have accomplished something. Because, if you have a huge area to census, what you are really trying to come out with is an estimate.

DAWSON: But from a probabilistic point of view, it is purely fortuitous.

FREESE: The other point that we were talking about was the time of day and that definitely did make a difference. We did not census from 12 noon until 2 in the afternoon, because the number of monkey sightings dropped.

MITTERMEIER: A one-shot census in any area is going to give you problems. You don't really have meaningful density data unless you go over an area several times. I have worked several sites where I would go out one day and see no monkeys and then the next day in the same area, I would see 4 or 5 groups. You have to cover a certain period of time to account for the ranging patterns. But I might pose another question. If the strip census technique doesn't work and if we want primate densities, what else are we going to use?

DAWSON: We are better off using what we have here as indices rather than absolutes.

MITTERMEIER: I don't think it's absolute, it is just that it is good for comparative reasons. If you go into several areas and do repeated censuses, you can say that one area has more monkeys of this species, that one area has comparatively higher densities. For example, the estimates in the Freezes' paper and my paper on individuals per square kilometer were based on estimates of group size. So, it's relative.

DAWSON: Let's say you have 23 per square kilometer and 30 per square kilometer elsewhere, and you say that one is more abundant so it has denser populations in that area. But you can't really say that. You can't compare these statistically, because you have no idea of the variance.

EISENBERG: Could I make a statement? You say that your standard deviation was greater than your mean. As I understand it, you did a standard walk of a certain distance. In this case it was 5 km, and each sample of 5 km was treated as an independent sample. Why did you pick 5 km?

DAWSON: It was convenient.

EISENBERG: There is a bias right away. Why didn't you take every 10 feet? Now, here is my point. Suppose, just as arbitrarily, I said that I am going to consider 20 km as my single sample space. Now I could do that, and I'll bet that, if I took a 20-linear km walk as a single sample, then maybe my standard deviation wouldn't be greater than my mean. I could make the standard deviation even worse if I took 10 feet as my sample distance. So where is the objectivity in this? It is possible, if you know something about a species' behavior in space, to pick a census method that gives you replicability and perhaps a mean value with a smaller standard deviation. Is that not true? So, therefore, we choose parameters for our census technique based on what we know about a species. The first preliminary efforts will always be fraught with difficulties because we set our sample space too small or too large. What ultimately determines a fair sample space, in this case linear distance walked, is some knowledge about the patchiness of the distribution of the appropriate home ranges and suitable habitat for the species. For species A, it may be distributed so patchily that the sample space would have to be much larger to give a representative value, when compared with species B in the same habitat which showed a uniform distribution, in which case one could use a smaller transect. These decisions are made all the time, in retrospect, based on an initial tryout. Then once a standardized technique has been developed, the accuracy of certain strip censusing procedures should be not bad at all, given that you make the appropriate assumptions for the differences in the structure of a habitat which determines the uniformity or regularity of distribution in space of the species in question. I grant you that's what we haven't done.

The other thing that a lot of us have failed to distinguish are the two elementary concepts common in the fifties that I hope we still recognize, crude density versus ecological density. Everything that has been reported today with some possible exceptions deals with crude density. Crude density is an estimate of the density of a species, as if the area were uniform without taking into account the fact that the resource distribution is patchy; ecological density takes into account the patchiness. So all the strip censuses will always come out maybe 5 to 10 times lower than a true ecological density for a population in balance.

DAWSON: I would like to qualify what I did on my strip census. The area in which I worked had been seriously disturbed, with the forest cut earlier in this century and the growth, except along forest streams, was fairly uniform. Given some idea about the species' biology and a rough estimate of what home-range size should be, I did consider, the practicability of taking a strip census, but 5 km was a long way for most marmoset home ranges, unless you were going along the river. Thus,it seemed that 5 km was an adequate strip census. I think the problem is that marmosets inhabit a very dense vegetation type and, because they are highly insectivorous, the home range is fairly large. Therefore,the probability of running into an animal is quite small. Now with howler monkeys which are fairly sedentary, visibility might be better, and the estimates would therefore probably be better and with less error.

EISENBERG: Well, that's true. The howler monkey lends itself in some kinds of habitat to a strip census far better than any other species. One other thing I would like to say is that a lot of assumptions are made about the width of a strip. It varies with the individual; eyes aren't the same, ears aren't the same. I think each individual must for himself honestly try to determine what is a reasonable width. Also, "visibility" varies from species to species. In some species, it is a fact that you have a bigger strip available for perceiving; in others, no. And that could have an important bearing on locating something like *Saguinus*.

THORINGTON: It seems to me that there is a common ground here. One, you know that your approximate densities are running in the order of 10 to 40 per km^2, if we look at all the marmoset densities given in these papers. Now, if you take an arbitrary length of strip and you make your estimate of the width, you know what size the area is that you are covering and you know what your mean is going to be, if your density is what you expect. On the basis of this, if your mean is 10 or less, then you should be approximating a Poisson distribution in which your standard deviation approximates your mean. And that's not what you want. So if you take a longer strip, then you will be approaching a normal distribution which is the point at which you can hope to compare the estimates from different habitats. This may give us trouble with marmosets because it may simply be too long a strip to be a reasonably homogeneous habitat. This is something that some of you may want to comment on. But, when you get into a normal distribution, your standard deviation is proportional to your mean, so you are going to have to push your mean up in order for your standard deviation to be a relatively small part of the mean.

EISENBERG: Yes, one would hope that that would happen if you were choosing a real sample space.

THORINGTON: I think the value of Gary's (Dawson) comments is that we are not out of the Poisson distribution at 5 km with marmosets in his kind of habitat. I think that's a very valuable observation.

MITTERMEIER: I would like to get back to what Gary (Dawson) was saying before about strip censusing technique. What do you see as an alternative?

DAWSON: Quite simply, there isn't another method available right now. I had hoped that using radiotelemetry and extrapolating from that would give us a better idea. As Thorington pointed out in one of his papers, if you know the detailed movements of a group you might be able to pick up small territories of single individuals, for example, inside the territory of a larger group. But, because there are habitat differences and territoriality is not a strict phenomenon, you can't extrapolate home-range size or territory size as was supposed to get a reasonable density estimate—at least in these species. Now, maybe there are other species where territories are strict and then you could extrapolate. In that case, I think it would be the best census method. But you would have to look at the other callitrichids first.

GREEN: I was doing censusing in an area of northern Colombia using two types of censusing methods; one was what we commonly call a transect. Since the habitat I was working in was fairly hilly terrain, I also utilized another method, called a point census, which in essence was sitting on one slope overlooking the other side of the valley which allowed observations of the tree canopy on the other side. Just to indicate what happens when you are censusing in this manner, the values that I came up with in my transect censusing in individuals per km^2 were 22 for howler monkeys, 33 for spider monkeys, 15 for *Cebus*, and 15 for *Saguinus leucopus*. However, with the point censusing there were 20 individuals/km^2 in howlers, 20 for *Ateles*, 4 for *Cebus*, and 2 for *Saguinus*. So, the behavior of the species we are observing, the variability of the habitat, as well as the acuity of the observer and the conditions under which we are observing all relate to the census result. And maybe point censusing is more accurate than strip censusing. How do we know?

MITTERMEIER: The differences that you get for different species may be somewhat accounted for by the technique that Curt Freese has developed which is an improvement over Southwick's and Kaufmann's technique, that is, using different auditory-visual fields for different species.

GREEN: I am just referring to differences in censusing methodology in the same area, same habitat, same conditions, and with the same observer.

MITTERMEIER: The point-to-point method might be better, but in the Brazilian Amazon, for instance, your nearest slope is somewhat different than in the *llanos*.

THORINGTON: I think we need to focus on these problems, but I don't want them to prevent us from looking at the densities which we actually have, by arguing too finely about the accuracy of our techniques. I think that if we have reasonable ideas of the order of magnitude of densities of some of these animals, then it should be possible to say something about the size of the area needed to protect a reasonably sized population. Would anyone care to address himself to some generalizations based on the data that have been presented here today which seem to run from 40 animals per km^2 down to about 5 per km^2. All of these were running a density of approximately 10 per km^2. Is this a reasonable estimate for most marmoset habitats that are not heavily exploited or are not grossly disturbed in some way? There are a few exceptions, such as *Cebuella*.

MITTERMEIER: This is kind of hard to say because with the same species you get quite different densities in different areas, probably depending on where the species is in its range and who it is competing with. For instance, in Surinam, *Saguinus midas* shows incredible densities whereas in an area that was about half as large in the Brazilian Amazon, we didn't encounter anywhere near as many, perhaps 10 percent as many groups as we encountered in Surinam.

THORINGTON: Your 10 percent estimate is still the same order of magnitude because you are talking about 40 as opposed to 4 per km^2. Whereas I am asking whether it is 1 per km^2, 10 per km^2 or 100 per km^2. I am wondering if in a lot of the habitats we don't have sufficient knowledge already to make some general estimates of what the population should be, that is how many km^2 we should maintain for a viable marmoset population. Perhaps let's choose some specific examples. Is 1,000 an adequate number for a viable population over a period of time? Do you think that the population fluctuation of marmosets is such that you would hate to stake "genetic heterogeneity" on 1,000 individuals?

MITTERMEIER: From Terborgh's recent paper on the abundance of birds in the Caribbean Islands and mainland areas, I would think that 1,000 animals would not really be enough. He seems to think that you need something like 1,000 km^2 in order to maintain any kind of diversity in an area over long periods of time and this is based on some fairly careful analyses of what happened to a fauna like Trinidad's and some of the smaller Caribbean Islands. Over a period of time, they just continue to lose species, after they have been cut off from the mainland where the original stock came from.

THORINGTON: This would seem to be the major problem when it comes to the larger species of primates. Maintaining diversity is presumably not the question we are talking about here, if we are dealing with specific species of marmosets. I am not sure that we should be, in talking about conservation. I think perhaps we should be talking more about the maintenance of diversity.

MITTERMEIER: I think we have to consider areas too. For instance, in southern Brazil, you can say 1,000 km^2 would be nice, but you have to take into account the practicalities; you might not be able to preserve more than about 30-40 km^2. In an area like the Brazilian Amazon where you still have vast untouched areas, I think the bigger the reserve, the better, and the more you can include in it, the better. In southern Brazil, even if you do manage to create some reserves, they may not preserve the species over the long run.

THORINGTON: Well, could we afford to make small reserves if they are only going to hold animals for a period of time before they become extinct?

KLEIMAN: If the Poço das Antas land were expropriated today, would it preserve *Leontopithecus rosalia*?

MITTERMEIER: That is dubious.

COIMBRA: The Poço das Antas Reserve is only going to prosper if they have very intensive human management of resources there. The current area which was decreed by the president has already been 70 percent destroyed. Recently Dr. Blood, C. Freese, Magnanini, and I flew over the area. We want to increase the area from 3,000 to 5,000 hectares since the area is so degraded already. There has to be reforestation and reintroduction of animals. I don't think that the animals can survive in the long run even in an area of 5,000 hectares. All natural enemies would have to be removed as part of the human management or the animals would not survive more than a couple of decades. Only about 10 percent of about 30,000 hectares can support the animals now. Most of it is alpine vegetation and it is beautiful, but it is not going to preserve animal life. There were 54 *L. rosalia* existing ten years ago in the park, and in 1973 there were no more.

In secondary growth, it is important to put in wooden boxes for sleeping. This comment was made this morning. Botanical studies have to be done now to find out which tree species are best for supporting the animals. The golden lion marmosets* captured recently are very small and have very poor coloring, so it is very important to quickly find out their needs and start managing the areas. I still don't know of an efficient censusing method. You can walk for a week without seeing anything; at other times you see a lot.

THOMAS: I would think that in trying to delineate specific reserve sizes, we are really not facing the political realities. The size will be dictated by the political reality of the situation balanced by the economics of what can be realistically set aside. Probably the most logical approach would be pinpointing those areas that are obviously successful for a variety of species, where you find high densities, and then deciding how much land you can get and hammering that through as a political and economic reality. There is no such thing as having too much land, but there certainly is such a thing as having too little. And even if you have to settle for too small a reserve, like apparently is going to be the case with the golden lion marmosets, it can be utilized to some degree and should not be overlooked. Even if it is nothing more than just a public showcase, you must try to get across the lesson you are trying to hammer home. So, I think trying to break down the reserve size by species is a little immaterial. One should try to go for as much as you can politically and economically.

MITTERMEIER: You have still got to know which areas to try to save.

THOMAS: Yes, you've got to pinpoint the successful areas and then include as much as you think is expedient.

COIMBRA: The Poço das Antas situation really doesn't depend on economics or on land. It is poor land, not good for agriculture, and the government had the money for taking the land, but the authorities never pressed the point. None of the presidents of Brazil have allowed the appropriation of the area; the responsible persons simply haven't presented the case well to the higher authorities.

THOMAS: It simply is not a political reality as yet. There has to be enough world pressure to make it a political reality.

FREESE: I want to bring up a point that Eisenberg was talking about last night. If callitrichids survive

*Editor's note: *Leontopithecus* has traditionally been referred to as a marmoset although it is a tamarin.

better in secondary forest and you declare an area a preserve, are the populations going to decline over a long period as the forest matures?

EISENBERG: Well, it is all relevant to the species that you want to conserve. If it happens to be a marmoset species that is adapted to secondary growth habitat, obviously any methods for preservation have to include management recommendations for the land. On the other hand, if *Leontopithecus* is in fact adapted to a primary or semi-climax and mature situation, then the management technique has to be tailored to that species' requirements.

PERRY: Assuming that the population is under some form of exploitation, and it is important for management purposes that you know whether the population is increasing, remaining stable, or decreasing, are the forms of censusing talked about here today adequate to determine this with some reliability? Is it a question of technique or how frequently, how many, and how extensively these censuses are taken? The second part of the question is then, are the limitations so severe in this form of censusing that we cannot realistically talk of calculating an optimum sustained yield?

THORINGTON: I would say that strip censusing is useful for a general estimate of how many animals there are in a forest, which is the first thing you need to know before deciding whether to protect that forest. But, to start looking at population fluctuations and considering how many animals might be taken or whether management strategies are successful or not, then let me say parenthetically that I don't know of a single case (except the most trivial ones) in which a primate population is being managed with conscious concern for the results. There is unconscious management of course. I think for determining the numbers that can be removed, one has to go into a far more intensive type of survey technique which may very well necessitate marking individual animals. That's another ball game altogether. It is not the kind of survey that is presently being encouraged by conservation money or any other money, for that matter.

MITTERMEIER: And which may take so long that by the time we get around to finishing it and finding out where we should preserve the areas, there won't be anything left to preserve. The advantage of the strip censusing technique is that it is fast and there are large areas to be covered. If it is not perfectly accurate, it is at least relatively accurate, and allows comparisons between areas that permit us to set priorities concerning which areas are the most important.

PERRY: Then what you are saying in regard to the second part of my question is that any calculation of an optimum sustained yield today is pure bull.

EISENBERG: I wouldn't take that extreme a position.

THORINGTON: Except that if we have a relative estimate of the population and some information about the breeding potential of that group, we can say something. Of course, we don't know our causes of mortality, and if we have a population that is just hanging on, you don't know whether the additional removal of animals is going to tip the scales. It depends on whether you are removing what would be lost anyway from mortality.

CICMANEC: It seems as if you are making a quantitative statement and there just isn't the basis to present a quantitative answer, even comparatively. If there is the right kind of fruit tree along the path where you are walking, you might find lots of a given species. Are other means of censusing being developed and, if there are, could people explain these?

EISENBERG: The answer to the fruit tree problem is that if you walk far enough, you'll pass enough fruit trees. It has to do with the length of the strip that you have taken as the unit. Given the amount of money that most of us work with (and in the foreseeable future it can only go down), we aren't going to come up with anything better than what we have now in time to make recommendations to set up sensible preserves. That's the way I see it. So we are in a dilemma!

CICMANEC: In a sense you are saying that other methods would be multiples in cost; they wouldn't be just a little bit more.

EISENBERG: You bet.

FREESE: I think you have the impression with strip censusing that we are just following one straight line. One day we would go in one direction, the next day another. So you don't have to worry about one fruit tree on one trail biasing your results. We moved in a web within an area and went out in multiple directions over different days. Every path was censused 2, 3 or 4 times with days in between.

CICMANEC: But are the paths chosen on the ease with which they could be walked which is determined by the vegetation?

FREESE: Unfortunately, that does happen. But we stayed on trails most of the time, because once you go off a trail, you make so much noise that you scare anything around you. So when there were existing trails, we used them. We could thus be quieter and have the opportunity of seeing more than if we whacked through with machetes, like a bulldozer.

MITTERMEIER: You don't always use established game trails. If you use trails just where the hunters have gone, you surely aren't going to see anything. And if you know that there is heavy hunting in an area, you should get off the trail.

EISENBERG: I think the staunchest critics of these techniques would have to admit that we can make a few generalizations about carrying capacities. We honestly can say certain things about certain species that are real and not just a lunatic fringe of estimation.

CICMANEC: I sympathize with your motivation and am happy for the effort that you are making. However, I feel that, once a quantitative statement is made, people feel that there is a great deal of knowledge behind the statement, such as government officials who would determine how many animals could be taken from a region. They would decide based on what they think is a quantitative answer and unfortunately they really don't have the answers to make the judgments. That's my concern as an end user of these animals; I sympathize with the problem.

THORINGTON: I think it very useful to have a quantitative answer to some people I discuss things with who are thinking in terms of hundreds or thousands of animals per km^2. I can disabuse them of that quantitatively in a hurry.

THOMAS: I would echo what John [Eisenberg] said. I think that the tools that we have at hand are adequate enough to pinpoint the areas that should be set aside. But no matter how much we hammer around the questions of how many animals per km^2, it still comes back to the political reality. Beyond that, we need mobilization of whatever political pressures can be brought to bear on a national or international basis to make it into a reality.

THORINGTON: Following through on that, let's take a look at the Poço das Antas Reserve with 5,000 hectares potentially holding, on the basis of our estimates of marmoset density, under good conditions, perhaps 500 animals. How hard do we work politically for 500 golden marmosets, or how hard do we work politically for a larger area in the Amazon? Just try to think in terms of game theory. What are the best strategies and where should the effort be put?

THOMAS: What you would have to ask is, is the golden lion marmoset affair a fait accompli? In other words, have the powers with control in the Brazilian government been hammered away so much that you have an impossible situation? You might as well realize that, if you throw everything you have into this reserve and go down in flaming defeat, it will hamper you for any other operation any place else. You have to make an assessment as to whether it is worthwhile to mobilize all effort to make that reserve a reality or whether you are better off in the long run to let that alone and maybe turn your attentions elsewhere. Because if you have a political impossibility, you will only end up destroying yourself in the process of trying to make something impossible happen.

THORINGTON: I think that is an excellent point. It is part and parcel of the strategy. If you can save a particular area, is it worth it biologically and is it worth it politically?

PERRY: I don't think that it is an impossibility, because I don't think that there are any substantial economic forces on the other side. What you have is a series of political hang-ups.

THOMAS: They can be very substantial.

PERRY: When I was down there in March, for example, everyone was assuring me that there was money in this year's budget for the acquisition of the land. I think right now that this has become such an international issue (there will undoubtedly be a resolution from the IUCN meeting in Zaire and Prince Bernhard is taking an interest, and so forth), that it would be a political disaster to write if off.

KLEIMAN: This was an international issue five years ago. What has happened in the last five years?

PERRY: A number of things have happened since 1969: the presidential decree, the legal work, the surveys, the delineation of the land; it is all now at the point where there is a legal basis for acquisition and action. My feeling is that it will take a relatively small shift of the political winds down there to make it happen because you do not have powerful economic forces on the other side. The land is not particularly valuable. It is not coveted for a hydroelectric dam or something of this sort.

COIMBRA: The greatest pressure can be made by the international press showing that the whole world is upset about the situation. The patron saints can't make the miracle. It will take people from outside to make the people in Brazil and the political leaders understand.

There is this tremendous dilemma to insure the survival of the species. If the Poço das Antas area is large enough, it will also protect the natural enemies of the animals; there are various predatory animals living there now, e.g., the hawk eagle. Since the callitrichids have only two young each year, there has to be control of the predators and good management. In Tijuca Park, the *Callithrix jacchus* is probably decreasing, because of the presence of *Felis wiedii* and *Eira barbara*.

PERRY: Is that part of the forest maturing? It was replanted in the 1890s.

COIMBRA: It is secondary forest and still in its initial stages.

THORINGTON: So we need letters to the *New York Times*, the *Washington Post*, and various other places.

PYLE (World Wildlife Fund): Has there been any concentrated effort, and I speak as a layman, to get the attention of the international press? My husband is a newsman and I can tell you that a good animal story (and I don't mean to be disparaging) is always good news. And have people who have been on projects in South American countries tried to get their stories published with the international wires or with the *New York Times* or *Washington Post*?

MITTERMEIER: I don't think that the international press has covered the situation enough. What was done at the last marmoset meeting was that a single letter was sent to the president of Brazil. A single letter like that can just be picked up and thrown away.

PYLE: One of the things that could come from here is a press release to the major wire services that a conference was being held now.

THOMAS: But it has got to be voluminous and continuous.

EISENBERG: One of the more effective things is to have a head of state make a statement to another head of state, which was done in South America by the president of Venezuela to the president of Peru and it had some positive repercussions.

COIMBRA: We must consider that it is not only the golden lion marmoset, but the golden-headed marmoset that is disappearing. They are cutting the forest regions around the area they inhabit and changing the microclimate for the cacao plantations. Those happen to be a great source of income for the region. The land is expensive and also rich in valuable wood, particularly in rosewood.

THORINGTON: Do you think that the same sort of problem exists in Colombia for the cotton-top marmoset? Is it basically a political problem?

COOPER: I think Pat [Neyman] could speak to the situation in the area where she worked which is apparently the largest remaining single tract of land in which the cotton-top marmoset exists.

NEYMAN: The owners of the land I worked on preserve the woods mainly for their own use. At the present time, they are selectively cutting the forest and have cut just about everything they can selectively cut, so the question of the future of the forest is somewhat up in the air. They are talking about seeding, but the land owner claims that INDERENA is not interested in doing anything positive only in regulating. The owner also is somewhat worried that I might try to suggest to INDERENA that the land be taken out of his hands. I have not tried to approach the owner one way or the other, because I was afraid I would prejudice him before really knowing how to properly deal with the question.

THORINGTON: How large is this woodlot?

NEYMAN: Slightly over 600 hectares. Even with that, according to what's been said here, it is not enough to sustain a viable population of *Saguinus oedipus*. Based on what I have seen there so far, that would have somewhat under 600 animals.

THORINGTON: I think one of the major problems that faces us with marmoset conservation and generally with ceboid conservation in South America is whether these small areas are really worth the political hassle or whether we should be going for the big areas in the Amazon which contain more of the species. We hate to see the golden marmoset go; we hate to see the cotton-top marmoset go, but are we really being practical in this?

MITTERMEIER: I think it is a matter of model

cases, too. I am in favor of trying to preserve any species, no matter how far gone, but I think with the golden marmoset, it is also bringing to the Brazilian government an awareness of the importance of conservation. Whereas a large area in Amazonia without such spectacular species, and something that the press can't identify with, might not do the same thing. I think that is one of the major things with the golden marmoset; it is something that can make the press.

THORINGTON: I have not been convinced of it yet. I just wonder if we are trying to save species and losing fauna.

MITTERMEIER: But can't we do both?

PERRY: I think there are two things happening. One, a slow and long-term shift for the better in the basic posture of the Brazilian government toward conservation. I think also we all ought to be looking at the program which is developing under Nogueira-Neto in the development of biological reserves because this is funded. The funding in the first couple of years has been fairly substantial. I have looked at the areas that have been delineated for acquisition in various ways, have gone through the sequence of the acquisition, and this program is still being shaped in such a way that inputs from field data could influence both the areas selected and the order of priority. None of the Amazonian areas that have been talked about today are presently high on the priority list.

MITTERMEIER: One or two of them are. Coimbra has some comments relative to the Nogueira-Neto program.

COIMBRA: The program of the Ministry of Mines and Engineering is choosing large areas in the Amazon for fauna and flora preservation. I have been invited to collaborate on the project. But this doesn't mean that we should not fight to the death for the three subspecies of the golden tamarin. If Magnanini and I drop the work, nobody else is going to do a damn thing about it. We are both making a lot of personal sacrifice to see to it that the work is done. I think that clarification of the issue in international journals, such as *Time*, rather than political pressure is going to show the importance of preserving primates. We can be pragmatic about it; we don't have to be poetic. We can talk about animals as a great biological value. Almost every politician in Brazil reads *Time* magazine. Relatively speaking, the position of Nogueira is that he hardly has any resources. He should have the resources that IBDF has, but unfortunately up to this point, he doesn't have. IBDF has the resources, but it doesn't have the philosophy for moving ahead in preservation of the biota. The human aspect has to be joined with the financial and material resources that IBDF has, to get the right people with the money: then there might be more hope. IBDF is transferring money to appropriate certain private enclaves within national parks in eastern Brazil. These are very expensive areas so there is a delay in setting aside land in the Amazon on the part of IBDF because they are concentrating elsewhere. There are great hopes for the Amazon because of a program called RADAN which might be the hope of preserves in the Amazon.

COOPER: At the risk of boring you, some of the perceptions that I came away with from my Peace Corps experience in Colombia involve as much things about the ecology and zoogeography of primates as some of the cultural differences. I think that we do not know the many informal assumptions which our culture makes about the world, and the way things are done. The same is true of others. But after awhile, it gets a little frustrating to see us speaking from these informal assumptions across barriers which we don't see, differences in thought pattern. For instance, it may be apparent to those of us who are products of the American culture that we think that the world is a place that is understandable if we devote enough resources, that is manageable if we want to make the effort, and that it is separate from us. My experience in Colombia leads me to suspect that my Colombian friends have a tendency to view the world as a place in which things happen more because they happen. I think the frequent saying is "if God wants, things will happen." This says a lot along those lines. It is a very important thing to understand in terms of the motivation of the people involved. I think that my Colombian colleagues and friends have intentions that are very much like mine; we are very sympathetic with each other in, for example, thinking that the cotton-top marmoset should be preserved. However, we perhaps differ in that we do not realize what that would imply, in terms of what we try to do and what it would be worth. I think that those Colombians with whom I find myself most philosophically in agreement would be those that had the same exposure to values and assumptions that I have, which are somewhat foreign to Latin culture at a broader level.

Also we must look at some of the best efforts of the agency with which we worked in Colombia,

INDERENA, to preserve areas that were considered to be biologically very important. La Macarena National Park is a large mountain that comes up out of the plains east of the eastern range of the Andes and is a rare situation where there are a number of agencies, universities, and people within Colombia who feel that it is incredibly important. Therefore, the most intensive program for the preservation of La Macarena has been under way for some years. But then one looks at the history of what happened. It started off as a very large reserve. The first thing INDERENA did there was to establish border stations, and they had to cut some trails. Because of some violence in Colombia 20 or 30 years ago, many people had immigrated to this area from the mountains. It happened to be fairly rich as such areas go, and there was a lot of colonization. The colonists were able to move where INDERENA moved because when INDERENA cut trails, the colonists followed them. To get back to the problem of culture. There are those who suggest that it is very difficult for Latins to be isolated policemen or to expect a guard to say, "Look, I'm protecting this point and you all must respect my position, so you're not going in." In my estimation, Latins cannot maintain that attitude and get out of life and the people around them the kind of emotional resources they need. Thus, there are cultural reasons why things cannot be guarded as well as economic reasons, such as when people are incredibly poor. At any rate, over a period of 5 or 10 years, the people went farther into the La Macarena reserve. In response, INDERENA, realizing that the reserve could not be maintained with people living in it, decided to chop off the area in which the people lived. Then, they felt that new stations needed to be established, because the original stations were inhabited and not within the new reserve. So they cut trails into the new reserve, and the colonists followed those trails. It is the best example of a sincere effort on the part of many people, and a rather rare one, and yet these efforts at preservation in my estimation have had more to do with allowing colonists to come in than anything and the future promises to be the same. That which will be protected is that which is most inaccessible, that which is most inhospitable. One other point. In looking at the history of science, biology, and conservation in Colombia, it looks as though they have been held up by individual people. Carlos Lehman, who died several years ago, is an example. He fought all his life for preservation of the Andean condor, to establish reserves, to establish the museum; yet, when that person dies there is no one to continue. Things that happen, happen because of individuals, because of people, not because of policies. I don't wish to be negative, imperialistic or judgmental, but it is somehow amazing that in meetings of this sort, when we are talking about different cultures, that these kinds of factors never seem to intrude, either formally by someone stating them or informally by the attitudes and understandings that people express.

THORINGTON: I hope we all appreciate how elitist conservation interests are and how unique our bias is from a broadly biological perspective. That, however, does not reduce my interest and concern about trying to do something. I am very well aware of the cultural differences and the cultural biases. They enter into my consideration, however, in reviewing the mechanisms which I think we are trying to do here. If there are those who have their doubts about whether conservation in other countries is something we should be dabbling in, I guess we are just on different wave lengths. I want to see some of these animals around in the future. I know I am going to see lots of humans around in the future. Reading Richard's article in *Scientific American* predicting that there will be no rainforest in the tropics in the year 2000, I find very frightening.

EISENBERG: The thing is, you are not alone. We should not fall into the trap of condemning ourselves for having a narrow special interest. In this room, we have people drawn from a number of different countries, mutually trying to work out a policy so it is not a one-sided interest at all. I am a one worlder, but, if I want to really get parochial and think about it, I feel that the differences in my own attitudes toward life and politics, with respect to fellow citizens of my country of origin, are as great here as it might be relative to any place I would want to transport myself on the globe.

COOPER: Dr. Coimbra is one of those people who is really laying it on the line and I think that, if we are going to do anything, it is going to be individuals and groups doing what they feel like doing. I think that is appreciated, because that is very much the way things have been. On the other hand, it would seem that we can't individually wash our hands of it because we talk to *Time* magazine, because we write letters, because we have done anything collectively.

MITTERMEIER: Well, it is less of a cultural thing

really than it is a matter of prestige and economics. If NSF suddenly started funding conservation research in South America and it became a prestigious thing to work in conservation biology, I'll bet you would suddenly find many people working in it. People who do have some interests in conservation right now just cannot work at it because, if they try to, they starve. I don't find the differences between ourselves and the Latins all that deep. I think it is just a matter of the level conservation has reached here and the level which has not reached there yet.

COOPER: I think that what you say is true, but I think there are qualitative differences. Think about the problems of La Macarena National Park. The basic reason that there are problems there right now is because in the 1920s and 1930s Rockefeller went in with his own cultural assumptions and values about disease and did things which made it possible for human populations to expand tremendously. So now you talk to the government and they say the population is doubling every 20 years, and there is not enough money. And the people who did it came from another culture and they had great intentions, but it was a projection of certain cultural values. We can see all around the world what happens when our assumptions about life, its length and quality, have been applied in the name of global health, to help people out.

BLOOD: I agree with Bob [Cooper] that we have to do a better job of understanding the way people in another country or another city see things and try to find answers that meet their interests and their needs as they see them. We are doing an injustice by simply projecting our own biases and our own interests into a completely different setting. I had the opportunity to be in Brazil very recently, and I have the feeling that there are some subtle but very important developments and progress. But I think we have to do a better job of doing our homework, if I may say it that way, of knowing exactly what it is that we need to do there, rather than to sit at this distance and say, "Let's find a quick and ready solution." There are many aspects of this that we need to look into before we find an easy remedy for it.

THORINGTON: I'm not sure that the best missionaries have followed your advice but that's another point. Pekka [Soini], having heard the problems elsewhere, with respect to conservation of primates and specifically of marmosets, would you like to comment on problems in Peru?

SOINI: Yes. Now that I have been listening to the problems in Brazil and Colombia, I feel even better about the situation in Peru, because something is being done, even though it is going slowly. We have one big national park already in Amazonia, Manu, that Marge [Freese] was speaking about. Callitrichids are pretty safe in there although that still leaves some species outside, but none of them seem to be in any immediate danger. *S. nigricollis* occupies a very small area outside any conserved area in Peru, but I don't think that it is in any real danger. And then the government is very enthusiastic about locating an area north of Manu National Park for the National Park of Loreto that we expect to be in existence in a few years. I think part of our mission (with the Proyecto Primates) is to locate and recommend a place for this park which would then protect several or probably all the remaining species. I don't see that any callitrichids in Peru are endangered at the moment. Of course, the government probably will have controlled harvesting of natural populations and also breeding in captivity, maybe managing islands or other isolated areas, but they certainly are interested in carrying out the studies first and in assessing the results of those studies before they take any action. As most of you know, there is no exporting of monkeys right now from Peru, and I expect that there won't be until some information is available concerning densities and poulations of animals. But the government's intention is eventually to exploit the fauna for the nation, to expolit it but with wise control.

HEARN: If we are trying to identify where practical action is possible, it seems there are two main areas. The first forms the basis of this discussion as to realities of political and economic affairs in these countries. The second is the international demand for primates. The demand is going to increase whether we like it or not and, at the present time, the survival of animals that are being exported is disastrously low. On a small survey of dealers in Europe, I get estimates of about 10 percent survival, from capture to eventual establishment in the laboratory. Once they arrive in the lab, quite a lot can be done by intensive care of animals. However, the losses before they leave South America and the losses between South America and Europe are probably in the region of about 60 to 70 percent. The essential thing seems to be the condition of the animal when it is exported, whether it can survive the trip. Perhaps we could turn some attention to government control of the state that animals are in when they leave South America.

PORTER: I am not acquainted with Europe; I am acquainted with the United States. In the United States, with the new CDC regulations eliminating the pet trade within 60 days, they will have cut down to about three or four parties importing into the United States. And I think all of us are now engaged in working or getting ready to go into South America to take over the supply itself down there, maybe when it first comes out of the jungle or through breeding farms. We have been forced into a corner. We already recognize the problem. I think we can keep these death losses way below what you are talking about. The 60-70 percent death loss is a horrible thing, but the people that were involved in it are fast being eliminated, and I think this is going to be a thing of the past within a year or so, as far as the United States is concerned.

HEARN: I am sure that will be the case. One point that many dealers make in Europe is that the pet trade absorbs a good deal more than research. I think this will be eliminated very quickly. However, it seems that, if a system can be built up where animals are exported in good condition, then demand will not be inflated by unnecessary deaths enroute.

PORTER: In Europe I know a couple of the different ports that they go into, but the ports they leave from in South America for the pet trade in Europe aren't necessarily in the country where the animals were first located. There is lots of contraband involved and, as a result, these animals are in very bad condition before they ever leave South America. The Department of Interior is moving to require permits to come in for research monkeys.

THORINGTON: Other comments?

COOPER: One of the things our agency was aware of in Colombia was that the total value of animals that supplied both the pet trade and the research trade was incredibly small. Ken Green published documentation of that. And while these animals are incredibly important to people who need them here in the laboratory, they are of little economic value to Colombia. There is no reason to expect that a great amount of effort would be or could be devoted to preserving, regulating, and managing these species, given the economic value that they have within the country. And that's a reality with regard to politics.While some of us were in Colombia, the pet trade was eliminated. A few animals were allowed out for research, with the exception of a couple of species from northern Colombia (the cotton-top marmoset which is greatly endangered), not because of research needs, but because of habitat destruction. The reasons for not exporting any more primates from Colombia for research, given the few thousand needed, are political; they have nothing to do with the status of these animals in nature and so somehow the research need (while it may sometimes be the tail that wags the dog because it provides money, provides pressure at least in countries where they seem to be important) is really a small part of the greater problem which we are considering, and that is the preservation of habitat.

THORINGTON: Let me review a few things here. It seems to me that in our discussions we have come to the conclusion that we have fairly good ways of determining relatively how many primates, marmosets particularly, there are in certain areas. We did not get very far into the discussion of how large marmoset populations should be to maintain themselves, but I would just point out that the literature is increasingly emphasizing the need for large habitats, if we want to maintain animals in the wild for extended periods of time. In the United States there are only about four parks that are really big enough to maintain the top carnivores and none of those are east of the Mississippi; they are all out in the southwest. So we really should be concerned about large areas and trying to maintain animals and habitats over an extended period of time. Let's consider our strategies carefully, what is possible politically and what is reasonable to do biologically.

I would like to have a couple of people follow through on some of the things that we have discussed. I hope that we will try to maintain a network of communication among each other. Being among the poorest of correspondents here, I hesitate to suggest this, but I do feel that it is important that our South American colleagues feel free to call on us to try to motivate things and to let us know what strategies may be the most successful in their particular countries.

SECTION III:

REPRODUCTION AND PROPAGATION OF THE CALLITRICHIDAE

Introduction

Studies of callitrichid reproductive cycles have been hindered by the apparent randomness and infrequency of copulatory behavior and the lack of visible morphological changes (e.g., in the genitalia and menstruation) during the ovarian cycle. With the use of two distinct hormone assays and standardized behavioral observations, it is now apparent that callitrichids are polyestrus, with a cycle of approximately 16 days. It is encouraging that similar results have been obtained in four species of three different genera, using different techniques.

Most callitrichid species observed in captivity can mate and conceive within the first week postpartum. Thus, continuous breeding is possible, although it seems to occur more commonly in marmosets (*Callithrix* spp.) than in tamarins (*Saguinus* and *Leontopithecus*). Both Gengozian et al. and Kleiman provide strong evidence for seasonality in tamarins while *Callithrix jacchus* appears not to be seasonal. *Callithrix* spp. also seem to have a higher reproductive output than tamarins in that they more commonly produce triplets and even quadruplets (see Gengozian et al. and Mallinson, Section V) than do *Saguinus* spp.

In both small and large colonies of callitrichids, a continuing problem has been the poorer reproductive success of captive-born animals as compared with their wild-caught parents. The reasons for this are multiple; captive-born animals may exhibit parental neglect and have stillbirths or abortions. As Gengozian et al. point out, very subtle differences in behavior and physiology (e.g., milk production) may be contributing to the poor survivorship of F_2 and F_3 young. The problem is usually not a failure of reproductive function.

There is a good birth rate, but high mortality in fetuses and infants before weaning. A resolution of these problems is necessary before the viability of captive colonies will be ensured. Again, there may be a difference in the survivorship of F_2 and F_3 *Callithrix jacchus* when compared with *Saguinus* spp. and *Leontopithecus*.

The Panel Discussion and papers by Kleiman and Jones, and Wilson point up the difference in approach between those breeding on a large scale and those dealing with small numbers of endangered species. Techniques which are routinely used on the common marmoset, *C. jacchus*, including frequent handling, vaginal smears, and bleeding, are infrequently performed on the endangered golden lion tamarin, *L. rosalia*. Concern over the resulting stress is a major reason; with less than 15 breeding females world-wide in scattered zoos, the concern is probably justified. In small scattered colonies, there is more interest in developing policies for the genetic management of species, yet because individual animals are owned by different institutions, long-range management programs cannot be instituted where they are most needed. Political factors seem to intervene.

In large colonies, reproductive success can be evaluated by the study of trends over a number of years. In small colonies, trends are difficult to establish and more emphasis is placed on individual animals and individualized solutions. Short-term changes in sex ratios or survivorship may have catastrophic effects on small colonies, but be insignificant in large ones.

Rothe's detailed description of parturition in *C. jacchus* is the first study of this phenomenon in a callitrichid. The limited care provided by the mother suggests that any weakness in the clinging ability of a neonate will result in its death. There are management implications in the results of this investigation; family groups should be provided with several nest boxes and several shelves so that parturient females are not forced to bear young on narrow branches where the likelihood of a neonate falling to the ground is greater.

JOHN P. HEARN
M.R.C. Unit of Reproductive Biology
2, Forrest Road, Edinburgh EH1 2QW.

The Endocrinology of Reproduction in the Common Marmoset, *Callithrix jacchus*

ABSTRACT

The levels of progesterone, estradiol 17β and LH measured by radioimmunoassays in the peripheral plasma of common marmosets (*Callithrix jacchus*) showed a clearly defined ovarian cycle of 16.4 ± 1.7 days. Marmosets do not menstruate, and in captivity they mate frequently throughout the cycle, so there is no behavioral index of estrus. Prealbumin proteins appear in the vaginal washings of common marmosets in an apparently cyclical fashion, but although their appearance correlates generally with the late luteal phase this method is not precise enough to determine accurately the stage of the cycle. Furthermore, vaginal cytology gives no indication of the time of ovulation. Hysterectomy during the luteal phase of the ovarian cycle does not prolong the life of the corpus luteum in the common marmoset, indicating that as in women and rhesus monkeys cyclical luteal regression is not under the control of uterine prostaglandins.

The gestation period is 148 ± 4.7 days as estimated by the time from the day of the last midcycle LH peak to the day of parturition. Preliminary results showed that luteectomy could be performed as early as the sixth week of pregnancy without causing an abortion. Progesterone or chorionic gonadotrophin assays could be used to diagnose pregnancy by two weeks, and pregnancy could be diagnosed by the fourth week using trans-abdominal uterine palpation. Commercial human chorionic gonadotrophin assay kits have so far proved unsatisfactory in marmosets, and immunoassays of common marmoset chorionic gonadotrophin show that this hormone is immunologically different from that in the human, although CG from both species show biological cross reactivity in standard mouse bioassays. Common marmosets continued to mate throughout pregnancy, although they did so less frequently after the first 6–8 weeks.

After birth, the young marmosets are carried by either parent or by older siblings of either sex. Neonatal male common marmosets showed high levels of plasma testosterone which dropped to basal levels by 70 days after birth and remained basal until the onset of puberty at 250 to 400 days of age. Young male common marmosets reached 'sexual maturity' before females as judged by the dates of their first fertile inseminations, and pubertal female common marmosets underwent a number of anovulatory cycles after the first appearance of progesterone in their plasma. There was a high rate of spontaneous abortion in pubertal common marmosets. Growth curves were constructed for common marmosets up to 600 days of age from measurements of head width, knee-heel length, body weight, pudendal pad width, vulva and testis size. These measurements were correlated with the first appearance of detectable levels of testosterone in males and progesterone in females, and the onset of reproductively directed behavior.

Adult males showed very high plasma levels of circulating testosterone, which was released in a markedly episodic fashion. Vasectomized males continued to mate frequently. The testis dimensions of adult males measured monthly through the year showed no seasonal variation in size.

The endocrinology of reproduction of male and female common marmosets shows that *Callithrix jacchus* is a suitable species for studies related to human reproduction. Their small size, ease of management, and rapid rate of reproduction make them excellent laboratory animals. The recent increase in basic information about the reproductive biology of marmosets promises a rapid increase in their use in the future. It is, therefore, essential that the exploitation of stocks in the wild should be carefully controlled over the next ten years until there is an adequate laboratory-bred supply, otherwise a serious depletion of the irreplaceable wild stocks will result.

Introduction

The great potential of marmosets and tamarins in biomedical research is only now becoming recognized. Consequently, there is a greatly increased demand for wild callitrichids in order to establish captive colonies. Simultaneously, they are becoming harder to obtain in the wild as deforestation and the expansion of human populations reduces the area of their natural habitat in the forests of South America. Several countries have banned the export of callitrichids, so exacerbating the demand on wild stocks in countries where no such prohibition operates.

Marmosets and tamarins have been used in research on immunology, virology, toxicology, and behavior for many years (for reviews see Gengozian, 1969; Deinhardt, 1970; Epple, 1970) but there is little knowledge of their reproductive endocrinology. Aspects of the endocrinology of reproduction of adult, neonatal and juvenile common marmosets are presented here. The studies were made on a group of 250 common marmosets, *Callithrix jacchus*, established in Edinburgh as a self-sustaining colony for basic studies of reproductive endocrinology and the development of immunological methods of contraception. A detailed report of the management of this colony has already been published (Hearn, Lunn et al., 1975). These studies are being pursued with two main objectives. Firstly, to evaluate the common marmoset as a suitable model for studies in human reproduction. Secondly, to learn from their reproductive endocrinology ways in which captive marmosets may be managed best to ensure optimal breeding. We hope that our results may prove useful in the future management of callitrichids in captivity and in the wild.

The Adult Female

The Ovarian Cycle

Marmosets and tamarins do not menstruate, nor do they show in captivity any externally obvious signs of estrus. There is no clear cyclicity in their vaginal cytology and although some cyclical variation in plasma diamine oxidase (Hampton and Hampton, 1975) and prealbumen proteins in vaginal washings (Hearn and Renfree, 1975) have been reported in callitrichids, there is no simple procedure by which the stage of the ovarian cycle can be determined. Since the common marmoset copulates throughout the cycle, mating is not an index of ovulation, although there may be an

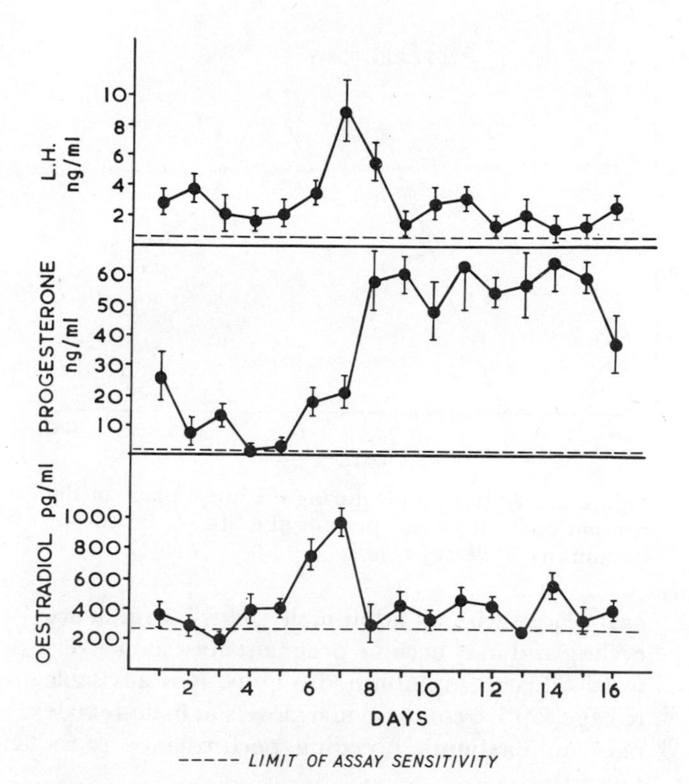

Figure 1. The peripheral plasma levels of LH, estradiol 17β and progesterone during the ovarian cycle of *Callithrix jacchus*. Each point represents the mean value (± S.E.M.) for nine ovarian cycles. Dashed line indicates the limit of assay sensitivity (From Hearn and Lunn, 1975).

increased incidence of mating around this time. There is a clearly defined ovarian cycle of 16.4 ± 1.7 days when monitored by plasma levels of LH, progesterone and estradiol (Figure 1) and simple progesterone radioimmunoassays are perhaps the most accurate method of following the cycle (Hearn and Lunn, 1975). This does, however, require frequent bleeding of animals from the femoral vein (Plate 1a).

Hysterectomy during the luteal phase does not prolong the ovarian cycle in common marmosets (Figure 2), implying that the corpus luteum is under luteotrophic control, as in women and rhesus monkeys and not under uterine luteolytic control as in the horse, cow, sheep, pig, guinea pig, and other species (for reviews see Short, 1969; Heap and Perry, 1974).

Several laboratory workers have suggested that in groups of marmosets and tamarins only the dominant female will produce young (Epple, 1970). When *Callithrix jacchus* females are caged together in the laboratory, the dominant animal continues to cycle normally but the others stop cycling. If a noncycling female is removed from the group

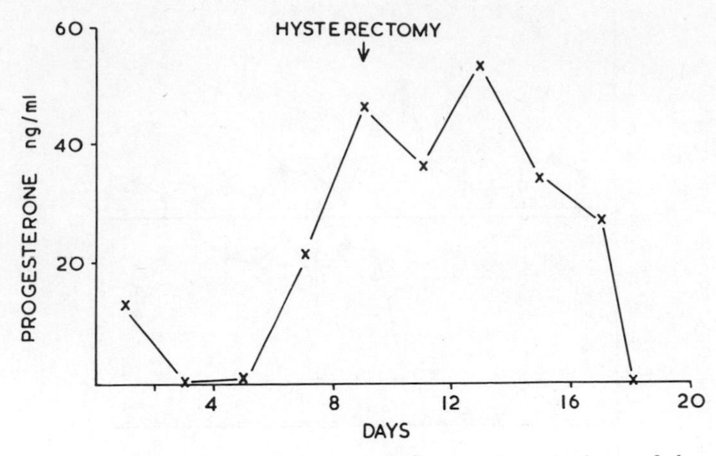

Figure 2. Hysterectomy during the luteal phase of the ovarian cycles does not prolong the life of the corpus luteum in *Callithrix jacchus*.

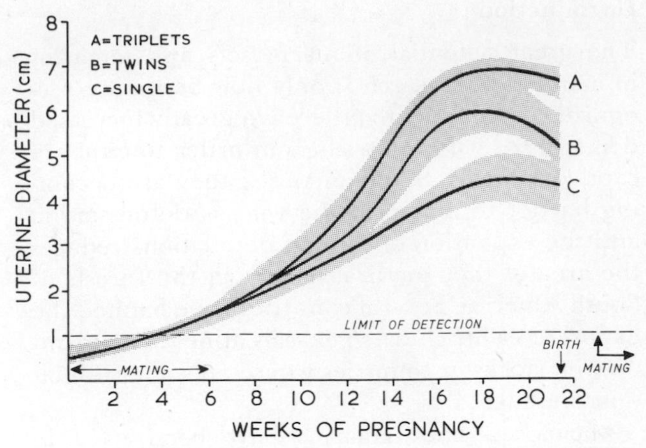

Figure 3. The diameter of the uterine fundus during pregnancies of single, twin, and triplet fetuses in the common marmoset. Mean and range values are given for 5-single, 20-twin, and 10-triplet pregnancies.

and placed with an adult male she will commence cycling and may become pregnant very soon afterward (Hearn, unpublished). Thus, it is advisable to cage captive common marmosets in male-female pairs if maximum breeding performance is required.

Pregnancy

Common marmosets have a gestation of 148 ± 4.3 days as calculated from the day of the ovulatory LH peak to the day of parturition, and will usually produce young (twins 80 percent, triplets 15 percent, single 5 percent, (n = 100) in our colony) twice a year when kept under suitable conditions. Pregnancy can be diagnosed by transabdominal uterine palpation from the fourth week of gestation and by the second week using radioimmunoassays of progesterone or chorionic gonadotrophin. It is difficult to diagnose pregnancy accurately using commercially available human pregnancy testing kits and the excretion of chorionic gonadotrophin even when measured by bioassay shows considerable variation between animals and even between sequential samples taken from the same animal (Hobson et al., 1976). Growth curves of the uterus of common marmosets carrying single, twin, and triplet young are shown in Figure 3, and in Figure 4 the plasma levels of progesterone, estradiol, and luteinizing hormone/chorionic gonadotrophin during pregnancy are shown.

In a preliminary study, it was found that the corpus luteum of pregnancy could be removed after the sixth week of gestation without disrupting the course of pregnancy (Figure 5). This is rather later than in women (Csapo et al., 1976) but the early stages of embryonic development occur more

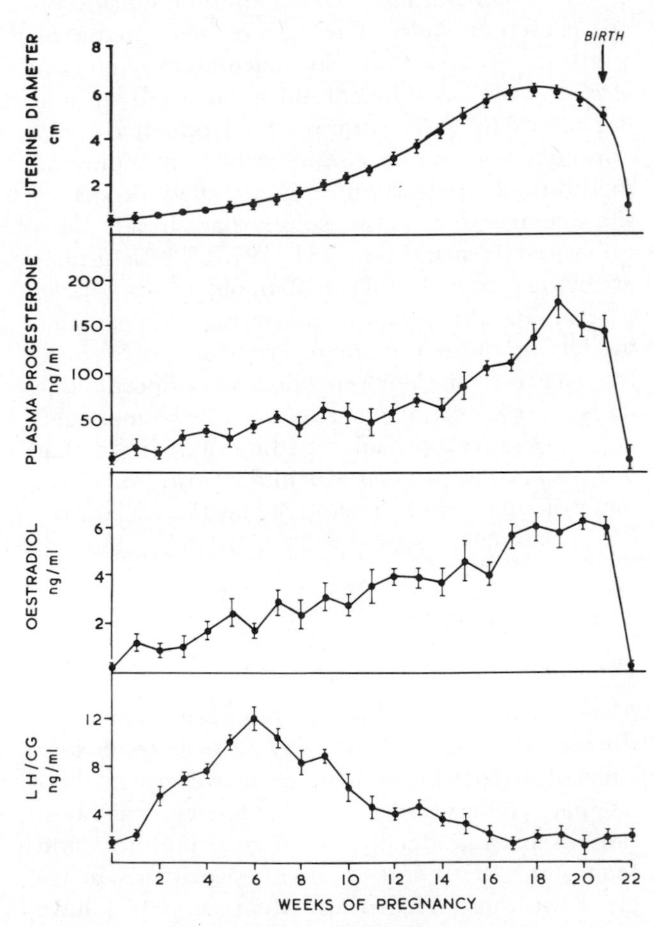

Figure 4. The uterine fundal diameter and the peripheral plasma levels of progesterone, estradiol 17β and LH/CG during pregnancy in the common marmoset. Each point represents the mean value (± S.E.M.) for seven pregnancies each of which produced twin young. (From Hearn and Lunn, 1975.)

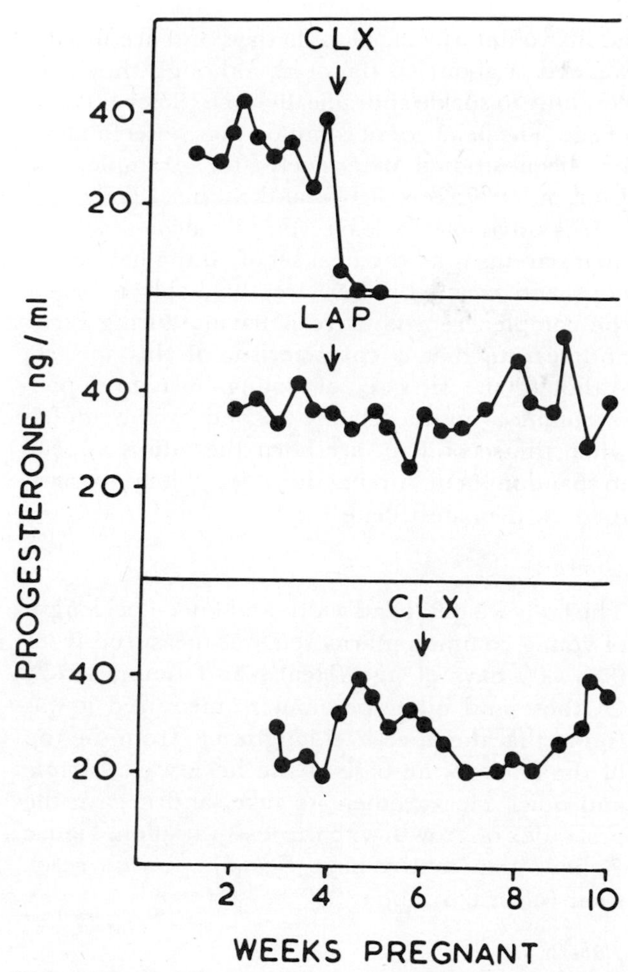

Figure 5. Peripheral plasma progesterone levels after luteectomy (CLX) performed at the fourth and sixth week of pregnancy in common marmosets, and in a control animal after laparotomy in the fourth week. After the sixth week, pregnancy can proceed without the corpora lutea.

slowly in common marmosets than in macaques and humans (I. R. Phillips, pers. comm.).

The time of parturition can be roughly predicted in *Callithrix jacchus* by transabdominal palpation or by radioimmunoassay of progesterone or estrogens. The fetal heads are found to position low in the mother's abdomen two to three weeks before birth, coincident with a reduction in uterine diameter (Figure 3) and a fall in the level of peripheral progesterone (Figure 4). In the last three to four days before birth, the levels of estrogens increase markedly (Hearn and Chambers, unpublished) to reach a peak at parturition, only to fall again to undetectable levels by the day after birth.

Postpartum Ovulation

The breeding records of many marmoset colonies show that the interbirth interval may be only a few days longer than the gestation period. Hearn and Lunn (1975) reported that common marmosets mate three to ten days after giving birth. Figure 6 shows that common marmosets commence cyclical ovarian activity immediately after parturition and may become pregnant on the first cycle after giving birth, although there is a considerable amount of early embryonic loss in the first two to four weeks of pregnancy. Recent data suggest that this may also be true for the human (Cutright, 1975; James, 1970). In *Callithrix jacchus*, lactation does not appear to inhibit ovarian cyclicity and ovulation, a factor that undoubtedly contributes to their high rate of reproduction.

The Adult Male

As far as is known, the common marmoset does not show any marked seasonality of birth in the wild. Environmental conditions are generally controlled in the laboratory, and most authors have reported that captive common marmosets breed throughout the year. The testes of captive adult males show no significant seasonal variation. The male common marmoset shows an episodic fluctuation of plasma testosterone levels similar to that found in many other species, but the concentrations are very high. This is not altered after vasectomy, and vasectomized marmosets continue to copulate frequently (Hearn and Corker, unpublished).

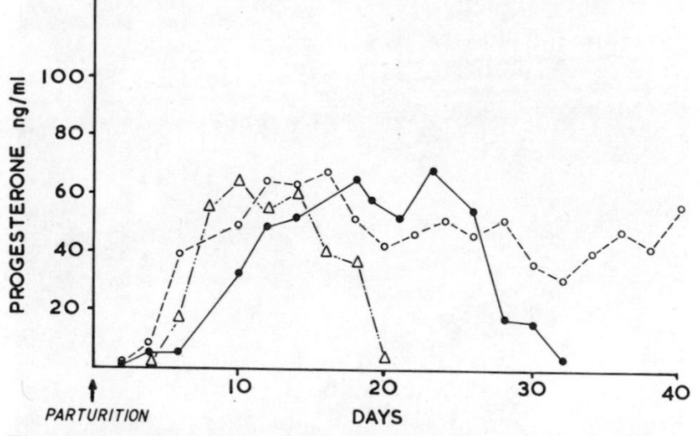

Figure 6. Common marmosets may commence cycling immediately after parturition and may become pregnant again in the first cycle while lactating. There may be considerable spontaneous fetal loss during early pregnancy in marmosets. △—△ postpartum nonfertile cycle; •—• early postpartum abortion; ○—○ postpartum pregnancy.

The Neonate and Juvenile

Behavior

After birth, the neonates are intensely cared for by both the parents. The young start to leave the

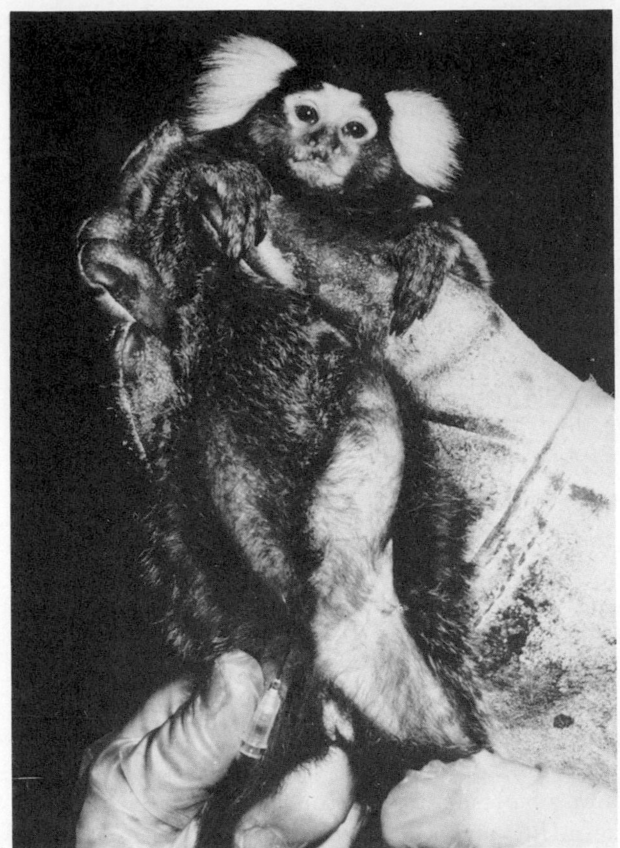

1a.

1b.

adults voluntarily at 20 to 25 days and are usually weaned at about 40 days old, although they may continue to suckle sporadically up to 80 or 90 days of age. The behavior of common marmoset families has been studied extensively (for example, see Ingram, 1977; Box, 1977; and Rothe, 1977).

It is advisable to leave young marmosets with their parents until the next set of young have been born and weaned. They are then able to learn the complex reproductive behavior during birth and rearing that is characteristic of this species. Although the first set of young born to a pair of common marmosets are carried by the adults, when the second set are born the adults appear to abandon them, forcing the older siblings to carry the new neonates (Plate 1b).

Growth

The body weight, head width and knee-heel length of young common marmosets was measured from 0 to 200 days of age (Hearn and Lunn, 1975). Of these and other parameters measured it was found that the knee-heel length, i.e. from the top of the knee to the base of the heel, was a simple and quick measurement to take, and it gave the best index of growth with the least variation. Figure 7 shows that captive-born *Callithrix jacchus* reach their full size by about 350 days of age.

Puberty

In the common marmoset, as in the human and other primates, puberty is a relatively lengthy process. Although *Callithrix jacchus* reach full body size at about one year old and they may show reproductively directed play and even copulation well before this age, gonadal size and the levels of progesterone in females, and testosterone in males, may not reach adult levels until 6 to 12 months later. Figure 8 shows that neonatal male common marmosets have high levels of plasma testosterone over the first 3 to 80 days of age. A similar neonatal testosterone rise has been reported in humans and although its significance is not understood (Forrest, et al., 1973), it is thought

Plate 1a. Blood samples may be taken easily from the femoral vein of *Callithrix jacchus* using a 27-gauge needle. The procedure causes little stress to the animals and if they are given 0.25 ml Fersamal Iron Syrup (Glaxo) it serves both as a reward and as a replacement of iron.

Plate 1b. A juvenile male *Callithrix jacchus*, aged 150 days, carrying two three-day old twins that have been passed on to him by the parents.

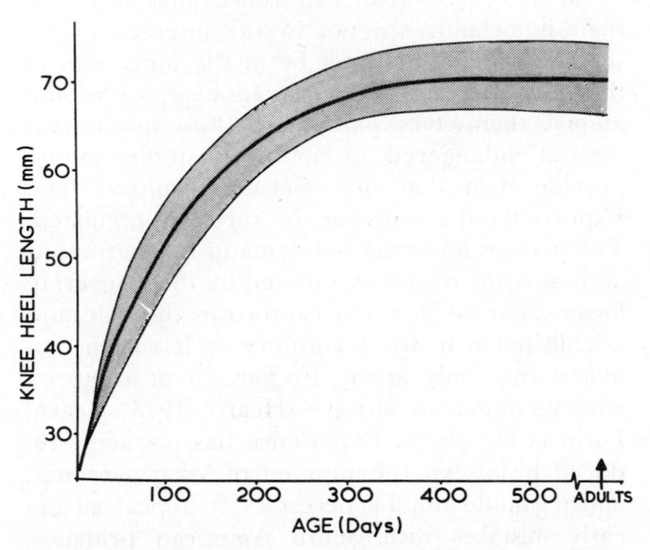

Figure 7. The knee-heel length of young common marmosets from 0 to 500 days old. The mean and range of values are shown.

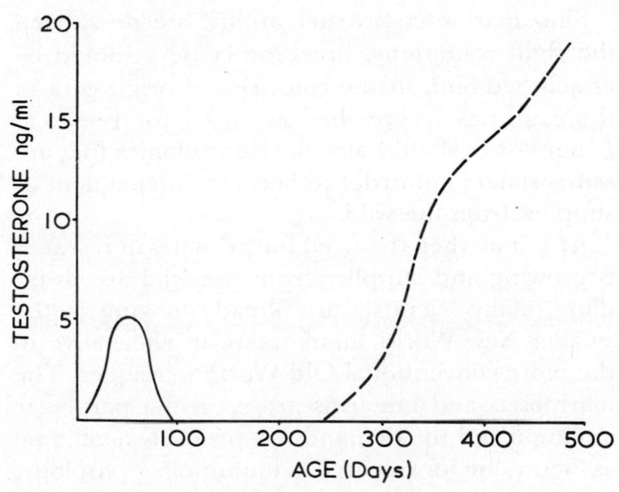

Figure 8. Peripheral plasma testosterone in common marmosets before and after puberty. There is an elevated level in the neonate between 3 and 90 days of age.

to be involved in the sexual differentiation of the brain.

Progesterone first appears in the plasma of female common marmosets at about 300 days, but the pattern does not suggest normal ovulation until 6 to 12 months later. There are verbal reports that females may become pregnant at 12 to 14 months of age and that they usually abort their first pregnancy or neglect and lose their first young, but our experience suggests that on average captive born females are not fully mature until 20 to 24 months old although occasionally individuals may become pregnant well before this age.

The onset of puberty may be advanced and the survival of first-born young improved if juvenile males or females are paired with reproductively experienced adults. Our experience suggests the following management program for rearing young common marmosets in order to achieve the best breeding results: they should be left with their parents until a second set of siblings has been born and reared; they should then (8 months old) be placed in peer groups where they learn to socialize with other animals until 12 to 18 months old; they can then be paired up with experienced adult partners to form breeding pairs.

Discussion

Captive common marmosets can be easily maintained in family groups and the contact that this allows between animals reduces the effects of boredom so often seen in colonies of larger monkeys that are caged singly. We have found, however, that the use of exercise cages can greatly improve the condition and breeding performance of the animals. An exercise cage placed in the center of each animal room, with the animals in each cage using it for 24 h periods in rotation, serves the dual purpose of giving each individual an occasional change of environment—a change is as important as extra space—and also provides a source of interest and distraction to the other animals in their cages around the room (Hearn, Lunn et al., 1975).

Callithrix jacchus adapts readily to captivity and their small size and ease of maintenance make it possible to establish self-sustaining breeding colonies relatively inexpensively when compared to the larger primates. As they usually produce twins, show no lactational anestrus, and become sexually mature at under two years old, the inherent rate of population growth is at least double that found in breeding colonies of the macaque monkeys that are more conventionally used in biomedical research. As a result, even small colonies of marmosets can become self-maintaining, thus minimizing the threat to wild populations from continued exploitation for research.

Our initial studies suggest that the common marmoset is an eminently suitable model for research in certain aspects of human reproduction. The endocrinology of the ovarian cycle, pregnancy, and puberty show close parallels to the human, and suggest that common marmosets can be used in studies of the control of the ovarian cycle and of early pregnancy. They are hardy animals and recover rapidly from major surgery. Their rapid rate of reproduction is particularly useful in studies of fertility, where the effects of immunological or endocrinological treatments can be seen relatively quickly (Hearn, Short et al., 1975).

In addition to the advantages that marmosets have as a model for the study of human reproductive physiology, they exhibit a number of unique features that make them an ideal group for basic research in immunology. They usually produce twin chimeric young as a result of placental anastomoses which also render the twins immunologically tolerant to each other. This creates many novel opportunities for the study of transplantation immunology and fetal endocrinology.

The major limitation of marmosets is their small size. This precludes the withdrawal of large quantities of blood (only 2 ml per week can be removed over long periods) and makes intricate transplantation and vascular surgery difficult. The lack of menstruation or any externally obvious signs of estrus or ovulation is also a great disadvantage in reproductive studies. Marmosets, in common with most South American primates, do not have spiral arterioles in the uterus.

Compared to man and the larger macaques, the common marmoset has a very short interval between birth and puberty. This makes it a suitable primate for the study of puberty and in particular the development of physical, endocrinological, and behavioral maturation; such studies can be completed in under two years, whereas the macaques require five, the baboons six, and the human up to sixteen years to complete pubertal development.

The common marmoset is perhaps the most amenable of all the marmosets and tamarins, and appears to have the best reproductive performance in captivity, although *Saguinus oedipus* and *Saguinus fuscicollis* also settle down and breed well in captive conditions. At present, all callitrichid species are threatened by the clearance of natural forest, and this lends urgency to research on marmosets and tamarins both in the laboratory and in the wild. The interests of conservationists, field and laboratory workers should be accepted by each other as complementary, in that any knowledge gained about these species will ultimately help in their management in captive or wild environments in the future. The endangered species of marmosets and tamarins already require careful management in the wild, and many more South American primate species may be in a similar predicament if the present rate of deforestation continues.

Sweeping bans on the export of callitrichids from their homelands are not in the interests of researchers. They may not be in the long term in the best interests of conservationists, or of the animals themselves, particularly those species that are not endangered, as laboratory studies should provide data that are useful in conservation. Exports should, however, be carefully regulated. The present international demand for marmosets and tamarins is grossly inflated by the staggering losses incurred between capture in the wild and establishment in the laboratory; it has been estimated that only about 10 percent of captured animals ultimately survive (Hearn, 1975; Hearn, Lunn et al., 1975). Experience has markedly reduced the losses in the shipment of Asian macaques, and it should not be necessary to repeat all the early mistakes with South American primates. Although the export trade is probably not a significant factor in the decline of callitrichid populations when compared to deforestation and the lack of conservation of the wild stocks, any improvement would help.

Since marmosets are such prolific breeders, given the right conditions, breeding centers should be established both in the countries of origin and in the countries where they are used for research. Laboratories should also develop colonies that are self-sustaining in order to become independent of supplies from the wild.

At a time when the need for primates in research is growing and supplies from the wild are dwindling, many scientists are already looking to the smaller New World monkeys as an alternative to the more conventional Old World macaques. The marmosets and tamarins are receiving particular attention and the demand will probably accelerate as their value for research in immunology, virology, teratology, and reproduction receives wider recognition. Their exploitation must be carefully managed in the next ten years until there is an adequate laboratory bred supply, because if the pattern of the last ten years is not changed there may be serious consequences for the wild populations.

Literature Cited

Box, H. O.

1977. Social interactions in family groups of captive marmosets (*Callithrix jacchus*). In *The Biology and Conservation of the Callitrichidae,* edited by Devra G. Kleiman, pp. 239–249. Washington, D.C.: Smithsonian Institution Press.

Csapo, A. I.; M. O. Pulkinnen; and H. K. Kaihola.
1976. The regulatory significance of the human corpus luteum. *J. Clin. Endocrin. Metab.*, (in press).

Cutright, P.
1975. Spontaneous fetal loss: A note on rates and some implications. *J. Biosocial Sci.*, 7:421–433.

Deinhardt, F.
1970. Use of marmosets in biomedical research. In *Medical Primatology,* edited by E. I. Goldsmith and J. Moor-Jankowski, pp. 918–925. Basel: Karger.

Epple, G.
1970. Maintenance, breeding and development of marmoset monkeys (Callithricidae) in captivity. *Folia Primatol.*, 12:57–76.

Forrest, M. G.; A. M. Caithard; and J. A. Bertrand.
1973. Evidence of testicular activity in early infancy. *J. Clin. Endocrin. Metab.*, 37:148–151.

Gengozian, N.
1969. Marmosets: Their potential in experimental medicine. *Ann. N. Y. Acad. Sci.*, 162:336–362.

Hampton, J. K., and S. H. Hampton.
1975. Some expected and unexpected characteristics of reproduction in Callithricidae. In *Breeding Simians for Developmental Biology,* edited by F. T. Perkins P. N. O'Donoghue. Laboratory Animal Handbooks, 6:235–240.

Heap, R. B., and J. S. Perry.
1974. The maternal recognition of pregnancy. *Brit. J. Hosp. Med.*, July 8–14.

Hearn, J. P.
1975. Conservation of marmosets. *Nature (Lond.)*, 257:358–359.

Hearn, J. P., and Lunn, S. F.
1975. The reproductive biology of the marmoset monkey, *Callithrix jacchus.* In *Breeding Simians for Developmental Biology,* edited by F. T. Perkins and P. N. O'Donoghue. Laboratory Animal Handbooks, 6:191–202.

Hearn, J. P.; S. F. Lunn; F. J. Burden; and M. M. Pilcher.
1975. Management of marmosets for biomedical research. *Lab. Anim.*, 9:125–134.

Hearn, J. P., and M. B. Renfree.
1975. Prealbumin proteins in the vaginal flushings of the marmoset, *Callithrix jacchus. J. Reprod. Fert.*, 43:159–161.

Hearn, J. P.; R. V. Short; and S. F. Lunn.
1975. The effects of immunising marmoset monkeys against the β subunit of HCG. In *Physiological Effects of Immunity against Hormones,* edited by R. G. Edwards and M. H. Johnston, pp. 229–247. Cambridge: Cambridge University Press.

Hobson, B.; Lunn, S. F.; and Hearn, J. P.
1976. Gonadotrophin excretion during pregnancy in the marmoset. *Folia Primat.*, (in press).

Ingram, J. C.
1977. Parent-infant interactions in the common marmoset (*Callithrix jacchus*). In *The Biology and Conservation of the Callitrichidae,* edited by Devra G. Kleiman, pp. 281-291. Washington, D.C.: Smithsonian Institution Press.

James, W. H.
1970. The incidence of spontaneous abortion. *Population Studies,* 24:241.

Rothe, H.
1977. Parturition and related behavior in *Callithrix jacchus* (Ceboidea, Callitrichidae). In *The Biology and Conservation of the Callitrichidae,* edited by Devra G. Kleiman, pp. 193–206. Washington, D.C.: Smithsonian Institution Press.

Short, R. V.
1969. Implantation and the maternal recognition of pregnancy. *Ciba Foundation Symposium on Foetal Autonomy,* edited by G. E. W. Wolstenholme and M. O'Connor, pp. 2–26. London: Churchill.

SUZANNE H. HAMPTON
KinetoMatic
Route 1, Box 63C
Templeton, California 93465
and
JOHN K. HAMPTON, JR.
California Polytechnic State University
San Luis Obispo, California 93407

Detection of Reproductive Cycles and Pregnancy in Tamarins (*Saguinus* spp.)

ABSTRACT

In the human, leucocyte alkaline phosphatase (LAP) activity is related to disease states, pregnancy, and stages of the menstrual cycle. LAP levels were measured in tamarin (*Saguinus* spp.) blood smears stained by the azo dye procedure. In *Saguinus fuscicollis*, at least 75 percent of all leucocytes were unstained, resembling the human case; in *S. oedipus* LAP activity was consistently three times higher than in *S. fuscicollis*. Females of both species exhibited peaks of LAP activity at intervals of approximately 14 days. A reproductive cycle of approximately 14 days has also been demonstrated by serial measurements of serum estrogen and progestin levels in tamarins. During pregnancy, LAP activity increased to levels well above that seen in nonpregnant tamarins.

Introduction

Members of the family Callitrichidae, like other New World primates, show no external cyclic menstrual bleeding and no sex skin is present (Lucas et al., 1937; Langford, 1963; Hampton et al., 1966; Epple, 1970). Serial vaginal smears from marmosets (*Callithrix jacchus*) show no cyclic changes in vaginal cytology, pH or peroxidase activity (Hearn and Lunn, 1975). Also, copulatory activity appears to occur randomly (Hearn and Lunn, 1975; authors' observations). Thus, marmosets and tamarins exhibit neither a conspicous estrous cycle, characteristic of lower mammals, nor a readily detectable menstrual cycle, characteristic of higher primates.

Because of the decline in populations of wild marmosets and the increasing difficulties in obtaining feral animals for research uses, it has become essential that captive breeding populations be established. Although several breeding colonies, large and small, are in existence (Levy and Artecona, 1964; Hampton et al., 1966; Gengozian, 1969; Deinhardt, 1970), in no case has breeding success reached the point of eliminating the need for occasional or regular additions of feral animals to replace colony losses or to provide for experimental uses. Probably a major contributing factor to this lack of highly productive breeding colonies is lack of knowledge regarding the reproductive physiology of marmosets and tamarins, especially methods of detection of a reproductive cycle or pregnancy.

It is the purpose of this paper to review recent work related to detection of reproductive cycles and pregnancy in the Callitrichidae and to present some preliminary new data concerning means of detection of ovulation and pregnancy.

The Reproductive Cycle

Because of difficulties in detecting an estrous or menstrual cycle by the usual clinical signs, the levels of serum estrogens and progestins have been analyzed as a measure of reproductive cyclicity in *Saguinus oedipus oedipus, S. fuscicollis* (Preslock et al., 1973), and *C. jacchus* (Hearn and Lunn, 1975). Blood samples were collected from five female *S. oedipus* and one female *S. fuscicollis* by venipuncture three times weekly for ten weeks. Serum estrogens were measured by radioimmunoassay using a modification of the dextran-coated charcoal method of Hotchkiss et al. (1971). Serum estrogens levels are characterized by an increase from a mean level of 2.3 ± 0.2 ng/ml to a mean peak level of 6.2 ± 0.8 ng/ml in approximately five days, followed by a rapid decline. The mean duration of the serum estrogen elevations was 9.0 ± 1.0 days, while the mean interim between successive elevations, during which time the estrogen levels were relatively constant, was 6.9 ± 1.0 days. The mean interval between successive estrogen peaks was 15.9 ± 1.0 days. Hearn and Lunn (1975) used the method of Linder and Bauminger (1974) to determine plasma estradiol-17β levels in five *C. jacchus* from which blood samples were taken on alternate days for approximately ten weeks. The time lapse between successive estradiol peaks was 16.4 ± 1.7 days with levels of estradiol of approximately 400 pg/ml between peaks and peak values of approximately 1000 pg/ml. Thus, although the interval between the estrogen peaks is similar in both *Callithrix* and *Saguinus*, i.e., 16.4 ± 1.7 days and 15.9 ± 1.0 days, respectively, *Saguinus* shows serum levels of estrogens several times greater than *Callithrix*. These differences in hormone levels may reflect differences in assay techniques.

Serum progestin levels were also determined in both species. Preslock et al. (1973), using the serum protein binding technique of Murphy (1967) as modified by Johansson (1969), found an interval of 15.5 ± 1.5 days between successive progestin peaks in *S. oedipus* and *S. fuscicollis*. The progestin levels began to increase shortly after the initiation of the estrogen rise, attained a mean peak value of 173.3 ± 17.5 ng/ml in approximately five days, and declined to levels of 40.3± 6.8 ng/ml before the ensuing rise in plasma estrogens. There was a lapse of approximately 48 hours between the estrogen and progestin peaks. The mean duration of the progestin elevation was 8.8 ± 0.8 days with 6.7 ± 0.7 days between successive elevations. Progesterone levels in *C. jacchus* were determined by Hearn and Lunn (1975) using a specific radioimmunoassay for progesterone (Neal et al., 1975). The interval between successive progesterone peaks was similar to that seen in *S. oedipus* as was the time lapse between estradiol and progesterone peaks. However, *C. jacchus* displayed much lower levels of plasma progesterone than did *S. oedipus*. Peak values for *C. jacchus* were approximately 60 ng/ml whereas in *S. oedipus* peak values approached a mean of 175 ng/ml. Again, these differences may reflect differences in assay techniques.

Although the reproductive cycle of callitrichids may be followed using radioimmunoassays of hormone levels, such procedures may be time-consuming. In the hope of developing a simpler method to follow the reproductive cycle, we have carried out preliminary studies to determine levels of neutrophilic granulocyte alkaline phosphatase (LAP) in *S. oedipus* and *S. fuscicollis*. It has been shown that, in the human, LAP activity is related to stages of the menstrual cycle (Polishuk et al., 1968). A rise in LAP activity is seen at the time of rise in basal body temperature corresponding to the time of ovulation. The method of Kaplow (1955) was used to determine LAP levels. Blood smears were fixed in 10 percent formalin in methanol and stained by the azo dye procedure, using propanediol buffer at a pH of 9.6. LAP activity was quantified by scoring 100 consecutive neutrophilic granulocytes from zero to four on the basis of intensity and appearance of the precipitated dye in the cytoplasm. The sum of the rating of 100 cells was considered the score for a particular sample.

Blood samples were obtained three times per week by venipuncture over a three-month period. Samples were obtained from seven *S. oedipus* (five female, two male) and six *S. fuscicollis* (four female, two male).

The serial levels of LAP activity in three *S. oedipus* females (#2109, #1839, #2115) are shown in Figures 1 and 2. LAP activity increased at consistently repeating intervals with a mean of 15.2 ± 1.6 days (Table 1). In all three cases, there were noncyclic levels of LAP activity during the initial 10 to 30 days of sampling. Hearn and Lunn (1975) have reported that, in *C. jacchus*, the first cycle during which animals were bled was usually longer (up to 26 days) than subsequent cycles. The LAP activity seen in two male *S. oedipus* (#2138, #2206) is shown in Figure 2. No cyclic increases in LAP activity are seen. LAP levels which did not demonstrate cyclic increases were observed in two female *S. oedipus* (Figure 3). The mean LAP score for all five *S. oedipus* females, a total of 162 samples, was 170 ± 43.1; the mean score for the two males, a total of 69 samples, was 168 ± 51.8.

Blood samples from all four *S. fuscicollis* females (#1663, #1859, #2369, #2375) showed cyclic increases in LAP activity (Figures 4 and 5). The mean LAP score for all *S. fuscicollis* females, a total of 121 samples, was 44 ± 34.5. Samples from two *S. fuscicollis* males did not show cyclic increases

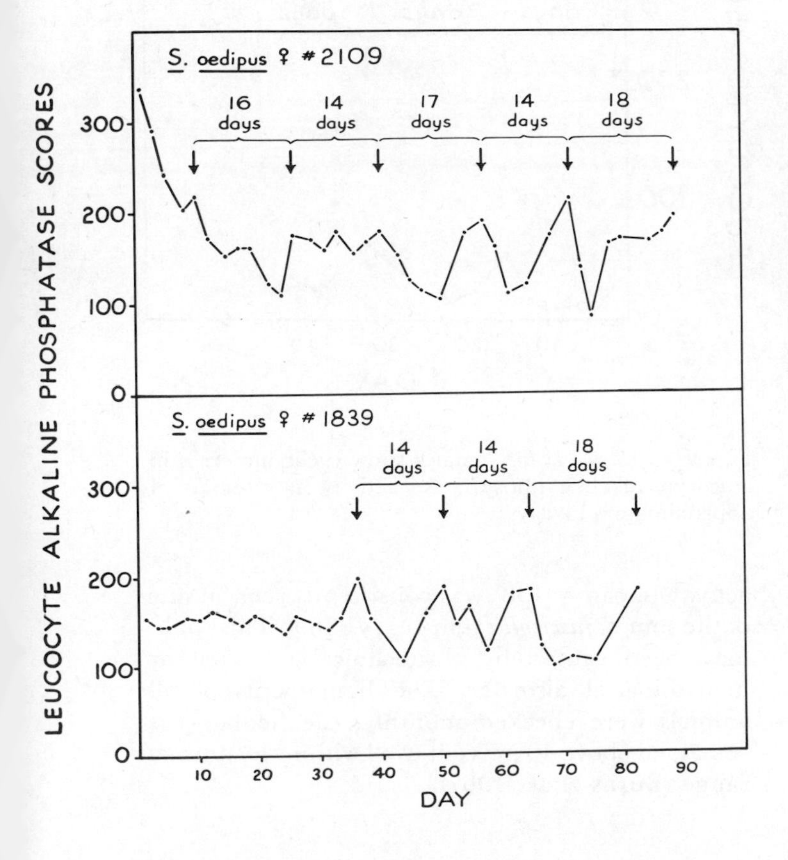

Figure 1 (top and bottom). Cyclic increases in leucocyte alkaline phosphatase activity in female *S. oedipus* occurred at intervals of approximately 14 days.

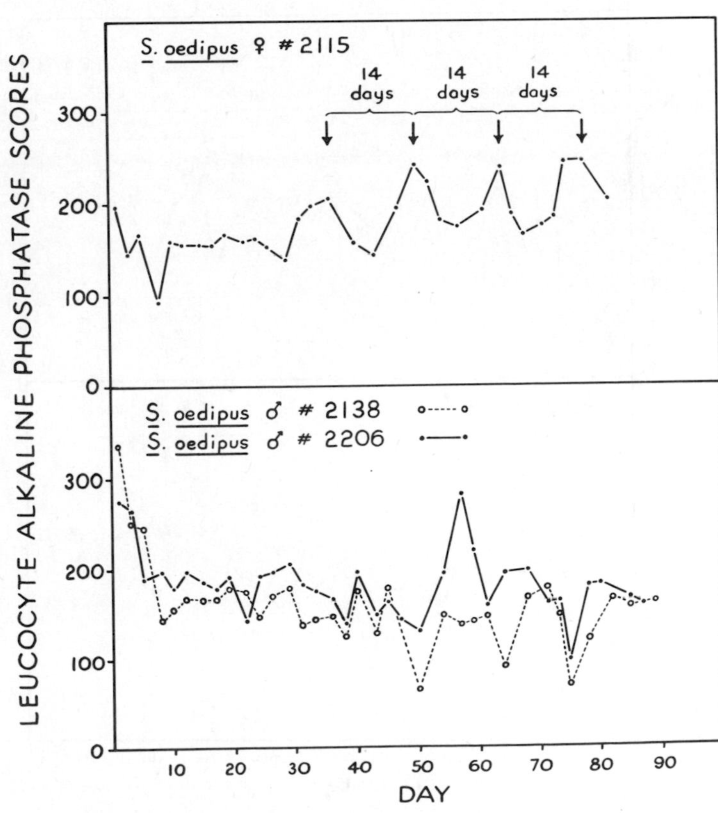

Figure 2. The 14-day cycle found in female *S. oedipus* (top) was absent in the two male *S. oedipus* studied (bottom).

Table 1. Estrous cycle in tamarins.

Species	*Mean (days)* *±Standard Deviation*	*Range (days)*	*Number of cycles Measured*
Saguinus oedipus	15.2 (±1.6)	14 – 18	11
Saguinus fuscicollis illigeri	14.6 (±1.4)	13 – 18	13

in LAP activity. Their mean LAP value, in 52 samples, was 84 ± 59.7.

In summary, in *S. oedipus* the mean cycle length in 11 cycles was 15.2 ± 1.6 days with a range of 14 to 18 days. In *S. fuscicollis*, the mean cycle length was 14.6 ± 1.4 days in 13 cycles with a range of 13 to 18 days. Although both species show a very similar length of reproductive cycle, there was a conspicuous difference in level of LAP activity between the two species. Samples from *S. oedipus* had a mean LAP value three times as great as that seen in *S. fuscicollis* females. Although the differences between the males of each species was not as great, this may be because of the small number of samples obtained. A high level of LAP

Figure 3 (top and bottom). Females which did not show a cycle of leucocyte alkaline phosphatase activity may be anestrus.

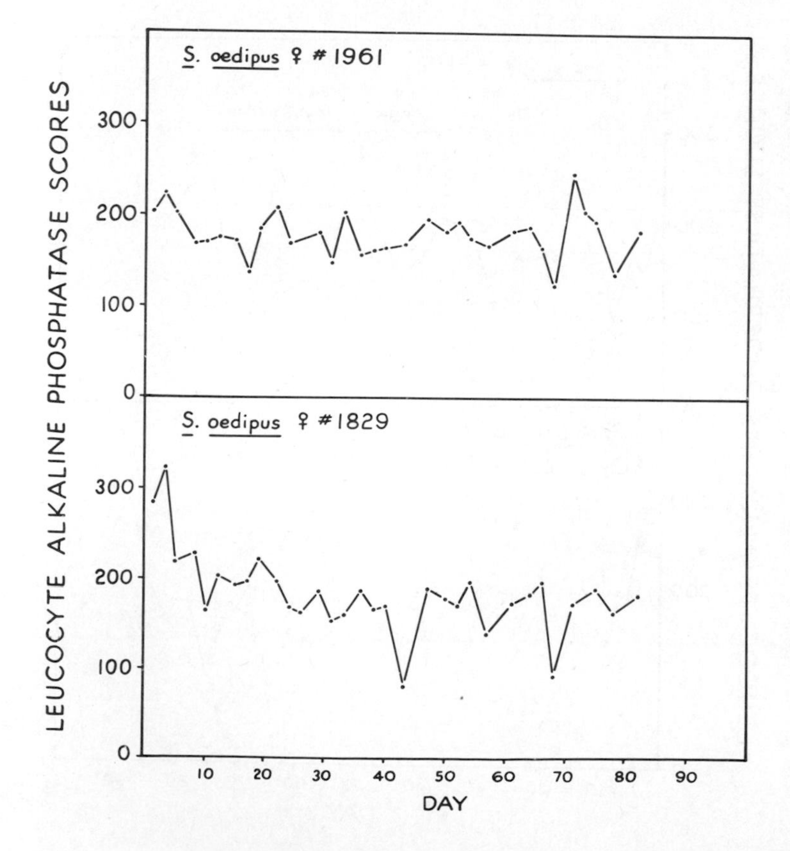

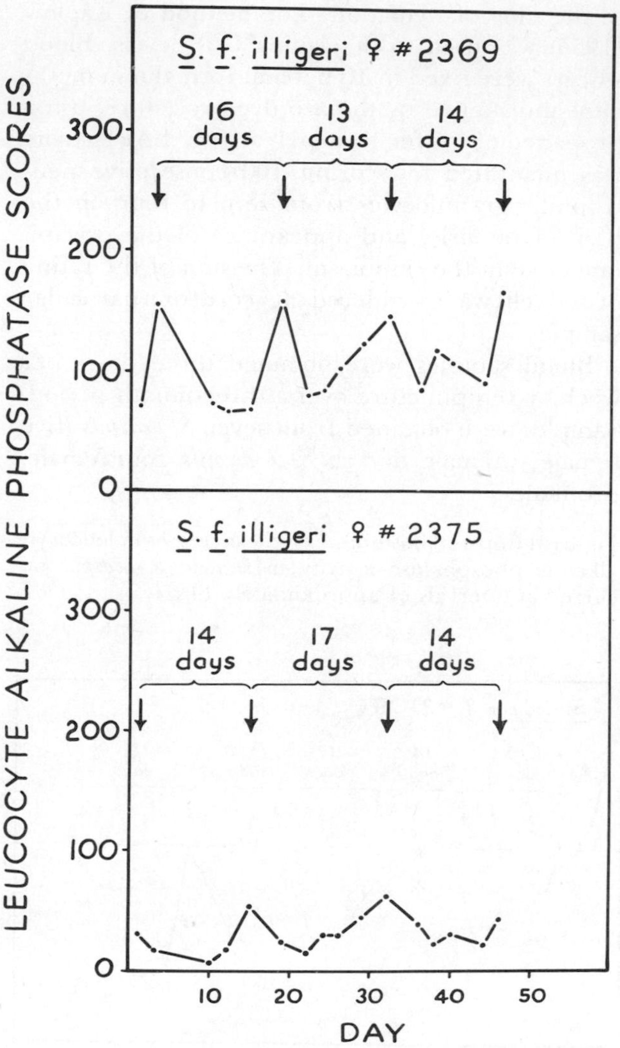

Figure 4. *S. fuscicollis* females show cyclic increases in leucocyte alkaline phosphatase activity at intervals of approximately 14 days.

activity (mean = 138) was consistently seen in one of the four *S. fuscicollis* females (#2369). This could have been caused by a subclinical infection or hematological disorder. The hematocrits of all animals were checked monthly; the hematocrits remained above 40 percent, well within the normal range (Burns et al., 1967).

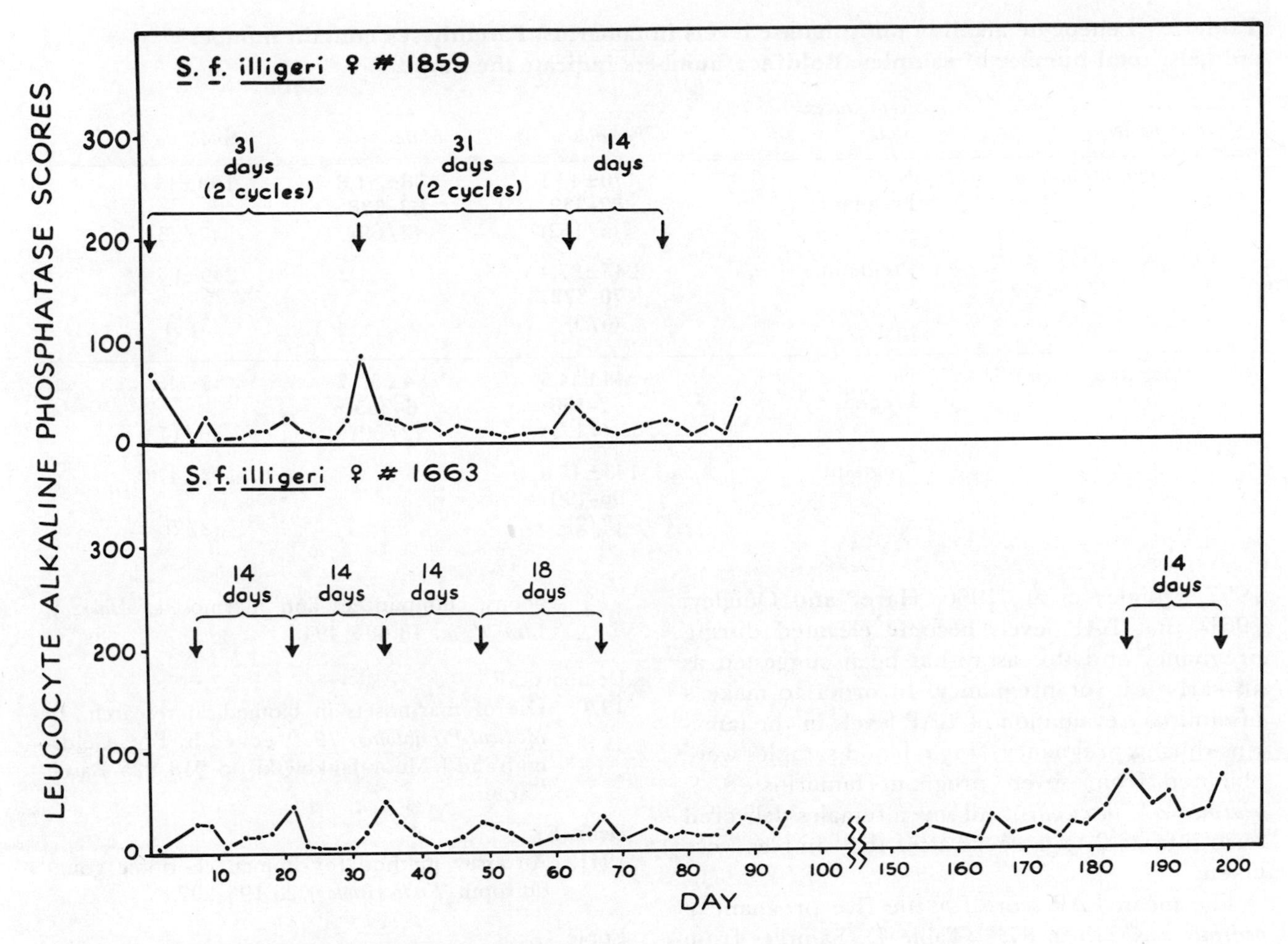

Figure 5. *S. fuscicollis* females show cyclic increases in leucocyte alkaline phosphatase activity at intervals of approximately 14 days.

Pregnancy

The ability to detect pregnancy in marmosets in its early stages is required both for the management of breeding colonies and for studies requiring interruption or manipulation of pregnancies. In 1969, we reported (Hampton et al., 1969) the results of a study in which chorionic gonadotrophin levels were determined for eight female *C. jacchus*. The method used was essentially that of Delfs (1941). Overnight urine collections were treated with cold acetone to precipitate gonadotrophins and the samples bioassayed in immature female rats. Using 20 weeks as a gestation period, excretion of chorionic gonadotrophin began at about two weeks after conception and continued at significant levels until 2 to 3 weeks before delivery. The data indicated that *C. jacchus* excretes a considerable amount of gonadotrophin which reaches its peak at mid-pregnancy. In some cases, gonadotrophin excretion was not correlated with confirmed pregnancy; pregnancy with fetal resorption or undetected abortion may be common. Hearn and Lunn (1975), using a radioimmunoassay for LH, and probably chorionic gonadotrophin, found levels of LH/chorionic gonadotrophin which peaked at 6 weeks gestation in *C. jacchus*. LH/chorionic gonadotrophin levels returned to nonpregnant levels at approximately 17 weeks gestation.

Gengozian et al. (1974) reported the use of uterine palpation to identify stage of pregnancy in *S. fuscicollis*. Uterine palpation has also been used by Hearn and Lunn (1975) to follow pregnancy in *C. jacchus* and they were able to detect pregnancy as early as the third week of gestation and distinguish singleton pregnancies from twin and triplet pregnancies between the 12th and 20th week of gestation.

In the human, it has been reported (Pritchard,

Table 2. Leucocyte alkaline phosphatase levels in tamarins. Parentheses contain number animals/total number of samples. Boldface numbers indicate the range.

Species	*Reproductive State*	*Females*	*Males*	*Total*
Saguinus oedipus	Not Pregnant	170±43.1 **82-339** (5/162)	168±51.8 **63-338** (2/69)	169±45.8 (7/231)
	Pregnant	245±87.4 **170-372** (5/5)		245±87.4 (5/5)
Saguinus fuscicollis illigeri	Not Pregnant	44±34.5 **2-166** (4/121)	84±59.7 **6-265** (2/52)	57±48.1 (6/173)
	Pregnant	143±47.0 **96-190** (2/2)		143±47.0 (2/2)

1957; Quigley et al., 1960; Harer and Quigley, 1961) that LAP levels become elevated during pregnancy and this assay has been suggested as an early test for pregnancy. In order to make a preliminary evaluation of LAP levels in the tamarins during pregnancy, single blood samples were obtained from seven pregnant tamarins (5 *S. oedipus*, 2 *S. fuscicollis*). All seven females delivered term infants 2 to 6 weeks after the samples were taken.

The mean LAP score for the five pregnant *S. oedipus* was 245 ± 87.4 (Table 2). Samples from the two pregnant *S. fuscicollis* gave a mean LAP value of 143 ± 47.0. The mean LAP values for pregnant females of both species were well above that seen in nonpregnant animals, but, because of overlap in the ranges, this test could not be used as confirmation of pregnancy from a single sample. It seems likely that the range of LAP values in the pregnant and nonpregnant state will not overlap in individual animals. We are now continuing this study to determine when, after conception, LAP values become elevated, and to follow levels of LAP activity throughout the course of pregnancy.

Acknowledgments

This investigation was supported in part by NIH research grant number HD08695 from the National Institute of Child Health and Human Development.

Literature Cited

Burns, K.F.; F. G. Ferguson; and S. H. Hampton.
1967. Compendium of normal blood values for baboons, chimpanzees and marmosets. *Amer. J. Clin. Path.*, 48:484–494.

Deinhardt, F.
1970. Use of marmosets in biomedical research. In *Medical Primatology 1970*, edited by E. I. Goldsmith and J. Moor-Jankowski, pp. 918–925. Basel: Karger.

Delfs, E.
1941. An assay method for human chorionic gonadotropin. *Endocrinology* 28:196–202.

Epple, G.
1970. Maintenance, breeding, and development of marmoset monkeys (Callithricidae) in captivity. *Folia Primat.*, 12:57–76.

Gengozian, N.
1969. Marmosets: Their potential in experimental medicine. *Ann. N. Y. Acad. Sci.*, 162:336–362.

Gengozian, N.; T. A. Smith; and D. G. Gosslee.
1974. External uterine palpation to identify stages of pregnancy in the marmoset, *Saguinus fuscicollis* ssp. *J. Med. Primat.*, 3:236–243.

Hampton, J. K., Jr.; S. H. Hampton; and B. T. Landwehr.
1966. Observations on a successful breeding colony of the marmoset, *Oedipomidas oedipus*. *Folia Primat.*, 4:265–287.

Hampton, J. K., Jr.; B. M. Levy; and P. M. Sweet.
1969. Chorionic gonadotrophin excretion during pregnancy in the marmoset, *Callithrix jacchus*. *Endocrinology*, 85:171–174.

Harer, W. G., Jr., and H. J. Quigley.
1961. Alkaline phosphatase activity in granular leucocytes as a test for early pregnancy. *Obstet. Gynecol.*, 17:238–242.

Hearn, J. P., and S. F. Lunn.
1975. The reproductive biology of the marmoset monkey, *Callithrix jacchus*. *Lab. Animal Handb.*, 6:191–202.

Hotchkiss, J.; L. E. Atkinson; and K. Knobil.
1971. Time course of serum estrogen and luteinizing hormone (LH) concentrations during the menstrual cycle of the rhesus monkey. *Endocrinology*, 89:177–183.

Johansson, E. D. B.
1969. Progesterone levels in peripheral plasma during the luteal phase of the normal human menstrual cycle measured by a rapid competitive protein binding technique. *Acta Endocrin.*, 61:592–606.

Kaplow, L. S.
1955. A histochemical procedure for localizing and evaluating leucocyte alkaline phosphatase activity in smears of blood and marrow. *Blood*, 10:1023–1029.

Langford, J. B.
1963. Breeding behavior of *Hapale jacchus* (common marmoset). *S. Afr. J. Sci.*, 59:299–300.

Levy, B. M. and J. Artecona.
1964. The marmoset as an experimental animal in biological research. 1. Care and maintenance. *Lab. Anim. Care*, 14:20–27.

Linder, J. R. and S. Bauminger.
1974. Production and characterisation of antisera to steroid hormones. In *Proceedings of the International Conference on Gonadotrophins and Gonadal Hormones*, edited by V. H. T. James and P. G. Crossignani, pp. 193–227. London and New York: Academic Press.

Lucas, N. C.; E. M. Hume; and H. H. Smith.
1937. On the breeding of the common marmoset (*Hapale jacchus*) in captivity when irradiated with ultra-violet rays. II. A ten years' family history. *Proc. Zool. Soc. Lond.*, 107:205–211.

Murphy, B. E. P.
1967. Some studies of the protein-binding of steroids and their application of the routine micro and ultramicro measurement of various steroids in body fluids by competitive protein-binding assay. *J. Clin. Endocrin.*, 27:973–990.

Neal, P.; T. G. Baker; K. P. McNatty; and R. J. Scaramuzzi.
(in press) Influence of prostaglandins and HCG on progesterone concentration and oocyte maturation in mouse ovarian follicles maintained in organ culture. *J. Endocrin.*,

Polishuk, W. Z.; H. Zuckerman; and Y. Diamant.
1968. Alkaline phosphatase activity in leucocytes during the menstrual cycle. *Fertil. Steril.*, 19:901–909.

Preslock, J. P.; S. H. Hampton; and J. K. Hampton, Jr.
1973. Cyclic variations of serum progestins and immunoreactive estrogens in marmosets. *Endocrinology*, 92:1096–1101.

Pritchard, J. A.
1957. Leucocyte phosphatase activity in pregnancy. *J. Lab. Clin. Med.*, 50:432–436.

Quigley, H. J.; E. A. Dawson; B. H. Hyun; and R. P. Custer.
1960. The activity of alkaline phosphatase in granular leucocytes during pregnancy and the puerperium: a preliminary report. *Amer. J. Clin. Path.*, 33:109–114.

DEVRA G. KLEIMAN
National Zoological Park
Smithsonian Institution
Washington, D.C., 20008

Characteristics of Reproduction and Sociosexual Interactions in Pairs of Lion Tamarins (*Leontopithecus rosalia*) During the Reproductive Cycle

ABSTRACT

Pairs, trios, and family groups of lion tamarins (*Leontopithecus rosalia*) were observed during an 18-month study to determine the species' basic reproductive characteristics. Changes in sociosexual interactions were then correlated with different phases of the behavioral estrous cycle. Results indicate that the lion tamarin is essentially a seasonal breeder, bearing most litters between January and June in the temperate zone, the reverse of the breeding season in Brazil. Gestation is the shortest recorded for a callitrichid, 128 days, and the estrous cycle is variable but averages two to three weeks in length.

Male attention to the female in terms of approaches, sniffing, and allogrooming is correlated with a peak in male sexual behavior. Male scent-marking frequencies, however, do not vary much during the female's estrous cycle. The female exhibits less interest in the male than the reverse through the cycle, but shows peaks in frequencies of sniffing, approaching, and grooming the male approximately two to four days before the male's peak in sexual behavior. Scent-marking frequencies of the female drop precipitously during peak estrus.

The reproductive characteristics and changes in sociosexual interactions during the reproductive cycle in lion tamarins are compared with other New World primates and discussed with respect to the monogamous social structure exhibited by callitrichids.

Introduction

Basic information on reproduction in New World callitrichids is scarce, as are data on all New World primates (Eisenberg, 1977). The determination of gestation, seasonality, and estrous cycle length has been hampered by the fact that many New World primates are not hardy in captivity and have bred poorly. Squirrel monkeys (*Saimiri sciureus*) are the most commonly used New World monkeys, yet considerable controversy and disagreement have resulted from studies of this species' reproduction (Rosenblum, 1968; Travis and Holmes, 1974; Lang, 1967). Moreover, until recently there were no data on the basic characteristics of reproduction in callitrichids because certain methods useful in the determination of reproductive status in Old World monkeys could not be applied successfully to the marmosets and tamarins. For example, there is no menstrual cycle, and no cyclical changes in the external genitalia during the ovarian cycle. Moreover, vaginal smears have not proved useful, and sexual behavior is reported to occur randomly. Only recent studies of hormonal changes in urine and blood have indicated the presence of an ovarian cycle (Preslock et al., 1973; Hampton and Hampton, 1977; Hearn, 1977).

In this study, the determination of the characteristics of the reproductive cycle in the golden lion tamarin (*Leontopithecus rosalia rosalia*) and the associated changes in sociosexual interactions was instigated because of the seriously endangered status of the species, both in Brazil and in captivity. The study was initiated in the hope that reproductive data could be applied to the problems of captive propagation. The status of the captive *Leontopithecus* population has been recently discussed by Kleiman and Jones (1977) and Kleiman (1977a).

In addition to the conservation function of this study, I was interested in determining whether there would be any unusual characteristics in the sociosexual interactions of lion tamarins, since they are thought to be monogamous like other callitrichids (Epple, 1975). Some of these findings have been summarized elsewhere (Kleiman, 1977b). Also, I was concerned with whether a successful pair bond could be identified behaviorally by comparing interactions of breeding pairs and nonbreeders and whether there would be differences in behavior between pairs with bonds of long duration when compared with newly paired animals. Such information would be useful in the captive propagation program, but would also improve our understanding of the development and maintenance of the pair bond in callitrichids.

Materials and Methods

Seven breeding age females were observed in pairs, trios, and family groups between October 1973 and March 1975. The females' reproductive and housing histories are presented in Table 1. New pairs were often established by allowing females to choose between two male partners. Once young were born, family groups were left intact, except that older offspring were removed from the family at 16 to 24 months of age.

The two older breeding females (33692 and 32721-B) were housed on exhibit in the Small Mammal House of the Zoo, while the remaining animals have been maintained in an off-exhibit facility. Small Mammal House facilities included a suite of 3 cages for each family group, measuring 6.1 m × 1.8 m × 2.5 m. In the off-exhibit facility, animals were housed in rooms measuring 4.6 m × 3.7 m × 4.6 m, with adjoining outside porches for use during the summer. The tamarin enclosures contained diagonal and horizontal limbs and branches and at least two nest boxes. Substrates were either soil or wood chips. Temperature was maintained at 70–85°F and, although the humidity varied considerably, it was always greater than 20 percent. All animals were exposed to natural light. The tamarins were fed twice daily with a prepared diet consisting of Science Diet, Marmoset Ration (Hill Packing Co., Riviana Foods), fruits, crickets, hairless mice, vegetables, and shrimp.

Manipulation of the tamarins was kept at a minimum; handling occurred only if an animal had to be treated by the veterinarian or transported to a new cage. Such infrequent manipulations were deemed necessary because of the endangered status of the species and our desire to observe naturalistic behavior in the absence of stress.

The pairs, trios, or family groups were observed daily for 30 minutes, usually for six days per week. Observations were conducted in the morning and early afternoon. Behaviors were recorded on a check sheet. The following patterns were studied carefully in the adult pairs.

1. Approach with contact: frequency recorded.

2. Huddle: when individuals rest together with large parts of the body surface in contact; frequency and duration recorded (Figure 1).

3. Sniff: frequency recorded.

4. Allogroom: involves nibbling and picking at fur of partner; frequency and duration recorded (Figure 2).

Table 1. Captive histories of adult females of *L. rosalia* at the National Zoological Park, October 1973 to March 1975.

Female Zoo Name or No. (Studbook No.)	*Arrival (A) or Birth Date (B)*	*Wild-caught (WB) or Captive-born (CB)*	*Length of Observations (months)*	*Cage Mates*	*Comments*
Evansville (64-C)	A 27 Apr 71	WB	9	1) LA ♂ + M00276 ♂ (2 mos) 2) LA ♂ (7 mos)	Possibly post-reproductive. Last litter born 30 Sept. 1969.
33692 (67-M)	A 16 Jan 69	WB	18	Family group, 30571A ♂ + juveniles	Breeding regularly.
32721-B (68-3b)	B 16 Feb 68	CB	18	Family group, 33691 ♂ + juveniles	Breeding regularly.
M00277 (72-1b)	B 15 Jan 72	CB	18	1) 35163 ♂ (6 mos) 2) M00068 ♂ (12 mos)	First litter born 15 May 1975.
M00320 (72-6b)	B 17 Apr 72	CB	16	1) M00319 ♂ + M00276 ♂ (6 mos) 2) M00276 ♂ (1 mos) 3) Family group, M00276 ♂ + juveniles	First litter born 5 Apr. 1974. Female died 11 Feb. 1975.
M00709 (73-7b)	B 11 June 73	CB	6	1) LA ♂ (2 mos) 2) M00716 ♂ + M00715 ♂ (4 mos)	First litter born 19 May 1975.
M00940 (73-6b)	B 10 June 73	CB	8	1) M00319 ♂ + M00716 ♂ (4 mos) 2) M00319 ♂ (4 mos)	First litter born 23 Apr. 1975.

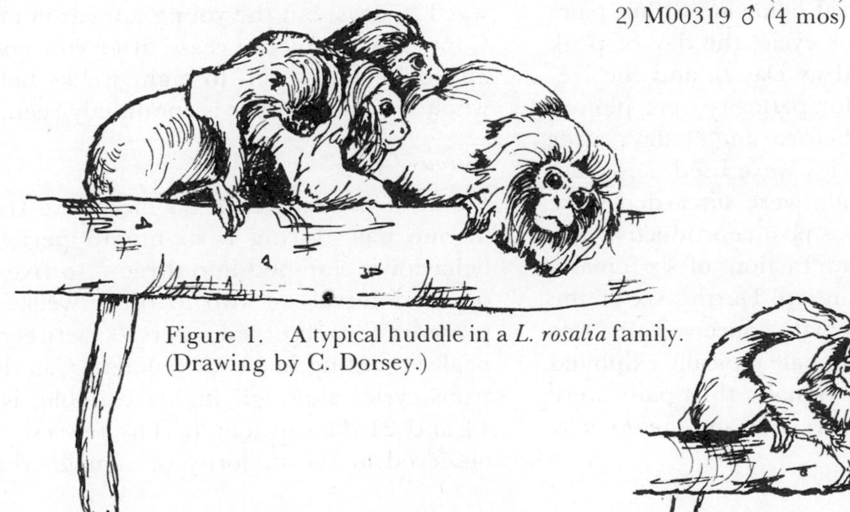

Figure 1. A typical huddle in a *L. rosalia* family. (Drawing by C. Dorsey.)

Figure 3. Scent marking by a *L. rosalia* male. (Drawing by C. Dorsey, reprinted from *Animal Kingdom*, February–March 1976).

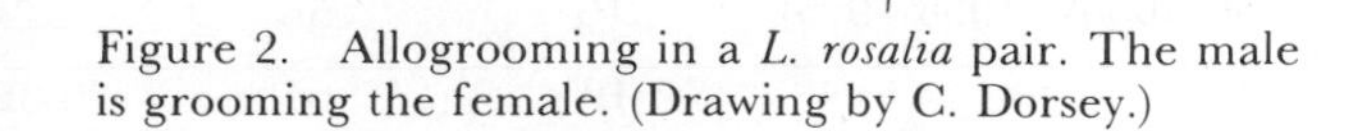

Figure 2. Allogrooming in a *L. rosalia* pair. The male is grooming the female. (Drawing by C. Dorsey.)

5. Scent-mark: total frequency of scent marking recorded including both sternal and perineal drags and rubs (Figure 3).

6. Mounting: mounting is of a typical primate form with the female standing quadrupedally and the male clasping her around the hindquarters; the male's hind legs remain on the substrate. Copulatory behavior includes multiple mounts with intravaginal thrusting. Ejaculation does not occur on every mount. Ejaculation is characterized by a prolonged intromission, a change in the tempo of thrusting, and a subsequent absence of sexual behavior with genital grooming by one or both animals. During ejaculation, the female may turn her head towards the male and/or vocalize. Frequencies and durations of mounts with and without thrusting recorded.

Other behavior patterns, such as wrestling, chasing, arch-walking, food-stealing, and tongue-flicking were also recorded, but will not be analyzed here. For a discussion of callitrichid behavior patterns, see Epple (1967, 1968, 1972, 1975) and Moynihan (1967, 1970, 1976).

For the determination of the estrous cycle, daily frequencies of male mounting were plotted for each female, and the interval between days of peak mounting activity was computed.

To determine how social behavior within pairs altered during the estrous cycle, the day of peak mounting was considered as Day 0, and the frequencies of other behavior patterns were plotted separately for 10 days before and 9 days after Day 0. Only clear-cut cycles were used, and data from the Evansville female were discarded since it was believed that she was post-reproductive (see Table 1). In total, social interactions of six females during 17 cycles were studied. During six of the 17 estrous periods, the female conceived. In trios or family groups, only one male typically exhibited sexual behavior with the female, thus pairs were considered to consist of the female and the sexually active male.

The basic data on reproduction presented here are derived from the International Studbook on the golden lion tamarin first completed by Jones (1973) and revised and updated by Kleiman (1976).

Results

Birth Season, Litter Size, and Gestation Period

Coimbra-Filho and Mittermeier (1973) report a birth peak in *Leontopithecus* in Brazil between September and March. The season is reversed in the north temperate zone with most births occurring between March and August (Table 2). Females which have two litters per year typically give birth first between January and April. There are 23 recorded cases of females bearing two litters in a calendar year as compared with 90 occurrences of single litters during a year. One female has given birth three times during a calendar year once, and another female twice.

Litter size is typically two, with 41 singletons, 116 twins, and 8 triplets being recorded through 1975 (Jones, 1973; Kleiman, 1976).

The gestation period of *Leontopithecus* is 128.6 days (range 125-132; n = 9), based on observed copulations in our colony. One interbirth interval was 126 days, but the young appeared premature. Copulations typically cease after conception until approximately seven to eight weeks before birth when sexual behavior is commonly seen.

Estrous Cycle

Figure 4 presents data on mounting frequencies in one pair during a six-month period. Sexual behavior is clumped into three- to five-day periods at intervals of two to three weeks. Table 3 presents the observed intervals between days of peak mounting in seven females. Clearly, the estrous cycle, although highly variable, is between 14 and 21 days in length. The few estrous cycles observed in the majority of females result from

Table 2. The distribution of births in captive *L. rosalia* in the Northern Hemisphere.*

	Months						
Source of Data	*J-F*	*M-A*	*M-J*	*J-A*	*S-O*	*N-D*	*Total*
National Zoological Park (1966-1975)	6	8	8	2	1	0	25
International Studbook (1958-1975)	10	56	24	27	9	10	136
Totals	16	64	32	29	10	10	

*Data from Jones (1973) and Kleiman (1976).

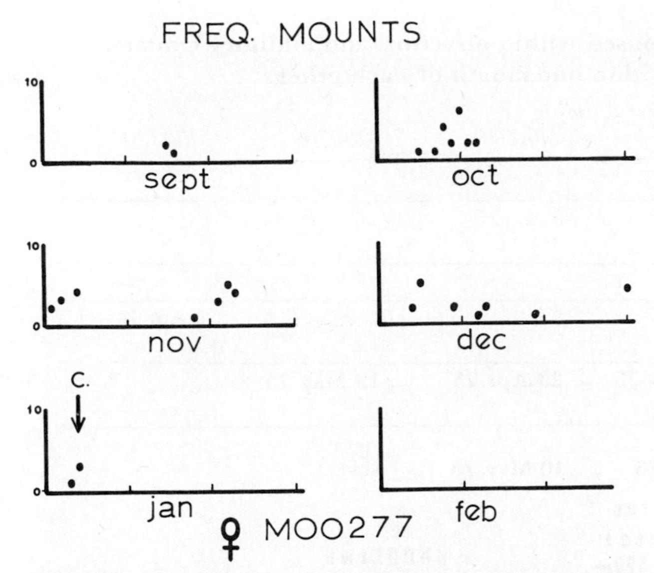

Figure 4. Frequency of mounting of the female (M00277) by the male in a pair of *L. rosalia* during a six-month period. C. indicates conception.

the fact that most females conceive on the first or second estrus at the onset of the breeding season in autumn. Sexual behavior is sporadic during the nonbreeding season in the summer and early autumn.

There was considerable variability between pairs in the frequency of sexual behavior, some rarely exhibiting mounting while others showing it almost on a daily basis. The differences in levels of sexual activity are, in part, a function of pair bond duration; pairs that have been cohabiting for long periods exhibit very infrequent sexual behavior, even when the female is in estrus (Kleiman, 1977b). In at least one case, a female conceived without any observations of copulatory activity. Levels of sexual activity may also be related to whether or not a pair has borne and reared offspring.

There appears to be a tendency for females in olfactory and auditory contact to synchronize their breeding such that they give birth nearly at the same time. Table 4 presents the birth dates for six different females housed in two separate buildings. The two females housed in the Small Mammal House have given birth within a month of each other during three of six years, while three females in the marmoset building all gave birth within one month of each other. Although tentative, these data do suggest that females in close contact stimulate each other reproductively as long as they have separate territories and are housed with an adult male.

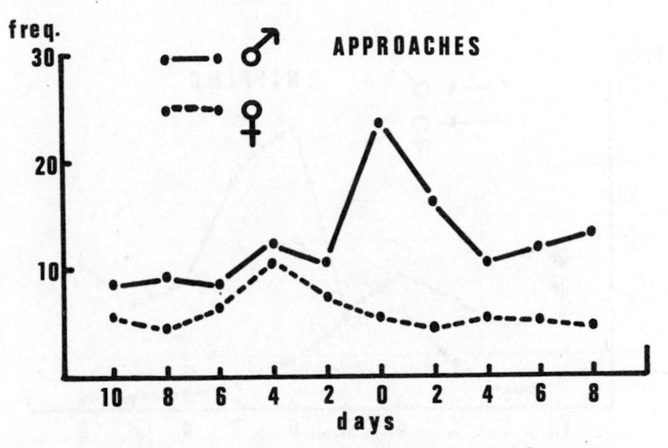

Figure 5. Mean frequency of approaches with contact by pairs of *L. rosalia* before and after the estrous peak in sexual activity (Day 0). The data are derived from 17 cycles in 6 different females. The means of 2-day blocks are presented.

Sociosexual Interactions

Figures 5 through 9 present the mean frequencies of approaches with contact, sniffing, and scent marking, and mean durations of huddling and allogrooming for ten days before and nine days

Table 3. Number of observed intervals between bouts of sexual behavior in 7 *L. rosalia* females, October 1973 to March 1975.

Female Name or Number	*Intervals (days)* <5	*6-9*	*10-13*	*14-17*	*18-21*	*22-25*	*26-29*	*29-60*	*N*
33692	0	0	0	0	1	0	0	1	2
32721-B	0	0	0	0	0	0	0	2	2
M00320	0	0	1	0	1	0	0	0	2
Evansville	0	1	2	4	3	0	0	0	10
M00277	0	2	2	4	5	6	0	1	20
M00940	0	0	0	2	0	0	1	0	3
M00709	0	0	1	1	0	0	0	1	3
Totals	0	3	6	11	10	6	1	5	42

Table 4. Breeding synchrony in *L. rosalia* females housed within olfactory and auditory contact, based on birth dates. Boldface type indicates births within one month of each other.

	Small Mammal House		*Marmoset Building*			
Year	*33692*	*32721-B*	*M00277*	*M00940*	*M00709*	*M00320*
1971	1 Jun 71	19 Mar 71				
1972	17 Apr 72	15 Jan 72 1 Jun 72				
1973	**11 Jun 73**	**16 Jun 73**				
1974	**15 Mar 74** **16 Aug 74**	**11 Feb 74** **4 Sept 74**				**5 Apr 74**
1975	26 Mar 75 30 July 75	30 May 75	**15 May 75**	**23 Apr 75**	**19 May 75**	
1976	**17 Jan 76** **18 Jun 76**	**20 Jan 76** **18 Jun 76**	**3 Jun 76**	**10 May 76**		

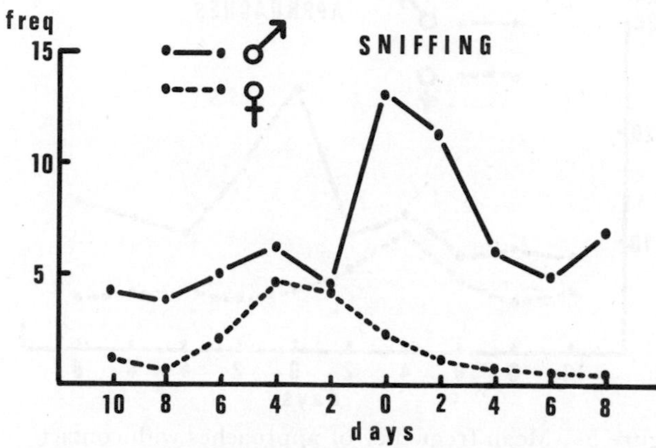

Figure 6. Mean frequency of body sniffing by pairs of *L. rosalia* before and after the estrous peak in sexual activity (Day 0). See legend for Figure 5.

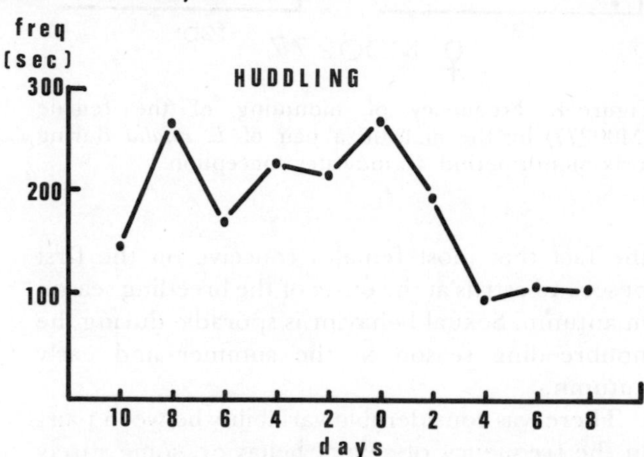

Figure 7. Mean huddling frequency (seconds) by pairs of *L. rosalia* before and after the estrous peak in sexual activity. See legend for Figure 5.

after peak sexual activity in six different females, representing 17 separate estrous periods.

Males approach females with contact more frequently than the reverse and almost double the frequency of approaches at the time of the peak mounting activity. Females approach males rarely but exhibit an increase approximately three to five days before estrus, as defined by male mounting (Figure 5).

Sniffing commonly occurs after an approach, but may also be seen during huddling, grooming, or other contact behaviors. An animal may sniff any section of the partner's body, but males more commonly sniff the rump and anogenital region. As with approaches, the sniffing frequency of the male more than doubles during the three- to four-day period when mounting is highest. Females sniff males infrequently, but the highest female sniffing frequencies occur during the three- to five-day period preceding the male's mounting peak (Figure 6).

Huddling behavior of long duration often occurs during the midday rest period, but also occurs frequently throughout the day. Huddling involves a pair or more of animals resting, either by sitting or lying in close body contact. Grooming may occur during huddling bouts, but the two behavior patterns can be mutually exclusive. In family groups, young may huddle between the adult pair, thus separating them, but all members of a family may be in body contact during huddling (Figure 1).

Huddling frequencies may increase under a variety of conditions, lowered temperatures, illness, or during pregnancy (Kleiman, unpublished). During the estrous cycle, huddling is most common during what is probably the follicular phase, but drops precipitously after the male's mounting peak (or probable ovulation) (Figure 7).

Allogrooming in lion tamarins is typically primate in form. The hands are used to part the fur with both the hands and mouth being used to pick out pieces of detritus (Figure 2). Grooming solicitation usually involves an individual approaching the partner and sprawling horizontally

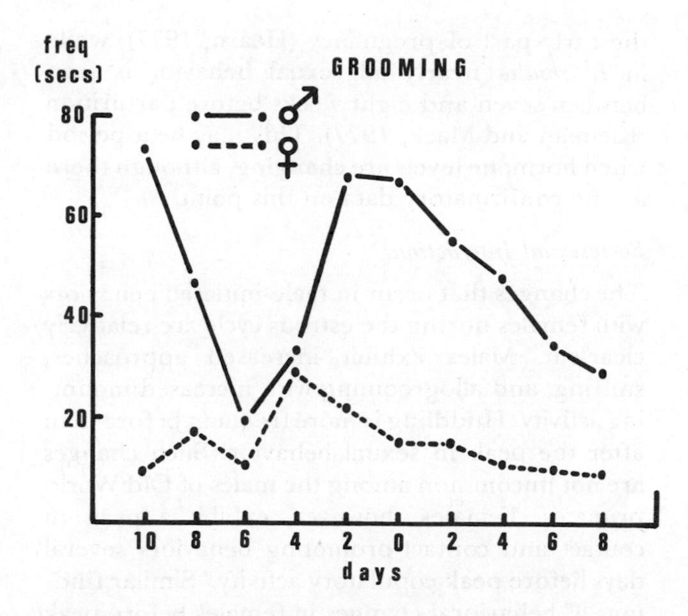

Figure 8. Mean allogrooming frequency (seconds) by pairs of *L. rosalia* before and after the estrous peak in sexual activity. See legend for Figure 5.

on a nest box or branch in either a supine or prostrate position. During huddles, grooming bouts are usually long, but short bursts of allogrooming may occur sporadically.

Males groom females more than the reverse at all stages of the estrous cycle and peak shortly before the period of greatest mounting activity. Females, on the other hand, exhibit their highest allogrooming frequencies three to five days before the male's peak in sexual behavior (Figure 8).

Lion tamarins have two glandular fields, one on the sternum immediately below the throat and the second in the area surrounding the anogenital region (Epple and Lorenz, 1967; Epple, 1972). A variety of scent-marking postures are used depending upon which glands are involved and where the scent mark is placed. The sternal gland is used on horizontal and vertical objects and is typically pressed on the site with the animal moving forward or upward, in a sternal rub (see Epple, 1972) (Figure 3). Sternal gland rubbing may lead into marking with the circumgenital glands (Mack and Kleiman, in prep.), especially when the animal is on a horizontal surface. In this case, the suprapubic area is usually dragged forward by the animal pulling itself forward using the forelimbs, followed by the animal returning itself into a sitting position. This is termed a ventral drag. The anogenital area may be dragged forward for marking or the animal may rub it back and forth or side to side, while in a sitting position. These are usually

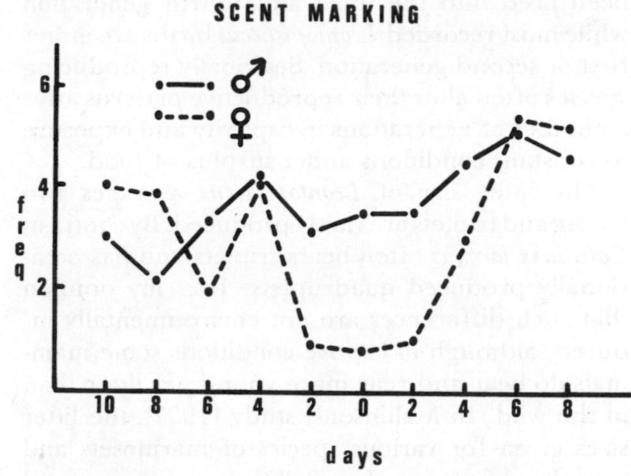

Figure 9. Mean frequency of scent-marking by pairs of *L. rosalia* before and after the estrous peak in sexual activity. See legend for Figure 5.

termed sit-rubs. Urination is common during scent-marking with both the sternal and circumgenital glands.

In the lion tamarin pairs, scent-marking frequencies vary, with some females marking more than males and vice versa. Female scent-marking tends to increase during pregnancy and prior to parturition (Mack and Kleiman, in prep.). During the estrous cycle, males, on average, scent mark more after, than before, the period of copulatory activity, but, in fact, do not vary much. Females exhibit a precipitous drop in scent-marking when mating activity is at its height and reach their highest levels about one week after sexual activity has peaked (Figure 9).

Discussion

Birth Season, Litter Size, and Gestation Period

Both field data and captive observations suggest that *L. rosalia* is highly seasonal in its breeding activity. This contrasts with observations by Epple (pers. comm.), Hearn (1977) and Rothe (1977) that *Saguinus fuscicollis* and *Callithrix jacchus* exhibit little seasonality in reproduction in captivity. The differences may be more apparent than real for two reasons. Most laboratory colonies of *Saguinus* and *Callithrix* are maintained indoors with no access to temperature, humidity, and light changes. *Leontopithecus* in zoos are more typically exposed to the elements, at least for part of the year. Secondly, both *S. fuscicollis* and *C. jacchus* have been bred into the third and fourth generation while most recorded *Leontopithecus* births are either first or second generation. Seasonally reproducing species often alter their reproductive patterns after a number of generations in captivity and exposure to constant conditions and a surplus of food.

The litter size of *Leontopithecus* averages two young and triplets are rarely produced. By contrast, *Callithrix jacchus* often bears triplets and has occasionally produced quadruplets. It is my opinion that such differences are not environmentally induced, although in captive conditions some mammals do bear and rear more young per litter than in the wild. In Mallinson's study (1977), the litter sizes given for various species of marmosets and tamarins do suggest that *Callithrix* spp. on average have larger litters than *Saguinus* spp.

Few gestation periods are known for marmosets and tamarins, but the available data suggest that *Leontopithecus,* the largest callitrichid, has the shortest gestation, 128 days. This is in contrast to many mammalian species where greater size usually results in a longer gestation among closely related species.

Estrous Cycle

The 14- to 21-day cycle in sexual activity found in *Leontopithecus* strongly suggests the presence of an ovarian cycle. These behavioral data are also in close accord with the ovarian cycle seen in *C. jacchus* of 16.4 days (Hearn, 1977) and that seen in *S. oedipus* and *S. fuscicollis* of 15.9 days (Hampton and Hampton, 1977), based on cycles of steroid hormones in the blood. It thus appears as though callitrichids have an estrous cycle of approximately 16 days which is less than the lunar cycle more typical of Old World monkeys (Hafez, 1971), but more than the very short 7- to 13-day cycle of another ceboid, the squirrel monkey (*Saimiri sciureus*) (Travis and Holmes, 1974).

Other investigators of callitrichids have not described cyclical sexual behavior in the species under study and have even suggested that sexual behavior occurs randomly (although Hearn, 1977, indicates that *C. jacchus* may exhibit a higher incidence of mating at mid-cycle). Sexual activity was not random in *Leontopithecus,* but the cycle was only apparent from quantitative differences in mating frequencies. Such differences would not, of course, be apparent without daily standardized observations of animals.

Sexual behavior during pregnancy has been reported for *C. jacchus* (Hearn, 1977; Epple, 1970). In *C. jacchus* copulatory behavior is restricted to the early part of pregnancy (Hearn, 1977), while in *L. rosalia* nearly all sexual behavior is seen between seven and eight weeks before parturition (Kleiman and Mack, 1977). This may be a period when hormone levels are changing, although there are no confirmatory data on this point.

Sociosexual Interactions

The changes that occur in male-initiated behaviors with females during the estrous cycle are relatively clear-cut. Males exhibit increased approaches, sniffing, and allogrooming with increased mounting activity. Huddling is more frequent before than after the peak in sexual behavior. Such changes are not uncommon among the males of Old World primates. Females, however, exhibit a peak in contact and contact-promoting behaviors several days before peak copulatory activity.[1] Similar findings of behavioral changes in females before peak estrus have not been reported for other primates, but may be more common than a review of the literature would suggest. Such behavior clearly indicates that females initiate interactions with males before estrus and thus have an important role in the development and maintenance of a consort relationship or pair bond. The female's role in the choice of a mate has been underplayed in many studies of mammalian reproduction, but recent publications (e.g., Doty, 1974) have begun to shift this emphasis.

The dramatic decrease in female scent-marking behavior during periods of increased sexual activity suggests that females are not advertising their reproductive state when they are most receptive. This may function to reduce the possibility that unattached adult males will interfere with the breeding of the bonded pair, since scent marking is used in inter-group communication (Mack and Kleiman in prep.).

The modal social system of a species is likely to influence the quality and quantity of male-female interactions (Eisenberg, 1967). *L. rosalia* is monogamous and therefore exhibits certain patterns of social interaction which differ from what is commonly described for polygamous primates. Firstly, frequencies of heterosexual contact and contact-promoting behaviors are high. Thus, adult males and females often huddle and groom. Mates are also conscious at most times of each other's location

[1]The females of pairs in which low levels of sexual activity are seen appear to exhibit more contact-promoting behaviors than do the females of pairs which show high frequencies of sexual behavior.

and activity, and many behaviors, such as Long Calling (McLanahan and Green, 1977) and scent marking (Mack and Kleiman, in prep.) are performed simultaneously or in alternation. In a variety of polygamous primate species, descriptions of social behavior suggest that adult males and females interact less with each other than with their own age and sex class or with young (e.g., *Ateles fusciceps,* Eisenberg, 1976; *Macaca mulatta,* Lindburg, 1973; *Presbytis johnii,* Poirier, 1970; *Papio ursinus,* Saayman, 1971; *Pan troglodytes,* Van Lawick-Goodall, 1968). Although one might argue that these differences exist because in polygamous primates living in large troops more partners are available for grooming and huddling, Rothe (1976) reports that the adult breeding pair in a *Callithrix jacchus* family group groom each other more with larger numbers of offspring present, regardless of the age of the young. (Rothe, 1976, has allowed some family groups to increase to 18 members.)

Mason (1974, 1975; Cubicciotti and Mason, 1975) has been examining frequencies of contact behaviors in isolated heterosexual pairs of the monogamous titi monkey (*Callicebus moloch*) and the polygamous squirrel monkey (*Saimiri sciureus*) and finds that titi monkeys spend almost twice as much time in contact as do squirrel monkeys, again supporting the notion that pair-bonding leads to high frequencies of contact.

Another characteristic of pair-bonded lion tamarins is that males groom females more often than females groom males regardless of the stage of the estrous cycle. Similar findings have been reported by Rothe (1976) for *C. jacchus.* In polygamous primates, the direction of most grooming is often the reverse, except at the time of estrus (e.g., *Macaca mulatta,* Lindburg, 1973; Michael et al., 1966). This difference between polygamous and monogamous primates in the direction and overall frequencies of heterosexual grooming appears to hold true for most primate species (see Kleiman, 1977b). Thus, the occurrence of monogamy in a primate species seems to result in a particular pattern and frequency of social interaction which can be contrasted with the behavior of primates exhibiting a nonmonogamous social structure.

Acknowledgments

This research was supported by NIMH R03MH 25242-01 and R01 MH 27241-01. I am grateful to J. Hitchcock, D. Mack, L. Dorsey, and R. Hoage for their aid.

Literature Cited

Coimbra-Filho, A., F. and R. A. Mittermeier.

1973. Distribution and ecology of the genus *Leontopithecus* Lesson 1840 in Brazil. *Primates,* 14:47-66.

Cubicciotti, D. and W. A. Mason.

1975. Comparative studies of social behavior in *Callicebus* and *Saimiri:* male-female emotional attachments. *Behav. Biol.,* 16:185-197.

Doty, R. L.

1974. A cry for the liberation of the female rodent: courtship and copulation in Rodentia. *Psych. Bull.,* 81:159-172.

Eisenberg, J. F.

1967. Comparative studies on the behavior of rodents with special emphasis on the evolution of social behavior, Part I. *Proc. U.S. Nat. Mus.,* 122(3597):1-55.

1976. Communication and social integration in the black spider monkey, *Ateles fusciceps robustus,* and related species. *Smithson. Contrbs. Zool.,* 213:1-108.

1977. Comparative ecology and reproduction of New World monkeys. In *The Biology and Conservation of the Callitrichidae,* edited by Devra G. Kleiman, pp. 13-22. Washington, D.C.; Smithsonian Institution Press.

Epple, G.

1967. Vergleichende Untersuchungen über Sexual und Sozialverhalten der Krallenaffen (Hapalidae). *Folia primat.,* 7:37-65.

1968. Comparative studies on vocalization in marmoset monkeys (Hapalidae). *Folia primat.,* 8:1-40.

1970. Maintenance, breeding, and development of marmoset monkeys (Callithricidae) in captivity. *Folia primat.,* 12:56-76.

1972. Social communication by olfactory signals in marmosets. *Internat. Zoo Yearb.,* 12:36-42.

1975. The behavior of marmoset monkeys (Callithricidae). In *Primate Behavior, Vol. 4,* edited by L. A. Rosenblum, pp. 195-239. New York: Academic Press.

Epple, G. and R. Lorenz.

1967. Vorkommen, Morphologie, und Funktion der Sternaldrüse bei den Platyrrhini. *Folia primat.,* 7:98-126.

Hafez, E. S. E.

1971. Reproductive cycles. In *Comparative Reproduction of Nonhuman Primates,* edited by E. S. E. Hafez, pp. 160-204. Springfield, Illinois: Charles C. Thomas.

Hampton, S. H. and Hampton, J. K.

1977. Detection of reproductive cycles and pregnancy in tamarins (*Saguinus* spp.). In *The Biology and Conservation of the Callitrichidae,* edited by Devra G. Kleiman, pp. 173-179. Washington, D.C.: Smithsonian Institution Press.

Hearn, J. P.
1977. The endocrinology of reproduction in the common marmoset, *Callithrix jacchus.* In *The Biology and Conservation of the Callitrichidae,* edited by Devra G. Kleiman, pp. 163–171. Washington, D.C.: Smithsonian Institution Press.

Jones, M.
1973. *Studbook for the golden lion marmoset,* Leontopithecus rosalia. Wheeling, West Virginia: American Association of Zoological Parks and Aquariums.

Kleiman, D. G.
1976. International Studbook, Golden Lion Tamarin, *Leontopithecus rosalia rosalia.* Washington, D.C.; National Zoological Park.
1977a. Progress and problems in lion tamarin, *Leontopithecus rosalia rosalia,* reproduction. *Internat. Zoo Yearb.,* 17:92–97.
1977b. Monogamy in mammals. *Quart. Rev. Biol.,* 52:39–69

Kleiman, D. G., and M. Jones.
1977. The current status of *Leontopithecus rosalia* in captivity with comments on breeding success at the National Zoological Park. In *The Biology and Conservation of the Callitrichidae,* edited by Devra G. Kleiman, pp. 215–218. Washington, D.C.: Smithsonian Institution Press.

Kleiman, D. G., and D. Mack.
1977. A peak in sexual activity during mid-pregnancy in the golden lion tamarin *Leontopithecus rosalia* (Primates: Callitrichidae). *J. Mammal.,* 58:657–660.

Lang, C. M.
1967. The estrous cycle of the squirrel monkey. *Lab Anim. Care,* 17:442-451.

Lindburg, D. G.
1973. Grooming behavior as a regulator of social interactions in rhesus monkeys. In *Behavioral Regulators of Behavior in Primates,* edited by C. R. Carpenter, pp. 124–148. Lewisburg, Pennsylvania: Bucknell University Press.

Mallinson, J. J. C.
1977. Maintenance of marmosets and tamarins at Jersey Zoölogical Park with special reference to the design of the new marmoset complex. In *The Biology and Conservation of the Callitrichidae,* edited by Devra G. Kleiman, pp. 323-329. Washington, D.C.: Smithsonian Institution Press.

Mason, W. A.
1974. Comparative studies of social behavior in *Callicebus* and *Saimiri*: behavior of male-female pairs. *Folia primat.,* 22:1-8.
1975. Comparative studies of social behavior in *Callicebus* and *Saimiri:* strength and specificity of attraction between male-female cagemates. *Folia primat.,* 23:113-123.

McLanahan, E. B., and K. M. Green.
1977. The vocal repertoire and an analysis of the contexts of vocalizations in *Leontopithecus rosalia.* In *The Biology and Conservation of the Callitrichidae,* edited by Devra G. Kleiman, pp. 251–269. Washington, D.C.: Smithsonian Institution Press.

Michael, R. P.; J. Herbert; and J. Welegalla.
1966. Ovarian hormones and grooming behaviour in the Rhesus monkey (*Macaca mulatta*) under laboratory conditions. *J. Endocr.,* 36:233-279.

Moynihan, M.
1967. Comparative aspects of communication in New World primates. In *Primate Ethology,* edited by D. J. Morris, pp. 236-266. London: Weidenfeld and Nicolson.
1970. Some behavior patterns of Platyrrhine monkeys. II. *Saguinus geoffroyi* and some other tamarins. *Smithson. Contrbs. Zool.,* 28:1-77.
1976. *The New World Primates.* Princeton: Princeton University Press.

Poirier, F. E.
1970. The Nilgiri langur (*Presbytis johnii*) of South India. In *Primate Behavior, Vol. 1,* edited by L. A. Rosenblum, pp. 251-383. New York: Academic Press.

Preslock, J. P., S. H. Hampton, and J. K. Hampton.
1973. Cyclic variations of serum progestins and immunoreactive estrogens in marmosets. *Endocrinology,* 92:1096-1101.

Rosenblum, L. A.
1968. Some aspects of female reproductive physiology in the squirrel monkey. In *The Squirrel Monkey,* edited by L. A. Rosenblum and R. W. Cooper, pp. 147-169. New York: Academic Press.

Rothe, H.
In press. Allogrooming by adult *Callithrix jacchus* in relation to postpartum oestrus. *J. Human Evol.*
1977. Parturition and related behavior in *Callithrix jacchus* (Ceboidea, Callitrichidae). In *The Biology and Conservation of the Callitrichidae,* edited by Devra G. Kleiman, pp. 193–206. Washington, D.C.: Smithsonian Institution Press.

Saayman, G. S.
1971. Grooming behaviour in a troop of free-living chacma baboons (*Papio ursinus*). *Folia primat.,* 16:161-178.

Travis, J. C. and W. N. Holmes.
1974. Some physiological and behavioural changes associated with oestrus and pregnancy in the squirrel monkey (*Saimiri sciureus*). *J. Zool., Lond.,* 174:41-66.

Van Lawick-Goodall, J.
1968. The behaviour of free-living chimpanzees in the Gombe Stream Reserve. *Anim. Behav. Monogr.,* 1:165-311.

CHARLES G. WILSON[1]
Zoological Curator
Oklahoma City Zoo

Gestation and Reproduction in Golden Lion Tamarins

A female golden lion tamarin (*Leontopithecus rosalia*) at the Oklahoma City Zoo produced what is believed to be a species record of 23 offspring in 11 births during 8 years in captivity. Only 9 lived more than one day; all 9 were males. Interbirth intervals for 5 consecutive births by this female, all producing living offspring, indicate a mean maximum gestation of 136.8 days, with a range of 133 to 143. The gestation was previously reported for this species as 134 days by Rabb and Rowell (1960), citing only one interbirth interval, and 132 to 134 days by Ulmer (1961). Jennison (1927) reported a gestation of 140 days although the source of the data is not given; however, a more realistic gestation of 126 to 128 days may be indicated since most mammals require 8 to 10 days for the uterus to recover from the birth and prepare for the next implantation. Immediate postparturient ovulation, fertilization, and implantation would be physiologically unlikely.

Further evidence indicates the 126- to 128-day gestation. At Oklahoma City Zoo, another pair of golden lion tamarins were observed breeding 7, 10, and 13 days after the female had given birth and was raising the twin young. Although the pair were not under continual observation, the observed breedings on these days correlates strongly with the 8- to 10-day recovery period most mammals require. This female gave birth to live twins 130 days after the last observed breeding, adding still further support to a 128-day gestation. Kleiman (1977) reports an average gestation of 128 days from observed copulations.

The oldest female, "Rosita" (Studbook No. 64-D), gave birth to triplets twice in 1972 after she had been in captivity for eight years. She had been

[1]Present address: Overton Park Zoo, Overton Park, Memphis, Tennessee, 38112.

received on June 22, 1964, as a presumed adult and died May 25, 1973. Lucas et al. (1937) indicate that the production of triplets may occur late in a marmoset's reproductive life. Epple (1970) speculates that triplets are caused by the highly nutritious captive diets. She cites body weights on a set of triplet neonatal *Callithrix jacchus* that were approximately equal in development. The fetuses weighed 27.5, 29.3, and 30.0 grams at an estimated one day before delivery. The Oklahoma City Zoo's female that produced triplets never really accepted the artificial marmoset diet (Marmoset Diet, manufactured by Riviana Foods) and subsisted for nearly nine years on crickets, apple, orange, cucumber, carrot, banana, raisins, milk with supplemental vitamins, and boiled egg.

In the second of Rosita's triplet births, the smallest neonate weighed 44.0 grams while the other two weighed 51.5 and 53.3 grams.[2] In the first birth, two of the triplets were apparently developed full term and one appeared markedly premature. The smallest neonate weighed 14.8 grams. Of the other two neonates, one is still living and the other was stillborn, weighing 60.0 grams. The neonate that lived was not weighed, but appeared to be normal in size and development.

The weight differences among the triplets may be explained in two ways: (1) superfetation, and (2) a minimum of three ova being released, fertilized, and implanted, but one being inhibited from developing to full term because of insufficient room or nutrient supply in the uterus. Superfetation, the phenomenon of successive fertilization of two or more ova of different ovulations resulting in the presence of embryos of unlike ages in the same uterus, has not been described in primates, although it is known to occur in a number of other mammals (e.g., cat, *Felis catus;* hare, *Lepus europaeus;* mink, *Mustela vison*).

It is unlikely, however, that this birth would indicate superfetation since most mammals in which superfetation has been observed are induced rather than spontaneous ovulators. Research into tamarin reproduction has indicated that induced ovulation is unlikely (Hampton et al., 1971).

[2] All Oklahoma City Zoo golden lion tamarin body weights are approximates of the actual weights at parturition as the neonates had been preserved in formalin prior to being weighed. Nonetheless, they do indicate proportionate differences.

Literature Cited

Epple, G.
1970. Maintenance, breeding and development of marmoset monkeys (Callithricidae) in captivity. *Folia Primat.,* 12:56-76.

Hampton, J. K., S. H. Hampton and B. M. Levy.
1971. Reproductive physiology and pregnancy in marmosets. *Medical Primatology,* pp. 527-535. Proc. 2nd Conf. Exp. Med. Surg. Primates. Basel: Karger.

Jennison, G.
1927. *Table of gestation periods and number of young.* London: A. and C. Black, Ltd. 8 pp.

Kleiman, D. G.
1977. Characteristics of reproduction and sociosexual interactions in pairs of lion tamarins (*Leontopithecus rosalia*) during the reproductive cycle. In *The Biology and Conservation of the Callitrichidae,* edited by Devra G. Kleiman, pp. 181–190. Washington, D.C.: Smithsonian Institution Press.

Lucas, N. C., E. M. Hume and H. H. Smith
1937. On the breeding of the common marmoset (*Hapale jacchus*) in captivity when irradiated with ultra-violet rays. II. A ten year's family history. *Proc. Zool. Soc. Lond.,* 107:205-211.

Rabb, G. B. and J. E. Rowell.
1960. Notes on reproduction in captive marmosets. *J. Mammal.,* 41:401.

Ulmer, F. A.
1961. Gestation period of the lion marmoset. *J. Mammal.,* 42:253-254.

H. ROTHE
Lehrstuhl für Anthropologie
Universität Göttingen
West Germany

Parturition and Related Behavior in *Callithrix jacchus* (Ceboidea, Callitrichidae)[1]

ABSTRACT

Nine *Callithrix jacchus* females have been observed during 15 deliveries. Beginning with the fourth month of pregnancy considerable abdominal swelling is seen in all pregnant females. This, however, does not allow a definite conclusion as to the stage of pregnancy. The same is true for the changes in form and size of the breasts and nipples. Feeding behavior provides little clue to gestation stage and parturition date. There are no indications of either changes in general activity or in spontaneous interactions of the pregnant females with other group/family members during gestation. The pregnant female leaves the common sleeping box about 5 to 90 minutes preceding the first clearly detectable signs of contractions and withdraws from her mate or group members. Preparation for birth (Phase I) nearly always starts at night (1930 to 2100 hrs). The beginning and the end of the dilatation period (Phase II) are very difficult to determine. The female's behavior during Phase I and Phase II is characterized by restlessness and/or by lethargy and specific body postures. During Phase II contractions occur in nearly regular intervals of 3 to 9 minutes. When expelling the fetus(es) (Phase III), the female assumes either a squatting or an 'arch posture' or sits upright. Expulsive contractions follow one

[1]The author is greatly indebted to G. Ch. Boswell for his help in translating this article into English.

another in intervals of a few seconds to about two minutes. With the exception of one infant, all live born have been presented in vertex position, both occiput anterior and occiput posterior. With two neonates, the mode of presentation was not clear. The interval between the expulsion of the fetuses ranges from 1 to 20 minutes. The placental stage (Phase IV) lasts on the average 15 to 25 minutes. The placenta is expelled by one or three contractions. Placentophagia takes 20 to 25 minutes. Normally, the newborn succeed in clinging unaided to the mother's fur within a few seconds after expulsion. The infants reach the nipples within 2 to 20 minutes. Maternal care for the neonates is confined to more or less intensive but random licking of the infants. Group members show great interest in the delivery process, and they participate in rearing the newborn. Births are spread over all months of the year. Deliveries occur for the most part in the evening (1900 to 2400 hrs). Births consisted of four cases of singletons, 22 of twins, 23 of triplets, and one set of quadruplets. The interbirth-interval usually was 150 to 155 days.

Introduction

Despite the fact that parturition is a fundamental biological process, this aspect of primate biology still lacks sufficient observational and quantified comparative data. This is especially true for South American primates with the possible exception of the squirrel monkey, *Saimiri sciureus*, (see Bowden et al., 1967; Hafner and Woodburne, 1964; Hill, 1962; Hopf, 1967; Takeshita, 1961/62). For most cebids and callitrichids, we have very incomplete information on the various aspects of sexuality and reproduction, e.g. endocrinology, estrous cycle, ovulation, mating and breeding season, and gestation length, to say nothing of parturition and pre- and postpartum behavior. In addition, the few data we have on parturitional behavior in some primate species are based on a few isolated observations (see Brandt and Mitchell, 1971), so that the danger of misinterpretation and unjustified generalizations is ever present.

The observations on parturition in *Callithrix jacchus* in this paper are far from complete, but will hopefully encourage other research in this field. There is a need for more data not only on the delivery process itself, but also on initial social bonds, e.g., early interactions of mother and infants(s) and their possible consequences for the development of the behavior of the neonate both as an individual and as a member of the group.

It is to be hoped that the study of puerperial mother-infant interactions will provide the key to a better understanding of some problems of marmoset social behavior not yet completely understood.

Material and Methods

Nine females of our *Callithrix jacchus* colony have been observed during 15 deliveries. Table 1 summarizes the histories of the breeding females. On the basis of more than 50 births we have been able to determine within ±2 days the parturition date. As a rule, we began our observations 149 to 150 days after the previous birth. Initially, we checked the prospective mother every hour from 0800 hrs in the morning until 1800/1900 hrs in the evening. For the remainder of the night, observation was continuous until 0600/0700 hrs the next morning. As soon as it became apparent that nearly all parturitions occurred at night between 1930 and 2400 hrs, we abandoned daytime checks and started the continuous observations at 1930 hrs in the evening finishing at 2330/2400 hrs if, up to this time, no indication of impending

Table 1. Life histories of breeding *Callithrix jacchus* females.

Female	*Origin*	*Age⁺ / Birth (years)*	*Deliveries in Captivity*		*No. of Singletons (S), Twins (Tw), Triplets (Tr), Quadruplets (Q), Uncertain (uc)*	*Observed Deliveries*
			Full-term	*Premature*		
0004	wb	appr. 3–4	7		7 Tw	
0007	wb	appr. 3–4	1		1 Tw	
0011	wb	appr. 4–5	15	1	2 S, 5 Tw, 8 Tr, 1 uc	4
0012	wb	appr. 7–8	11	4	6 Tw, 6 Tr, 3 uc	3
0014	wb	appr. 2–3	1	1	1 Tw, 1 Tr	1
0015	wb	appr. 2–3	1	1	1 Tw, 1 Q	2
0020	wb	appr. 4–5	1		1 S	1
0032	cb/F_1	Mar. 3, 1970	3		3 Tr	1++
0045	cb/F_1	Jan. 2, 1971	3		1 S, 2 Tr	1
0049	cb/F_1	June 19, 1971	2		2 Tr	1
0060	cb/F_1	Aug. 30, 1972	1		1 Tr	1

cb = captive born
wb = wild born
⁺ = when arrived at laboratory 1968
⁺⁺ = observed partially

birth was to be seen. Seven deliveries have been observed with room lights off or with sparse illumination from an infrared bulb. Street lighting through the windows was sufficient to permit exact observations. With six deliveries, we kept the room lights switched off until the beginning of Phase I (preparation for birth) of the parturition process. As soon as the pregnant female left the common sleeping box, we switched on the lights. Two deliveries occurred in daytime.

Results

Physical and Behavioral Indications of Pregnancy; Stage of Pregnancy

The female's physical appearance alone does not permit a foolproof diagnosis of pregnancy. This is especially true for the first three months of gestation. Beginning with the 13th to 15th week of pregnancy, we observed a considerable abdominal swelling in all pregnant females. Due to the variability in the abdominal swelling not only between different females but also from one gestation to the next in a single animal, we are unable to use abdominal swelling to draw a definite conclusion as to the stage of pregnancy. This result is in accordance with the observations of other workers (Lucas et al., 1927; Marik, 1931; Phillips and Grist, 1975).

The abdominal distension seems to be more pronounced in nulliparous and primiparous females and those who have given birth less than five times than with females who have had five and more deliveries. On the other hand, there is no clear relation between abdominal swelling and the age of the female. Moreover, abdominal enlargement is not always correlated with the number of fetuses. Females pregnant with a single fetus may show more pronounced enlargement than those with twins or triplets.

Regular weekly weighing (three females, six pregnancies) did not provide a reliable measure of postconception period, or even of pregnancy, since body weight may fluctuate by a mean value of 15 g from week to week.

With primiparous and pluriparous females the enlargement of the breasts and nipples is of little value as an indicator of pregnancy and stage of gestation. Most females are lactating for the major part of the subsequent pregnancy (3 to 3.5 months). Therefore the changes in form and size of the breasts and nipples are to a great extent influenced by lactation.

Nulliparous females show slight breast enlargement during the last four to six weeks of pregnancy (see Lucas et al., 1927; Marik, 1931). However, this is not a good indication of impending birth, especially since nonpregnant females and even males may exhibit spontaneous breast enlargement at irregular intervals. It should be noted that some males may even exceed lactating females in this respect. According to Phillips and Grist (1975), a grey secretion can be expressed from the nipples less than one week before parturition and is a good indicator of forthcoming delivery.

Swellings and other changes of the genital region are very seldom observed and, as a rule, appear within one to two days preceding parturition. To some degree the swellings are more intense and occur more frequently, with nulliparous and primiparous females than with pluriparous ones. This is a subjective impression, however, and is not useful for predicting the date of birth. Only one of our females had a conspicuous vulval swelling during the last four days of pregnancy (F 0012, tenth pregnancy). This female gave birth to triplets, one of which was presented in breech position. We have no indication of vaginal outflow during gestation as described for some other marmoset species (Benirschke and Richart, 1963; Christen, 1974).

Some recent papers on marmoset biology describe an external uterine palpation method to determine the course of pregnancy (Gengozian et al., 1974; Phillips and Grist, 1975; Poswillo et al., 1972). At present, transabdominal palpation seems to be the only method providing a reliable measure of stage of pregnancy in marmosets and some other platyrrhines, for instance *Callicebus.*

Two to three weeks preceding parturition, fetal movements can be seen when the female sits in an upright posture or clings to a vertical perch. Such movements are, however, inconspicuous and are not always detectable.

About one month before parturition, pregnant females, especially the primiparae and nulliparae, show slightly increased appetite; however, this increase can only be roughly correlated with the stage of gestation. Fluctuations of appetite are normal in both females and males and occur in the former throughout gestation. During the fifth month of pregnancy, the amount of drinking may be three times the level observed from month one to four. The nulliparous and very old females (for instance F 0012) tend to drink more than the others toward the end of gestation. In short, feeding behavior provides little cue to gestation stage and parturition date.

There were no fundamental changes in the general activity of pregnant females with the exception of two females who during four gestations (F 0011, ninth pregnancy; F 0012 13th, 14th, 15th pregnancy) slept more frequently and for longer periods each day during months three to five of pregnancy than during months one to two (see Grüner and Krause, 1963; Mallinson, 1965; Marik, 1931). During the fifth month of pregnancy, however, we observed in all females a significant increase in the frequency of some body postures and comfort activities, such as upright sitting, leaning against perches, sprawling, lying stretched out on the flanks, and hanging by arms.

There were no indications of either qualitative or quantitative changes in spontaneous interactions of the pregnant females with other family members (e.g. grooming, greeting ceremonies, huddling, initiating contact) during the course of gestation with the exception of some periods when the females are in postconception estrus. Pregnant females living in family groups do not become more aggressive as gestation proceeds. This is true up to the parturition day (see Lucas et al., 1937; Mallinson, 1965). On the other hand, the pregnant α-female of 'artifical groups' shows more aggression toward lower ranking females, and the group may split apart if the breeding pair is not removed.

Family members, especially the α-male and adult sons and daughters, initiate more interactions with the pregnant female a few days preceding parturition than they do in earlier stages of gestation. This increased interest in the α-female is expressed by more frequent active grooming, greeting, huddling, and close contact, as well as following behavior, sniffing, licking, and/or tasting the female's urine marks, sniffing, touching, and licking of her genital region. As a rule, the female tolerates intensive physical contact such as licking and touching of her genital region only by the α-male, but there is a great variability of tolerance among females and in the same female from gestation to gestation. It should be noted that the α-female of small families (up to 8 to 10 members) is more tolerant in this respect than are those of bigger families. In general, we observed the trend that the α-male and sons are rather more tolerated than are adult daughters when they try to lick, sniff, or touch the mother's genitals.

We have paired pregnant females at various stages of gestation (three weeks postconception up to less than one week preceding parturition) with an unfamiliar male. In all cases, the female gave birth at the forecasted date and reared the infant(s) with the help of the male though he was not the father.

Delivery Process

The classification of the delivery process into different phases is far from being standardized in the primatological literature. In the present paper, I follow the classification by Naaktgeboren and Slijper (1970) since it has proved to be the most applicable to the study of parturitional behavior in the marmoset. Other classifications are given by Bo (1971), Bowden et al. (1967), Lindburg and Hazell (1972) and Takeshita (1961/62).

PHASE I: PREPARATION FOR BIRTH

Phase I includes all behaviors of the pregnant female related to birth which are to be seen until contractions begin with Phase II (dilatation period). In primates, Phase I may last from a few minutes up to some hours or even days; however, its beginning and its end, or its transition to the dilatation period, are very hard to determine.

With 12 of 15 observed deliveries, the pregnant female leaves the common sleeping box about 5 to 90 minutes preceding the first clearly detectable signs of labor pains and withdraws from the group members or mate to a distance of between 0.8 m and 5 m depending on cage size. At this time, the abdominal swelling is visibly more caudal than during earlier stages of gestation or even some hours ago (Plate 1a). Pluriparous females and those who are living in families of more than 8 to 10 members withdraw earlier and farther from the others than do nulliparae, primiparae, or those females who live in smaller families or in pairs. It may happen that the female returns to her group members in the nest box, but leaves again some minutes later.

As a rule, the group members or the male take no notice of the withdrawal of the parturient female if the cage room is kept in darkness. We once observed the eldest daughter awaken and follow the mother by arch walking for about 2 m, and then immediately return to the nest box.

Parturition probably never occurs in the sleeping box since we have never found traces of blood in the nest box, but always outside on a perch or board.

Preparation for birth always started in the evening or at night between 1930 and 2100 hrs except for two deliveries which occurred in daytime. The female's behavior in Phase I is characterized by two extremes: some females, especially pluriparous ones, are extraordinarily restless. They walk for some minutes along the perches with no interruption, anal rub, urinate, defecate, scratch, and eat and drink more frequently than usual. Other females, for the most part nulliparae and primiparae, behave almost apathetically. They sit or lie for extended periods, move slowly and cautiously, occasionally scratch and groom, ruffle the coat and doze (Figures 1 and 2). Sometimes, the females give very faint contact calls which are not answered by group members.

Plate 1 a. Abdominal swelling in a *C. jacchus* female two hours preceding expulsion of the fetuses (F 0011, gestation 16).

b. "Defecation posture" of a laboring *C. jacchus* female during Phase II of the delivery process (F 0012, birth 10).

c. "Sphinx-posture" of a laboring *C. jacchus* female during Phase II of the delivery process (F 0011, birth 16).

d. Modified "marking-posture" of a laboring *C. jacchus* female during Phase II of the delivery process (F 0011, birth 16).

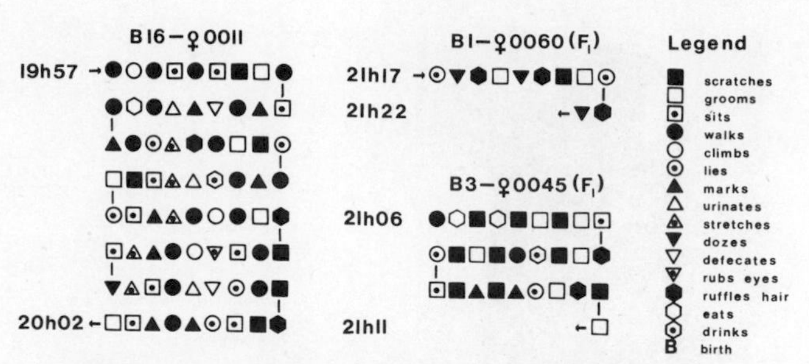

Figure 1. Sequences of behaviors of three *C. jacchus* females during Phase I of the delivery process.

PHASE II: DILATATION PERIOD

This stage extends from the beginning of the first contractions until the onset of the definitive birth contractions (Phase III: Expulsive stage). The exact beginning of Phase II is difficult to determine since the first contractions are weak and may be overlooked during the early dilatation period. In any case, the end of Phase I and the beginning of Phase II overlap. The same is true for the transition from Phase II to Phase III. In order to get a useful measure, we take the beginning of Phase II from the moment the first contractions are clearly detected. This does not exclude the possibility, however, that labor and consequently Phase II may have started earlier.

The body postures and behaviors of the laboring female in Phase II are very characteristic. These include: freezing of locomotion, lifting of the abdomen and tail, adduction of the flanks and thorax, ruffling of the hair, heavy breathing, closing of the eyes, and opening of the mouth. Some females have vaginal outflow which moistens the proximoventral part of the tail (Plate 1b). A moistened tail base combined with the above mentioned body postures and behaviors is a very good indicator of an immediately impending expulsion of the fetus(es), within 2.5 hours at the latest. The end of Phase II is characterized by contractions following each other at shorter intervals and which are increasingly more convulsive; further, by the rupture of the amnion sac, and again by characteristic postures and behaviors.

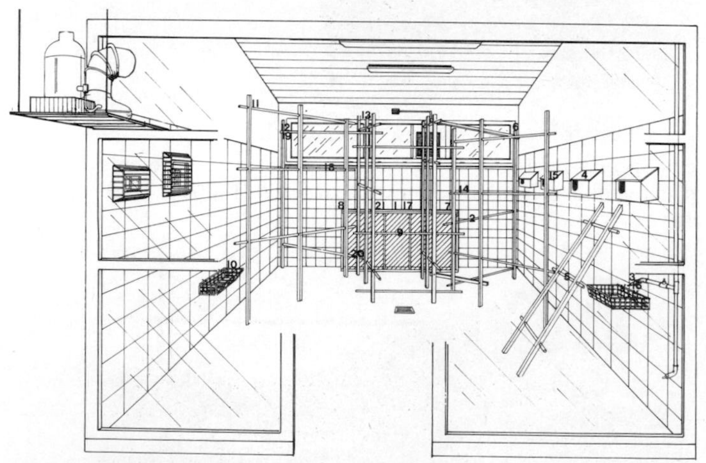

Figure 2. Drawing of the cage of a *C. jacchus* family. The figures 1 to 21 mark the places of the cage the laboring female has been during a five-minute period during Phase I of the delivery process (F 0011, birth 16).

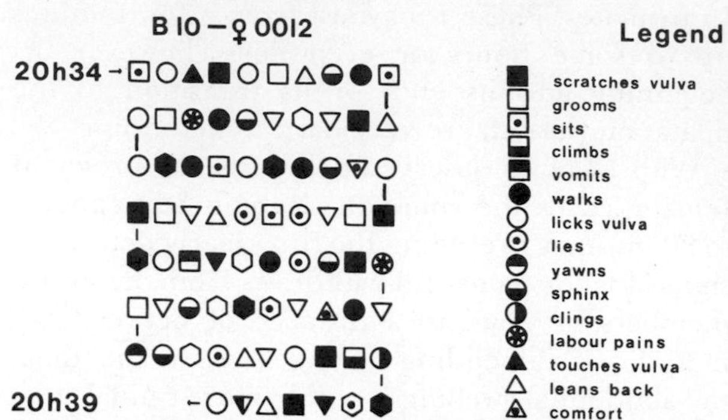

Figure 3. Sequence of behaviors of a *C. jacchus* female during Phase II of the delivery process (F 0012, birth 10).

During Phase II, the female appears extremely restless (Figure 3). She sits or lies only for short periods (some seconds) in "sphinx-posture" (Plate 1c) or "sliding-posture" and when laboring, in "defecation-posture" (see Plate 1b). She licks, touches, and very often scratches the genital and vulval regions, licks her hands, feet, and tail, rubs her eyes, and slides or even creeps in a modified "marking posture" (Plate 1d) 1 m to 1.5 m along a perch or shelf.

With the continuation of dilatation, the female often lies stretched out on her belly, sits in an upright posture on her hind legs and tail, leans back against a perch or the wall of the cage, or clings to a perch with ruffled hairs and tufts, sometimes closing the eyes and opening the mouth.

During dilatation, contractions follow each other in nearly regular intervals of about 3 to 9 minutes and for a mean duration of 2 seconds. It is interesting to note that the time span between contractions varies among the females; however, they are very constant in duration for each female during successive deliveries.

Duration of Phase II may vary from less than 5 minutes to more than 90 minutes. We could not observe any correlation between length of dilatation period and either the age of the females, number of previous births, or number of fetuses born. With the beginning of expulsive contractions, i.e., the end of Phase II/beginning of Phase III, the female remains in one place which is sometimes a small perch and sometimes a shelf. This varies from female to female and from birth to birth.

PHASE III: EXPULSIVE STAGE

Phase III begins with the onset of expulsive contractions and ends with the complete expulsion of all fetuses. The expulsion of the fetus(es) starts with strong contractions of 10 to 20 seconds duration each and following each other at intervals of a few seconds to 2 minutes. The intervals between contractions are variable and extend from between 0.5 to 2.5 minutes. The parturient female assumes either a squatting- , an "arch-posture" or sits upright and grasps a vertical perch with her hands and leans back. During labor the females utter groans, adduct their flanks, breathe rapidly and heavily, ruffle their coat, tremble, open their mouths, usually close their eyes, and sometimes drag the body along the support in prone position.

As a rule, the amnion sac tears after one to two expulsive pains. The female licks the amnion fluid from her legs, tail, hands, and from the birthplace. Between contractions, the female licks, touches, and very often scratches her genital and vulval regions, legs and tail—in general exhibits an increased cleaning activity.

The duration of expulsion of the first fetus ranges from one to 25 minutes. The number of expulsive contractions for the first fetus ranges from one to more than 20. The expulsion of the first fetus takes more time and more contractions than that of the following infants(s). Normally, the interval between expulsions of the fetuses in multiple births is no more than 4 to 7 minutes; however, it ranges from less than one to more than 20 minutes.

Fetus two and three are expelled with a mean of eight contractions each, again with a range of one to 20. As soon as the head of the infant has appeared (see Plate 3), one more uterine contraction suffices to complete expulsion. The duration of expulsive stage of all fetuses sums up to about 20 to 25 minutes. There are no definite relations between duration of expulsion and either the age of the mother, the number of previous deliveries, or the number of the fetuses. During the intervals between expulsions, the mother thoroughly cleans the birthplace as well as her legs, tail, and genital region. With the exception of one infant (see Plate 2a) all live born have been presented in vertex position, occiput anterior and occiput posterior. With two neonates the mode of presentation was not clear (possibly face presentation). The umbilical

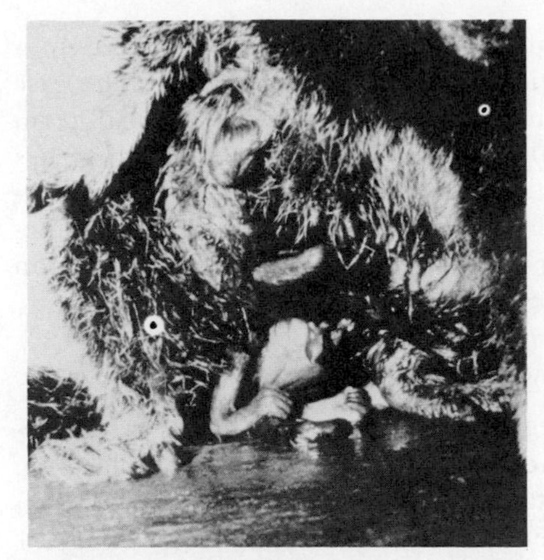

Plate 2. a. Breech presentation of a *C. jacchus* neonate (F 0012, birth 10).

b. *C. jacchus* female 30 s after expulsion of the placenta. The neonates are still connected to the placenta by the umbilical cord (F 0012, birth 9).

c. Placentophagia. The female pulls loose shreds of the placenta by her teeth (F 0012, birth 9) (from Rothe, 1973).

cords are ignored by the mother until after expulsion and eating of the placenta except in one case (F 0049, birth two) (Plate 2b). With full-term deliveries, discharge of blood regularly occurs during expulsion of the placenta. Bleeding preceded placental discharge in only two deliveries. Females never used their hands to assist expulsion of the fetuses.

PHASE IV: PLACENTAL STAGE AND PLACENTOPHAGIA

Phase IV begins when all fetuses have been delivered and ends with the expulsion of the placenta. The placenta appears on the average 15 to 25 minutes after delivery of the last fetus (range, 1 to 45 minutes). It is expelled by one to three contractions following each other at intervals of a few seconds.

After a short time (less than one minute), during which the female licks and cleans her genital and vulval regions, thighs, feet, and tail, she starts to eat the placenta pulling loose small shreds with her teeth (Plate 2c). She eats the umbilical cord starting with the placental end. Only a short piece (0.5 cm to 1.5 cm) of the umbilical cord is left close to the infant's body. Placentophagia takes about 20 to 25 minutes. Subsequently, the female thoroughly cleans the birthplace, and her hands, legs, and mouth. Placentophagia seems to be a regular element of the parturitional behavior of the common marmoset.

Behavior of the Newborn and Mother-Infant Interactions

In some earlier publications (Rothe, 1973, 1974, 1975) the author has referred to the fact that newborn *Callithrix jacchus* get little maternal care immediately following delivery and, further, that the chance for survival depends largely on the infant. Additional observations of the parturitional behavior of the common marmoset have confirmed this result.

The most critical time for the newborn is immediately after expulsion when they must cling unaided to their mother within a few seconds (see Plate 3). As soon as the head and arms are free

a.

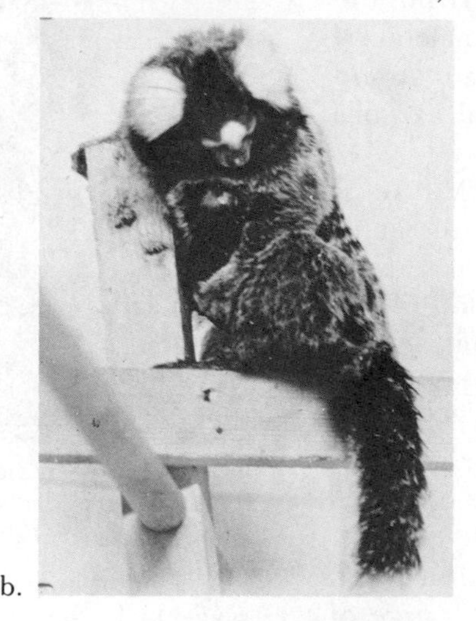
b.

c.

d.

e.

a and b. The head of the infant is expelled. Note the opened eyes of the infant.

c. 23 seconds later, the upper part of the infant's body has appeared; the infant grasps the mother's fur with its right hand.

d. 16 seconds later, the infant clings to its mother, the lower part of the infant's body is not yet expelled.

e. 3.5 minutes later, the infant is completely expelled, and clings to its mother's left thigh (F 0045, birth 3).

Plate 3. Expulsion of a *C. jacchus* fetus (second born of a set of triplets).

of the birth canal the infant tries with rowing, rotating, and grasping arm movements and pendulous movements of the upper part of its body to grasp the mother's fur, to cling, and to climb on her back and belly. This may occur even before the infant has been completely expelled (see Plate 3). As a rule the infant succeeds within some seconds or after 0.5 minutes at the latest in grasping the fur of its mother. This success is independent of the mode of presentation, provided that the infant has been born in good physical health.

The body posture of the mother during birth and the birthplace (i.e. whether perch or board) has some influence on the part of the mother's body to which the infant clings first. If the mother gives birth on a perch and expels nearly the entire body of the infant before the latter has been able to cling, the baby may not succeed in climbing on its mother, but dangles beneath her connected by the umbilical cord. Such situations occur very rarely and are quickly mastered by the baby, provided the mother does not move at the moment of expulsion. The infant always orients itself head forward on the mother's body.

Within 2 to 20 minutes after the infant has a firm hold on its mother, it reaches the nipples unaided using pendulous breast seeking movements while still connected to the placenta by the umbilical cord. The time needed to reach the nipples depends on where on the mother's body the infant first clung and on the amount of maternal care it has received (i.e., how often it has been licked and cleaned by the female).

The tail of the newborn has no stabilizing function within the first minutes following expulsion. It lies rather limply on the mother's fur or hangs loosely. Afterward, the distal third is curled up and pressed against the mother's body.

Shortly after delivery or as soon as the female has licked the newborn's face, the neonate opens its eyes (see Plate 3). Healthy neonates which are able to cling to their mother seldom vocalize during the first minutes of their extrauterine life.

Maternal care for the newborn immediately following expulsion is limited, being confined to two to nine lickings and cleanings of the face, head, and shoulder region, which take no more than a few seconds. The mother does not actively help the baby grasp her fur, nor does she grasp the baby even if it is dangling beneath her connected by the umbilical cord. The female does not support the infant as it attempts to reach the nipples. At best she may abduct her arms to facilitate access. Sometimes she supports the baby with one or both hands while licking it; however, in many cases the mother does not even touch the infant.

The extent of maternal care depends essentially on whether the neonate climbs via the belly or the back to the nipples. Infants who climb via the back are at a decisive disadvantage and get almost no maternal care save one to two lickings upon reaching the nipples. Our observations indicate that the firstborn infant receives more attention from the mother than the following one(s), provided it does not cling to its mother's back.

Infants born in poor physical condition receive more pronounced care from their mother only if they move and especially if they vocalize (distress and contact calls). The mother's behavior, however, depends essentially on that of the infant's. If the neonate does not vocalize or if it lies motionless beside or under its mother, she shows no further interest.

In summary, if the newborn *Callithrix jacchus* succeeds in clinging unaided to its mother immediately after expulsion, it has a fair chance of survival. Infants who fail to cling have no chance or at best a very reduced one.

First Hours of Puerperium

A few minutes following the eating of the placenta and the thorough cleaning of the birthplace, the mother repeatedly drinks rather eagerly, licks and cleans her genital region, feet, tail, and hands, sprawls sometimes, and falls asleep within 30 to 45 minutes. The female does not return to her group members, but spends the birth night sleeping alone with her newborn. Normally, the female is rather tired following delivery, but by no means lethargic, provided she has been in good health during gestation and the delivery was normal.

During the first night postpartum, the newborn rest more or less motionless at the mother's nipples. In the case of triplets, one infant does not get access to the nipples for the first hours of its extrauterine life. From time to time during the night it tries to reach a nipple, but since both are occupied by its sibs it climbs back to its mother's neck. The mother does not react to the infants' movements on her neck, shoulder, or breasts.

Primiparae tend to behave more cautiously and passively and move more slowly when carrying their infants the first morning postpartum. Furthermore, they carry their babies for longer periods than do pluriparae before trying to get rid of them.

All group members participate in rearing the newborn, but with varying intensity, frequency,

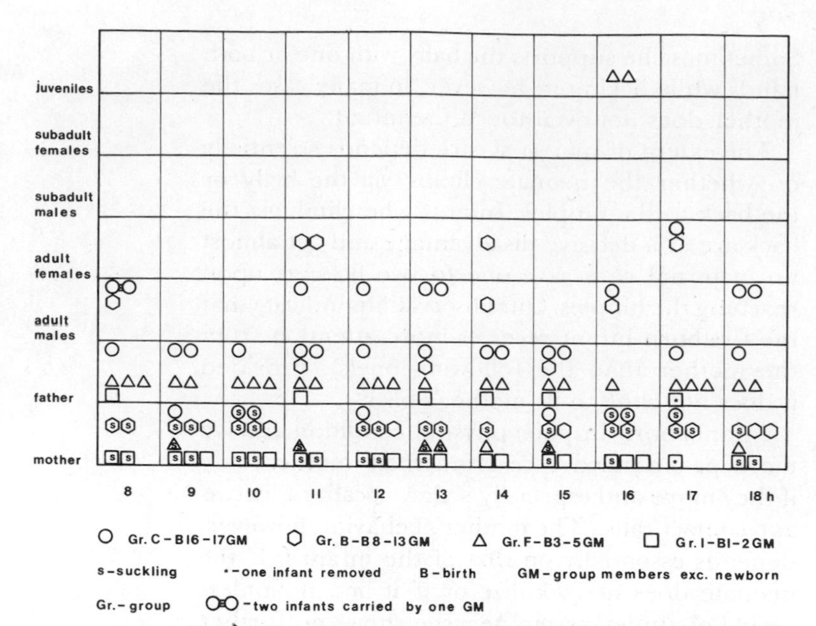

Figure 4. Group members carrying the newborn (○ ⬡ △ □) at the first day postpartum. Hourly checks.

and duration (see Figures 4 and 5). During the first week of life and even later, the newborn show pendulous breast-seeking movements when being carried by their father and elder brothers and sisters. These group members even tolerate nipple contact for considerable time without trying to get rid of the infants. This behavior is only seen in groups of about twelve or more individuals.

We could not ascertain regular suckling intervals. Suckling frequency and duration varies among different females and from litter to litter in a given female (see Figure 4 and 5). Some mothers, especially those who have given birth less than three to four times, suckle their infants nearly the whole day, or at least at very short intervals for the first days following parturition whereas others do so less than four to five times a day. The newborn are carried by group members other than the mother during the night starting with the second night postpartum.

The larger the group (about 18 to 20 animals), the more problematic becomes the rearing of the infants, e.g., suckling frequency decreases in comparison to smaller groups. In large groups, the interval between nursing may exceed five hours; therefore, the infants become weak rapidly and would die within three days postpartum if not removed from the group and artificially reared. However, these observations must be confirmed by future results.

Group Members Attending Parturition

Some deliveries have been observed with artificial lighting with group members present to study reactions to the newborn and interactions between the mother and the attending group members.

Starting with Phase I (Phase II at the latest) the α-male and the adult sons and daughters show increased interest in the α-female. They exhibit more frequent and, to some extent, more intense grooming, greeting, huddling, and contact behavior, and sniff and lick more than usual the marking spots of the α-female. They also try to lick, sniff, and touch her genital region which she does not

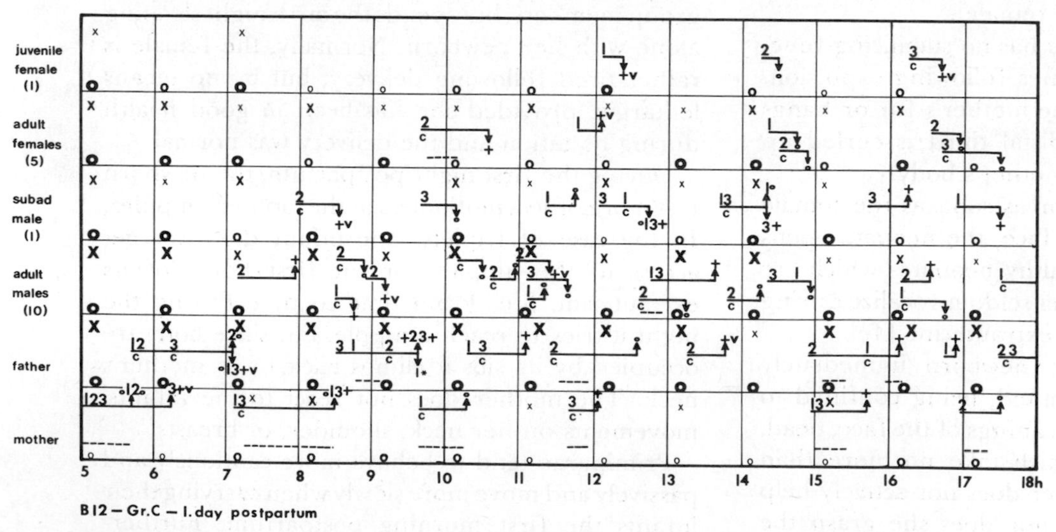

B12 – Gr.C – 1. day postpartum

—— duration of carrying newborn ----- nipple contact 123 - nr. of triplet ↑ - GM tries to dislodge newb.

↟ - GM takes newb. ↑ - newb. changes GM spotaneously c - newb. is cleaned v - newb. vocalizes(distress)

Ooo - frequency of infant-directed behavior (grooming, licking, manipulating, sniffing, touching) per GM, o - 1 to 5, o - 5 to 10, O - >10 X X x - GM tries to take newb., x - 1 to 3, X - 4 to 6, X - >6 per GM (group member) B - birth

Figure 5. Frequency and duration of carrying and suckling the neonates by group members at the first day postpartum. Frequency of interactions.

always allow, especially in the case of her daughters.

The parturient female is surrounded by nearly all group members who try to investigate the infants as soon as they have been born (Plate 4a and 4b). Altogether, the spectrum of behaviors shown by the group members that is initiated by the appearance of the infants is manifold and includes curiosity, manipulation, vocalization, grooming, licking, touching, play, and, in the case of the youngest family members (five months old) a kind of aggressive fear. All things considered, the behavior of the attending group members is friendly and by no means aggressive; however, during the excitement of the birth process the somewhat rough treatment (e.g., chewing of the infant's tail or extremities) by some group members may endanger the neonate(s).

The behavior of group members toward the infants varies both quantitatively and qualitatively from group to group and from litter to litter. The behavior of the α-male ranges from extreme indifference to well-developed paternal care. There is a distinctive trend, however, toward more care for the newborn by both the α-male and the elder sons and daughters than by the α-female.

The mother at first tries to fend off group members (Plate 4c), but later she usually tolerates the neonates being investigated and cared for by family members. On only one occasion the α-male succeeded in taking an infant from the mother before parturition had ended. We did not observe active help by family members during delivery as described by Langford (1963).

The newborn are in danger of being injured by group members especially at the moment when the placenta is expelled, because the family members try eagerly to grasp and consume the placenta or parts of it. As a rule, the group members succeed within a few minutes in getting the placenta (Plate 4d) and the umbilical cord separates close to the infant's body. Sometimes the newborn even loses hold of its mother and is dragged across the cage; it is then in great danger of being eaten, at least partially (head or tail).

The female interferes only if the newborn vocalizes. Otherwise, she remains indifferent.

Nonviable infants are also in danger of being eaten by group members, especially if they are unable to move or vocalize or if they assume unnatural postures.

Following consumption of the placenta and cleaning of blood and amnion fluid from the birthplace and perches, the group members return to their sleeping boxes leaving the mother alone.

Plate 4. Group members attending parturition.

a. The father (on the left) investigates the first-born infant. The mother (on the right) is laboring (F 0012, M 0013, birth 10).

b. Sub-adult male (on the right) sniffs second-born neonate (F 0045, M0078, birth 3).

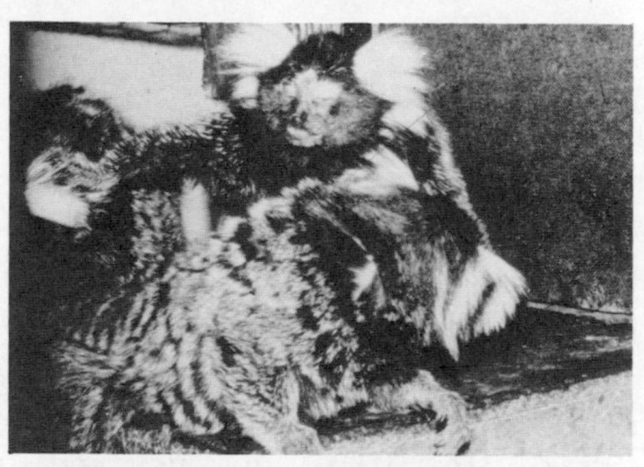

c. Laboring *C. jacchus* female fends off obtrusive family members (F 0012, birth 9).

d. *C. jacchus* female eats the placenta (F 0032, daughter of F 0012, birth 10).

In two parturitions which we observed by light, the father stayed with the mother till the next morning but showed no reaction to the newborn during the night.

Month and Hour of Birth, Number of Fetuses, Sex Ratio, and Interbirth-Interval

Births are spread over all months of the year (Table 2). Fifty-one of 53 deliveries took place in the evening and during the night. Thirty-two were completed before 2230 hrs, 13 before 2400 hrs, 1 before 0300 hrs. None began before 1900/2000 hrs. Only 2 deliveries occurred in daytime, one between 0600 and 0700 hrs, in the morning. The other was an incomplete and complicated birth, during which the uterus ruptured, that lasted more than 26 hours. With 5 deliveries, the hour is uncertain. We could not observe any correlation between age of the female, number of previous deliveries, and hour of birth.

Litters included 4 singletons, 22 twins, 23 triplets and 1 set of quadruplets. Three deliveries were miscarriages, and the number of fetuses could not be determined (Table 3).

The sex ratio is slightly in favor of males; however, there are significant differences in the sex ratio of our groups. The number of stillborn males and the death rate of males of triplet deliveries within the first week postpartum equals that of the females.

The interval between two full-term deliveries has been 153 days, eleven times; 151 days, six times; 152 and 155 days, four times; 154 days, three times; 150 days, two times; and 156, 160, 165, 184, 187, 267 days, one time.

Table 2. Month of year of full-term births in *Callithrix jacchus*.

	Month											
Female	*J*	*F*	*M*	*A*	*M*	*J*	*J*	*A*	*S*	*O*	*N*	*D*
0004		1		1	1		1⁺		1	1		1
0007							1⁺					
0011	1	1	2	1	2⁺	1	1	1	1	2	1	1
0012	1	1	2			1	1	3		1		1⁺
0014									1⁺			
0015						1⁺						
0020							1⁺					
0032			1					2⁺				
0045			1		1⁺					1		
0049		1⁺					1					
0060				1⁺								
total	2	4	6	3	4	3	6	6	3	5	1	3

⁺ = first full-term delivery in captivity

Conclusions

To date, behavioral data do not permit the diagnosis of pregnancy or stage of pregnancy in the common marmoset; however, the palpation method developed by Poswillo et al. (1972) and improved

Table 3. Sexes of singletons, twins, triplets, and quadruplets of *Callithrix jacchus*.

	Sex of Litter										
Female	*M*	*MM*	*MMM*	*F*	*FF*	*FFF*	*MF*	*MMF*	*MMMF*	*MFF*	*uc*
0004		1			3		3				
0007					1						
0011	1	3		1		1	2	4		3	1
0012			1		2		4	1		3	4
0014		1								1	
0015		1							1		
0020				1							
0032								3			
0045	1							1		1	
0049								1		1	
0060								1			
Total	2	6	1	2	6	1	9	11	1	9	5

uc = uncertain

by Gengozian et al. (1974) and Phillips and Grist (1975) does seem to provide exact data. Unfortunately, frequent catching of pregnant females living in family groups may cause disturbances which result in unpredicticable behavior that may be difficult to control. As a result, the group structure may be altered or the group split into segments.

In comparison to the difficulties in diagnosing pregnancy and stage of pregnancy, the recognition of the onset of the delivery process is relatively easy. Characteristic behaviors and body postures of the pregnant female provide the cue. Precise determination of the phases of parturition described herein, especially the beginning and the end of Phase II and the transition of Phase II to Phase III, is more difficult.

One striking aspect of the parturition of the common marmoset is the astonishingly abrupt onset of the delivery process and the relatively short duration of the complete parturition, when compared with other primates for which data are available. Until the beginning of Phase I, the behavior of the pregnant female is not very distinguishable from her behavior during earlier stages of pregnancy or the behavior of nonpregnant females.

According to our observations, the female's withdrawal from her group members when parturition begins and the subsequent occurrence of birth in the absence of family members seem to be regular behavioral elements of the parturient female. This behavior combined with the fact that parturition nearly always occurs in the evening or at night, may have at least two functions: (1) protection against obtrusive group members; and (2) the possibility of recovering from the strains of parturition alone until the next morning.

In the author's opinion, the most fascinating aspect of delivery is the need for the newborn to cling unaided to the mother and the near lack of maternal care. The nature of maternal care in the common marmoset gives rise to various hypotheses. (1) The exceptionally prolific reproduction of marmosets may compensate for the potential loss of one or more newborn immediately following their expulsion due to lack of pronounced maternal care. (2) The sparse initial contact (quantitatively and qualitatively) and interactions of the mother with her infants during and immediately after expulsion provides the minimum of such essential behaviors as carrying, feeding, and cleaning of the infants, and it may not be sufficient to establish very close mother-infant bonds. This lack, however, is then compensated for by the involvement in rearing by group members other than the mother which may even include non-nutritive nipple contact and sucking of the father and elder brothers and sisters. (3) In combination with hypothesis two, the rearing of the newborn by all family members has a positive effect on socialization of the infants, and on stability and growth of the group.

The above hypotheses are highly speculative and other explanations may have the same or even a higher probability of being correct; however, the following observations may strengthen the above hypotheses.

1. Endangered neonates are paid little attention by their mothers and maternal care must be elicited to a great extent by the infants themselves.

2. During the weaning period and even some weeks later, *Callithrix jacchus* infants flee not always to their mother when frightened but to the nearest group member. This observation is an indication of the possibility that the mother-infant bond is little developed; the biological value of this behavior is evident.

3. Family members other than the mother engage in pronounced infant care except, of course, nursing.

Many more observations are needed of parturition not only of the common marmoset, but also of other callitrichids. The study of the early postpartum period should receive great emphasis.

Literature Cited

Benirschke, K., and R. Richart
1963. The establishment of a marmoset breeding colony and its four pregnancies. *Lab. Anim. Care,* 13:70–83.

Bo, W. J.
1971. Parturition. In *Comparative Reproduction of Nonhuman Primates,* edited by E. S. E. Hafez, pp. 302–314. Springfield: Charles Thomas.

Bowden, D.; P. Winter; and D. Ploog
1967. Pregnancy and delivery behavior in the squirrel monkey (*Saimiri sciureus*) and other primates. *Folia primat.* 5:1–42.

Brandt, E. M. and G. Mitchell
1971. Parturition in primates. Behavior related to birth. In *Primate Behavior,* volume 2, edited by L. A. Rosenblum, pp. 178–223. New York and London: Academic Press.

Christen, A.
1974. Fortpflanzungsbiologie und Verhalten bei *Cebuella pygmaea* und *Tamarin tamarin. Fortschr. der Verhaltensforschung* 14: Berlin and Hamburg: Parey.

Gengozian, N.; T. A. Smith; and D. G. Gosslee
1974. External uterine palpation to identify stages of pregnancy in the marmoset *Saguinus fuscicollis* ssp. *J. Med. Primat.* 3:236–243.

Grüner, M., and P. Krause
1963. Biologische Beobachtungen an Weisspinseläffchen, *Hapale jacchus* (L. 1758). *Zool. Gart. (N.S.)* 28:111–114.

Hafner, E., and L. S. Woodburne
1964. Breeding *Saimiri sciureus. Lab. Primate Newsletter* 3:15–16.

Hill, W. C. O.
1962. Reproduction in the squirrel monkey, *Saimiri sciureus. Proc. Zool. Soc. Lond.* 138:671–672.

Hopf, S.
1967. Notes on pregnancy, delivery, and infant survival in captive squirrel monkeys. *Primates* 8:323–332.

Langford, J. B.
1963. Breeding behavior of *Hapale jacchus* (Common marmoset). *S. Afr. J. Sci.* 59:299–300.

Lindburg, D. G., and L. D. Hazell
1972. Licking of the neonate and duration of labor in great apes and man. *Amer. Anthrop.* 74:318–325.

Lucas, N. S.; E. M. Hume; and H. H. Smith
1927. On the breeding of the common marmoset (*Hapale jacchus* Lin.) in captivity when irradiated with ultra-violet rays. *Proc. Zool. Soc. Lond.* 30: 447–450.
1937. On the breeding of the common marmoset (*Hapale jacchus* Lin.) in captivity when irradiated with ultra-violet rays. II. A ten years' family history. *Proc. Zool. Soc. Lond.* 107:205–211.

Mallinson, J.
1965. Notes on the nutrition, social behavior and reproduction of Hapalidae in captivity. *Internat. Zoo Yearb.* 5:137–140.

Marik, M.
1931. Beobachtungen zur Fortpflanzungsbiologie der Uistiti (*Callithrix jacchus* L.). *Zool. Gart. (N.S.)* 4:347–349.

Naaktgeboren, C., and E. J. Slijper
1970. *Biologie der Geburt. Eine Einführung in die vergleichende Geburtskunde.* Berlin and Hamburg: Parey.

Phillips, I. R., and S. M. Grist
1975. The use of transabdominal palpation to determine the course of pregnancy in the marmoset (*Callithrix jacchus*). *J. Reprod. Fertil.* 43:103–108.

Poswillo, D., W. J. Hamilton, and D. Sopher
1972. The marmoset as an animal for teratological research. *Nature, Lond.* 239:460–462.

Rothe, H.
1973. Beobachtungen zur Geburt beim Weissbüscheläffchen (*Callithrix jacchus* Erxleben, 1777). *Folia Primat.* 19:257–285.
1974. Further observations on the delivery behaviour of the common marmoset (*Callithrix jacchus*). *Z. Säugetierk.* 39:135–142.
1975. Influence of newborn marmosets' (*Callithrix jacchus*) behaviour on expression and efficiency of maternal and paternal care. In *Contemporary Primatology*, edited by S. Kondo, M. Kawai, A. Ehara, pp. 315–320. Basel: S. Karger.

Takeshita, H.
1961/62. On the delivery behavior of squirrel monkeys (*Saimiri sciureus*) and a mona monkey (*Cercopithecus mona*). *Primates* 3: 59–72.

N. GENGOZIAN,
J. S. BATSON AND
T. A. SMITH
Medical and Health Sciences Division
Oak Ridge Associated Universities
Oak Ridge, Tennessee 37830

Breeding of Tamarins (*Saguinus* spp.) in the Laboratory[1]

ABSTRACT

Reproductive data from a colony of acclimated and laboratory-bred *S. fuscicollis* ssp. are presented. Information pertaining to the following aspects of breeding has been included: time in residence of imported animals until first pregnancy; a "seasonal" breeding pattern in the laboratory; reproductive performance of imported and laboratory-bred animals and the yearly survival rate of their offspring; interbirth intervals among established breeders; estimation of stages of pregnancy and time of parturition; and twinning frequency. It is concluded that greater information pertaining to the behavioral characteristics of animals in a captive colony environment relative to that in a more natural setting is required in order to realize fully the potential already demonstrated in a laboratory whose major research efforts have been directed toward immunology and not reproductive physiology.

[1]This work was supported in part by United States Public Health Service Grant ROI AM09289-05-09 AIB from the Division of Arthritis and Metabolic Diseases, and by Grant 9 ROI AI 12007-09AI from the Division of Allergy and Infectious Diseases, National Institutes of Health.

Introduction

In 1961 we initiated a program involving utilization of marmosets and tamarins as laboratory primates for studies in radiation biology. A one year's evaluation prompted us to suggest that marmosets and tamarins could well replace other species of primates as experimental subjects, offering certain advantages not observed with other species (Gengozian et al., 1962). In particular, we had noted successful breeding of these animals even within the first year of receipt, without any special efforts by us toward reproduction. During the past ten years our studies have been directed toward a better understanding of the hematologic and immunologic consequences of the natural blood chimerism that occurs in these species (Benirschke et al., 1962; Gengozian et al., 1964; Gengozian, 1969; Gengozian et al., 1969; Gengozian, 1972; Gengozian, 1971; Porter and Gengozian, 1969; Porter and Gengozian, 1973; Barnhart and Gengozian, 1975.) This goal necessitated greater attention to breeding itself and indeed a more conservative utilization of these animals by 'non-terminal' experimentation.

The recent exportation bans of callitrichids by several South American countries places successful breeding of these animals at a high priority if we are to realize the specific contributions they are to make in various biomedical disciplines. The present report, therefore, in addition to providing reproductive data derived from a captive, research laboratory environment, describes some pertinent observations made in the establishment of our colony which may be of value to those presently engaged in attempting to breed callitrichids on a large or small scale.

Materials and Methods

Animals

The predominant callitrichid species maintained and bred in our colony has been *Saguinus fuscicollis* ssp. (Hershkovitz, 1966). A few *S. oedipus oedipus* have also been bred and information pertaining to this group is specifically noted in the text.

The term "imported" breeders indicates wild-caught animals received from various dealers that were acclimated to the laboratory and ultimately produced the first (F_1) lab-bred tamarins. Second (F_2) and third (F_3) generation animals are those subsequently derived from the preceding generation that have reached reproductive age.

Caging and Maintenance

Our general husbandry procedures were detailed in 1969 and these have not changed significantly since that time (Gengozian, 1969). In brief, caging has been done in units fabricated from pressed galvanized metal and measuring approximately 18 × 20 × 42 inches. Designed to hold six animals, particularly adolescents who take advantage of the large vertical space for exercise, these units can readily be subdivided into two cages measuring 18 × 18 × 20 inches. The majority of our paired breeders are now housed in the latter size cages. Each cage has a wooden perch and is also provided with a wooden nesting box at the side for free access by the animals.

The basic diet of the animals has consisted of a primate commercial chow (Teklad) supplemented with hard-boiled eggs and a multivitamin preparation. Fed twice a day, the animals are also offered a different fruit (apple, orange, banana) three times a week at an evening meal with the primate chow. In addition to the vitamin D_3 contained in the chow the animals are exposed daily to sunlamps for approximately 30 minutes in the morning and afternoon. No special dietary formulas or regimens have been used for the breeding females or their young following weaning.

All young were separated from their parents at approximately three months of age and either maintained as co-twin pairs or placed in the larger cage units with other infants of the same age. These were then subsequently paired for mating at about ten months of age, most with other unrelated animals, but in a few cases with full sibs or co-twins.

Results and Discussion

As noted above, the tamarins in our colony have been utilized in research related to the hematology and immunology of blood chimerism. While a significant breeding nucleus has been developed, complete isolation and separation of these animals from our research activities have not been possible. For example, although all adult animals (imported or lab-bred) are paired for possible mating, their inclusion in a variety of experiments before any pregnancy occurs is a common occurrence. The majority of our studies has utilized in vitro procedures requiring blood at scheduled intervals, but in a few instances some in vivo 'non-terminal' experimentation has also been performed. If during this preconception period a pregnancy does occur, the experiment is interrupted and the animals used only after delivery of the young if warranted by the study. Most often this entails additional bleeding for serum or cells; subsequently, the male and female are set aside as a breeding pair and bled only if there is a specific need, and are not involved in any further long-term study.

Specific reproductive data forthcoming from our colony concerning certain aspects of breeding under the conditions described above are as follows.

Time to Pregnancy of Imported Animals

A frequent question asked by individuals interested in establishing a breeding nucleus of marmosets and tamarins has been "When can one expect breeding to occur after receiving the animals?" Sufficient numbers of *S. fuscicollis* ssp. have been received to provide a reasonably accurate estimate for this species. In analyzing our data, we included only those females that had always been caged with an adult male (received in the same shipment) since their arrival into the colony. Of 93 females, 64 (69 percent) became pregnant within the first year, the average and median times in the colony being approximately four months. The earliest pregnancy occurred within 26 days after receipt. During the second year better than 90 percent of this group had experienced at least one pregnancy. In contrast, of 14 *S.o. oedipus*, only five or 36 percent became pregnant within the first year, but by the second year the frequency approached that of *S. fuscicollis*.

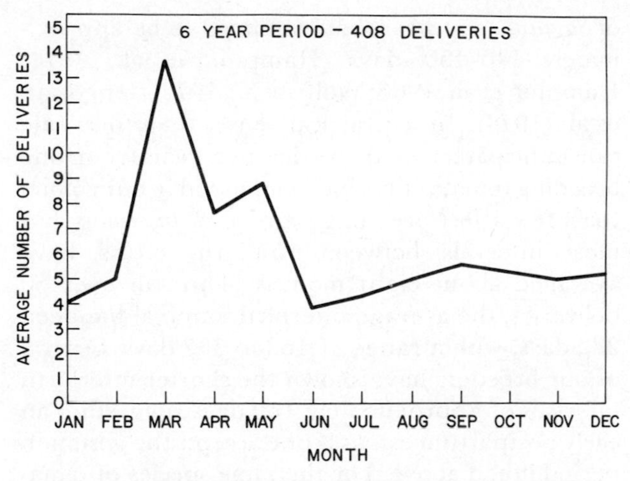

Figure 1. Average number of deliveries per month in *Saguinus fuscicollis* throughout a six-year period. Greater than 90 percent of the females yielding these data were wild-caught, laboratory-acclimated tamarins with the balance being laboratory-bred animals.

A "Seasonal" Breeding Pattern in the Laboratory

During the first few years of our attempts at establishing breeding pairs, it became apparent that even under laboratory conditions, a seasonal breeding pattern similar to that occurring in the wild took place. Figure 1 shows the average number of deliveries (full-term) per month for a period of six years. Although deliveries occurred every month, peak numbers appeared in March, April, and May, indicating most active breeding during the late fall months. It should be noted, however, that for each month there was a wide range in the number of deliveries for the different years.

Reproductive Performance of Imported and Lab-bred Breeders

Reproductive data over a five-year period for *S. fuscicollis* imported and lab-bred breeders are given in Table 1. Three points of comparison have been made for each group: the frequency of live births; the number born premature; and the first-year survival rate. Differences between the two groups are apparent for each criterion, particularly the frequency of premature deliveries and the yearly survival rate. Classification of a birth as "premature" was based on the size of the animal (weight and crown-rump length) and general appearance, notably the density of the pelage (specific data on these parameters of age will be presented elsewhere). None of the premature infants survived more than one week, even those we attempted to hand-rear following what appeared to be "parental neglect" (see below). Therefore, if we eliminate this group of infants and consider only those that were identified as full-term at birth, the first year survival rate was 65 percent for full-term animals derived from imported breeders (F_1 offspring) and only 49 percent for those from lab-bred breeders (F_2 offspring). When the mortality pattern of young obtained from these two groups of breeders was examined at bimonthly intervals throughout the first year (Figure 2), it became apparent that the greatest differences occurred within the first two months after birth. Both groups showed the highest loss within the first seven days, but this was a reflection primarily of the death of the premature infants. Excluding these animals, the percent mortality of full-term F_1 and F_2 offspring was quite low the first week with only a slight difference between the two groups, 7 percent versus 4 percent (obtained by subtracting the premature deaths from the total percentage

Table 1. Comparative reproductive data for imported and lab-bred breeders in *Saguinus fuscicollis*.

Breeders	*Total Animals Delivered**	*Number Born Alive*	*Number Premature*	*Number Survive One Year*
Imported	584	395	45	227
	(F_1)	(68%)	(11%)	(65%)**
				(58%)***
Lab-bred	198	115	25	44
	(F_2)	(58%)	(22%)	(49%)**
				(38%)***

*Five-year period.

**This value does not include premature animals which died within a few days after delivery.

***This value considers all live animals regardless of state of maturity at birth.

shown in Figure 2). During the next two months, however, the percentage mortality increased in each group, with 16 percent of the F_2 dying as contrasted to only 8 percent of the F_1; subsequent mortality for the balance of the first year was not significantly different between the two.

The reasons for the higher mortality during the first two months among the F_2 offspring are not readily apparent, but one observation concerning parental care may be relevant. This is the frequent "parental neglect" demonstrated by the lab-bred (F_1) breeders, a phenomenon rarely observed with the wild-caught imported breeders. Indeed, neglect by the imported breeder for its offspring invariably reflects a premature delivery or some other form of physiological distress which leads to death of the young. Numerous attempts to hand-rear such offspring have failed. In contrast, parental neglect among the lab-bred (F_1) breeders occurs with healthy, full-term young and hand-rearing of these is usually met with success. This characteristic occurrence of parental neglect among these animals may be a possible factor in the high mortality of the F_2 offspring in the first two months. Thus, although we can recognize when the parents obviously refuse to take care of their young, there may also be a more subtle form of inadequate parental care which has gone unnoticed by us, e.g., inability or failure to nurse properly. During this early critical period, such improper care could easily lead to death of the young even though they were healthy, full-term infants at birth. Hampton et al. (1971) have also reported parental neglect in *S.o. oedipus*, this being more prevalent with lab-bred than with wild-caught breeders.

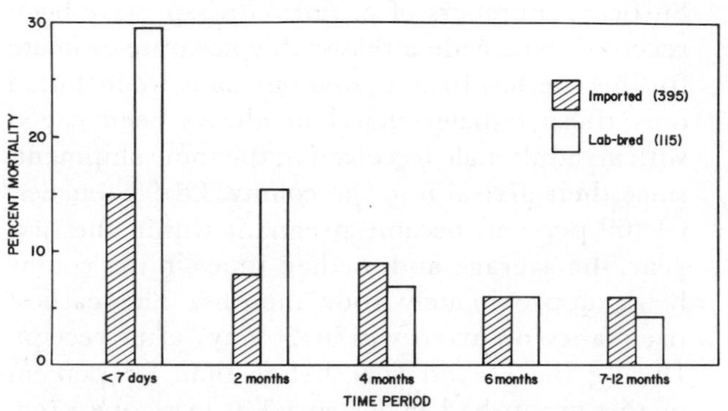

Figure 2. One-year mortality record of *Saguinus fuscicollis* offspring from imported (F_1) and lab-bred (F_2) breeders. The numbers in parentheses indicate the total number of live animals for each group under analysis (see Table 1). The percentage mortality at each designated time period was calculated on the basis of the number of animals delivered live regardless of whether they were classified as full-term or premature at birth.

Interbirth Intervals

Based on a few timed pregnancies, but primarily on the interbirth intervals, the gestation period of *Saguinus* spp.has been estimated to be approximately 140–150 days (Hampton et al., 1971; Hampton et al., 1966; Wolfe et al., 1972; Gengozian et al., 1974). In a practical sense, therefore, one can anticipate two deliveries per year from one breeding female. This has been noted in our colony for a few select breeding pairs of *S. fuscicollis*, but most intervals between full-term births have averaged about eight months. Thus, in over 60 deliveries, the average interbirth interval has been 222 days, with a range of 163 to 367 days. Several of our breeders have shown the shorter interbirth intervals of approximately 160 days suggesting an early postpartum estrus if one accepts the gestation period noted above. For the same species of tamarin, Wolfe et al. (1972) note, however, that early removal of the infants (within 48 hours) from the parents will shorten the interval significantly, to an average of 170 days (range 134–210, 40 intervals) as compared to 233 days (range 173–397, 20 intervals) when young are removed at 3 to 4 months. Although we have not had an opportunity to analyze this variable, we have noted that individual breeding females will show their own characteristic breeding pattern. Thus, one female in over 12 deliveries (with the same mate) has shown an average interbirth interval of 188 days, while another female (also with one mate) had an average interval time of 235 days for eight full-term deliveries. In *S.o. oedipus*, Hampton and Hampton (1965) reported an average interbirth interval of 240 days in 21 deliveries, slightly longer than in *S. fuscicollis*, an observation which agrees with our limited experience with these animals. A significant generic variation is apparent, however, when one examines the data of Hiddleston who reports an average interbirth interval of 178 days (S.E. ± 3 days) for 146 deliveries of *Callithrix jacchus* (Hiddleston, pers. comm.). This remarkably consistent productive performance has not been described for any other species of callitrichid in captivity; Lucas et al. (1937) also noted an average interval of only 162 days for eight interbirth intervals for a single *C. jacchus* female.

Estimation of Parturition Time

Early in the development of our breeding program, it became apparent that the ability to detect pregnancy and to estimate the days postconception would be a distinct advantage not only for colony management but also in experimental studies where fetal tissues of known ages may be required. At that time, the lack of any definitive information on the estrous cycle of these animals which would permit timed matings or hormonal assays to follow the course of pregnancy prompted us to explore external uterine palpation as a means of identifying pregnant animals. Although our methods and resulting data have been published in detail (Gengozian et al., 1974), a brief account is given to indicate the simplicity and accuracy of the technique. All paired females were routinely palpated each month in the region of the uterus by applying slight pressure with the thumb and forefinger on the lateral aspects of the abdominal wall immediately above the pubic glandular pad. The earliest palpable pregnant uterus feels like a firm rigid sphere, measuring approximately 0.5 cm in diameter. This sphere increases in size and moves toward the mid-abdominal area as time passes. Palpation of 92 pregnant females in varying stages of pregnancy permitted us to obtain graphic data relating the size of the uterus (to the nearest 0.5 cm) to remaining days to delivery of full-term twins. From an equation derived from this data, we then subsequently attempted to predict the parturition time of an additional 172 animals palpated in varying stages of pregnancy. Of these, 12 or 7 percent delivered on the day predicted, 108 or 61 percent delivered within 5 days (±) of the anticipated parturition date and 142 or 83 percent within 7 days (±). As noted above, the earliest palpable pregnant uterus was 0.5 cm, which, if one accepts a gestation time of 145 days, revealed this measurement to detect pregnancy within the first 15 days after conception. Table 2 shows the uterine measurements and the actual days to delivery for the 270 animals that were involved in this study. Since our report, similar palpation techniques for *C. jacchus* have also been described, with measurements not too different from those found by us for *S. fuscicollis* (Phillips and Grist, 1975). In vitro and in vivo assays for hormones either excreted in the urine (chorionic gonadotrophin) or found in the serum (luteinizing hormone, progesterone or estradiol) have also been used to diagnose pregnancy in *S. o. oedipus* and *C. jacchus*. (Hampton et al., 1969; Hearn et al., 1975). Although these methods have detected early pregnancies (within 14 to 20 days), it does not seem that the alterations in the levels of the hormones are of sufficient magnitude to provide a predictive tool throughout all stages of pregnancy.

Twinning Frequency and Sex of Offspring

A high frequency of fraternal twinning among marmosets was suggested by Wislocki (1939) based on analysis of uteri and corpora lutea of pregnant animals and from other data in the literature. Since

Table 2. External uterine palpation throughout pregnancy to estimate parturition time.*

Uterine Diameter (cm)	*Mean Days to Delivery***	*Number of Animals*
0.5	131	25
1.0	96	34
1.5	80	67
2.0	70	57
2.5	51	47
3.0	42	29
3.5	29	11

*Data adapted from Gengozian et al. (1974).

**All deliveries were live full-term twins.

Table 3. Frequency of single and multiple births in *S. fuscicollis* ssp.

Litter Size	*Offspring Sex**
Single	65 (20.4%) 25m, 24f
Twins	238 (74.6%) 57mm, 107mf, 40ff**
Triplets	16 (5%) 6mmf, 5ffm, lfff,lmmm

*Sex of all offspring could not be determined due in some cases to early abortus material and in others because of partial cannibalism of a still-born fetus.

**The subdivision of sexes does not deviate significantly from the expected 1mm:2mf:lff ratio for fraternal twinning.

then several reports on the twinning frequency in a colony environment have appeared (Hampton and Hampton, 1965; Wolfe et al., 1972). In our laboratory, twinning has occurred at a rate of approximately 75 percent, with the sexes approximating that expected for fraternal twinning (Table 3). Further data supporting fertilization of two ova were obtained by identification of red blood cell chimerism in isosexual co-twin pairs (Gengozian, 1972). The frequency of single and triplet deliveries has averaged approximately 20 percent and 5 percent, respectively, which compares to that reported by Wolfe et al. (1972) for the same species and also by Hampton et al. (1971) for *S.o. oedipus*. In contrast, Hiddleston (pers. comm.) reports for *C. jacchus* a twinning frequency of 67 percent, triplets 25 percent and only 5 percent singletons in 662 parturitions. A high frequency of triplets (17 percent of 142 deliveries) in *C. jacchus* has also been reported by Hampton et al. (1971).

Initially, our interpretation of single births among *S. fuscicollis* was intrauterine death and resorption of one of the twin fetuses (Gengozian et al., 1969); however, an extensive analysis of blood chimerism in over 50 recorded single births suggested that this occurs relatively infrequently and that approximately 80 percent of the single births are a consequence of only one ovum being fertilized at conception (Gengozian and Batson, 1975). Whether the same probability would apply to *C. jacchus* where multiple young are the rule is not known. With respect to single births, however, an interesting and quite unexpected observation made in our laboratory was the increase in number of such offspring with age of the breeding female. Thus, among 12 breeding *S. fuscicollis* females, the frequency of single deliveries changed from 14 percent during the first five years in the colony to greater than 33 percent during the second five years of residence (Gengozian and Batson, 1975). This observation contrasts to the suggestion of others that aging may lead to excessive fecundity (Epple, 1970).

Conclusions

The reproduction of callitrichids in a research laboratory environment can be achieved with little difficulty even in the absence of any concerted effort or means of allowing for a strict geographic and experimental isolation of these animals. Although the percentage live births among the wild-caught imported breeders was only 68 percent, these data included several animals which for unknown reasons have never produced live young, even after five or six deliveries with different mates; in contrast, some breeders have consistently yielded greater than 90 percent live deliveries. In the absence of definitive answers concerning the reasons for the poor performance of some animals, a simple culling procedure to utilize these in experimental studies rather than retain them as breeders would appear to be expedient if we are to realize successive generations of lab-bred offspring. In this context, we have already noted that animals derived from good imported breeders in turn have a high probability of becoming successful breeders, while the performance of those derived from females showing a high abortion rate appears to be less promising. Whether this reflects genetic factors, as suggested, or as yet unrecognized physiologic problems related to stress of a colony environment can only be answered in future studies. The low percentage of live offspring from the lab-bred breeders in comparison to that observed with imported breeders appears to be a separate problem in that significant improvement has been obtained with multiparity. The high mortality attendant during the first two months among the F_2 offspring would appear to be correctable once we have established that factors other than parental neglect can be excluded. Finally, greater information concerning the behavioral aspects of breeding with respect to a laboratory environment as contrasted to a more natural setting is needed to fully exploit the potential already demonstrated.

Literature Cited

Barnhart, D. D., and N. Gengozian
1975. An evaluation of the mixed lymphocyte culture reaction in marmosets. *Transplantation* 20:107–115.

Benirschke, K.; J. M. Anderson; and L. E. Brownhill
1962. Marrow chimerism in marmosets. *Science* 138: 513–515.

Epple, G.
1970. Maintenance, breeding, and development of marmoset monkeys (Callithricidae) in captivity. *Folia primat.* 12:56–76.

Gengozian, N.
1969. Marmosets: Their potential in experimental medicine. *Ann. N. Y. Acad. Sci.* 162:336–367.
1971. Male and female cell populations in the chimeric marmoset. In *Medical Primatology 1970,* pages 926–938. Selected papers from the 2nd Conference on Experimental Medicine and Surgery in Primates, held in New York in 1969. Basel: Karger.
1972. A blood factor in the marmoset, *Saguinus fuscicollis.* Its detection, mode of inheritance, and species specificity. *J. med. Primat.* 1:272–286.

Gengozian, N., and J. S. Batson
1975. Single-born marmosets without hemopoietic chimerism: Naturally occurring and induced. *J. med. Primat.* 4:252–261.

Gengozian, N.; J. S. Batson; and P. Eide
1964. Hematologic and cytogenetic evidence for hematopoietic chimerism in the marmoset *Tamarinus nigricollis. Cytogenetics* 3:384–393.

Gengozian, N.; J. S. Batson; C. T. Greene; and D. G. Gosslee
1969. Hemopoietic chimerism in imported and laboratory-bred marmosets. *Transplantation* 8:633–652.

Gengozian, N.; J. S. Batson; and T. A. Smith
1962. *Tamarinus nigricollis* as a laboratory primate. *Bone Marrow Therapy and Protection in Irradiated Primates.* Proceedings of a Symposium, August 15–18, 1962, Radiobiological Institute T.N.O. 27 pages.

Gengozian, N.; T. A. Smith; and D. G. Gosslee
1974. External uterine palpation to identify stages of pregnancy in the marmoset, *Saguinus fuscicollis* ssp. *J. med. Primat.* 3:236–243.

Hampton, J. K., Jr., and S. H. Hampton
1965. Marmosets (Hapalidae): Breeding seasons, twinning, and sex of offspring. *Science* 150 (3698): 915–917.

Hampton, J. K., Jr.; S. H. Hampton; and B. T. Landwehr
1966. Observations on a successful breeding colony of the marmoset, *Oedipomidas oedipus. Folia primat.* 4:265–287.

Hampton, J. K., Jr.; S. H. Hampton; and B. M. Levy
1971. Reproductive physiology and pregnancy in marmosets. In *Medical Primatology 1970,* pages 527–535. Selected papers from the 2nd Conference on Experimental Medicine and Surgery in Primates, held in New York in 1969. Basel: Karger.

Hampton, J. K., Jr.; B. M. Levy; and P. M. Sweet
1969. Chorionic gonadotrophin excretion during pregnancy in the marmoset, *Callithrix jacchus. Endocrinology* 85:171–174.

Hearn, J. P.; R. V. Short; and S. F. Lunn
1975. The effects of immunizing marmoset monkeys against the β subunit of HCG. In *Physiological Effects of Immunity against Reproductive Hormones,* edited by R. G. Edwards and M. H. Johnson, pp. 229–247. Cambridge: Cambridge University Press.

Hershkovitz, P.
1966. Taxonomic notes on tamarins, genus *Saguinus* (Callithricidae, Primates), with description of four new forms. *Folia primat.* 4:381–396.

Lucas, N. S.; E. M. Hume; and H. H. Smith
1937. On the breeding of the common marmoset (*Hapale jacchus* Linn.) in captivity when irradiated with ultraviolet rays. II. A ten years' family history. *Proc. zool. Soc. Lond.,* Ser. A. 107:205–211.

Phillips, I. R., and S. M. Grist
1975. The use of transabdominal palpation to determine the course of pregnancy in the marmoset (*Callithrix jacchus*). *J. Reprod. Fert.* 43:103–108.

Porter, R. P., and N. Gengozian
1969. Immunological tolerance and rejection of skin allografts in the marmoset. *Transplantation* 8: 653–665.
1973. Immunological responsiveness and tolerance of marmoset lymphoid tissue in vitro. *Transplantation* 15:221–230.

Wislocki, G. B.
1939. Observations on twinning in marmosets. *Amer. J. Anat.* 64:445–483.

Wolfe, L. G.; J. D. Ogden; J. B. Deinhardt; L. Fisher; and F. Deinhardt
1972. Breeding and hand-rearing marmosets for viral oncogenesis studies. In *Breeding Primates,* edited by W. B. Beveridge, pp. 145–157. Basel: Karger.

DEVRA G. KLEIMAN
National Zoological Park,
Smithsonian Institution
Washington, D.C. 20008
and
MARVIN JONES
San Diego Zoological Garden
San Diego, California 92112

The Current Status of *Leontopithecus rosalia* in Captivity With Comments on Breeding Success at the National Zoological Park

ABSTRACT

The trends in reproduction and survivorship in captive lion tamarins, *Leontopithecus rosalia*, are evaluated. An increase in numbers from 69 to 74 specimens between December 1972 and December 1975, although reversing a decline, is far below potential. In the captive population, survivorship of adults remains poor, mainly due to viral and bacterial diseases. Survivorship of young is also low, with maternal neglect and stillbirths common. There may be a tendency for older females to over-produce, by bearing triplets and/or two litters a year. Pubertal females have not been paired early enough, thus reducing recruitment. An analysis of the age classes and sex ratio in the current population suggests that the lion tamarin population is now at a crucial stage in its captive history.

The precarious status of the golden lion tamarin, *Leontopithecus rosalia rosalia*, has been recognized for at least a decade; numerous individuals and institutions have committed themselves to the preservation of this rare and beautiful monkey since its endangered status became evident. Magnanini and Coimbra-Filho (1972), Magnanini (1977), and Coimbra-Filho and Mittermeier (in press) have reviewed progress in the Brazilian efforts to establish a reserve for the subspecies while Bridgwater (1972a), Perry (1973), and Perry, Bridgwater, and Horseman (1973) have discussed the long-term prospects for the captive population. Jones (1973) prepared the first international *Studbook* for the golden lion tamarin and acted as studbook keeper until April 1975. The purpose of this paper is to discuss recent trends in reproduction of the captive population.

In 1972, D. Bridgwater and J. Perry organized a conference which was held at the National Zoological Park in Washington (Bridgwater, 1972b) and cosponsored by Wild Animal Propagation Trust (now disbanded), National Zoological Park, and New York Zoological Society. Specialists in callitrichid biology exchanged ideas and made recommendations for improving the management and husbandry of the lion tamarin. At that time, there were several apparent trends in the captive population which were pointed out by Perry et al. (1973). The total numbers of captive lion tamarins had decreased from a peak in 1968, just before the total ban on importation of this species was effected. There was also a steady increase in the percentage of captive-bred animals in the population, suggesting poor survivorship among the imported specimens. For example, between 1969 and 1971, the percentage of captive-bred lion tamarins increased from 23 to 51 percent (Perry et al., 1973). This trend has continued with captive-born animals currently (December 1975) accounting for about 80 percent of the population.

Another trend apparent since 1969 is a steady decline in the number of institutions maintaining this species. Between 1960 and 1969, 36 institutions or private individuals maintained lion tamarins. By 1973, less than half of these locations still had specimens (Jones, 1973). Currently, lion tamarins are found at 17 sites, but two of these were only recently uncovered (in Pretoria, South Africa) and one location, the Tijuca Biological Bank in Rio de Janeiro, has mostly wild-caught specimens which have been brought into captivity only recently (Magnanini et al., 1975). Moreover of the 17 sites, 5 have no breeding potential in that they possess only single specimens.

The decrease in the number of institutions with lion tamarins is partly due to a voluntary relocation of single animals to colonies with breeding potential. However, there are still periodic and sudden losses within colonies, usually caused by viral or bacterial outbreaks. Currently, the best method of reducing these losses is the housing of different family groups in separate facilities, a policy being followed by several institutions.

The trends in the captive population can only be understood in relation to the reproductive potential of the species (see Eisenberg, 1977). Twins are typical in tamarin births, and females occasionally may have two litters per year (Hampton et al., 1966; Epple, 1967). If the survivorship of young were optimal, and the adult lifespan approximated the probably 10-year longevity, it is conceivable that a captive population would double within a few years.

Captive collections of lion tamarins, however, have not begun to exhibit this trend. In Table 1 data are presented on the captive population in December 1972 (from Jones, 1973) and December 1975.

The net increase was five specimens, far below the potential. Males still outnumber females and the total number of pairs has remained stable.

Table 1. The status of the captive population of *Leontopithecus rosalia rosalia* on December 31, 1972 and December 31, 1975.*

					Wild-born		*Captive-born*			*Adult*		*Young < 16 mos*		
Year	♂	♀	?	*Total*	♂	♀	♂	♀	?	♂	♀	♂	♀	?
1972	46	23	0	69	14	11	32	12	0	37	19	9	4	0
1975	42	30	2	74	7	8	35	22	2	34	26	8	4	2

*Does not include two colonies in Pretoria, South Africa, nor individuals in the Tijuca Biological Bank, Rio de Janeiro, Brazil.

Table 2. Number of captive female *L. rosalia* giving birth each year.*

1960	3	1968	12
1961	3	1969	15
1962	3	1970	12
1963	2	1971	14
1964	2	1972	9
1965	2	1973	9
1966	5	1974	12
1967	10	1975	14

*Excluding colonies in Pretoria, South Africa, and Tijuca Biological Bank, Rio de Janeiro, Brazil.

Table 3. Survivorship of young *L. rosalia*, 1972–1975*

Year	*Number Litters*	♂	♀	*?*	*Total*	*Number Still-born or <30 Days*	*Number Surviv-ing >30 Days*
1972	13	14	6	5	25	15	10
1973	12	9	10	1	20	6	14
1974	16	14	14	0	28	13	15
1975	20	21	9	9	39	28	11

*Excluding colonies in Pretoria, South Africa, and Tijuca Biological Bank, Rio de Janeiro, Brazil.

There has been, however, an increase in numbers for the first time since a ban on importation was instituted. Moreover, the increase has comprised mainly females, thus there should be a rise in the number of breeding pairs in the near future. In fact, there has been an increase recently in the number of females which are actually breeding. Of 26 females over 16 months old, 14 had young in 1975 compared with 9 of 20 females in 1973. The numbers of females bearing young each year since 1960 is presented in Table 2.

The status of lion tamarins in captivity has not improved significantly because (1) adult mortality has remained high and (2) the survivorship of young has not been optimal. Table 3 presents a summary of reproduction in captive *L. rosalia* between 1972 and December 1975. Survivorship of young beyond 30 days averaged 55 percent until 1975 when it dropped precipitously to about 28 percent. The causes of infant mortality are numerous, and a review of these in the National Zoological Park colony may point up some of the problems.

In 1974, three reproductive females bore 9 young, of which 8 survived, giving a survivorship of 89 percent. Two of the females bore two litters during the year, and had been regularly reproducing (usually annually) since 1971. The third female bore her first litter and reared the young.

In 1975, six sexually mature females became pregnant, three for the first time. The 1974 primiparous female aborted twins and died soon after, possibly from enteritis of viral origin. One of the older females gave birth in March to triplets, of which two survived. This female became pregnant immediately postpartum and bore two young (one stillborn, one live-born which died) in July, with an interbirth interval of 126 days. The second older female bore triplets, all of which were stillborn and slightly underdeveloped, in May, one month after the death of her mate.

The reproductive history of the older females points up one problem of captivity—overproduction. It is unlikely that lion tamarins would produce more than one litter a year in the wild (Coimbra-Filho and Mittermeier, 1973) and triplets would be a rare occurrence. In recent years, several zoos have reported cases of females bearing two litters a year, including triplet births (e.g., Wilson, 1977). Such output is certain to impose a terrific burden on a female, and the result appears to be poor survivorship eventually. If such occurrences increase in the future, it may become necessary to break up pairs by removing the adult male for six months, thus allowing females a reproductive rest.

The three other primiparous females exhibited three different patterns. One female bore a single young and reared it. The second bore twins of a large size which died during delivery. The third gave birth to healthy twins, but rejected them on the first and second day, respectively. This female had had limited experience of infants as a juvenile, as had her mate, and initially nursed both young adequately; however, she appeared eager to transfer the young on the first day, and began scratching and rubbing them off. The male did not respond by carrying the young and they eventually fell off and were killed by the parents. Whether this pair will eventually exhibit appropriate parental care, given their lack of experience as juveniles, is open to question. In captive-born females from other collections, young are also often lost from stillbirths and maternal neglect.

Thus, of 15 young born in 1975 at the National Zoological Park, three survived past 30 days of age. In 1976, the National Zoo will have two more females reach puberty, totaling seven females of

Table 4. Age of first conception in female and first insemination in male *L. rosalia*.

	Age in Months						
Sex	*<24*	*25-36*	*37-48*	*>48*	*X̄*	*n*	*Range*
Males	10	5	4	1	28.7	20	(9-54)
Females	8	4	5	1	31.3	18	(14-52)

reproductive age; however, of these only one pair will be proven breeders, proven parents, and with no potential reproductive problems. Thus, with the largest captive population of lion tamarins outside of Brazil, the National Zoo is not secure in its breeding potential.

A final problem which has contributed to less successful reproduction in this species in captivity is the apparent delay in pairing up pubertal females. Table 4 presents the distribution in age of first conception in females and first successful insemination in males. Eight females have conceived before 2 years of age, suggesting that puberty may be reached between 16 and 20 months old. Between 1969 and 1973 most females were conceiving for the first time at a much later age, between 3 and 4 years old. Although this delay is being reduced, any delay is intolerable, given the scarcity of breeding age females and the surplus of males in the population.

The trends in reproduction in the captive lion tamarin population between 1972 and 1975 suggest that we are at a crucial stage in its history. Births and breeding females have increased in number, and the decline in numbers of animals, constant since 1968, has been reversed. Population increases have been minimal, however, and survivorship of both adults and young remains poor. Clearly, unless there is a population spurt over the coming two to three years, the population is unlikely to become self-sustaining and will thus eventually decline and disappear.

Literature Cited

Bridgwater, D.

1972a. Introductory remarks with comments on the history and current status of the Golden Marmoset. In *Saving the Lion Marmoset,* edited by D. Bridgwater, pp. 1-6. Wheeling, West Virginia: Wild Animal Propagation Trust.

1972b. (Editor) *Saving the Lion Marmoset.* Wheeling, West Virginia: Wild Animal Propagation Trust.

Coimbra-Filho, A. F., and R. A. Mittermeier.

1973 Distribution and ecology of the genus *Leontopithecus* Lesson 1840 in Brazil. *Primates,* 14:47-66.

in press. Conservation of the Brazilian lion tamarins (*Leontopithecus rosalia* ssp.). In *Primate Conservation,* edited by Prince Rainier of Monaco and G. Bourne. New York: Academic Press.

Eisenberg, J. F.

1977. Comparative ecology and reproduction of New World monkeys. In *The Biology and Conservation of the Callitrichidae,* edited by Devra G. Kleiman, pp. 13-22. Washington, D.C.: Smithsonian Institution Press.

Epple, G.

1967. Vergleichende Untersuchungen über Sexual- und Sozialverhalten der Krallenaffen (Hapalidae). *Folia primat.,* 7:37-65.

Hampton, J. K.; S. H. Hampton; and B. T. Landwehr.

1966. Observations on a successful breeding colony of the marmoset, *Oedipomidas oedipus. Folia primat.,* 4:265-287.

Jones, M. L.

1973 *Studbook for the Golden Lion Marmoset, Leontopithecus rosalia.* Wheeling, West Virginia: American Association of Zoological Parks and Aquariums.

Magnanini, A.

1977. Progress in the development of the Poço das Antas Biological Reserve for *Leontopithecus rosalia rosalia* in Brazil. In *The Biology and Conservation of the Callitrichidae,* edited by Devra G. Kleiman, pp. 131-136. Washington, D.C.: Smithsonian Institution Press.

Magnanini, A. and A. F. Coimbra-Filho.

1972. The establishment of a captive breeding program and a wildlife research center for the lion marmoset, *Leontopithecus,* in Brazil. In *Saving the Lion Marmoset,* edited by D. Bridgwater, pp. 110-119. Wheeling, West Virginia: WAPT.

Magnanini, A.; A. F. Coimbra-Filho; R. A. Mittermeier; and A. Aldrighi.

1975. The Tijuca Bank of Lion Marmosets, *Leontopithecus rosalia:* a progress report. *Internat. Zoo Yearb.,* 15:284-287.

Perry, J.; D. D. Bridgwater; and D. L. Horseman.

1973. Captive propagation: a progress report. *Zoologica,* 57:109-117.

Wilson, C.G.

1977. Gestation and reproduction in golden lion tamarins. In *The Biology and Conservation of the Callitrichidae,* edited by Devra G. Kleiman, pp. 191-192. Washington, D.C.: Smithsonian Institution Press.

Panelists:
W. Thomas (chairperson),
D. Kleiman, R. Lorenz,
J. Mallinson, C. Wilson.

PANEL DISCUSSION:
AUGUST 20, 1975

Reproduction of Small Colonies of Endangered Callitrichids

THOMAS: This panel is about reproduction in small colonies of endangered marmosets. In the past, this meant golden lion marmosets. But now I find to my chagrin that we are faced with another one virtually as bad, the cotton top. This is ironic in that, in years past, it was the single most available marmoset in this country. It hasn't been very long ago that cotton tops sold for as little as $15 a piece. Now, we find that it is a seriously endangered species.

We have several groupings of golden lion marmosets in which we are getting second- and third-generation births. Some discussion should be given to some of the consequences of starting with a very small gene pool. As yet, have there been any serious genetic defects reported with golden lions? When animals are born dead, it is hard to tell whether you might have a lethal mutation or a lethal characteristic causing the death in utero or up to the time of delivery. But I think we can only speculate on that; we haven't any data to support it positively or negatively.

KLEIMAN: Since we are just getting into second generation and maybe one case of third generation offspring, we really haven't had to deal with the problem of inbreeding and genetic factors. In fact, very few of the institutions that have had second-generation young have had brother-sister, father-daughter matings, etc. So it is not a problem now, but it is obviously something that has to be thought about very seriously for the future of the captive populations.

THOMAS: You have to come up with policy, should you have an animal born with an obvious genetic defect, particularly when it is an endangered species. Do you allow it to continue to survive (perhaps there are certain ones that could be surgically repaired)? What are you really doing when you save such an animal? Granted, an Indian rhino with a genetic defect that I could correct surgically, I would be hard-pressed to destroy, particularly if it was a female. But, by the same token, I might be thinking that perhaps I could play the numbers game until I could get enough animals and perhaps breed this characteristic out. Is that really the way one should look at endangered species or should one take a much more cold-blooded attitude? I suppose it would vary according to the species and the numbers you are dealing with.

THORINGTON: And also the genetic defect you had. You might want to breed a few of them and sell them for use in medical research. (Laughter.)

KLEIMAN: It has been suggested. At the AAZPA meeting in 1973, it was suggested that, for species where no more imported animals will be available, we had better rapidly inbreed and find out what genetic defects there are prior to making decisions about the ultimate breeding program. This doesn't go over very well in zoos, because inbreeding is thought of very negatively even at the second-generation level.

WILSON: At Oklahoma City, two of the existing breeding pairs are half-brother–half-sister combinations, and we have had no unusual circumstances surrounding any birth. We had one stillbirth, but all of the other offspring lived, so I don't see any problem at this stage with this type of inbreeding.

LORENZ: I have one case of inbreeding in *Callimico* which resulted in four offspring. One is still alive which was mated with a wild-born male as far as I can tell, and there are three surviving young from that pairing, so this is three generations from the inbred pair.

THOMAS: Obviously, the important thing is what genes you start with. If you have a good pair to start and you keep repeating good characteristics, you will stay out of trouble for quite some time.

KLEIMAN: But nobody (and it is not only true for zoos but for the research laboratory) has developed a breeding policy that I know of. Whatever our beliefs about the good or bad consequences of inbreeding, we are exchanging animals and breeding animals with no set policy. And if we cannot get new blood, what kinds of characteristics are the golden marmosets going to have without a breeding policy?

EGOSCUE: (National Zoological Park): In this book edited by Martin on breeding rare and endangered species, there is not a single reference to genetics.

KINGSTON: Well, there is in the laboratory field: Mice are the easy ones. There is Falconer, 1963. If you wish to retain an adequate non-inbred pool of any species, you must start with 20 unrelated breeding pairs and you must continue by having your F_2s equally drawn from each of those 20 unrelated pairs. You have then only .05 percent inbreeding which is as highly heterozygous as you can get. This is a printed established figure: 20 unrelated breeding pairs.

EISENBERG: Yes, it has all pretty much been worked out. I think we should strongly bear in mind that the structure of natural populations often is a mosaic of subpopulations which are highly inbred. Now, in a rare and endangered species, even if you were to get permits to bring out species X as a founder stock and you went into an area and removed that stock, you probably wouldn't come out with 20 unrelated pairs. The other side of it is that all that inbreeding can possibly do for you, depending on your stock, is bring out in a phenotypic expression deleterious mutants that are masked because they are recessive to a dominant in the paired condition. It may well be that the finest thing you can do is to find out how many lethals are floating in your population by doing a little fast inbreeding the first go-round and see how many phenotypic variations show up that are undesirable. If they don't show up, then you can feel a little more secure. Maintenance of heterozygosity deliberately is, in fact, a tactic that you must resort to if you do have a fairly high load of lethals in whatever your founder population is.

THOMAS: Are you suggesting then that we set up a pilot group of golden lion that would be purposely inbred?

EISENBERG: I don't think we can afford that. It is a hypothetical proposition, that's all.

THORINGTON: Also Hardy-Weinberg long ago established that, if you are breeding, you are going to maintain the same gene frequency; you are not going to have any increase in frequency of deleterious genes until you start selecting. When they started to breed laboratory mice, they bred for the best breeders, but what they were also breeding for was poor eyesight and poor hearing so the animals would be undisturbed under most laboratory conditions. It is that sort of selective factor that we have to worry about. Any inbreeding you do will bring out those things which are deleterious when they are homozygous, but they may not be deleterious in the heterozygous situation. Therefore, you may want to retain them in the population.

BLOOD: Wouldn't you want to avoid a rule, regulation, or policy at this stage in order to keep the options open? The general approach would be to avoid inbreeding if possible, but if necessary go ahead with it.

KLEIMAN: In the case of *Leontopithecus,* there is a committee which was supposed to inform institutions that had colonies of where other animals were and perhaps recommend changes. But, if you have a committee, from what basis are they going to make the recommendation? Obviously, if you have a pair of animals that haven't bred in five years, you want to break that pair up and give the female another experienced male; that's one recommendation you can make. But deciding which male from which colony is another question entirely. It seems to me one ought to try and develop a basic strategy that you follow.

EGOSCUE: Did anybody really look to see why they hadn't bred? Could it possibly have been nutrition, for example?

KLEIMAN: They bred once and then quit. But to find out why means going into another institution's management policies.

WILSON: But the criteria at this stage of the game will always be subjective, with this small a population. You only have so many males to pick from that are of the appropriate age.

LORENZ: May I please quote the breeding history of the *Callicebus* group which was established at Delta (Regional Primate Center). In 1965 until 1969 roughly 300 *Callicebus* were purchased, including a few *C. torquatus,* which never bred. Out of the *C. moloch,* I think there were three subspecies, most of them *cupreus.* There was a very poor breeding record until about 1970 at which time I started what I have termed selective pairing. I started out intuitively but from 8 pairs set up in 1970, we had 8 babies and 6 were raised. One was not accepted and we tried hand-raising it unsuccessfully. I can't recall the eighth one. The breeding record has kept at a level of 6 plus or minus one young per year. We have gone back and compared the breeding record of the animals that bred before and it turned out that they had spent more time with more animals but had not achieved the breeding that they did achieve after we remated them. We have come to a hierarchical key of features according to which we set up cage partners. I started out with subspecies to begin with, wild- or captive-born origin second. Size was important or weight (which of course is correlated) and the last is a very subjective term which I call temperament, the emotional disposition of the animal. The colony figured about 45 animals in 1969 when I started. We lost a few more of the wild animals and, in 1972, we had 24 adults left of which I sent 11 to Dr. Mason at Davis Primate Center. Of the remaining animals, the colony has rebuilt to 40 now, in spite of the relocation from Delta to Göttingen. I think what we essentially have found (I have just been talking in reproductive terms) are the components of the pair bond which is a very cumulative thing.

THOMAS: There are a couple of other things. I find that zoos tend to concentrate in too confined an area all of their potential. When I went to Los Angeles, I found a whole collection of golden lion marmosets in a single building, and in and out of that building came every baby on a bottle and so on.That was subsequently changed as quickly as I could do it. A singular policy should be: don't concentrate them all at one place.

COOPER: It appears to me that, while we are not at the stage where one can say too much about the genetic problems, one very immediate one seems to be the tremendous numbers of male offspring. This is a phenomenon which has been anecdotally described in a variety of vertebrates and related to such vague things as stress. Is there a way to determine exactly what is causing it? Is it possible to compare management in the National Zoo colony, which at this time is presumably not having that problem, with Oklahoma City and Monkey Jungle?

KLEIMAN: I think it was almost a freak coincidence because not only did it affect Monkey Jungle and Oklahoma, but it affected us (NZP) during a certain two-or three-year period. Because we are dealing with so few animals, a surplus of males is obviously going to seem like a much greater effect than it really is. Over the past two years, the sex ratio has evened out, and some zoos are even producing sets of twin females, for example.

COOPER: In identical circumstances in which they were producing males?

KLEIMAN: Yes. I think Rainer (Lorenz) will back me up on this. In *Callimico,* the problem is the exact reverse; there is a surplus of females and there are insufficient males to pair them with.

LORENZ: I am still searching for a place where more than one pair of *Callimico* within auditory reach of each other have been bred. I think that *Callimico* is a species that basically is set for an extremely low population density. I have the impression that there is a rank order between pairs within reach and the dominant pair will inhibit the breeding of the socially inferior pair by calling. In Göttingen we are getting the animals out of hearing range of each other in the hope that we will have more than one breeding group. I have talked to Frank Dumond about whether that could be the case with his goldens and he did not think so. As a matter of fact, he had used *Callimico* in between pairs of goldens, as a biological insulator, so to speak.

THOMAS: We probably do have a little better survival rate of males in captivity than you would see in the wild.

KLEIMAN: That is something that could be looked into. That is one point that Pekka (Soini) brought up. We could examine whether our survivorship is different based on sex.

WILSON: In Oklahoma City, we had a preponderance of males earlier. I was just checking the total record and there were 13 males and 14 females born in the last 11 years, so it is 50:50. Survival wasn't even, but total births were.

THOMAS: We have a major problem facing us with cotton-top marmosets. Unfortunately, the situation in the wild is so desperate that we have to take some major action with what is in captivity. I would guess that there have to be several hundred scattered around in the United States. Now, I propose that through the AAZPA we establish a Studbook, secondly, that we get the word out to every collection through the Simian Society, AAZPA, and any other source that we may have, to try to gather the animals into some meaningful breeding situation. I am sure that there are some stuck away as pets. We should institute and pursue with vigor a positive breeding program for cotton tops. Any comments or suggestions?

MALLINSON: In England we have two zoo bodies, one is the Federation of Zoos and the other is the Association. At the moment there is aggressive interaction between the two bodies. As far as the breeding committee of the Federation, we are trying to get four or five different places to take responsibility for particular species. The situation with the cotton top is amazing to me. At the next meeting of the Federation, I will do everything I can to see about exchanging specimens.

KLEIMAN: I talked to Gisela (Epple) shortly before she left about cotton tops. I felt that we can deal with the zoos, we have a newsletter to go through. But how are we to find the *oedipus* in university psychology departments, primate centers, or pharmaceutical companies?

THORINGTON: Nan (Muckenhirn) collected all that data. Just put it in the Primate Newsletter. Give us the name of the person who is going to put together the Studbook. Keith Hobbs did the same for Europe just a couple of years ago and should still have a pretty good handle on who might have cotton tops in Europe.

BLOOD: Speaking only secondarily about research needs and on the importance of this animal as a model for some kinds of cancer work, we have to first of all face the concern for maintaining stocks for zoos. If there is a consortium approach to pool whatever is available now, and animals are maintained for zoo purposes, then, if possible, from the production, we might have some animals for research purposes. We should even think of a third possibility; if this is really successful, there might be some animals that could go back to their native habitat and help repopulate a reserve, if that ever becomes possible. But I think if something is put together that makes good sense and there is honest-to-goodness collaboration between the institutions that now have these animals, then there is a very good chance of funding to make this possible with those very objectives that I mentioned and in that order.

THOMAS: Now we have all these elements together, who is going to do the Studbook. (P. Scollay volunteers.)

KLEIMAN: How are centers going to be designated?

THOMAS: I think we are first going to have to find out where the animals are, who is willing to serve, and who can serve.

BLOOD: And, who is likely to be most acceptable.

KLEIMAN: Did you mention the word money before? (Laughter.)

BLOOD: Didn't I say something like support?

KLEIMAN: Seriously, for zoos, when we go into breeding in reasonable numbers, we have always got to think about building off-exhibit facilities. We have to think about four to six pairs of animals per zoo. The same would be true for any research institution that had *Saguinus.* If they had to put a lot of space into a breeding facility and could not use the animals initially, there would have to be some financial support from some agency to keep the colony going.

BLOOD: I am confident that, if there is effort made by different institutions by contributing their animals or loaning them or otherwise, this in itself would make this very attractive to funding institutions.

POOK: I think it is worth considering breeding policies at this stage. I see that there are two alternatives. Either you can go for mixing and matching through the whole population or you can have inbreeding only within zones. There must surely be some one who knows enough about genetics to tell you which is the best policy.

THORINGTON: I would suggest that perhaps we could put together something on this: what the options are and what the problems are. In most lab colonies when you inbreed them for eight or nine generations, you go through a bottleneck effect. There is a real problem getting that tenth generation, if you are doing brother-sister matings. Now if it is outbred slightly more than that, then it takes longer. But I think the experience of the lab animal people could be brought together in a fairly brief synopsis in terms of the options and the problems that may come up under different breeding strategies.

THOMAS: Who can we have do that now?

THORINGTON: I was going to recommend Devra (Kleiman). (Laughter.)

THOMAS: I thought she was going to recommend you.

KLEIMAN: But who makes the ultimate decisions?

THOMAS: You can offer the alternatives, but whoever owns the animals is going to have to make the decision.

KLEIMAN: But then you have already lost the possibility of a policy.

THORINGTON: If you know what the different options are, you can inbreed for a certain period of time (zoo directors are only going to outlive so many generations of marmosets) and then you can outbreed at that particular point that you get in a congenial guy. You may have to adopt your policy to the political realities. You can use a mixed strategy here of outbreeding between the marmosets owned by compatible people, keeping track of how much inbreeding is going on, so you know how much outbreeding you want subsequently.

KLEIMAN: At the moment, Bridgwater from Minnesota is very involved in a new zoo inventory system called ISIS. It is being organized by Dr. U. S. Seal in Minnesota. I gave Seal the *Studbook* shortly before this meeting. They ran it through their computers and found very little existence of inbreeding to date in the golden marmoset populations, which is not, of course, surprising. But if we put the *Studbook* material on a computer and we have programs for determining levels of inbreeding over the years, we ought to be able to make decisions like that. If we have to make our decisions politically, we can, I suppose.

THOMAS: Who here today has a sizable representation of cotton tops?

CICMANEC: We have twelve pairs with inbreeding in three or four. But we would want to do serology tests before we would reintroduce any into the general population of cotton tops.

THOMAS: But the point is, you would be willing to join in some sort of a joint effort on their propagation.

CICMANEC: Let me explain. In contract research, it is really the National Cancer Institute that owns the animals, but I guess by fudging the reports, they may go out and see a white-lipped marmoset that they think is a cotton top. (Laughter.)

THOMAS: That sounds like an old zoo ploy.

CICMANEC: We have learned something at this meeting. (Laughter.)

KLEIMAN: But, even with a contract, I would have thought you could justify cotton-top breeding because if those animals are needed in cancer research, it is to your benefit to ensure that a viable population develops.

CICMANEC: That is exactly right. But their position is, let's do the cancer research and get the supposed answers now rather than think about the availability of the species or see that that species continues to exist. It is basically a matter of carrying these arguments to the director.

NICOLL: My cotton-top group is certainly not all that large. Right now I only have five breeding pairs with two more ready to be set up. But I don't have facilities for handling more than twelve breeding pairs.

THOMAS: If there could be brought together a really sizable group, at a zoo, university or lab, or whatever; just so that their future is guaranteed with very carefully drawn guidelines on what happens to them.

BLOOD: I would like to be educated just a little more. What is the procedure for those who have golden lion marmosets in terms of whether or not they collaborate toward the breeding. I gather there are some institutions who do and some do not. Do they just go their merry way or is there some place where the whole world can see that some people are working hard and others are not taking any part. Because, if there is a lesson to be learned from that, then maybe we can put that into the program for the cotton tops in order to avoid this.

KLEIMAN: Everything is voluntary. There is no way to force any zoo (and I'm sure it is going to be true for the research institutions and universities as well) to exchange, loan, or concentrate their animals. There never has been.

BLOOD: Well, where could I look in print that would say these are the institutions having animals and these are the institutions that have chosen not to collaborate? Or that these are the institutions that have them and do positive things?

KLEIMAN: Because the committee has been so inactive over the past three years, nothing has been drawn up. There is an AAZPA meeting in Calgary to which Warren (Thomas) is going. I am going to give him all the information on where animals are and recommendations for moving them. Within the AAZPA meetings, some statement has got to be made publicly about those institutions that are not helping out.

THORINGTON: I would like to make a plea that when these rare and endangered species die, that they go some place where they will be retained. I think this is very desirable from the standpoint of the biologist, because then we have more material to work with. It can also in the future be very desirable to know whether morphological traits you suddenly discover in your colony did exist in the founder stock. There are some retrospective questions that you may want to ask. So anything that is left over after autopsy, any carcasses, putrifying specimens that you didn't discover in time, let's try to save everything we can of these. I am willing to accept them for the Smithsonian's Natural History Museum and I am sure lots of other people are.

NICOLL: How do you want them saved?

THORINGTON: Most of them I am saving in 10 percent formalin and the better preserved they are, the more they can be used. If they can be perfused with 10 percent formalin, they can be used by the pathologists or the anatomists. If you can't do that and you can freeze them, then I'll preserve them and they can be used by the anatomists. If they can't be preserved, then we will save the skeletal parts or anything. Yes, on these endangered things, I will settle for anything.

COOPER: It has always occurred to me that it is worth looking at stillborn and aborted fetuses. For those of you who cannot do anything else with them, I would recommend measuring, weighing, and photographing them. Dr. Gengozian had said that, if golden stillborn or aborted fetuses are put in 10 percent formalin (10 volumes of formalin to one volume of meat) and sent to his lab, he and his associates will screen them histopathologically. They will do a very thorough job. I don't think that you should expect that it will tell you a whole lot. It may end up telling you that there is nothing histopathologically abnormal, and that is worth knowing. Two more points also with regard to fetuses. One, the problem with sexing. Everyone I have worked with has had problems in sexing infants. I think those kinds of things would be overcome by specimens getting to histopathologists; they see the gonads. The other thing is in determining whether a fetus is, in fact, stillborn or whether it breathed and you just found it dead. You can open the chest cavity right down the sternum, cut out a little piece of the apical lung and drop it in fluid. If it floats there was some ventilation; if it sinks, there was not. And if you send Dr. Gengozian these babies in formalin, it would be a good idea anyway to open the rib cage for penetration of formalin.

KINGSTON: In the extreme case of the golden lions, there seem to me three things which ought to be thought about. If pairs fail to breed, sometimes the males were thought to be infertile. A sperm count is easy to arrange, is atraumatic so far as the animals are concerned with electro-ejaculation. In an extreme instance of this kind, can't you do that?

THOMAS: I would be very hesitant to do that with them unless I had a lot of males.

KINGSTON: Secondly, there have been triplets in which the third young has been lost. Experience has shown that marmosets never manage to rear the third young unaided. You leave them in the hope that they will get the colostrum. But if you leave that third animal too long, you will find it on the floor and not be able to save it by handrearing. If you deliberately take it away after about 24 hours, you have a very good chance. They are perfectly easy (I'm not saying lion marmosets because I haven't had any experience, but I'm talking of four other species) to hand-rear. There is hardly any overtime work involved. If they have one feeding in the evening at about eight o'clock, then they survive perfectly well.

THOMAS: For the best return in numbers, then, you are saying that you would arbitrarily pull a third animal.

KINGSTON: I would take it away after 24 hours or earlier, if it was obviously in distress. Thirdly, there is ample evidence with other species that, if you are not concerned with behavioral research and family grouping, if you take the first born young away prior to the subsequent birth, it increases the likelihood of a postpartum conception very considerably. In an animal as scarce as this, I should have thought that production is surely our whole objective at this moment.

KLEIMAN: I think there is only one case where a lion marmoset has been hand-raised and that is in Oklahoma.

WILSON: They have hand-raised two.

KLEIMAN: We don't feel comfortable about experimenting unless we absolutely have to. Now in the case of triplets, I don't think there are any cases where three were born and found alive the next day. We haven't even had the opportunity to make a decision about it, about pulling young or not.

LORENZ: At the 1972 conference, I think it was said that in Los Angeles, three were actually reared by the parents.

WILSON: At Oklahoma, the first hand-raised one was pulled on the second day. There have been several that were found weak on the cage floor that we attempted to hand-raise, but their survival in the nursery was 30 minutes to a few hours. The second one, a female, was pulled about the third day, and was successfully hand-raised and lived at least for a year. There is a complete report of the hand-rearing of this first golden lion at Oklahoma City written three years ago. It was printed, but I don't know whether it has ever been distributed.

KINGSTON: You must use incubators to hold the temperature up and never use a teat or a cloth. Feed them only from a 1-ml syringe. Do not put the syringe into their mouth, but form a meniscus on the tip of the syringe and just put it to their lips. We have done this with four species; it is very simple and very rarely fails.

THOMAS: I have the records here for the set of triplets at Los Angeles; they were born May 10, 1971, the female died September 22, 1972; another one died February 6, 1974; and the third is still living, so that pair successfully raised the triplets.

KINGSTON: But, it is extremely rare, and really just a question of mathematics.

LORENZ: The Delta Primate Center has quite an experience with electro-ejaculation. They might make information available. Regarding hand-rearing, I think the Hamptons, the Deinharts, and others have comparative experience, and even if it might not exactly be applied to the golden lion I think this should be considered.

SCOLLAY: You should be able to check the urine for sperm very successfully. Secondly, since triplets have a low chance of all three being raised, why not take a male away and try to learn how to hand-raise it.

CICMANEC: I wonder if the hand-rearing case isn't perhaps overstated or stated too optimistically. We are working with a handicap in taking those infants that have been rejected. The marmosets know a great deal more than we do about what the chances are for raising infants. I can't remember the numbers of successfully hand-reared infants, but it is not that optimistic. Dr. Levy's slide that was presented yesterday afternoon showed that it was the fourth year before we saw even 50 percent survival of *C. jacchus* to a full year. In the initial years when they had had experience, they were still only seeing about 20 percent survival in the first year. It is not as easy as you think.

KINGSTON: We have hand-reared perhaps 40 animals of 4 species. We started off with teats and with doll bottles at first. In the early losses, they all choked because we put the teats in their mouths. But using a syringe and feeding them every 4 hours for 5 feedings a day, their survival is good. Dr. Ingram who was here this morning has a *C. argentata* that was hand-reared who is all right and a little over two years old. We saw no difference in survival rates with hand-reared animals.

KLEIMAN: I don't think the question is whether or not to hand-rear animals. If babies are rejected, we have to try to hand-rear. But it has only worked maybe 50 percent of the time in goldens. The question really is whether you take babies and hand-rear them as a policy. I don't think anybody with lion marmosets would go for that kind of approach because of the associated behavioral problems.

KINGSTON: I am talking about triplets and not leaving the third animal to die with the parents.

KLEIMAN: A comment on sperm counts in urine. When I used the term fertility, I wasn't necessarily referring to the presence or absence of sperm. We also have animals with behavioral problems. One of our males was doing a lot of mating, but I had the feeling that, if he was ejaculating, he was not doing so in the female. With that kind of problem, electro-ejaculation and sperm in urine are not going to help.

SCOLLAY: It's just that Kurt (Benirschke) has done electro-ejaculation with so many different animals. It is so easy that it is something that people should know about.

INGRAM: In John Hearn's colony, they regularly do vaginal smears and see sperm at all times of the cycles. That would be a better way of seeing if they were actually mating successfully.

WILSON: With lions, in general, we hate to handle the animals, and especially not the females.

MALLINSON: As far as the zoo world is concerned, so much relies on the compatability of directors for long-term breeding programs. We have four or five excellent illustrations of animals that have been lent to us at Jersey (on the endangered list) that are breeding well. I am not talking about marmosets, but animals ranging from the orangutan to the spectacled bear and the barefaced ibis. And so we feel that we have quite a long-term future as far as these species are concerned. It is very regrettable that it is compatibility between zoo directors which is so important as opposed to compatibility amongst the animals.

THOMAS: That is not a totally European malady.

SECTION IV:

SOCIAL BEHAVIOR OF CALLITRICHIDS

Introduction

Numerous behavioral studies of callitrichids in captivity have been conducted, and many of the essential characteristics of communication, social and parental behavior have already been described. The papers in this section, however, indicate that behavioral studies of callitrichids are entering a new phase. Whereas, previous investigations were almost entirely descriptive and minimized interspecific differences, the new trend emphasizes the collection of quantitative data, the development and testing of hypotheses, and the examination of inter- and intraspecific differences.

The relative roles of the adult male and female and older juveniles in parental care are detailed in Ingram and Hoage, and both report sex and age differences in the carrying of offspring for two species. Moreover, there is an apparent tendency for animals to carry members of their own sex preferentially, a finding which has implications for the ontogeny of social behavior and role differentiation. Both Ingram and Hoage find that captive-born parents differ from wild-imported ones in the quality and quantity of their parental care. Hoage emphasizes that he was observing primiparous parents which might have produced the difference, but Ingram indicates that a captive-born pair rejected young more and carried them less subsequent to their first litter. If such behavioral differences persist in captive-born young even after several litters are born, this obviously has implications for multi-generational breeding programs, and suggests that captive-born callitrichids may be different enough from imported animals to make comparisons between the two groups unwise.

McLanahan and Green emphasize the quantitative description of vocalizations and an analysis of the contexts in which distinct calls are given in lion tamarins. Their results suggest that some call types have been under selective pressure for context specificity, i.e., they are restricted to special situations, while others occur in a variety of contexts. Pook examined the use of contact calls in a variety of experimental conditions in two species, and found species differences which may reflect broader differences in social behavior.

The studies of Box and Epple are both examples of long-term studies of social behavior; in the first case, a non-manipulative approach and, in the second, an example of one part of a series of social experiments to delineate the characteristics of the pair bond.

Epple shows that, in general, female *S. fuscicollis* tend to be more aggressive than males toward intruder conspecifics and to scent mark more during and after encounters. She suggests that the female may be primary in maintaining the pair bond.

Box's observations suggest that it is important to follow individual differences in behavior over the long term. She also points out that there may be major differences within a species based on the housing of family groups. For example, from several studies in this volume, it appears that the levels of certain forms of territorial behavior, e.g., scent marking and vocalizing, are directly related to the degree of proximity of other unrelated conspecifics. When family groups are housed in close proximity, more scent marking and territorial behavior are seen. Box also mentions differences in the frequency of play behavior in her colony versus other colonies. The cause of this is unclear, but again may relate to differences in housing.

The variety of approaches reflected in all of these chapters suggests that behavioral studies of captive callitrichids will make important contributions to our future understanding of callitrichid social structure and behavior, as they have done in the past.

GISELA EPPLE
Monell Chemical Senses Center
University of Pennsylvania
Philadelphia, Pennsylvania 19104

Notes on the Establishment and Maintenance of the Pair Bond in *Saguinus fuscicollis*

ABSTRACT

The interaction of permanently mated male-female pairs of *Saguinus fuscicollis* with strange conspecifics was studied. Twelve pairs of subjects were each given ten-minute aggressive encounters with ten strange males and ten strange females. The subjects aggressed against the strangers, female-stimulus animals receiving more overt aggression than male-stimulus animals, particularly from the female subjects. The role of overt aggression and of scent-marking behavior in enforcing the pair bond and group integrity is discussed and reviewed with reference to earlier studies.

Introduction

During the past ten years my studies on several species of the Callitrichidae have shown that reproductive success of females maintained under laboratory conditions is strongly influenced by their social status within the group (Epple, 1967, 1970, 1975). In *Callithrix j. jacchus, Saguinus oedipus geoffroyi* and *Saguinus fuscicollis* groups consisting of several adults of both sexes, there was only one breeding female. In family units which had grown from a breeding pair and contained mature female offspring, only the mother continued breeding. In artificially formed groups, only one of the females, the dominant one, reproduced. Adult females, once removed from their parental group or from the presence of a dominant female and paired with a mature male often reproduced within a few months. The same happened when the dominant female was removed and one of the other female group members obtained the α position.

Other researchers have confirmed this observation (DuMond, 1971; Rothe, 1975). Moreover, Dawson's (1977) field study on *Saguinus o. geoffroyi* suggests that even wild groups of this species have only one breeding female.

All of these findings indicate that the dominant female or mother in some way inhibits reproduction in her daughters and in nonrelated subdominant females, either behaviorally or physiologically or both. Obviously, the mechanisms involved in this inhibition may be extremely complex.*

Under laboratory conditions *Saguinus fuscicollis* as well as *Callithrix j. jacchus* show a strong tendency to establish long-term pair bonds (Epple, 1975). The establishment and maintenance of a permanent pair bond might well be one of the necessary prerequisites for successful reproduction. Its inhibition, on the other hand, might directly influence reproductive physiology or might be the cause of behavioral and physiological events which, in turn, affect reproductive success. Since we feel that under the conditions prevailing in our laboratory pair bonding is not only an important factor in reproductive behavior but is also reasonably open to experimental and observational studies, we have looked at various possible mechanisms which are important in establishing and maintaining a pair bond in *Saguinus fuscicollis.*

**Editor's note:*
Hearn (this volume, 1977) indicates that in *Callithrix jacchus* the inhibition is physiological in that such subdominant females do not exhibit hormonal changes indicative of an ovarian cycle.

Intragroup Social Interactions

Family groups and artificial groups containing several nonrelated adults of both sexes have been under long-term observation. Family groups tend to be much more stable than artificial groups. Patterns of aggressive behavior, such as threatening or attacking, are very rarely seen in family groups while they occur more frequently in artificial groups. Moreover, aggressive interactions among group members appear to be more frequent in artificial groups containing several mature, nonrelated females than in groups containing one female and several males. It appears that one female actively and often, by means of overt aggression, inhibits pair formation between the other females and any male (Epple, 1975). This seems to be true even if females are outnumbered by males.

Artificial groups containing only one adult female, but several males, on the other hand, are much more stable. Very little aggression between the males is seen in these groups. Moreover, the existence of pair bonds within the group is not as obvious as in groups containing several females. Quantitative studies on groups containing two nonrelated males and one female, however, demonstrated the existence of a pair bond between the female and one of the males. It was expressed in a higher frequency of sexual and positive social interactions such as contacting, huddling, and grooming between the female and her "mate" than between the female and the second male (Epple, 1972). Sexual attachment to one partner, however, was not exclusive, since the female mated with both males. Moreover, the pair bond, though stable over extended periods of time, is not a lifetime attachment. In two of the groups under observation the females switched mates, changing from permanent association with one male to permanent association with the second male. In one of these groups the switch was associated with overt aggression between the males, making the removal of the loser of the fight necessary (Epple, 1975).

Intergroup Social Interactions

Methods

In another series of experiments the interactions of permanently mated, adult pairs with adult conspecifics which were introduced to the pairs for 10-minute test periods were studied. Each pair lived alone in a small testing room (approx. 10 × 10 feet) which was equipped with a home cage containing natural branches and a nest box. The subjects (called the "residents") were not confined to their home cage but had access to the entire test room. The animals were studied under three conditions.

1. Undisturbed pairs were observed for a series of twenty ten-minute tests (trial-free condition).

2. Strange conspecifics were introduced to each pair for a series of twenty ten-minute tests. The stranger, called the visitor, was wheeled into the test room in a wire mesh cage (2 × 3 × 4 feet) and remained confined to this transport cage throughout the test, while the subjects had the free run of the room. This was done to prevent excessive injury to the animals during the encounters. Ten male visitors and 10 female visitors were introduced to each resident pair.

3. Five minutes after the visitor had been removed from the test room, the residents were observed for a period of 10 minutes.

Under all conditions, the test period of 10 minutes was divided into 40 intervals of 15 seconds each. For each interval the subjects received a score of 1 for the performance of a variety of predetermined behavioral patterns. The experiment is still in progress with additional pairs being added. Data on overt aggression shown toward the strange visitors, on the amount of contact between mates and on scent-marking behavior under the various conditions were obtained from 12 pairs and are reported here.

We computed a cumulative score for overt aggression by summing the scores for each 15-second interval of attacking the visitor, chasing it, and of mutual fighting. During attacks the attacker attempted and/or succeeded in reaching through the wire mesh of the transport cage, scratching and clawing at the opponent. Occasionally, it succeeded in grabbing hold of a limb or tail and biting it. Attacks which resulted in contact by scratching and/or biting were not scored differentially from those in which the recipient avoided contact by running away.

A chase developed when the recipient of an attack ran away and the aggressor pursued it repeatedly, attempting to renew the attack. Most frequently the residents chased the visitors around inside their mobile cages, sometimes creating panic in the visitor. Occasionally, a visitor chased a resident along the length of the transport cage. Fighting was scored when both opponents engaged in mutual grabbing, clawing, and biting without one of them attempting to withdraw from the first attack. Fighting was an infrequent response since most attacks were answered by escape responses rather than counterattacks.

Contact scores between mates included any kind of contact. With the exception of one pair (see below), these contacts were almost exclusively friendly or nonspecific and provided a good measurement of social proximity between mates. Male and female *Saguinus fuscicollis* possess large specialized skin glands in the circumgenital area and above the sternum which they use in scent marking, rubbing them against the substrate or even the body of a social partner (c.f. Epple, 1972). When the circumgenital glands are used, not only glandular secretions are applied, but also a few drops of urine, and in the case of the female, vaginal discharge might be included. The use of the circumgenital glands is more frequent than that of the sternal gland and the data presented here refer to circumgenital marking only.

Results

1. *Aggression.* The pairs exhibited frequent overt aggression during their interactions with strange males and females; however, the degree to which the residents engaged in overt aggression varied strongly among individuals. While some pairs showed very little overt aggression toward strange conspecifics, others were highly aggressive. Within pairs, the aggression scores obtained by the mates showed a significant correlation, i.e., if one partner received a low score, its mate tended to receive a low score also (Spearman rank correlation coefficient $r_s = 0.81$; $p = 0.01$;[1]). Table 1 demonstrates the individual variability of the aggression scores and their correlation within the pairs.

Figure 1 shows that the total mean scores for male residents and the total mean scores for female residents do not differ significantly from each other. It becomes evident from Table 1, however, that there is a tendency for females to score higher in overt aggression than males and that the mean for males is strongly influenced by the high scores of 3 individuals (♂♂ 35, 70, 115). Thus in 8 out of 12 pairs the females were more aggressive than the males, in 3 pairs this relationship was reversed and 1 pair received equal scores. When the scores for males and females are considered with respect to the sex of the visitor (e.g., males against males, females against females), it becomes clear that

[1]All statistical tests follow Siegel (1956).

Table 1. Mean scores for overt aggression of individuals during encounters with both strange males and strange females. Pairs ranked according to increasing aggression scores of females.

Pairs	*Aggression Scores*	
	Females	*Males*
1 (♂ 71 × ♀ 81)	0.9	0.9
2 (♂ 55 × ♀ 6)	2.7	3.9
3 (♂ 10 × ♀118)	4.2	0.7
4 (♂ 73 × ♀ 40)	4.3	3.2
5 (♂ 89 × ♀ 19)	5.7	1.9
6 (♂ 68 × ♀ 79)	6.0	0.2
7 (♂100 × ♀103)	6.7	4.2
8 (♂ 75 × ♀102)	8.4	6.4
9 (♂ 44 × ♀ 39)	10.2	11.6
10 (♂ 35 × ♀131)	17.7	22.7
11 (♂ 70 × ♀ 80)	17.9	17.7
12 (♂115 × ♀ 7)	25.3	18.8

visiting females are the recipients of considerably more overt aggression than visiting males. This difference is highly significant in the scores of the resident females ($p = 0.011$, Walsh test) while male subjects show more individual variability. Eleven

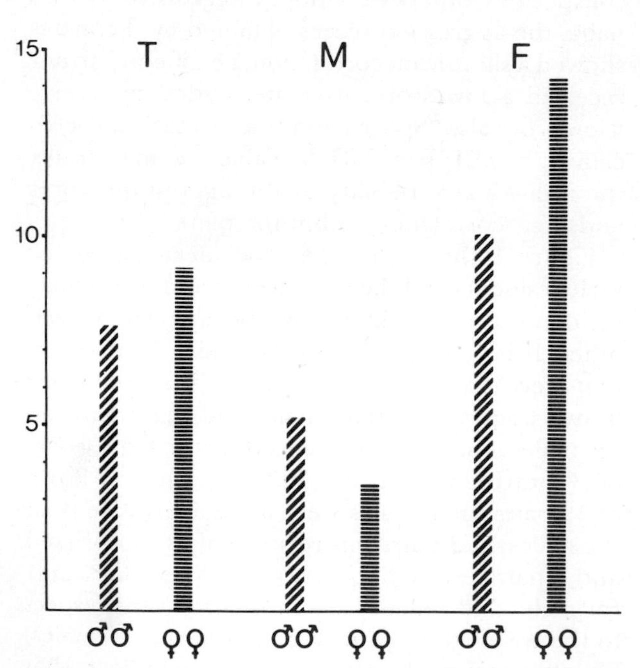

Figure 1. Mean scores for overt aggression of males and females during total encounters (T), during encounters with males (M), and during encounters with females (F).

Table 2. Mean scores for overt aggression shown toward visitors and toward the mate by ♂ 35 and ♀ 131 (pair 10).

Pair 10	*Aggression score shown toward*			
	strange ♂♂	*mate*	*strange ♀♀*	*mate*
♂ 35	7.60	10.1	37.80	2.9
♀ 131	0.89	0.7	32.8	0.2

out of 12 females aggressed more strongly against females than against males, while 4 males out of 12 were more aggressive against strange males. These results suggest that females show specifically strong aggression against conspecifics of the same sex.

While there was a considerable amount of overt aggression between the resident pair and the visitors, only one pair showed any appreciable amount of aggression toward each other. Table 2 presents the mean aggression scores of pair 10, directed at visitors and at each other. The high aggression score of the male toward his female in the presence of strange males may be redirected aggression.

2. *Contacts.* The scores for total contacts between partners (Figure 2) reflect almost exclusively

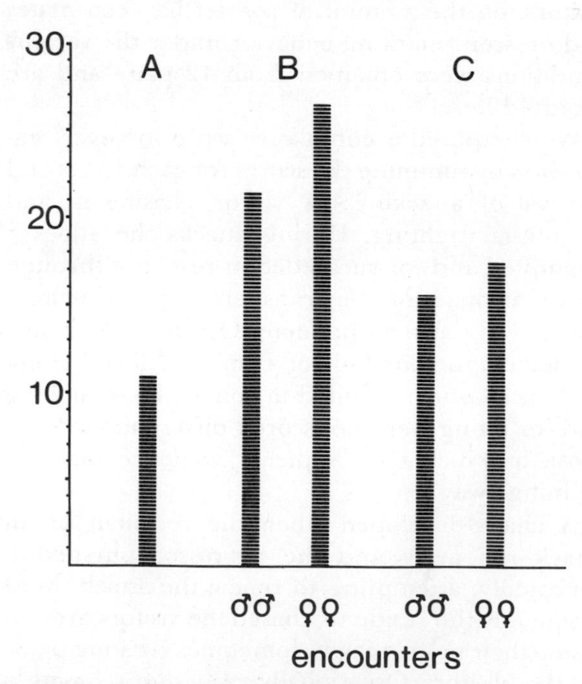

Figure 2. Mean scores for total contacts between partners under trial-free conditions (A), during encounters (B) with males and with females, and following encounters (C) with males and with females.

friendly social and sexual contacts as well as non-specific contacts except for the particular pair mentioned above. Interestingly, the contact scores between partners were significantly higher during aggressive encounters than during trial free conditions (p = 0.002, Mann Whitney U-test). Moreover, partners had more contacts with each other in the presence of strange females than in the presence of strange males (p = 0.048, Walsh test). Partner contacts, recorded for 10 minutes after the removal of the visitor, remained elevated above the level of the trial free condition (p = 0.02 Mann Whitney U-test). Contacts following encounters with females tended to be higher than contacts following encounters with males, but this difference was not statistically significant. An increase in the amount of contact between both mates in the presence of strangers probably is the result of a certain coordination in the aggressive actions of the pair, which causes both partners to be at the same location (e.g., the cage containing the visitor) in the test room more frequently than under trial-free conditions. This interpretation, however, only applies to partner contacts occurring in the presence of strangers, but does not explain the elevated scores following aggressive encounters. One wonders, therefore, whether the increase in partner contacts during, as well as after, aggressive encounters reflects a causal relationship between aggressive motivation directed at strangers and attraction between familiar partners.

3. *Scent marking.* The mean scent-marking scores of the 12 resident pairs are shown in Figure 3. As the figure shows, both partners marked relatively infrequently under trial-free conditions, but the introduction of a visitor strongly stimulated marking in males and females (p = 0.01, Walsh test). In females, marking remained elevated above trial-free levels during the 10-minute test following the removal of the visitors (p = 0.01, Walsh test). Under all conditions, whether strange males or strange females were introduced, and also during the first 10 minutes following removal of the visitors, the resident females showed more marking than the resident males (p = 0.02, Mann Whitney U-test). Male residents tended to mark more frequently in the presence of male visitors than in the presence of female visitors, while female residents increased their marking significantly in the presence of strange females (p = 0.001, Sign test).

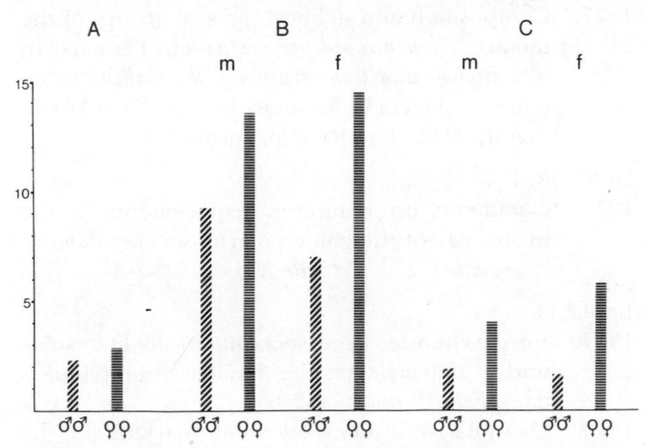

Figure 3. Mean scores for genital marking by males and females under trial-free conditions (A), during encounters with males (B-m) and females (B-f), and following encounters with males (C-m) and females (C-f).

Discussion

From the results reported above, it appears that in *Saguinus fuscicollis* the pair bond is based on social and sexual attraction between mates and is reinforced by sexual behavior and patterns of social behavior such as huddling and grooming. However, mutual sexual and social attraction is not the only mechanism involved in the maintenance of the pair bond. Dominant females and sometimes also males seem to use overt aggression, particularly directed at a member of the same sex, to prevent potential sexual competitors from *within* the group from establishing close relationships with their mates and maybe even with any other group member of the opposite sex. Against intruders from *outside* the group, a mated pair attacks in cooperation.

Under the specific conditions of testing used in this experiment, the mated pairs prevented positive social and sexual contact with conspecific intruders by overt aggression and females tended to be most active in doing so, particularly when the intruder was a female. It was my subjective impression that during the encounters resident males and females were quite frequently enticed into attacking a visitor by the behavior of their mates. As soon as one of the partners tried to establish contact with the stimulus animal, usually by mutual sniffing, the other partner approached, made contact with its mate and attacked the visitor. Quite frequently, the partner's interference appeared to excite the previously peaceful subject, who would then attack the stimulus animal. During a typical episode of this kind, one of the residents, frequently the male if the visitor was a female and the female if the visitor was a male, would exchange friendly sniffs with the stimulus animal. This behavior immediately caused the second subject to rush to the scene, squeeze itself between visitor and mate, and attack the visitor. If the visitor did not move away, an immediate attack by both partners frequently followed.

These findings suggest a basic role of aggressive behavior directed towards sexual competitors in the enforcement of the pair bond and maybe also in the maintenance of group integrity. It has to be kept in mind, however, that aggression might only be used under extreme conditions such as the testing conditions in our laboratory may represent. In a natural environment, where recipients of potential attacks are able to avoid the aggressor, overt aggression might rarely occur, and the monkeys might use more subtle behaviors in warding off sexual competitors such as long distance threat signals. Data on scent-marking behavior obtained during the social-encounter tests suggest that this behavior pattern, under certain conditions might serve as a signal of aggressive threat and as such, substitute for overt aggression.

In the present study, aggression strongly stimulated scent-marking behavior. Comparable data were obtained in a previous study of three large groups of *Callithrix j. jacchus* (Epple, 1970). Marking frequencies per hour during trial-free conditions were highest in the α-male of one group which contained more than one adult male and highest in the α-female in two groups containing only one adult male but two adult females. Ten-minute aggressive encounters with a strange male strongly stimulated scent marking in the α-male of each group, while interactions with strange females stimulated it in the α-female. Subdominant group members did not generally show an increase in marking (Epple, 1970).

The results of both studies suggest that one of the functions of scent marking in *Saguinus fuscicollis* and in *Callithrix j. jacchus* is the communication of aggressive motivation. The quality or the quantity of the odors produced during the aggressive marking might communicate the motivation of the marker to group members and strange conspecifics.

Previous studies have shown that the scent marks of *Saguinus fuscicollis* contain information on sex and individual identity. Moreover, *Saguinus* discriminated the scent marks of socially dominant males from those of subdominant males, whether on the basis of the quantity of odor deposited or on the basis of its quality or both (Epple, 1972, 1973, 1974a,b,c).

With these results in mind, it is easy to envision how scent might serve as a long-distance threat signal in intragroup and intergroup communication. A high frequency of marking, regardless of its motivation, automatically results in a large amount of this particular individual's scent in the environment. The scent identifies the individual, its sex, maybe even its social status and advertizes its presence throughout the living space, even if the animal itself is not present. The monkeys sniff the marks of conspecifics very frequently. Therefore, any social partner, be it a group mate or a stranger, must be aware of the amount and identity of the odor of an active marker throughout most of the area in which the group lives. A submissive member of the group or an intruder, after experiencing defeat by one or all of the group members, might be quite strongly affected by exposure to the superior's odor, either behaviorally or even physiologically. In this way, the scent marks of aggressive, dominant monkeys might function as a substitute for overt aggression, maintaining the hierarchy within groups and enforcing the pair bond as well as group integrity.

Acknowledgments

The study was supported by research grant No. GB 33104X from the National Science Foundation and by Research Career Development Award No. 5 K04 HD 70,575 from the National Institutes of Child Health and Human Development.

Literature Cited

Dawson, G.

1977. Composition and stability of social groups of the tamarin *Saguinus oedipus geoffroyi* in Panama. In *The Biology and Conservation of the Callitrichidae*, edited by Devra G. Kleiman, pp. 23–37. Washington, D.C.: Smithsonian Institution Press.

DuMond, F.

1971. Comments on minimum requirements in the husbandry of the golden marmoset (*Leontopithecus rosalia*). *Lab. Primate Newsl.* 10:30–37.

Epple, G.

1967. Vergleichende Untersuchungen über Sexual- und Sozialverhalten der Krallenaffen (Hapalidae). *Folia Primat.* 7:37–65.

1970. Quantitative studies on scent marking in the marmoset. (*Callithrix jacchus*). *Folia Primat.* 13:48–62.

1972. Social communication by olfactory signals in marmosets. *Internat. Zoo. Yearb.* 12:36–42.

1973. The role of pheromones in the social communication of marmoset monkeys (Callithricidae). *J. Reprod. Fertil. Suppl.* 19:447–454.
1974a. Primate pheromones. In *Pheromones,* edited by M. C. Birch, pp. 366–385. New York: North Holland/American Elsevier.
1974b. Pheromones in primate reproduction and social behavior. In *Reproductive Behavior,* edited by W. Montagna and W. A. Sadler, pp. 131–155. New York: Plenum.
1974c. Olfactory communication in South American primates. *Ann. N. Y. Acad. Scien.* 237:261–278.
1975. The behavior of marmoset monkeys (Callithricidae). In *Primate Behavior, Vol. 4,* edited by L. A. Rosenblum, pp. 195–239. New York: Academic Press.

Rothe, H.
1975. Some aspects of sexuality and reproduction in groups of captive marmosets (*Callithrix jacchus*). *Z. Tierpsychol.* 37:255–273.

Siegel, S.
1956. *Non-parametric Statistics for the Behavioral Sciences.* Toronto and London: McGraw-Hill.

HILARY O. BOX
Department of Psychology
University of Reading
Reading, England, U. K.

Social Interactions in Family Groups of Captive Marmosets (*Callithrix jacchus*)

ABSTRACT

The paper summarizes some observations on social play and allogrooming and presents quantitative data on scent marking within captive groups of common marmosets, from January 1973 until June 1975. Three family groups and one male-female pair were observed. The main purpose of the study was to examine changes in behavior over relatively long periods of time, and to examine changes in grooming and marking behavior relative to the reproductive cycle of two females. With respect to the quantitative data, adult pairs of marmosets appear to mark differently in established family groups than they do either as single pairs or in groups composed of several unrelated adults. Moreover, there appear to be predictable trends in marking behavior, especially of adult females during different stages of pregnancy. However, these trends appear to change as the size of the family increases and/or some forms of social contact (e.g. auditory and olfactory) with other groups are enhanced or relatively restricted.

Introduction

My interests in common marmosets (*Callithrix jacchus*) include a variety of aspects of social behavior, but especially, as a long term goal, their capacities for socially mediated learning. The marked variation in behavior of individual monkeys, and of family groups, however, makes long-term observations of at least some aspects of behavior a prerequisite to experimental manipulations. Hence, I have been concerned so far with sampling spontaneously occurring behavior within social groups.

In this context I have demonstrated gross individual, gender, and age-class differences for some categories of behavior, both among and within captive groups (Box, 1975a). I have also examined the ontogeny of those categories of behavior within a family group (Box, 1975b). The following data relate to some aspects of social play, allogrooming, and especially scent marking in several family groups.

Methods

One family group of marmosets was housed in a wooden cage which measured 6 ft × 4 ft × 4 ft (1.8 m × 1.2 m × 1.2 m). Two other families lived in wooden cages which measured 3 ft × 2 ft × 7 ft (0.9 m × 0.6 m × 2.1 m). A single pair of marmosets lived in a cage which measured 2.5 ft × 2.5 ft × 7 ft (0.75 m × 0.75 m × 2.1 m). These cages were furnished with an ample supply of branches, platforms, and two nest boxes. The groups were often housed in different rooms, but where there was more than one group per room, they were visually isolated. Food was available ad lib., and the light:dark cycle was 10L:14D. Temperature was maintained at 72°F–76°F and the humidity at 55–60 percent.

For recording, groups of animals were scanned visually. Coded observations were spoken into one channel of a stereo-tape recorder, whilst the other channel was marked at ten-second intervals by an electronic timer. Taped observations were later transferred to data sheets. Gross results were computed initially by counting the number of ten-second intervals in which particular events occurred. Most data recordings were made with the observer sitting in the room with the marmosets.

The studies were divided into different phases. In Phase I the categories of behavior were sampled over relatively short periods of time. There were six periods of observation during Phase I; in each of these, every hour of the marmosets' ten-hour day, was sampled three times, by a series of ten-minute observation sessions. The order of observing the different social groups, as well as the hourly periods of the day (e.g., 0900–1000 hrs., 1400–1500 hrs.) were randomized as far as possible. The observation periods were distributed from early January to early August 1973. During these periods, four groups of marmosets were observed for marking, allogrooming, and social play (Box, 1975a). At the beginning of Phase I, each social group included the following animals.

Group A contained a wild-caught pair approximately five years old, an unrelated male of twenty-four months old, and a nine-month old female of the pair. The younger male and female were removed from the group before Phase II. Group B contained a wild-caught pair of approximately five years old, their male and female twins of eight months old, and their female twins of three months old. Group C also contained a wild-caught pair of some five years of age, their male and female twins of seven months, and their single male of two months. Group D was made up from a wild-caught, previously mated male approximately three to four years old, and a laboratory-bred unmated female of two years old.

In Phase II (Box, 1975b), one of the four groups (Group A) was observed again for the same categories of behavior between January and June 1974. The family then consisted of the adult pair and two sets of male and female twins, of five and ten months of age. Phase II included a period of short observation sessions immediately before and after the birth of a third set of twins so that I could examine behavior with reference to the reproductive cycle.

When the third set of twins was four weeks old in February 1974, observations were increased to thirty-minutes duration, and frequently took place once a day. These extended observation sessions were continued until after the birth of a fourth set of twins in June 1974. During this period, a total of eighty observation sessions was accumulated with a maximum, but rare, gap of five days between any two of them. Observations were conducted between 1000 and 1500 hrs. The length of observation sessions was increased so that patterns of behavior of long duration but which occurred infrequently, such as social grooming and social play, could be sampled adequately.

After Phase II, a more detailed study (Phase III) of scent marking was carried out in relation to the reproductive cycle of two adult females within two family groups (A and C). For Group

A, eighty observation sessions were accumulated from July 17, 1974, until the birth of a single infant on November 30, 1974. For Group C, observations were begun on October 8, 1974, when the family consisted of an adult pair seven years of age and their male twins of nearly eight weeks of age. Observations continued through the birth of male twins on January 22, 1975, and a further birth of male twins on June 28, 1975. There was a total of sixteen sets of ten, twenty-minute observations sessions. During Phase III, observation sessions were limited to twenty minutes and occurred mainly between 1000 and 1500 hrs.

The statistical test of significance used in this paper was the Terry-Hoeffding Normal Scores test (see Bradley, 1968). The test is particularly robust and was used because the samples were small and the underlying population distribution and variance were unknown; hence, the use of common t-tests would be unreliable (Bradley, 1976).

Results

Play

General statements about the importance of social play in the normal development of monkeys and apes are familiar in behavioral science (e.g., Loizos, 1967). Moreover, some studies, such as that by Harlow and Harlow (1962) have shown that different components of play behavior such as rough and tumbling and chasing, are diagnostic of age and sex membership in captive rhesus monkeys. As I have pointed out elsewhere (Box, 1975a), the study of social play in marmosets is interesting because the sexes are relatively monomorphic physically and behaviorally, and results can be compared eventually with play in species which exhibit sexual dimorphism from an early age. Moreover, marmosets are typically reared with a twin of the same or opposite gender, a fact which might have implications for social development.

Play behavior was divided into categories. All behavior involving close physical contact such as wrestling, biting, and rough and tumbling was scored as "Rough and Tumbling." Chasing behavior, and various forms of mutual hide-and-seek, were scored as "Chasing." "Soliciting" play included instances where one animal touched and ran away from, and/or exhibited a visual hide-and-seek with another, nonparticipating marmoset.

During the last five observation periods of Phase I, data were collected on the three categories of play behavior for Groups A, B, C and D. Data for social play were also collected for Group A during Phase II. All these data may be inspected from previous publications (Box, 1975a, 1975b) for details of variation in play behavior among individuals and groups of marmosets. The following points emerged from these data. Adult marmosets never played with each other, and they played less with their offspring than the young played with each other; however, two particular adult females did play occasionally with juveniles, especially between the births of infants. "Rough and Tumble" play occurred more frequently than "Chasing." "Soliciting" play was seen least frequently. Individuals tended to spend more time playing within their own age groups; however, there was also some evidence that individuals preferred certain play partners.

Thus, social play was overwhelmingly a juvenile activity. Moreover, there were no gender differences observed in the frequencies of the three play categories. For example, both females and males exhibited chasing and biting behavior on different occasions. An important general point here is that both the quantity and quality of patterns of social play may reflect specific environmental conditions. For example, "Rough and Tumble" play may well be facilitated by the provision of a number of large flat surfaces within the cage; thin branches may be less conducive to this kind of behavior. Comparative data are not yet available.

Play was first recorded in juveniles at thirty days of age; the tendency of twins of the same litter to play together began at five to seven weeks. All three categories of play behavior apparently developed at the same rate. The persistent tendency to engage in more "Rough and Tumble" could be quantitatively demonstrated from the age of between five and seven weeks.

In these studies, no attempt was made to estimate nonsocial play; however, it was noticed that singletons, i.e., young without litter mates, were much more active in nonsocial play, particularly in the manipulation of inanimate objects, when compared with twin marmosets. They also "Solicited" play more often than young reared with siblings, and this included soliciting all members of the family group, including the parents.

Finally, it should be stated that juvenile common marmosets play frequently when living in settled family groups which are maintained in large and appropriately designed cages, with plenty of branch and platform space. Juveniles may even vigorously chase and wrestle while carrying infants. Hence, some of the more frequent patterns of social behavior to which infants are exposed, at least under some captive circumstances, are those of social play. My findings are most useful in indicating the need for a thorough examination of social play in young marmosets in which the behavior is studied with respect to age, gender, degrees of biological relationship, and social context.

Allogrooming

Quantitative data on allogrooming have been published (Box, 1975a, 1975b). The main findings are as follows. There was considerable variation in allogrooming scores both within and among social groups. However, adult pairs predominated in allogrooming activities, although there was variation among the pairs as to which adult was primarily involved. A principal "groomer" tended to groom all other marmosets more than the mate, and juvenile involvement in allogrooming both as a recipient and as a participant increased with age. Allogrooming frequencies appeared to increase before the birth of infants, but, given the variation of scores among groups of marmosets, more evidence is required to substantiate this finding. Finally, young marmosets were regularly observed to groom themselves from two to three weeks of age, and brief allogrooming was informally observed from the seventh week onward.

Scent Marking

Common marmosets exhibit high-intensity circumgenital marking, but a low frequency of sternal marking. My observations of scent marking began with four groups of marmosets (A, B, C and D) in a series of observation sessions (Phase I, Box 1975a), which was followed by a more detailed investigation of one family (Group A—Phase II, Box, 1975b) in a series of short observation sessions before and after the birth of a third set of twins (periods a and b, respectively), and in a subsequent period (c) in which eight sets of ten, thirty-minute observation sessions were accumulated from the time that the third twins were four weeks old until the birth of a fourth set of twins in June 1974.

The results of Phase I (Box, 1975a) indicated that individual marmosets varied in scent-marking frequencies. The highest overall scores were clearly made by adult males, but there was considerable variation within the adult pairs over periods of observation. Further, younger animals of equivalent age groups exhibited similar scent-marking scores independent of gender differences. In Group A during Phase II, the marking of both members of the adult pair was elevated during the two-week study period before the birth. Both older juveniles also showed more marking during the same time period, but there was no detectable differences in frequencies between them.

During period c of Phase II there were again clear differences between the marking of the adult male and his mate. In general, the adult male marked relatively less during this period than he had done previously. By contrast, the adult female showed marking scores which were similar to those in the four weeks after her previous birth, but notably increased her marking score six weeks before the birth; this continued, with some decline, until the birth. A similar but not so large increase in scent marking before the previous birth, which was only observed during the last two weeks, has already been mentioned.

The scent-marking scores of the oldest juveniles were similar across sets of observations for most of the time, and resembled scores seen during the first four weeks after the birth of the last infants. In the two weeks before the birth, however, the marking of both the male and his twin sister increased to a level which approximated their pre-birth scores for the previous birth. These juveniles had passed the age of approximate physical maturity during these observations. The other juveniles did not increase their scent marking this time, and their marking was similar to that of the older juveniles, for much of the time.

The variability of marking scores among individuals within this family group encouraged more detailed investigation. Hence, the family was observed during a further eight sets of ten observation sessions, which were limited this time to twenty minutes per session (Phase III).

During Phase III, after the first four sets of observations, the whole family was transferred to a much larger cage, 8.5 ft × 9 ft × 8 ft (2.6 m × 2.7 m × 2.4 m), in another room, but with their own nest boxes, perches and platforms. Thus, marking scores were compared under two conditions of housing. In essence, the observed levels of marking for the adult pair and their oldest male offspring altered between the first and second four sets of observations (Table 1). The adult male showed a significant decrease in marking ($p = .02$), the adult female also decreased her marking significantly ($p = .05$) whereas that of the ♂1 significantly increased ($p = .02$). There was no significant difference in marking over the first four and second four sets of observations for the oldest female offspring ♀1. Moreover, the adult male showed a further decline in marking relative to

Table 1. Scent marking scores for each member of Group A during the eight sets of observations in Phase III.

	Sets of Observations							
Animals	*1*	*2*	*3*	*4*	*5*	*6*	*7*	*8*
M	38	31	34	49	6	8	3	3
F	60	87	84	85	26	58	55	62
♂ 1	21	11	9	15	108	100	81	51
♀ 1	18	6	20	29	15	50	46	33
♂ 2	6	—	2	11	2	4	6	4
♀ 2	5	2	3	6	4	11	9	5
♂ 3	3	10	4	5	9	2	12	10
♀ 3	3	0	0	4	6	3	3	2
♂ 4	1	0	0	6	4	1	3	2
♀ 4	1	1	1	0	0	0	0	0

the previous eighty observation sessions, i.e., those as reported in Box (1975b) for eighty 30-minute observation sessions between February 21, 1974, and June 27, 1974 (Phase II, period c). During that series of observations his total average percent of marking frequencies[1] was 4.3 with a range across the eight blocks of ten observation sessions of 2.9 to 5.6. In this present series of observations, however, the overall average percent of his marking frequency over the first four sets (Table 1) was 3.2 with a range of 2.6 to 4.1 across the sets, and in the second four sets (Table 1) the total average percent was 0.43, with a range across them of 0.3 to 0.6. By contrast, the adult female sustained a relatively high marking level during the first four sets of Phase III (average percent 6.6). After set 4, however, her marking scores also dropped, but then were higher over sets 6, 7, and 8 toward the birth of an infant at the end of set 8. Her marking before the birth was equivalent to that of previous observations. For example, her average percent marking in set 8 before the previous birth was 5.1 (Box 1975b), and was also 5.1 in set 8 of the present series (Table 1). The significant increase in marking of the ♂1 over the second four sets of observations during this phase was particularly interesting in that it began shortly after the family was transferred to a new cage; moreover, his marking also declined toward the birth of the next young, to be below that of the adult female in set 8, whereas it had been more than that of the female in sets 5, 6, and 7 (Table 1). Further, ♂1 continued to mark more than the adult male after the birth of the next infant.

The decrease in scent marking by the adult male during Phase III was followed by other behavioral changes. For example, he made few visual threats (whereas his son did) and was generally timid; moreover, his play behavior with younger juveniles increased. On the other hand, the behavior of the adult female did not alter. Interestingly, there was no apparent friction between the adult male and his son ♂1; in fact, they were often observed in close proximity, and frequently nuzzled against one another. Friction did arise, however, between ♂1 and his younger brother ♂2, to such an extent that ♂1 was removed from the group in March 1975 when he was twenty four months old.

In Phase III, more detailed information on scent marking was collected. I first examined whether the adult pair marked in close temporal conjunction, defined by the adult male or female scent marking after the mate within three, ten-second intervals (thirty seconds). Since marking loci were coded, it was also possible to indicate whether the adults marked the same, or different loci, in relatively close temporal conjunction (Table 2).

These data were organized over the same sets of observations as in Table 1, but include the slightly higher *total* marking frequencies, rather than the number of ten-second intervals in which marking occurred. They are, however, expressed as a percentage of the latter to indicate the number of markings in temporal and physical conjunction with respect to the more general marking scores shown in Table 1. The Table also shows the percentage proportions of sternal marking that were observed for this group during Phase III of the study.

With regard to sternal marking, it was clear that all this marking was done by the adult male, that it occurred only in the first four sets of observations, and that it increased notably over sets 3 and 4.

From these data also there was no significant difference ($p = .20$) in the frequencies of the temporal contiguity of marking for either the adult male or female in the first four sets compared with the second four sets of observations. The frequency with which the male marked within thirty seconds of the female was, however, significantly greater than the other way round ($p = .02$). The adult male also frequently marked the same loci as the female in temporal contiguity. No other

[1]This measure is derived by dividing the marking scores (number of 10-second periods in which marking occurred) by the total number of 10-second periods per set of observations, e.g., in Table I. There are 1,200 10-second periods in ten 20-minute observation sessions (one set of observations), while in Phase II, period c, there are a total of 14,400 10-second periods during eighty 30-minute observation sessions.

Table 2. The percentage scores of the total scores (Table 1) per set of observations that each adult of the pair of Group A marked within thirty seconds of the other, together with a figure in brackets which shows the proportion of the unbracketed figure in which each animal was observed to mark on the same locus as the other. The percentage proportion of sternal marking is also given for each adult during each set of observations.

Sets of Observations	*Adult Female After Adult Male*	*Adult Male After Adult Female*	*Frequency of Sternal Marking*	
			Adult Male	*Adult Female*
1	3.3	23.7 (23.7)	5.3	0
2	5.7 (2.3)	25.8 (22.6)	3.2	0
3	3.6	41.2 (29.4)	8.8	0
4	9.4	24.5 (16.3)	20.4	0
5	7.7	33.3 (33.3)	0	0
6	0	0	0	0
7	0	0	0	0
8	0	0	0	0

dyad of marmosets in the family marked as much in close physical conjunction. On the other hand, the ♂1 was observed to mark more in physical contiguity with respect to the adult female during the first four sets of observations than during the second four sets ($p = .05$—Table 3). No other comparisons in this table were significant.

Observations of marking loci were conducted not only to observe the relative physical contiguity of marking between specific marmosets, but also to find out whether different substrates within the captive environments were marked preferentially. For this family, and for the first four sets of observations, a total of sixteen specific loci was recognized, in addition to a general category of "central branches" in which marmosets marked branches, but not at frequently used loci. Marking of the wires of the cage was never included in the results.

Table 4 shows the combined scores for the 17 marking locations for all the family over the first four sets of observations, and Table 5 the equivalent data for 19 marking locations over the second four sets of observations. From these data the following general observations can be made. The marmosets marked each other rarely; they also infrequently marked nest boxes, and nonspecific loci on branches. Food bowls, food platforms, and food itself were marked considerably more than those locations already mentioned. In the first four sets but not in the second four sets of observations, the top platform of the cage was frequently marked. Indeed, most scent marking took place on specific loci on branches that were mainly horizontal or, less frequently, one that was at 45° in the cage. Hence, the marmosets marked locations specifically, even the younger marmosets with relatively low overall scores.

From time to time new scent-marking loci were established by much chewing. Also, particular established loci were marked more frequently at some times than at others. The cause of this variability is not known, but further studies would be useful, especially with reference to which member(s) establish new loci.

After the removal of ♂1 from Group A, the family was observed to determine if there would be changes in marking behavior (Table 6). These data refer to the number of ten-second intervals in which marking occurred in twenty-four, ten-minute observation sessions which were distributed between March 17, 1975, until the birth of new young on May 6, 1975; they were divided into four sets of six observation sessions per set in order to emphasize their variability.

The marking of the adult male remained low and actually decreased during this stage compared with his marking during the last four sets of observation as shown in Table 1. During those last four sets of observations his total average percent marking was 0.43, whereas his total average percent marking following the removal of the ♂1 was 0.3. The adult female also decreased her marking with an average total percent of 3.8 after the removal of ♂1 compared with 4.2 previously. The ♀1 also decreased her marking with a total average percent of 1.3 compared with a total average of 3.0 during the last four observation sets of Phase III. There was, however, no substantial change in marking frequencies for the adult pair or their oldest ♀1 offspring, after the removal of the ♂1, compared with their marking frequen-

Table 3. The percentage proportion of the total scores (Table 1) per set of observations that particular combinations of marmosets marked within thirty seconds of the other, together with the proportion in brackets, of the unbracketed figure, in which each animal was observed to mark on the same locus as the other.

Sets of Observations	*♂ 1 After Female*	*Female After ♂ 1*	*♀ 1 After Female*	*Female After ♀ 1*	*♀ 1 After ♂ 1*	*♂ 1 After ♀ 1*
1	14.3 (4.7)	8.3 (1.7)	5.6	3.3	5.6	9.5
2	0	1.1	0	1.1	0	0
3	22.2 (11.1)	3.6	15.0	6.0	5.0	0
4	13.3 (6.7)	2.4	24.1 (6.9)	2.4 (1.2)	3.4	6.7
5	7.4 (2.8)	19.2	0	0	20.0 (6.6)	0.93
6	10.0 (4.0)	20.7 (5.2)	10.0	6.9	10.0	10.0 (2.0)
7	6.2 (1.2)	18.2 (1.8)	10.9	5.5	2.2	8.6 (2.5)
8	9.8 (5.9)	3.2	15.2	8.1	9.1	7.8

Table 4. The frequencies of marking for each member of Group A on seventeen marking locations during the first four sets of observation during Phase III. Loci 1, 2, 3, 7, 10, 11, 12, 13, 15, and 16 were at specific places on branches, locus 14 was the general category "central branches." Loci 4, 5, and 6 were on the top platform of the cage. Loci 8 and 9 were the food bowl and the food platform, and locus 17 was "other monkeys."

	Scent-marking Loci																
Animals	*1*	*2*	*3*	*4*	*5*	*6*	*7*	*8*	*9*	*10*	*11*	*12*	*13*	*14*	*15*	*16*	*17*
M	26	12	49	6	5	8	4	0	3	5	4	19	0	2	8	6	1
F	37	27	67	25	48	17	20	1	16	2	2	42	0	7	4	23	1
♂ 1	7	1	1	1	0	4	0	3	2	8	1	12	2	3	2	9	0
♀ 1	9	4	4	3	5	16	1	1	1	2	0	7	0	8	3	10	0
♂ 2	3	0	0	0	1	4	0	0	0	1	1	1	0	4	1	3	0
♀ 2	1	1	0	2	0	6	0	1	0	0	0	0	1	1	0	3	0
♂ 3	4	7	4	0	1	2	0	0	0	0	1	3	0	0	1	1	0
♀ 3	0	0	1	0	0	2	0	0	0	1	1	2	0	0	0	0	0
♂ 4	2	1	3	0	0	1	0	0	0	0	0	0	0	0	0	2	0
♀ 4	0	0	0	0	0	1	0	0	0	0	0	0	0	2	2	0	0

Table 5. The frequencies of marking for each member of Group A on 19 marking locations during the second four sets of observations during Phase III. Loci 1, 2, 3, 4, 5, 6, 7, 8, and 13 were on branches, and locus 14 was the category "central branches." Loci 9, 10, and 18 were the food platform, food on the food platform, and the food bowl. Locus 12 was the top platform. Loci 15 and 16 were two separate nest boxes. Locus 17 was the floor of the cage, and locus 19 was "other monkeys."

	Scent-marking Loci																		
Animals	*1*	*2*	*3*	*4*	*5*	*6*	*7*	*8*	*9*	*10*	*11*	*12*	*13*	*14*	*15*	*16*	*17*	*18*	*19*
M	5	2	0	0	0	2	3	0	0	0	2	0	6	0	0	0	0	0	0
F	15	15	2	1	3	14	61	27	13	7	4	0	42	0	1	0	0	6	0
♂ 1	24	83	1	1	13	7	57	71	15	4	16	0	43	2	1	0	0	13	0
♀ 1	5	24	0	1	9	12	18	20	1	2	16	0	39	0	2	2	0	0	0
♂ 2	0	0	1	0	0	2	3	2	2	0	0	0	4	0	1	0	2	0	0
♀ 2	3	2	1	0	1	2	1	0	8	5	5	0	0	0	0	0	2	3	0
♂ 3	7	6	0	0	2	0	2	4	4	1	1	0	7	0	1	0	0	1	0
♀ 3	1	2	0	0	1	0	1	1	2	0	2	0	3	0	1	0	1	0	0
♂ 4	0	1	0	0	3	1	0	1	2	2	0	0	1	0	0	0	0	1	0
♀ 4	0	0	0	0	0	0	0	0	0	0	0	0	0	0	0	0	0	0	0

Table 6. Frequencies of marking for members of Group A after the removal of the ♂ 1.

Animals	*Sets of Observations* 1	2	3	4
M	0	1	1	4
F	21	25	26	18
♀ 1	3	9	8	11
♂ 2	4	11	2	3
♀ 2	0	3	2	1
♂ 3	3	3	9	0
♀ 3	0	2	0	4
♂ 4	0	3	6	0
♀ 4	0	0	0	1
♂ 5	0	0	0	0

cies during other sets of observation; hence, for these animals there was a *general* decline in marking as the project continued with no evidence that it was particularly affected for any of them by the presence or absence of the ♂1.

By contrast, the behavior of the ♂1 himself changed considerably when he was paired with an unmated adult female. Apart from frequent initial attempts to copulate, he was not observed to make visual or vocal threats. Moreover, the relative frequency of marking declined with a total average percent of 3.5 in these observations (Table 7) compared with a total average percent of 7.1 during the last four observation sets in which he lived with his family. Again, he consistently marked less than his cage mate. This latter observation was one that we have noted informally for a number of new male-female pairs of marmosets, but had observed formally for only one other group (Box, 1975a). Table 7 shows data which were obtained to sample the marking behavior of the ♂1 of Group A from the first day in which he was paired with a female on March 17, 1975, until May 27, 1975. Again, ten-minute observation sessions were divided into a number of sets in which there were six observation sessions per set.

The data for Group A encouraged the collection of additional marking data for Group C (Phase III). Beginning in October 1974, marking was examined for the four members of this group up to, and immediately after, the birth of second twins (Table 8) and again up to the birth of a third set of twins (Table 9).

Figure 1 shows day by day marking scores for the adult male and female both before and after the second set of twins on the thirty-first day. From Table 8, it can be seen that the adult female changed from a marking level below that of the male in the first four sets to a much higher level of marking in sets 5 and 6. In set 5, the sudden increase in her marking also coincided with a drop in marking by the male. The marking of the female dropped considerably the day of the birth of twins, and remained low during set 8; by set 9 (Table 9) her scores were again similar to the first four sets in Table 8. During sets 9 to 16, the family was housed alone in a separate room and although the adult female's marking again increased to a level a little above that of her mate during the later stages of her pregnancy (set 15), the increase was smaller than before. However, statistical tests on the marking frequencies for both the adult male and female

Table 7. The frequency of marking for the ♂ 1 (M) and a new female in three sets of six ten-minute observation sessions.

Animals	*Sets of Observations* 1	2	3
M (♂ 1)	23	20	21
F	67	43	42

Table 8. The frequencies of marking for each member of Group C during the first eight sets of observations of Phase III.

Animals	*Sets of Observations* 1	2	3	4	5	6	7	8
M	117	105	84	80	53	97	85	63
F	37	50	33	47	142	134	71	10
♂ 1	3	1	3	5	6	7	6	12
♂ 2	—	3	2	7	3	8	7	5

Table 9. The frequencies of marking for each member of Group C during the second eight sets of observation during Phase III.

Animals	*Sets of Observations* 9	10	11	12	13	14	15	16
M	74	92	71	83	86	72	54	57
F	38	38	30	43	42	42	67	25
♂ 1	5	13	19	19	11	18	9	11
♂ 2	4	11	18	29	27	33	24	32
♂ 3	—	1	1	4	4	6	5	1
♂ 4	—	6	6	13	14	25	28	11

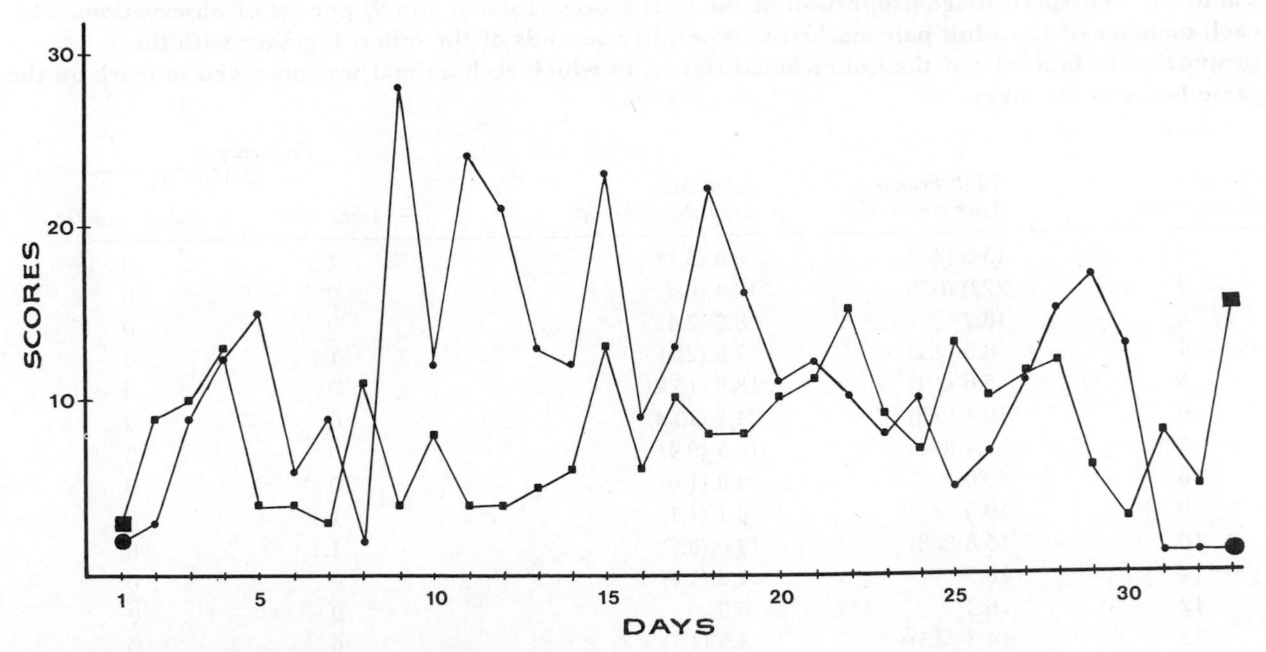

Figure 1. Individual marking scores for the adult male and his mate of Group C during a period of 30 days before the birth of young. The birth occurred on Day 31. (Solid squares, adult male; solid circles, adult female.)

over the first eight sets of observations, compared with the second eight sets, did not show significant differences.

The frequencies of marking for each of the older juveniles significantly increased over the second eight sets of observations compared with the first eight (for ♂1 $p < .01$, for ♂2 $p < .002$). It was also interesting to note that in the later sets of observations, i.e., from set 12 onward, one young male of each twin pair consistently marked more than the other.

Table 10 indicates the proportion of times, per set of observations, that the adult pair marked within thirty seconds of one another, together with the proportion of times that each marmoset was observed to mark on the same locus as the other. As for Group A, the percentages in this table were calculated from the slightly higher total marking frequencies represented in terms of the marking frequencies in Tables 8 and 9.

As in Group A, the adult male tended to scent mark more in temporal and physical conjunction with his mate than the other way around. Again, this trend was less noticeable during the second eight sets of observation than during the first eight sets. Statistical tests, however, showed only one comparison to be significant, namely, that the adult female marked more in temporal contiguity after the male during the second eight sets of observation than during the first eight sets ($p < .02$). Comparisons were made for both marmosets in the first and second set of observations as well as between them over the same periods, and included estimations of both temporal and physical contiguity between the pair.

The marking of specific loci can be seen in Table 11 which shows the total number of times that each family member was observed to mark thirteen identifiable loci, and a general category of "central branches." Like Group A, the marmosets of Group C rarely marked other monkeys; they also infrequently marked nonspecific loci on branches, or the floor. Like Group A during the first four sets of observations when they lived in similar circumstances to Group C, the adults of Group C marked substantially more on the top platform of their cage than at other locations. The same trend was also found for the older twin offspring of Group C. The loci connected with food, i.e., the food bowl, food platform, and food itself on the food platform, were scent marked relatively often, but they were most frequently marked, in relative terms, by the younger twins, and especially by the ♂4. As in Group A, particular loci were marked preferentially; the social/spatial significance of such a general finding is not known.

Table 10. The percentage proportion of the total scores (Table 8 and 9) per set of observations that each member of the adult pair marked within thirty seconds of the other, together with the proportion in brackets, of the unbracketed figure, in which each animal was observed to mark on the same locus as the other.

Sets of Observations	*Adult Female After Adult Male*	*Adult Male After Adult Female*	*Frequency of Sternal Marking*	
			Adult Male	*Adult Female*
1	13.5 (2.7)	6.8 (5.1)	4.3	0
2	22.0 (6.0)	12.4 (6.7)	0	0
3	18.2	8.3 (2.4)	0	0
4	8.5 (2.1)	7.5 (2.5)	1.3	0
5	5.6 (2.1)	18.8 (13.2)	0	0
6	10.4 (5.2)	24.7 (15.4)	0	0
7	14.1 (2.8)	16.5 (9.4)	0	0
8	40.0	4.8 (1.6)	0	0
9	10.5	8.1 (4.1)	1.4	0
10	15.8 (5.3)	13.0 (8.7)	1.1	0
11	16.7	2.8 (1.4)	0	0
12	16.3	6.0	0	0
13	14.3 (2.4)	3.5 (1.7)	0	0
14	19.0	11.1	0	0
15	9.0	18.5 (1.9)	0	0
16	20.0	1.8	0	0

Discussion and Conclusions

My observations have been mainly concerned with broad categories of behavior such as social play, allogrooming, and scent marking in common marmosets. With respect to social play for example, some few other workers have made observations, yet those of Stevenson (pers. comm.) are considerably more detailed than others. There is general agreement about our findings, however, with the notable exception that I have observed comparatively little social play which involved adult pairs, and especially adult males. With respect to scent marking, a general interest of my results is that I have examined behavior over the reproductive cycle of my breeding females, in established family groups, and over successive births. Previous studies of scent marking in marmosets, by other workers and especially Epple (e.g., 1972) have demonstrated, for instance, the importance of scent marking in the regulation of social relationships involving status among adults. My results are com-

Table 11. The total frequency of marking for each member of Group C on 14 specific loci during the 16 sets of observations during Phase III. Loci 1, 2, and 3 were on the top platform of the cage, loci 4, 5, and 6 were the food platform, food bowl and food on the food platform, respectively. Loci 7, 8, 9, 10, and 11 were loci on branches, whereas locus 12 was the general category "central branches." Locus 13 was the floor of the cage, and 14 was "other monkeys."

	Scent-marking Loci													
Animals	*1*	*2*	*3*	*4*	*5*	*6*	*7*	*8*	*9*	*10*	*11*	*12*	*13*	*14*
M	270	220	174	53	116	115	65	43	16	45	85	20	7	9
F	232	173	85	38	110	83	95	7	6	7	31	1	5	3
♂ 1	7	14	26	21	—	21	19	10	8	7	8	10	—	2
♂ 2	16	25	66	2	1	31	14	10	3	6	22	21	—	3
♂ 3	—	—	3	1	—	12	1	1	—	1	1	2	—	—
♂ 4	1	1	8	26	—	53	—	—	6	3	—	5	—	—

plementary to those of other people, and may provide hypotheses about additional functions of scent marking in marmosets. For example, it may be that increases in scent marking by adult females several weeks before a birth is connected with an increase in territorial behavior. Again, the relative decline in marking among adult pairs when housed with their families, separate from other groups, may indicate some aspects of intergroup functions such as territorial defense prompted by the presence of other groups. The relative lack of sternal marking in established family groups living separately may also lend support to this idea. On the other hand, it may be that as the family grows in size, increasing behavioral activity by juvenile members alters the relative roles which specific members, such as the adult pair, play. The family may act increasingly as a "unit" to achieve ends which were previously undertaken by a smaller number of individuals. This would seem to apply to a variety of aspects of behavior including caring for the young (Box, 1977). Further, the behavioral changes associated with young adult marmosets may provide hypotheses about the way in which they split from their families. For example, the relationship of the young adult to his siblings may be more important in some cases than an "inevitable" eviction by the parents. Actually, I have observed amicable relations between young adults and their parents in other families within our colony as well as some of strife!

All this aside, however, it is important that the variability of a wide variety of categories of behavior among different social groups of marmosets be recognized. Reliable tests for the robustness of any apparent trend require the observation of a large number of social groups of animals. Questions as to the robustness of our speculations about behavioral function await observations of our animals in nature. It is also the case that there are additional types of study that can be undertaken in captivity, which may help to elaborate our hypotheses. A good example here is the collaboration between people working on hormonal aspects of the reproductive cycle and those interested in specific kinds of behavioral changes during the reproductive cycle.

Acknowledgments

I am grateful to Jill Pook, Hazel Lee, and Gillian Cole for their care in looking after my marmosets; Joan M. Morris for much general assistance; B. R. Singer for statistical advice; and Mrs. Audrey Conner for typing the manuscript.

Literature Cited

Box, Hilary, O.
1975a. Quantitative studies of behaviour within captive groups of marmoset monkeys (*Callithrix jacchus*). *Primates,* 16(2): 155–174.

Box, Hilary, O.
1975b. A social developmental study of young monkeys (*Callithrix jacchus*) within a captive family group. *Primates,* 16(4): 419–435.

Box, Hilary, O.
1977 Quantitative data on the carrying of young captive monkeys (*Callithrix jacchus*) by other members of their family groups. *Primates,* 18: 475–484.

Bradley, J. V.
1968. *Distribution-free Statistical Tests.* Prentice-Hall.
1976. *Probability, Decision, Statistics.* Prentice-Hall.

Epple, G.
1972. South American primates in captivity: Social communication by olfactory signals in marmosets. *Inter. Zoo Yearb.,* 12: 36–41.

Harlow, H. F. and Harlow, M. K.
1962. Social deprivation in monkeys. *Scient. Amer.* 207: 136–146.

Loizos, C.
1967. Play behavior in higher primates: A review. In *Primate Ethology,* edited by D. Morris, pp. 176–218. London: Weidenfeld and Nicolson.

E. B. McLANAHAN
National Zoological Park
Washington, D. C. 20008
and
K. M. GREEN
National Zoological Park
Washington, D. C. 20008,
and Johns Hopkins University
Baltimore, Maryland 21205

The Vocal Repertoire and an Analysis of the Contexts of Vocalizations in *Leontopithecus rosalia*

ABSTRACT

The vocal behavior of three pairs and one trio of golden lion tamarins was analyzed with respect to structure and context. A total of seventeen call types was distinguished and grouped into five sound classes: trills, whines, clucks, nontonal sounds, and Long Calls (combinations of elements from each of the tonal sound classes). These call types were uttered within twenty-five separate contexts, divided under four general headings: Solo-activity, Foraging, Vigilance, and Contact. Each sound class was correlated with one of these general contexts at a high level of significance: trills with Solo-activity, whines with Contact, clucks with Foraging, and Long Calls with Viligance. Sexual dimorphism with respect to vocalizing rate and context was minimal, and this is thought to reflect the similarity of male and female roles in *Leontopithecus.* Most call types tend to occur in a variety of contexts, but those thought to relay vital information are linked to a specific context, are associated with medium to high arousal, and also show a considerable degree of stereotypy. Data on Long-Call duetting within pairs indicate that it may be instrumental in harmonizing male and female social roles, meanwhile strengthening the pair bond.

Analyses of the physical structure of the repertoire show vocal communication in the golden lion tamarin to be a mixture of discrete and graded systems.

Introduction

In its basic form, vocal communication involves a communicator, a signal, and a recipient. While the signal, primarily its physical measurements, has been the focus of many studies on primate communication, signal context and function as well as features of the communicator and recipient have been given less attention (Eisenberg, 1976). Few studies of primate vocal communication consider the interplay between the communicator and recipient. For an understanding of auditory communication, a consideration of vocalizations from both a behavioral and structural standpoint is the most comprehensive approach. To this end, we present a quantitative analysis of vocal signals in *Leontopithecus rosalia rosalia* as well as the behavioral contexts in which they occur.

We believe that information on age and sex of vocalizers, a close consideration of vocalization context—of behaviors preceding and following vocalization—and of the environment in which the vocalization occurs is prerequisite to an ascription of function. Our preliminary analysis of the vocal repertoire of *L. r. rosalia* attempts to fulfill this objective.

Little is known about the social and sexual behavior of this species in the wild, but captive studies indicate that tamarins form permanent pair-bonds and that males as well as females take an active role in caring for the young (Epple, 1967). Little sexual dimorphism in external morphology and familial roles has led us to ask if it also characterizes certain aspects of vocal behavior.

Methods and Materials

This study was concurrent with one of reproductive behavior of *Leontopithecus rosalia* conducted by Dr. D. G. Kleiman at the National Zoological Park in Washington, D. C. (see Kleiman, 1977). Subjects were maintained in either family, pair, trio (one female and two males), semi-solitary, or solitary situations, depending upon their age, sex, and length of time in the reproductive study.

Eight animals, four adult females and four adult males were selected from a potential twenty-four subjects for our study. Although subsidiary data were collected from the nonsubject individuals, these eight proved most suitable to our study in that they had formed pair-bonds.

In the Marmoset Building, subjects were maintained in housing units measuring approximately 4.6 m × 3.7 m × 4.6 m. All units were equipped with numerous branches, a feeding platform, and two or more nest boxes. Glass doors permitted access to porches also equipped with branches and artifacts for climbing and running. The ventilation system facilitated the transfer of vocal as well as olfactory information between groups. Such conditions played a significant role with respect to vocal chorusing and rate of scent marking (Mack and Kleiman, in press).

A total of 4,373 vocalizations representing 77 hours of observation was collected from the entire colony during the months of June through October 1974. Of this, 2,782 vocalizations representing 53 hours of observation were analyzed contextually by means of a 10-minute check sheet on which the behaviors of the vocalizer prior to and succeeding emission were recorded.

These data were transferred daily to a master sheet whereon sex of vocalizer, number of vocalizations, and the contexts in which they were emitted were matched. When a vocalization occurred in several contexts, as for example if an animal was grooming and huddling, the call was scored under both activities. The criteria employed in the allocation of a vocalization to a particular context included consideration of facial expression, posture, orientation of animal, intensity of locomotion, sequence of behaviors, inferred relationship between vocalizer and referent (Eisenberg, 1976), and group social relationships.

In addition, observations totaling 251.5 hours of vocal chorusing and duetting were studied six days a week between July and November 1974. This study involved the aforementioned groups, in addition to two additional pairs that were maintained in family groups and were on exhibit to the public in the Small Mammal House. These two pairs had been at the National Zoo for over four years and all were wild-born except for one female. They were housed in opposite ends of the Small Mammal House and consequently lacked the olfactory stimuli from other family groups available to the pairs in the Marmoset Building. A false ceiling in the exhibit hall, however, aided the transmission of vocal information. The other four pairs in the Marmoset Building had been formed no more than four months prior to the beginning of the study and were all captive-born except for one female.

Approximately five hours of *Leontopithecus* sounds were recorded on a Uher tape recorder (4000 Report L) at tape speeds of 7-1/2 inches per second using a Dan Gibson parabolic reflector and microphone (Model 200 E.P.M.). Spectro-

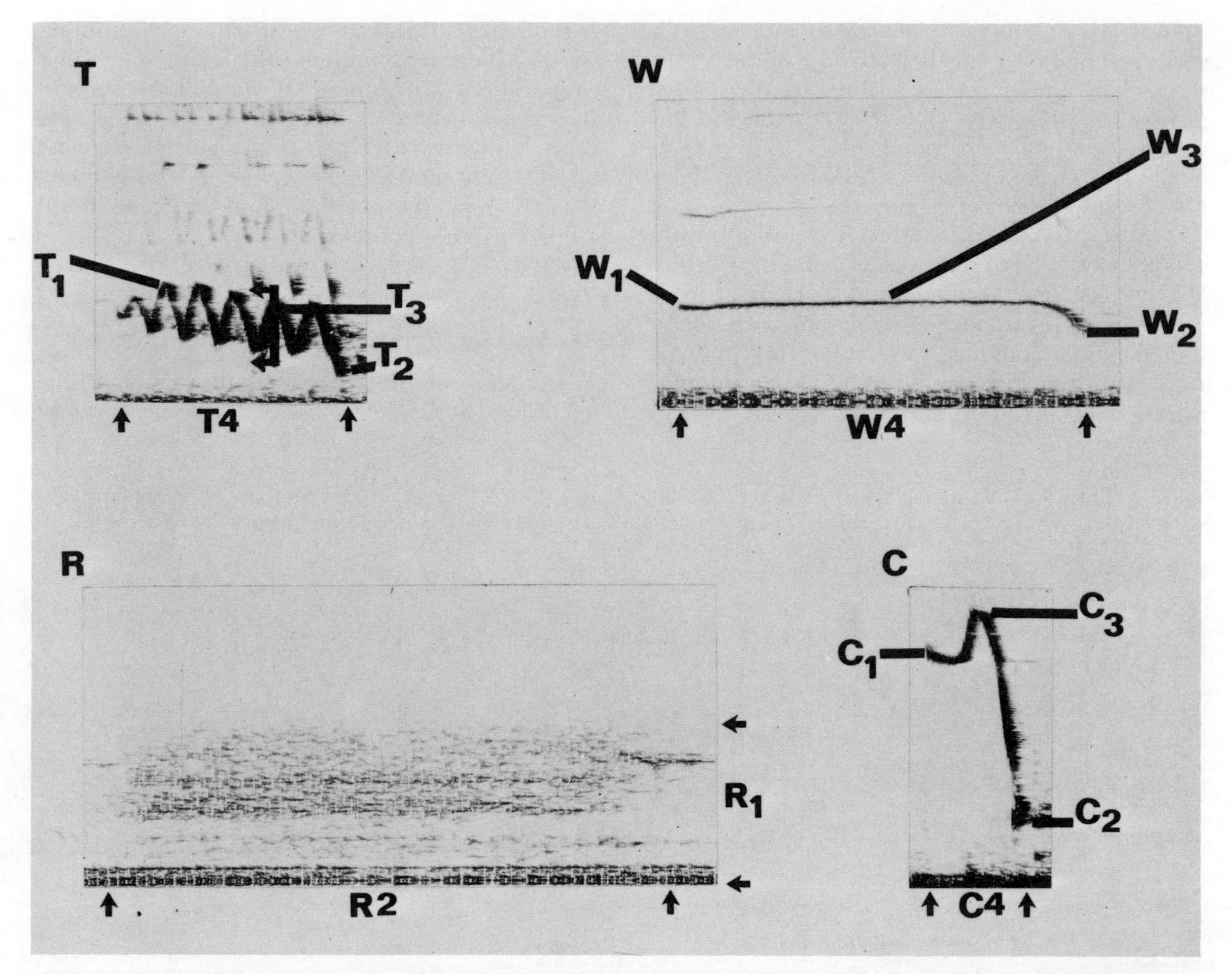

Figure 1. The above spectrograms illustrate the basic structural measurements that appear throughout the paper. The vertical axis represents frequency in kHz and the horizontal axis indicates time in seconds. The solid black bar is a time marker = 0.5 seconds. Unless noted, the narrow band filter is used.

(1) Trill. T1 and T2 indicate the dominant frequency in kHz at the start and end of the syllable, respectively. T3 indicates the sweep range of an oscillation near the mid-duration point. T4 is the time measure on the horizontal axis and represents the duration in seconds. (2) Whine. W1 and W2 indicate the dominant frequency at the start and end respectively. W3 is a measure of the frequency at the midpoint. W4 is the duration of the syllable. (3) Cluck. C1 and C2 measure the beginning and end frequency, respectively. C3 represents the upper frequency at a point of inflection, if present. C4 is the duration in seconds. (4) Nontonal sounds, represented by "R," a rasp, are noisy call notes composed of sound that is continuously developed over a wide range of frequencies. R1 represents the frequency range. R2 is the duration.

grams were prepared for analysis with a Kay Missilyzer (Model 675) using both the wide band (600 Hz) and narrow band (60 Hz) pass filters. Structural spectrographic measurements were taken with a transparent plastic grid overlay in the manner illustrated in Figure 1, following Struhsaker's (1967) nomenclature.

Sound recordings were made of 6 males, 5 females, and 2 infants in April, June, September, and October 1974. Certain observations by EBM and recording sessions by KMG occurred simultaneously, but the majority of the data are from independent sessions.

Results

Based on physical characteristics, seventeen different call types could be distinguished which are grouped into four major sound classes: trills, whines, clucks, and nontonal sounds. The fifth sound class or Long Call contains elements from each of the tonal sound classes. The vocalizations were classified by ear according to their basic physical similarities. These occurred within twenty-five contexts, divided under four general headings: (1) Solo-activities—autogroom, solitary play, scent mark, arch posture or arch walk (Moynihan, 1966), locomotory hesitation, activity completion; (2) Foraging—scan, probe, selection of food from tray, feeding, food-share, food-steal; (3) Vigilance—response to flying objects, orientation to visual stimulus, examination of stimulus from a distance, fixation upon visual stimulus during low, medium, and high arousal, alarm-flight; and (4) Contact—approach, touch, huddle, wrestle, allogroom, mutual sniff, and mount. The proportional distribution of vocalizations in the four general contexts was: Solo-activities, 9.4 percent; Foraging, 20.5 percent; Vigilant behavior, 50.7 percent; and Contact behavior, 15.4 percent. Table 1 shows the distribution of the vocalizations in the different contexts.

Trills

1. TEMPORAL PATTERN AND PHYSICAL STRUCTURE

A summary of the spectrographic characteristics of trills is available in Table 2 and examples of

Figure 2. Trills. (A) Rapid trill, (B) Trill A II, (C) Trill A descending, (D) Trill B II, (E) Trill C, (F) three notes of a Trill B 1. Frequency scales are in 3 kHz intervals. The solid black bar is a time marker = 0.5 seconds.

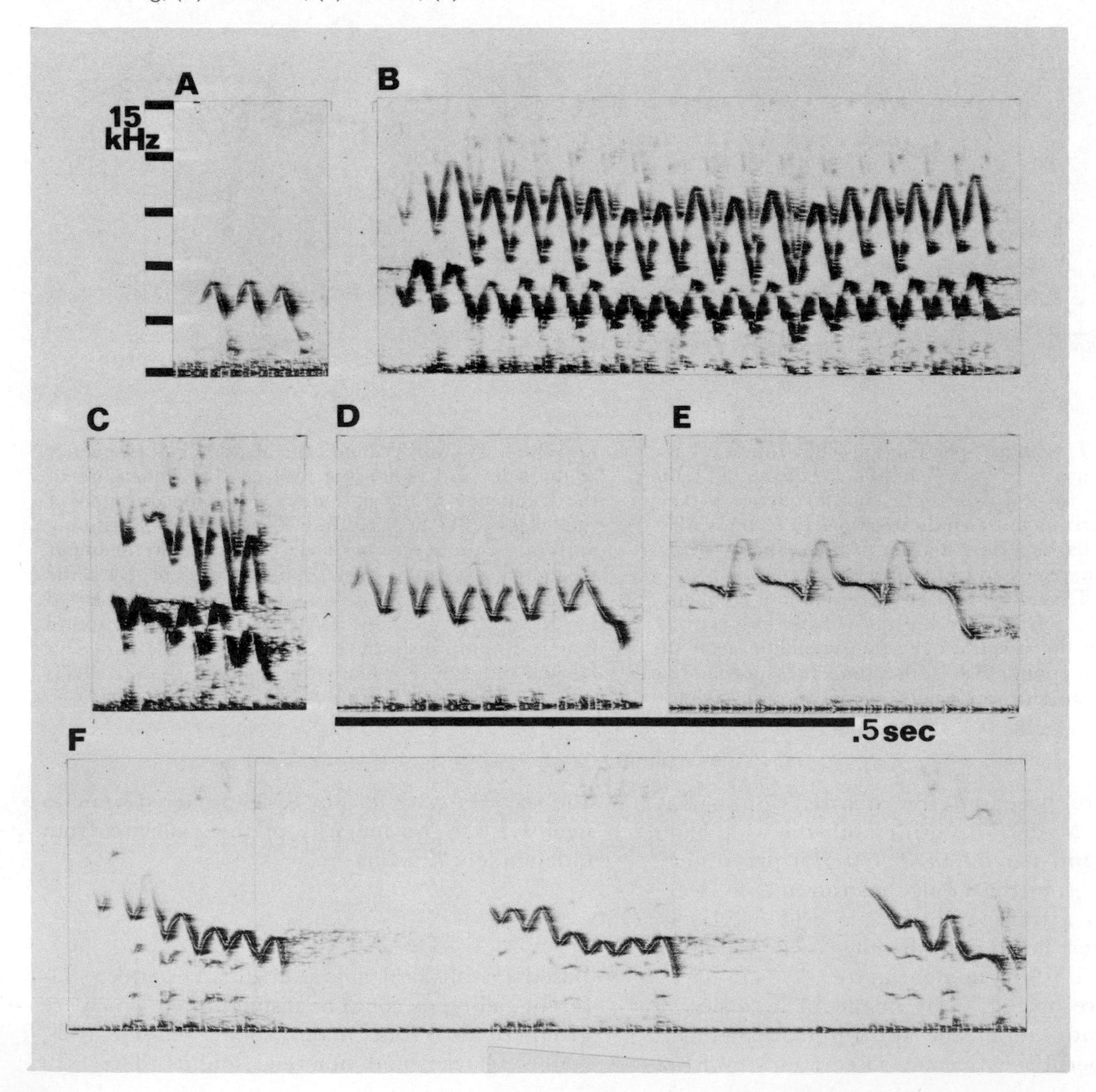

spectrograms in Figure 2. Most trill types, with the exception of Trill C, may be given in sequence or associated in phrases with clucks. Duration of the specific syllable types varies from .18 to .44 seconds. Generally, these elements are composed of a rhythmic sequence of sharply ascending and descending modulations with bands of energy uniformly spread across the spectrogram. The number of oscillations per call type is variable but in Trill C the sound structure is distinguishable since each oscillation is, on the average, longer than any other trill type. Furthermore, the first part of each oscillation descends in pitch like other trills, but then, at a point of inflection, has a more gradual frequency decline (Figure 2E). Trill B, which is highly correlated with situations of stress, has a much greater mean-sweep range and like the descending Trill A, exhibits greatest frequency changes from mean-top to mean-bottom frequency (Figure 2B, 2C, 2D).

2. CONTEXTS AND ASSOCIATED BEHAVIORS

Table 3 shows the relative distribution of male and female trills in the main contexts. Males trill significantly more than females ($p < .05$) and the data indicate that this significance is attributable to differences in all contexts except contact. Almost half of all male and female trills are in the context of vigilance.

Trills are the most frequently uttered sound in the repertoire of *L. rosalia rosalia.* Although variants occur through a wide range of contexts, most trilling is associated with vigilant behavior. Sexual differences are attributable to males vocalizing more than females in the context of fixating upon visual stimuli. The lowest percentage of trills are emitted during contact behavior.

The Rapid Trill (Figure 2A) is one of the few vocalizations in the repertoire occurring in a specific context, i.e., when the animal lands on a surface during locomotory activity. By emphasizing individual movements within the cage, the vocalization may be instrumental in spacing. This trill is given about 50 percent more frequently by males than females.

Trill A (Figure 2B, 2C) occurs in all general contexts, but predominates in association with vigilant behavior. Mostly, it is given with a completely closed mouth, but occasionally with a partially opened mouth.

No noticeable change in facial expression is observed during emission. In the context of Foraging, Vigilance, and Solo-activity, the male vocalizes between 50 and 80 percent more than the female, but the female produces Trill A twice as often as the male during contact behaviors.

Trill B (Figure 2D) has been recorded from individuals under stress, most often when being caught or handled by a keeper.

Trill C (Figure 2E) is associated with high arousal vigilant behavior and occurs most commonly as a preface to Long Calling (see Figure 7). Initiated by one animal, the trill is gradually picked up by adjacent groups until one eventually breaks into the Long Call and a chorus begins. The emission of the Trill C does not necessarily elicit a trill, but may also elicit a crescendoing whine or Wah-Wah, the preface element of the Long Call. The trill is occasionally emitted while the animals are searching and probing for food.

Whines

1. TEMPORAL PATTERN AND PHYSICAL STRUCTURE

The frequency range of the whine calls is stereotyped and differences in the categorization of these syllables is based on the degree of variability in duration and intensity as indicated in Table 4. The Wah-Wah is the most staccato, or shortest, but often is emitted sequentially. In such a phrase, the notes become longer in duration as well as higher in pitch. Figure 3 illustrates such development in a solitary animal and Figure 7 shows a corresponding situation among a pair of golden lion tamarins. The Soft Whine is shorter in duration and of less intensity than the Whine. The latter call can often be emitted in excess of one second.

2. CONTEXTS AND ASSOCIATED BEHAVIORS

Table 5 shows the relative distribution of male and female whines in the main contexts. Though there is not a significant sexual difference in whining rate or context, the data suggest that females whine slightly more than males in the area of foraging. This is the only tonal sound class which females emit more than males.

Whines comprise the second largest percentage of the repertoire. From Table 5, it can be seen that they are associated primarily with vigilance and secondarily with contact behavior. Female whines in solo, foraging, and contact contexts occur during autogrooming, food-selecting, approaching, huddling, and allo-sniffing. Male whines in the context of vigilance often occur during fixation upon visual stimuli at medium to high arousal.

The Wah (not analyzed spectrographically) occurs in foraging and vigilance contexts. In relation to food, it is given most often by females

Table 1. The distribution of vocalizations in different contexts in *Leontopithecus rosalia.*

Vocalizations	*Solo-activities*					*Foraging*					
	Groom	*Play*	*Mark*	*Loc. Hes.*	*Action Comp.*	*Scan*	*Probe*	*Select*	*Feed*	*Share*	*Steal*
Trills											
1. *Rapid Trill*				XXX							
2. *Trill A*		X	X			XX	XX				
3. *Trill C*						XX	XX				
Whines											
1. *Soft Whine*	XX	X	X			X	X	X	X	X	
2. *Soft Whine Series*											
3. *Wah*						X	X	X	X		X
Clucks											
1. *Cluck*	X	X	X			X	X	X	X	X	X
2. *Tsick*							X	XXX			
3. *Feeding Chatter*						X	X	X	XX	X	X
4. *Contact Chatter*											
5. *Burst*											
Long Calls											
1. *Long Call*											
Nontonal Sounds											
1. *Snough*					XX						
2. *Rasp & Squeal*											X

Horizontal axis: contexts in which vocalizations are emitted.
Vertical axis: Vocalizations arranged by class.
Loc. Hes. = locomotory hesitation
Action Comp. = action completion
R.F.O. = response to flying object
O.V.S. = orientation to visual stimulus
Exam. = examination of visual stimulus from a distance
Fix. M = fixation upon visual stimulus during low to medium arousal
Fix. H = fixation upon visual stimulus during high arousal
A-F = alarm-flight
App = approach
T = touch
Hud = huddle
Wr = wrestle
G = allogroom
Sn = allosniff
Mt. = mount
X = 1-29 vocalizations
XX = 30-60 vocalizations
XXX = 61-150 vocalizations
XXXX = 151-600 vocalizations

Table 2. Physical characteristics of trills.

Type		*Sample Size*	*Mean Top Frequency (kHz)*	*Mean Bottom Frequency (kHz)*	*Mean Sweep Range (kHz)*	*Mean Duration (sec.)*	*Mean No. of Oscillations*	*Mean Duration of One Oscillation (sec.)*
Rapid Trill		11	7.41 (±2.77)	3.55 (±1.62)	3.45 (±1.13)	.18 (±.06)	3.55 (±.93)	.05 (±.02)
Trill A	I	7	5.75 (±1.04)	1.20 (±.76)	3.17 (±.52)	.36 (±.12)	9.86 (±4.10)	.04 (±.01)
	II	15	5.47 (±1.48)	1.87 (±.64)	2.83 (±.56)	.37 (±.12)	10.20 (±3.08)	.04 —
	III	18	7.00 (±1.06)	2.31 (±.57)	2.33 (±.59)	.44 (±.11)	13.40 (±3.33)	.03 —
	Descending	10	6.17 (±.26)	2.00 (±.05)	2.42 (±.38)	.19 (±.10)	5.90 (±3.0)	.03 —
Trill B	I	5	10.2 (±1.64)	2.60 (±.55)	5.60 (±1.56)	.21 (±.03)	4.80 (±.84)	.05 —
	II	13	8.92 (±1.59)	2.81 (±.48)	4.00 (±.96)	.22 (±.04)	5.90 (±.95)	.04 —
	III	4	7.25 (±.96)	2.50 (±1.68)	1.75 (±.65)	.25 (±.04)	7.25 (±.96)	.03 —
Trill C		22	10.64 (±1.47)	4.34 (±.99)	4.84 (±1.30)	.24 (±.09)	3.09 (±1.34)	.08 (±.01)

Table 1. The distribution of vocalizations in different contexts in *Leontopithecus rosalia* (Continued).

	Vigilance			*Fix.*			*Contact*						
Vocalizations	*R.F.O.*	*O.V.S.*	*Exam.*	*M*	*H*	*A-F*	*App*	*T*	*Hud*	*Wr*	*G*	*Sn*	*Mt.*
Trills													
1. *Rapid Trill*													
2. *Trill A*		X	X	XX	X	XX	X			X	X		
3. *Trill C*		X	X	XX	XXX								
Whines													
1. *Soft Whine*		XX	XX	XXX			XX	XX	XX	XX	XX	X	
2. *Soft Whine Series*		X		XX									
3. *Wah*		X	XX	XX		XX							
Clucks													
1. *Cluck*		X	XX	X	XX								
2. *Tsick*		X	X	X	X								
3. *Feeding Chatter*													
4. *Contact Chatter*									X	X	X		X
5. *Burst*	XX												
Long Calls													
1. *Long Call*					XXXX								
Nontonal Sounds													
1. *Snough*													
2. *Rasp & Squeal*								X	X				

Table 3. The distribution of male and female Trills in the four behavioral contexts. Number of vocalizations (in parentheses) given in particular context. Numbers in columns represent proportion of total number of vocalizations given in particular context by males or females.

Sex	*Solo-activities*	*Foraging*	*Vigilance*	*Contact*	*Total*
Males	.21 (100)	.23 (108)	.49 (232)	.07 (34)	474
Females	.22 (62)	.21 (60)	.44 (123)	.13 (37)	282
Total	$n = 162$	$n = 168$	$n = 355$	$n = 71$	$N = 756$

$\chi^2 = 7.97$; d.f. = 3; $p < .05$.

Table 4. Physical characteristics of Whines.

Type	*Sample Size*	*Mean Frequency Start (kHz)*	*Mean Frequency Midpoint (kHz)*	*Mean Frequency End (kHz)*	*Mean Duration (sec.)*
Wah Wah	41	6.58 (±1.85)	6.68 (±1.54)	6.27 (±1.22)	.19 (±.11)
			Range (kHz)		
Soft Whine	13		3–6		.34 (±.06)
Whine	34		3–6		.42

Table 5. The distribution of male and female Whines in the four behavioral contexts. Number of vocalizations (in parentheses) given in particular context. Numbers in columns represent proportion of total number of vocalizations given in particular context by males or females.

Sex	*Solo-Activities*	*Foraging*	*Vigilance*	*Contact*	*Total*
Males	.05 (16)	.11 (40)	.54 (189)	.30 (107)	352
Females	.06 (24)	.15 (57)	.48 (179)	.30 (114)	374
Total	n = 40	n = 97	n = 368	n = 221	N = 726

$\chi^2 = 4.41$; d.f. = 3; n.s. at <.05 significance level.

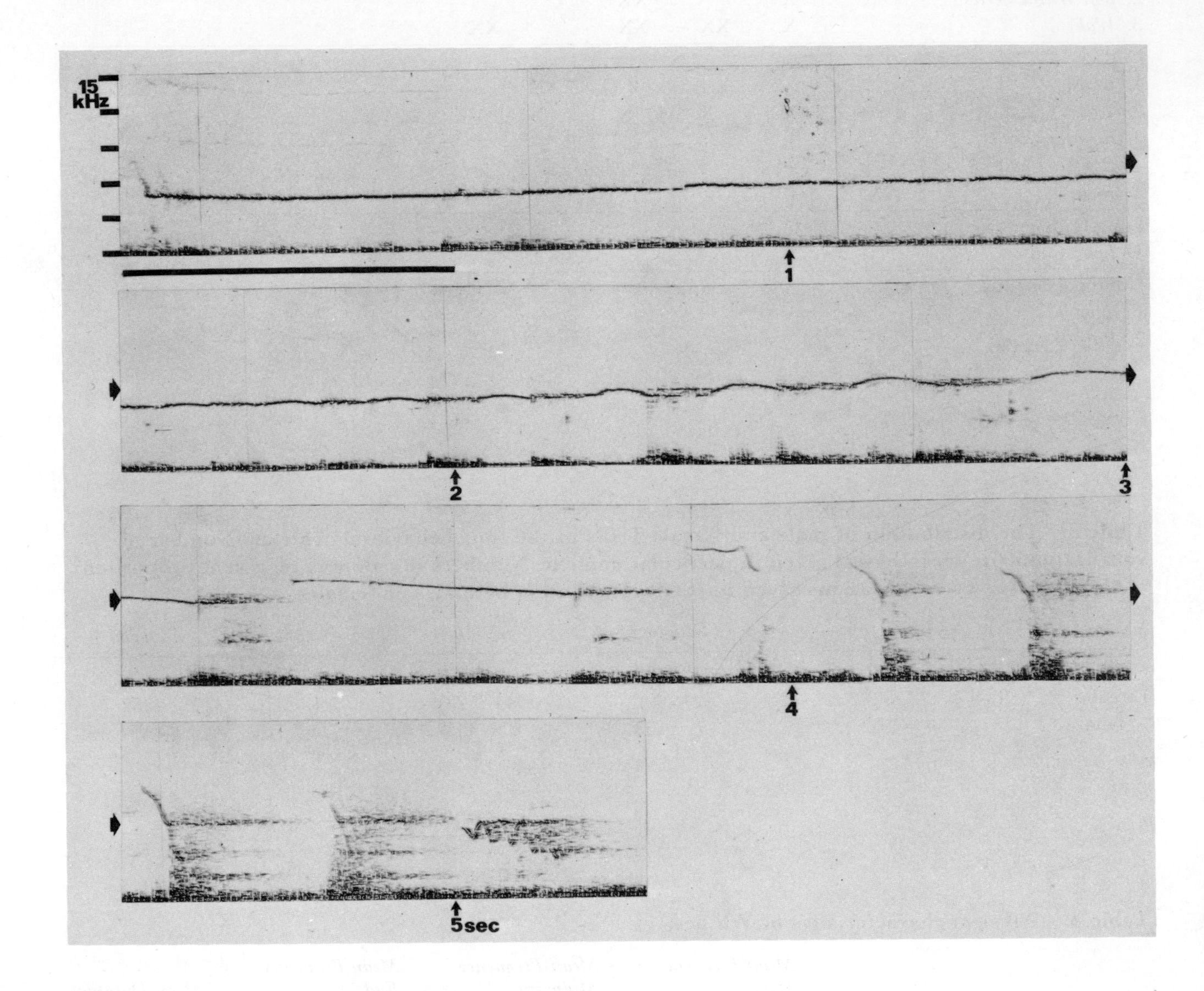

Figure 3. A whine call series emitted by Los Angeles male. Duration of bout is 5.25 seconds. In the first second, the whine is clearly distinguishable and grades into Wah-Wah notes while increasing in frequency. At the third second, the Wah-Wah has transformed into a descending whine; by the fourth second, clucks are emitted. Frequency and time are indicated.

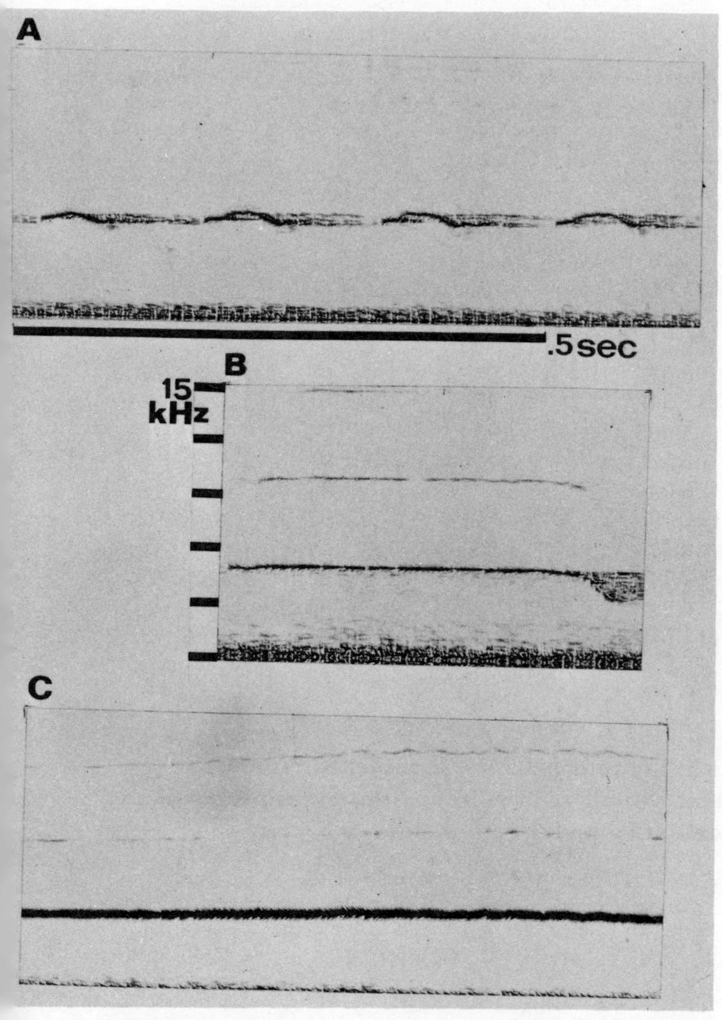

Figure 4. Whines. (A) Wah-Wah, (B) Soft Whine, (C) Whine. Frequency and time as in Figure 2.

from whom food is being stolen or females alarmed by a live insect while probing in crevices for food items. In the context of vigilance, it is given during alarm-flight sequences. During such a behavioral episode, an animal may flee while uttering Trill A, and at a safe distance from the fearful stimulus, may turn toward the stimulus and emit a Wah. The vocalization is given by both sexes during examination of foreign stimuli. It can be elicited by the observer's approach in which case piloerection and sideward darting movements accompany emission.

The Wah-Wah (Figure 4A) is a structural element of the Long Call, and thus its contextual analysis is subsumed under that of the Long Call.

The Soft Whine (Figure 4B) is the most common vocalization in the entire repertoire. Like the Trill A, it has a wide contextual distribution but is most often associated with vigilant behavior. In almost all contexts it is emitted more frequently by the female than the male, especially during vigilant behaviors where the female whines more than twice as often as the male during orientation to and examination of visual stimuli.

The Whine (Figure 4C) is similar to the Soft Whine, except that the sound intensity is greater, the call is longer in duration, and it is emitted under conditions of higher arousal than is the Soft Whine.

Clucks

1. TEMPORAL PATTERN AND PHYSICAL STRUCTURE

Syllables of this category are characteristically short notes, in most cases less than 0.1 second in duration. This staccato feature, together with the concentration of sound energy results in most Cluck types appearing very sharp to the ear (Table 6). The Peep syllable has little frequency change between the start and end of the call, but several forms, i.e., Cluck, Feeding Chatter I and II, and Pü exhibit a prominent inflection point (Figure 5A, 5E, 5F). Like trills, certain call notes of this group are emitted in phrases.

2. CONTEXTS AND ASSOCIATED BEHAVIORS

Table 7 shows the distribution of male and female clucks in the general contexts. While sexual differences in calling rate and context are not significant, males do produce clucks more than females, particularly during foraging.

Clucks show a clustering in foraging and vigilance contexts. The calls are recorded in all foraging contexts, but particularly probing, food-selecting, and feeding. In vigilance they are emitted during examination of and fixation upon unfamiliar stimuli.

The Cluck (Figure 5A) is evenly distributed among foraging contexts: scanning and probing for food, selecting items from the food tray, food-sharing, and food-stealing. Given in the context of vigilant behavior, it is closely associated with examination of and fixation upon foreign objects. Males cluck about 20 percent more than females in both contexts.

The Tsick (Figure 5C and 5D) is heard during feeding on live prey such as insects and small rodents. When we introduced a live bull snake

Table 6. Physical characteristics of Clucks.

Type		*Sample Size*	*Mean Frequency Start (kHz)*	*Mean Frequency End (kHz)*	*Mean Frequency Inflection (kHz)*	*Mean Duration (sec.)*	*Mean Internote Interval (sec.)*
Cluck		50	10.41 (±.35)	10.68 (±2.60)	5.95 (±1.64)	.09 (±.01)	.09 (±.05)
Tsick		50	7.28 (±1.80)	4.36 (±1.61)		.04 (±.02)	
Sharp Cluck		7	4.44 (±.42)	6.21 (±.81)		.06 (±.02)	
Feeding Chatter	I	11	5.50 (±.45)	6.27 (±.34)	4.95 (±.47)	.06 (±.01)	.36 (±.06)
	II	10	5.50 (±.64)	5.90 (±.52)	4.90 (±.45)	.04 (±.01)	
	III	6	7.50 (±1.79)	6.00 (±2.10)		.04 (±.02)	.18 (±.02)
Pü		31	10.46 (±1.00)	8.77 (±2.14)	4.89 (±1.26)	.07 (±.02)	.12 (±.05)
Peep		19	6.45 (±2.06)	6.32 (±1.81)	.13 (±.05)	.09 (±.05)	

Table 7. The distribution of male and female Clucks in the four behavioral contexts. Number of vocalizations (in parentheses) given in particular context. Numbers in columns represent proportion of total number of vocalizations given in particular context by males or females.

Sex	*Solo-activities*	*Foraging*	*Vigilance*	*Contact*	*Total*
Males	.02 (8)	.52 (180)	.38 (133)	.08 (28)	349
Females	.02 (5)	.42 (104)	.47 (116)	.08 (20)	245
Total	$n = 13$	$n = 284$	$n = 249$	$n = 48$	$N = 594$

$\chi^2 = 5.48$; d.f. = 3; n.s. at <.05 significance level.

into the housing unit of one pair, and recorded the response of another pair to the presence of an armadillo, *Cabassous centralis,* we heard this vocalization uttered frequently in rapid series. Alternating movements to and from the snake indicated conflicting motivational tendencies.

The Feeding Chatter (Figure 5E) is heard during live feeding and also food-stealing and food-sharing, especially between parents and offspring. Retraction of the corners of the mouth and chattering of the teeth occur. In addition, one of the subtypes has been recorded from males during medium to high arousal contact behaviors such as wrestling, allogrooming, and mounting.

The Cluck Burst is given equally by males and females in response to unexpected flying objects, such as birds, falling leaves, and large insects. Because of the rarity of this vocalization, no tape recording was made for structural analysis.

Clucks may be combined with other vocalizations. For example, the Cluck-whine is uttered most often by males during examination of foreign objects and when probing for food items. The Cluck-trill (Figure 5H) is frequently heard during medium to high arousal vigilant behavior, and is more often uttered by the male than female. Intense locomotor activity interspersed with scanning of the outdoors often occurs during emission.

Nontonal Sounds

1. TEMPORAL PATTERN AND PHYSICAL STRUCTURE

In these sounds, the note duration is relatively

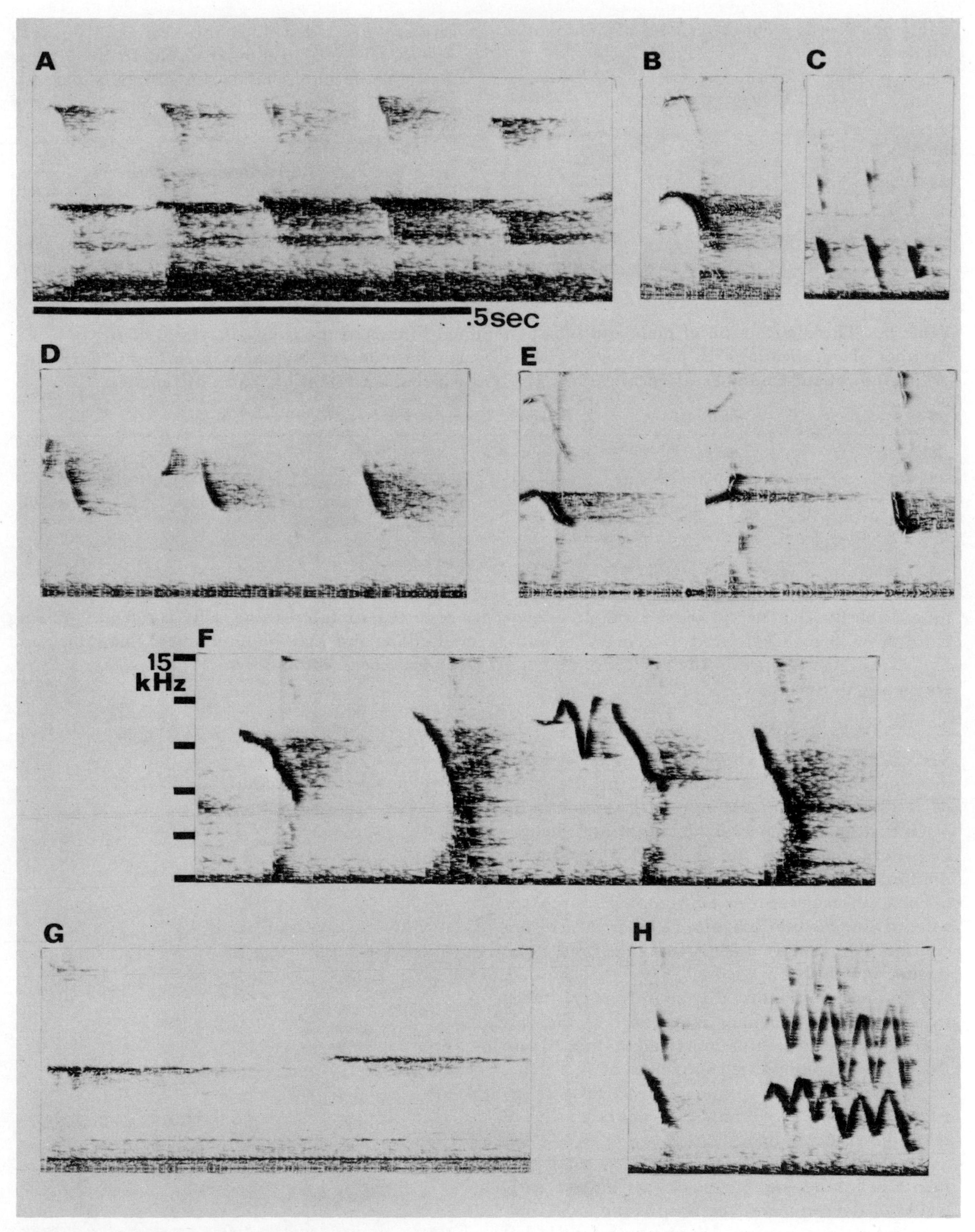

Figure 5. Clucks. (A) Cluck, (B) Sharp Cluck, (C) and (D) Tsick, (E) Feeding Chatter, (F) Pü, (G) Peep, (H) Cluck-trill. Frequency and time as in Figure 2.

Table 8. Physical characteristics of nontonal sounds.

Type	*Sample Size*	*Range (kHz)*	*Mean Duration (sec.)*
Snough	5	0–6	.32 (±.07)
Screech	4	0–7	.23 (±.15)
Infant Rasp	11	.36–8.45	.64 (±.2)

ears are pressed against the head and one can observe piloerection about the head. In the carrying position, the infant emits this sound, apparently when it is uncomfortable or desires a change of position. In adults, the Rasp occurs more frequently, but both the Rasp and Squeal occur in the same contexts. Both notes are given most often by females in unfamiliar and seemingly stressful situations. They are occasionally uttered by subadult animals when they are the victims of food-stealing. Females, approaching parturition, are reported to utter these calls frequently during huddling and wrestling in the nest box (Mack and Dorsey, pers. comm.).

Table 9. The distribution of male and female nontonal sounds in the four behavioral contexts. Number of vocalizations (in parentheses) given in particular context. Numbers in columns represent proportion of total number of vocalizations given in particular context by males or females.

Sex	*Solo-activities*	*Foraging*	*Vigilance*	*Contact*	*Total*
Males	.70 (21)	.30 (9)	0 (0)	0 (0)	30
Females	.37 (26)	.16 (11)	0 (0)	.47 (33)	70
Total	$n = 47$	$n = 20$	$n = 0$	$n = 33$	$N = 100$

$\chi^2 = 0.58$; d.f. = 2; $p < 0.01$.

long (Table 8). All note types have a wide frequency band from 0 to 7 kHz and are "noisy" sounds, as contrasted to the previously discussed call notes which are all tonal.

2. CONTEXTS AND ASSOCIATED BEHAVIORS

Sexual differences in vocalizing rates and context distribution are highly significant in this class ($p < 0.01$). Females use nontonal sounds more often than males, especially during contact behavior where males were never heard to produce a nontonal sound (see Table 9).

Forty-seven percent of all nontonal sounds are uttered during solo-activities, 33 percent during contact, 20 percent during foraging, and none during vigilant behavior.

The Snough (Figure 6A) sounds very similar to a sneeze. Its frequent occurrence at the end of feeding and locomotor bouts leads us to postulate that it may communicate completion of an activity.

The Screech is only emitted in contexts of extreme stress, during pursuit by keepers, or during occasions of being hand-held.

The Rasp (Figure 6B) and Squeal are commonly heard in infants and juveniles. During the early weeks of development, the Rasp is the most common infant vocalization. When the infant is left alone on a substrate in the absence of either parent, the call is given repeatedly. The mouth is open,

Figure 6. Nontonal sounds. (A) Snough, (B) Rasp. Frequency and time as in Figure 2.

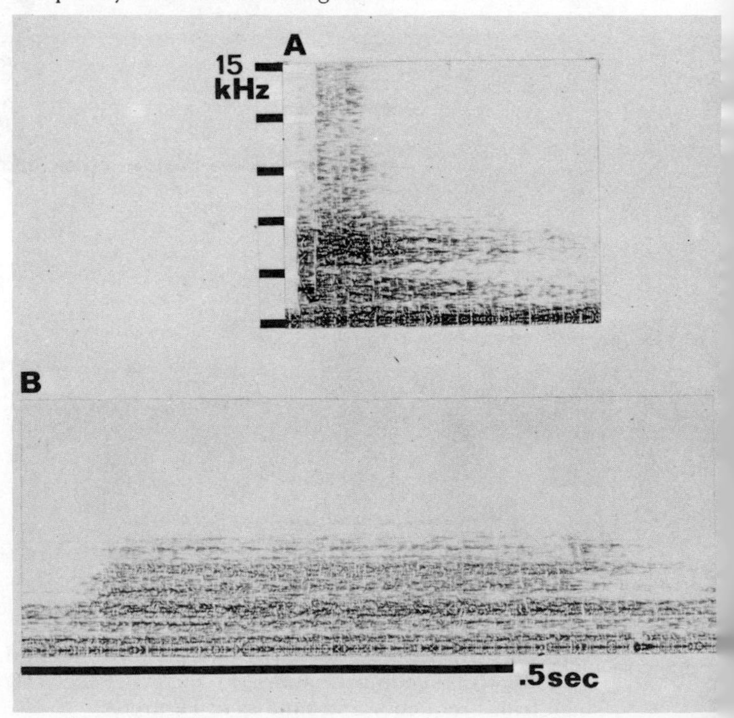

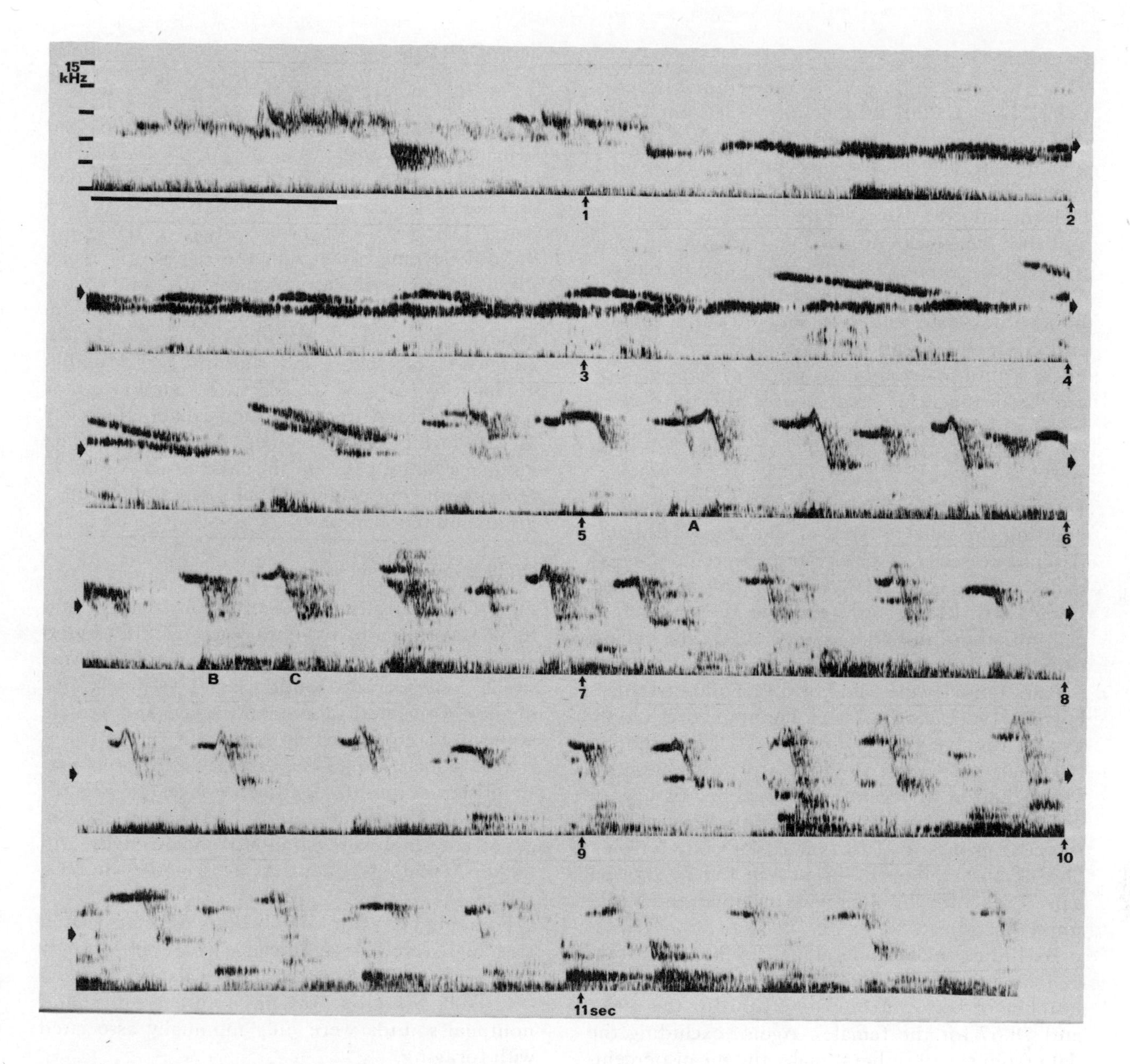

Figure 7. A Long Call duet recorded from Los Angeles male and Evansville female of almost 12 seconds duration. The Trill C is shown as the preface during the first 0.5 seconds. During the second second, one individual is emitting whines at a rather constant frequency of 6 kHz while the other animal is giving Wah-Wah calls at a slightly higher frequency. By the third second, the Wah-Wah calls change in structure and become descending whines and at the five-second mark both animals are giving clucks. A indicates two clucks, emitted by different individuals almost simultaneously; B and C represent the same sound type, but as distinct non-overlapping notes. The spectrograph is made with a wide-band filter.

Long Calling

1. TEMPORAL PATTERN AND PHYSICAL STRUCTURE

A combination of elements from the tonal sound classes—Trills, Whines, and Clucks, is characteristic of the Long Call bout. The Trill C may serve as a preface and often is followed by a Whine or Wah-Wah series which increases in intensity as well as duration. Commonly, if the stimuli are great enough to keep the vocalizers aroused, the sounds become harsher and eventually grade into Clucks and sporadic Trill A's (Figure 7). The total duration of this bout can vary from several minutes up to in excess of ten minutes.

2. CONTEXTS AND ASSOCIATED BEHAVIORS

There was no significant difference in the proportion of Long Calling by males ($n = 336$ Long Calls) and females ($n = 270$ Long Calls).

Long calls are emitted only during high-arousal vigilant behavior and most often are accompanied by locomotor displays, piloerection about the head and shoulders, scanning of the outdoors, fixation upon visual stimuli, and defecation and urination. Most of these vocal displays occur in the morning hours. Long Calls tend to be initiated by one animal and gradually spread to pair, family, and adjacent groups until many animals from different groups are vocalizing. This vocal bout is the third most frequent sound of the repertoire.

Duetting

Duetting occurs when one individual is joined in emitting the syllable types of the Long Call bout. A separate series of observations were conducted to determine (1) the relative rate of male and female Long Calls, (2) sex differences in the number of call initiations and chorusing, and (3) the calling rates of different pairs.

Pairs 1 and 2 were wild-born, except for a female born in 1968, and had been maintained as family groups for five years. Pairs 3 and 4 were captive-born and had been paired for at least six months. Pair 5 had been recently established, and Trio 6 was in the final stages of a "pair-preference" test (i.e., the grouping of two males with one female). During observations of Trio 6, the female showed a preference for the older male (Kleiman and Mack, unpublished).

Excluding the peripheral male in the trio (group 6) from our calculations, we find that the mean number of bout initiations is 91.00 for the males and 78.67 for the females. Again, excluding the data from the peripheral male, the mean percentage of chorusing is 35.6 percent for the males and 53.1 percent for the females (Table 10). These data suggest the following: Both sexes initiate approximately the same number of bouts, but females join in more than males. In long-established pairs, females initiate more Long Calls than the males, whereas the opposite is true in the newer pairs. The female in Trio 6 duetted more with the central male than with the peripheral male; this was correlated with greater contact and sexual behavior. Thus, Long Calling may be functional in developing and maintaining the pair bond. The additional olfactory and auditory information available to pairs in the Marmoset Building may have contributed to the greater calling frequency in that facility by both males and females. The greater calling rate in the Marmoset Building indicates the importance of the Long Call in intergroup communication.

Summary of Contexts

Figure 8 illustrates the distribution of male and female vocalizations within each of the sound classes within the four general contexts. It shows which contexts and sound classes are most frequently associated, also taking male and female vocalizing frequencies into account.

The association of solo-activity with trills was significant at the .05 significance level ($\chi^2 = 8.78$; d.f. $= 3$). Nontonal sounds and whines showed an equally low association with solo-activity and higher female frequencies were noted in both classes.

Foraging behavior and clucking were associated at a high level of significance ($p < .001$; $\chi^2 = 18.51$; d.f. $= 3$). A moderate association between trills and foraging was apparent. Whines and nontonal sounds were only minimally associated with foraging.

Table 10. A comparison of Long Call duetting by males and females in six pairs of *L. rosalia*.

<table>
<tr><th rowspan="3">Activity</th><th colspan="2">Small Mammal House Pair Nos.</th><th colspan="3">Marmoset Building Pair Nos.</th><th colspan="2"></th></tr>
<tr><th>1</th><th>2</th><th>3</th><th>4</th><th>5</th><th colspan="2">6*</th></tr>
<tr><th></th><th></th><th></th><th></th><th></th><th>CM</th><th>PM</th></tr>
<tr><td>Male initiates</td><td>15</td><td>17</td><td>175</td><td>195</td><td>69</td><td>75</td><td>87</td></tr>
<tr><td>Female joins in (%)</td><td>33.3</td><td>59.2</td><td>60.3</td><td>65.4</td><td>59.2</td><td>41.1</td><td>7.0</td></tr>
<tr><td>Female initiates</td><td>89</td><td>55</td><td>138</td><td>129</td><td>30</td><td colspan="2">31</td></tr>
<tr><td>Male joins in (%)</td><td>25.1</td><td>11.1</td><td>46.3</td><td>60.2</td><td>70.7</td><td>0</td><td>10.5</td></tr>
<tr><td>Observation hours</td><td>53.5</td><td>53.5</td><td>48.5</td><td>53.5</td><td>11.5</td><td colspan="2">31.5</td></tr>
</table>

*Trio with one female and two males. CM is preferred male, PM is peripheral male, based on female's contact behavior.

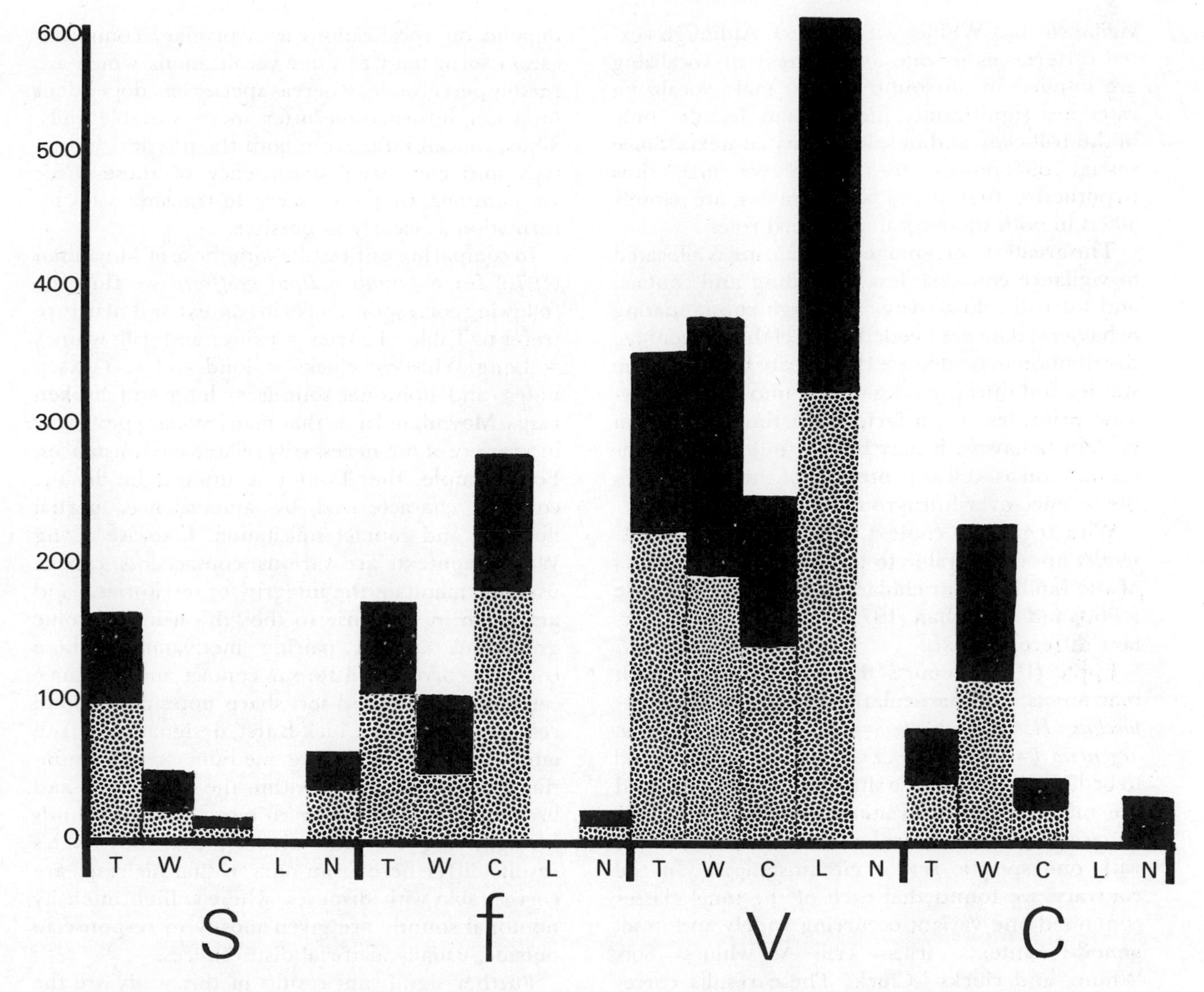

Figure 8. The distribution by sex of the five sound types in the four behavioral contexts in *L. rosalia*. Vertical axis: number of vocalizations; horizontal axis: sound classes divided among the four behavioral contexts. (S = Solo-activities, F = Foraging, V = Vigilance, C = Contact, T = Trill sound class, W = Whine sound class, C = Cluck sound class, L = Long Call sound class, N = Nontonal sound class.) Black bars represent the percentage of vocalizations uttered by females and stippled bars the percentage of vocalizations uttered by males.

The association of vigilant behavior with Long Call emission, particularly by males was highly significant ($p < .001$; $\chi^2 = 16.54$; d.f. = 3). Trills and whines were associated about equally with vigilant behavior, although about half as much as the Long Call. Clucks showed even less association and nontonal sounds were not associated at all. Long Call emission occurred only in vigilance contexts.

Contact behaviors and whine emission were associated at the .001 significance level ($\chi^2 = 31.65$; d.f. = 3). Sex differences in contact-related whining were minimal. Trills, clucks, and nontonal sounds showed a relatively small association.

Discussion

The vocal repertoire of the golden lion tamarin consists of four distinct sound classes, three tonal and one nontonal, in addition to a Long Call comprising elements from each of the tonal classes. Vocalizations within these classes have a characteristic association with one or more specific contexts in the categories of solo-activity, foraging, vigilance, and contact. Each of the tonal sound classes as well as the Long Call shows a significant association with one general context: Trills with solo-activity; Clucks with foraging; Long Calls with vigilance; and Whines with contact. Although sexual differences in rate and context of vocalizing are implied in all sound classes, male vocalizing rates are significantly higher than females only in the trill class and in solo-activity contexts. Since sexual differences are minor, we may thus hypothesize that males and females are closely allied in both their vocal and social roles.

The greatest percentage of vocalizing is allocated to vigilance contexts, less to feeding and contact, and least to solo-activity. Although substantiating behavioral data are needed, we feel this percentage distribution may indicate the activity budget of the species and thus provides insight into social behavioral priorities. If, in fact, much time is spent in vigilant behavior, it may be that intergroup communication as well as protection of the group takes precedence over intragroup interactions.

With regard to context, the vocalizations of *L. rosalia* are comparable to those of other members of the family Callitrichidae, as discussed by Epple (1968) and Moynihan (1970), though some important differences exist.

Epple (1968) reports that most calls of adult marmosets, and particularly *Hapale* (= *Callithrix*) *jacchus*, *H. leucocephala* (= *C. geoffroyi*), *Mico a. argentata* (= *C. a. argentata*), and *L. rosalia*, tend to be linked to a specific situation. While we found the majority of vocalizations to cluster within a general context, very few of them were associated with one specific set of circumstances. On the contrary, we found that each of the tonal classes contained one variant occurring widely and in all general contexts: trills—Trill A; whines—Soft Whine, and clucks—Cluck. These results correspond with Epple's description of "monosyllabic calls given in loose visual contact" by *Oedipomidas* (= *Saguinus*) *oedipus* and *O. spixi* (= *S. oedipus geoffroyi*). In these species, according to Epple, the calls are easily elicited by slight disturbances within the group and may serve to inform members of slight motivational shifts. Similar calls given by *L. rosalia* in an array of contexts may serve to transmit finer details of information at the intragroup level, where olfactory, visual, and tactile cues are also operative.

The tonal vocalizations which tended to be associated with specific contexts and also showed a considerable degree of stereotypy were the Rapid Trill, at the end of locomotor bouts; the Cluck Burst in response to flying objects; and the preface Wah-Wah element of the Long Call, given during high arousal, multi-participant vocal sessions. Moynihan (1964) has suggested that species which depend on vocalizations as a primary communication form tend to utter vocalizations which are readily perceivable, whereas species less dependent on vocal information utter more variable calls. Thus, we can infer from both the physical stereotypy and contextual consistency of these three vocalizations, that they serve to transmit vital information as clearly as possible.

In comparing our results with those of Moynihan (1970) for *Saguinus oedipus geoffroyi*, we find the following correspondences in context and structure (refer to Table 11): trills = twitter and trill; whines = Long Whistles; clucks = loud and soft sharp notes; and nontonal sounds = long and broken rasps. Moynihan finds that many vocal types occur in a variety of not necessarily related circumstances. For example, the Twitter is uttered in distinct contexts characterized by ambivalence, partial hostility, and contact solicitation. Likewise, Long Whistle contexts are various: contact loss, can be used to "maintain the integrity of territories," and are given in response to the "thwarting of some gregarious and/or pairing motivation." These contexts correspond to our contact and vigilance categories. Loud and soft sharp notes correspond respectively to our Cluck Burst, designating alarm, and Cluck, given during medium arousal, nondirectional movements within the cage. Long and broken rasps are similar to our nontonal sounds in signifying extreme distress, but Moynihan's results differ here from ours in that his rasps are correlatable with disputes, whereas high intensity nontonal sounds are given mostly in response to outside, usually artificial disturbances.

Further significant results of this study are the implications Long Call duetting has for pair formation, maintenance, and development. The fact that the female in the trio duetted most with the "consort" male as well as the trend for females in long-standing pairs to take greater initiative in Long Calling, all point to the Long Call's significance in bringing the male and female social roles into relative harmony as well as cementing their bond.

The task of comparing the results of the structural analysis of this investigation to other related research among the Callitrichidae is not easy. Both Epple (1968) and Moynihan (1970) categorize the vocalization types in reference to the contextual framework of the sound emissions. Since several sound or call notes are classified among several contexts, close scrutiny is required to differentiate these names and classify according to the basic syllable notes found in this study. Furthermore,

Table 11. A comparison of vocalizations in marmosets, tamarins, and *Callimico goeldii*.*

Note Type	*Cebuella pygmaea*	*Callithrix jacchus*	*Callithrix argentata*	*Leontopithecus rosalia*	*Saguinus oedipus geoffroyi*	*Saguinus oedipus oedipus*	*Callimico goeldii*
Trills	open-mouth trill[5]; closed-mouth trill[5]	twitter[1,2]	trills[2]	trills[2]; twitter[1,2]	trill[2,3]; twitter[3]; pulses[4]; structured bursts[4]	trill[2]; twitter[1]; chirp[2]	trills[2]
Whines	alerting call Type A[5]; alerting call Type B[5]	"phee"[2]		"whee"[2]; "phee"[2]	mono call in isolation[2]; long call[4]; long whistle[3]; bar whine[4]	warning call[2]	mono call when threatened by observer[2]
Clucks	adult J-call[5]; "tsick"[5]; "phee"[5]; infant J-call[5]; warning call[5]; tonal-squeak click[5]; chatter[2]	"tsick"[2]; mono contact call (close visual contact)[2]; "egg" (crackle)[2]; "tsee"[2]; warning call[2]; chatter[2]; alarm call[6]; mobbing call[6]	"tsick"[2]; "egg" (crackle)[2]; "ock" (cough)[2]	"pe"[2]; "Pu"[2]; cluck[2]; "tsick"[2]	mono call in loose contact[2]; sharp notes (loud, soft sneezing)[3]; chevrons[4]; slicing chirp[4]; complete chitter[4]	mono call in isolation[2]; "te"[2]; mono call in loose contact and when disturbed[2]; "tsick"[2]	"tsick"[2]; "te"[2]; "tschong"[2]; "tsee"[2]; "phee"[2]
Nontonal	squeal[5]; click[5]; Type A screech[5]; Type B screech[5]; intergroup screech[5]; excitement squeal[5]	"Nga"[2]; screams[2]; squeal[2]		"screech"[1]; squeal[1,2]; long squeak[1]; wavering squeak[1]	slicing scream alarm[4]; persistent infantile patterns[3]; long broken rasp[3]; "mrou"[4]; "kneah"[4]	scream[2]; squeal[2]; rasping screech[1]	
Combination		mono contact with "phee"[2]; shrill[6]; mobbing call "tsick" with crackles and coughs[2]		long call whine and intro-slicing chirp[4]	long call whine and intro-slicing chirp[4]		

[1]Andrew, 1967.
[2]Epple, 1968.
[3]Moynihan, 1970.
[4]Muckenhirn, 1967.
[5]Pola and Snowdon, 1975.
[6]Pook, 1977.

*Since the preparation of this Table, Moody and Menzel (1976, *Folia primat.* 25:73–94) have published on *Saguinus fuscicollis* vocalizations.

Note type: Trills are called trills, twitter-hooks.
Whines are called mew, chee, chip.
Clucks are called short call, sudden call, yips.
Nontonals are called rattle, squawk, snarl, sneeze.
Combinations are called long call.

structural characteristics of the sound types presented in sonograms, the sample size and the amount of variability are not always available. Often, only line drawings representing the original sonogram are included, which adds to the difficulty of identifying the call note. However, Muckenhirn (1967) provides both sample size, original sonograms, and discusses gradations and variability. Comparisons between studies and species is available in Table 11.

Muckenhirn (1967) used five basic sound categories: pulses, structural chirps, whines, chevrons, and noises. Based on the sample sonograms, this classification corresponds closely to the present *Leontopithecus* classification: trills = pulses and structural chirps; whines = whines; clucks = chevrons; and nontonal = noise.

Moynihan (1970) identifies nine main sound types "some of which intergrade with one another through more or less continuous series of intermediates." He adds, however, "persistent infantile patterns" as another type of vocalization, thus, in effect having a total of ten sound types. Review of this classification suggests that these can be grouped structurally as: trills = twitter and trill; whines = Long Whistle; clucks = sharp, loud, soft, and sneezing sharp notes; and nontonal = long and broken rasps, persistent infantile patterns.

Epple (1968) places the vocalizations of several callitrichid species into eight broad contextual divisions. This work presented major difficulties with regard to classifying the sounds according to the basic vocal types in this paper, and it is best to review the numerous call notes in Table 11.

It is instructive to note that these vocalization studies of taxonomically related species may be classified broadly into similar categories. Of more importance is the fact that many of these related vocalizations are found to occur in similar contexts. For example, "screams" and "screeches" in *H. jacchus* (= *C. jacchus*), *Leontopithecus,* and *S. oedipus* (Epple, 1968), Moynihan's "broken rasps," Muckenhirn's slicing scream alarm, and the present study's nontonal squeal, all occur in the context of extreme stress and during pursuit by keepers.

The nontonal structural quality of this alarm note is interspecifically stereotyped. The similarity in both structure and function of at least this syllable type suggests that equivalent selective pressures have led to this homologous call note (Eisenberg, 1974; Marler, 1967, 1973; and Moynihan, 1967, 1970).

On the other hand, many sound emissions belonging to the cluck type, i.e., Pü, Tsick, and Peep are more variable both in structure and context. Such variable sound elements therefore correspond to the characteristics of a graded sound system (Gautier, 1974; Green, 1972; Marler, 1969, 1970, 1973). We would stress that the communicatory system considered from the standpoint of vocal, olfactory, tactile and/or visual signals tends to be graded, even though several discrete communicatory displays may persist.

Acknowledgments

Part of this research was funded by a work-study program to KMG through the Johns Hopkins University and Smithsonian Institution. Discussions on sound structure with Eugene Morton and Rebecca Field were helpful. EBM would like to thank Wolfgang Dittus, Charles Roberts, and Christopher Hewat for their comments on analysis of the contextual data. Both of us are deeply grateful to John Eisenberg and Devra Kleiman for the stimulation and guidance they always provided. In addition, R. Hoage, D. Mack, and C. Dorsey aided us in further understanding the behavior of *Leontopithecus.* Finally, we appreciate all assistance by the personnel of the Office of Zoological Research.

Literature Cited

Andrew, R. J.
1963. The origin and evolution of the calls and facial expressions of the primates. *Behaviour,* 20:1–109.

Eisenberg, J. F.
1974. The function and motivational basis of Hystricomorph vocalizations. *Symp. Zool. Soc. Lond.,* 34:211–247.
1976. Communication mechanisms and social integration in the black spider monkey, *Ateles fusciceps robustus,* and related species. *Smithson. Contrbs. Zool.,* 213:1–108.

Epple, G.
1967. Vergleichende Untersuchungen über Sexual- und Sozialverhalten der Krallenaffen (Hapalidae). *Folia Primat.,* 7:37–65.
1968. Comparative studies on vocalization in marmoset monkeys (Hapalidae). *Folia Primat.,* 8:1–40.

Gautier, J. P.
1974. Field and laboratory studies of the vocalizations of Talapoin monkeys (*Miopithecus talapoin*). *Behaviour,* 51:209–273.

Green, S.
1972. Communication by a graded vocal system in Japanese monkeys. Ph.D. Thesis, Rockefeller University, New York.

Kleiman, D. G.
1977. Characteristics of reproduction and sociosexual interactions in pairs of lion tamarins (*Leontopithecus rosalia*) during the reproductive cycle. In *The Biology and Conservation of the Callitrichidae,* edited by Devra G. Kleiman, pp. 181–190. Washington, D.C.: Smithsonian Institution Press.

Mack, D. and D. G. Kleiman.
(in press). Distribution of scent marks in different contexts in captive lion tamarins, *Leontopithecus rosalia* (Primates). In *Biology of the Callitrichidae,* edited by H. Rothe. Göttingen: University of Göttingen.

Marler, P.
1967. Animal communication signals. *Science,* 157:769–774.
1969. Vocalizations of wild chimpanzees. *Rec. Adv. Primat.* 1:94–100.
1970. Vocalizations of East African monkeys. I. Red Colobus. *Folia Primat.*, 13(2–3):81–91.
1973. A comparison of vocalizations of red-tailed monkeys and blue-tailed monkeys, *Cercopithecus ascanius* and *C. mitis* in Uganda. *Z. Tierpsychol.*, 33:323–347.

Moynihan, M.
1964. Some behavior patterns of platyrrhine monkeys. I. The night monkey (*Aotus trivirgatus*). *Smithson. Misc. Colls.*, 146(5):1–84.
1966. Communication in the Titi monkey, *Callicebus. J. Zool. Lond.*, 150:77–127.
1967. Comparative aspects of communication in New World primates. In *Primate Ethology,* edited by D. Morris, pp. 236–266. London: Weidenfeld and Nicolson.
1970. Some behavior patterns of platyrrhine monkeys. II. *Saguinus geoffroyi* and some other tamarins. *Smithson. Contrbs. Zool.*, 28:1–77.

Muckenhirn, N. A.
1967. The behavior and vocal repertoire of *Saguinus oedipus* (Hershkovitz 1966) (Callithricidae, Primates). M. Sc. thesis, University of Maryland, College Park.

Pola, Y. V. and C. T. Snowdon.
1975. The vocalizations of pygmy marmosets (*Cebuella pygmaea*). *Anim. Behav.*, 23:826–842.

Pook, A. G.
1977. A comparative study of the use of contact calls in *Saguinus fuscicollis and Callithrix jacchus.* In *The Biology and Conservation of the Callitrichidae,* edited by Devra G. Kleiman, pp. 271–280. Washington, D.C.: Smithsonian Institution Press.

Struhsaker, T.
1967. Auditory communications among vervet monkeys (*Cercopithecus aethiops*). In *Social Communication Among Primates,* edited by S. A. Altmann, pp. 281–325. Chicago: University of Chicago Press.

A. G. POOK
Jersey Zoo, Trinity, Jersey,
Channel Islands

A Comparative Study of the Use of Contact Calls in *Saguinus fuscicollis* and *Callithrix jacchus*

ABSTRACT

In a laboratory study of the vocalizations of *Saguinus fuscicollis* and *Callithrix jacchus*, it was found that contact calls are the vocal signals most frequently used by adult pairs of animals under normal laboratory conditions. In both species, contact calls are more frequent at higher activity levels. However, preferences for different types of contact call at different levels of activity are found in *S. fuscicollis*, but not in *C. jacchus*.

Further contextual and sequential analyses reveal other differences in the use of contact calls, which suggest that *S. fuscicollis* generally maintains close contact by vocal means, whereas *C. jacchus* uses a combined visual and vocal system.

The significance of these and other observations on social behavior, in relation to the ecology and hypothesized social organization of the two species in the wild, is discussed.

Introduction

There is as yet very little published information concerning the social behavior of marmosets, but those studies that have been carried out have indicated the importance of vocal communication (Epple, 1968; Muckenhirn, 1967; Moynihan, 1970). Altmann (1967) states that vocal communication is probably more important for arboreal primates, such as marmosets, because of the limitations imposed on visual communication by dense foliage.

All species of marmosets possess a large and varied vocal repertoire. The use of these repertoires, though basically similar, differs in various ways from species to species. It may be that such differences reflect variations in the social organization and ecology of the species in the wild.

This paper presents some of the results of a comparative study carried out in the laboratory of the use of vocalizations in the saddle-back tamarin (*Saguinus fuscicollis leucogenys*) and the common marmoset (*Callithrix jacchus*). The main aims of this study were to establish the vocal repertoire of each species and to try to account for any differences that were found, either in structure or use.

Methods

Although general observations were made on a variety of social groupings of both species, the studies described here were carried out on two adult mated pairs of each species. This enabled accurate quantitative analyses to be made of vocalizations and their contexts. Some of the subjects were captive-born, and all were over eighteen months old and had been paired for at least two months before the study began. During the course of these studies three of the four females became pregnant.

For each study, one pair of subjects was observed at a time, isolated in an experimental room, in cages measuring 0.6 × 1.2 × 1.8 m high and equipped with wooden perches, nest boxes, and platforms. Between 16 and 32 hours of observation were made on each pair for each study. Observation sessions lasted for one hour and were spread evenly over most of the day. Data recording was done by hand, using prepared data sheets, during three to four recording periods in each observation session.

The walls of the experimental room were hung with cardboard egg trays to reduce the level of acoustic reverberation. Recordings were made with a Revox A77 tape-recorder at a speed of 38 cm / sec or, less often, 19 cm/sec. A B&K Type 4134 condenser microphone, coupled with a B & K Type 2619 preamplifier were used. The frequency response of this system was flat within the range 30 Hz–20 kHz.

Results

General Observations

There is a close correspondence between some sections of the vocal repertoires of the two species, although they represent different genera. For example, the intense alarm call of *C. jacchus* (Figure 1c) is almost identical to the intense alarm call

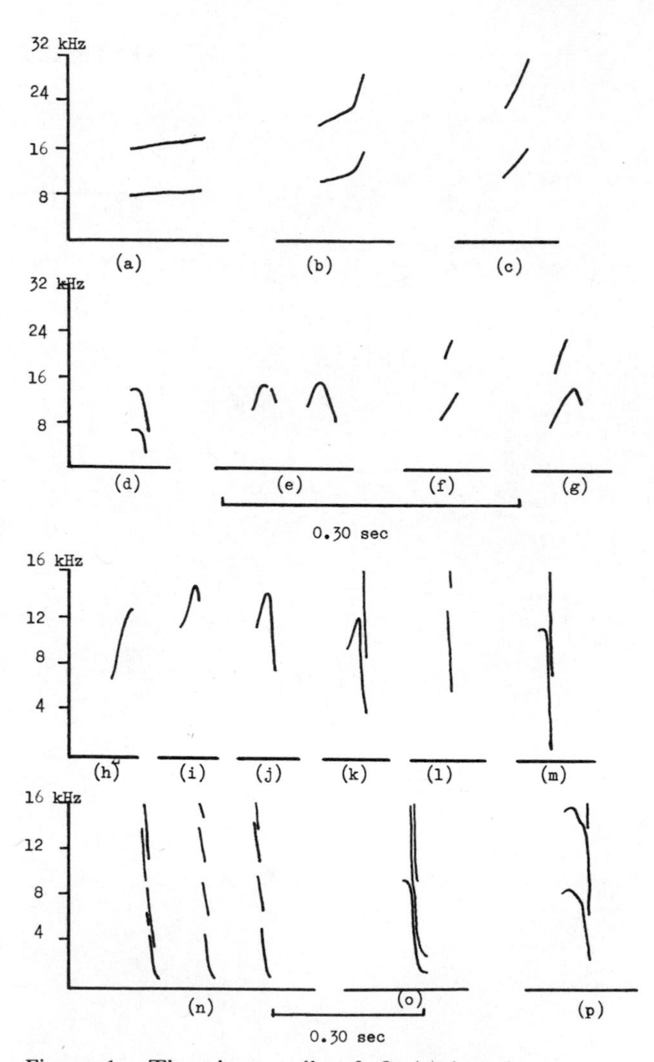

Figure 1. The alarm calls of *C. jacchus*, (a)–(c), and *S. fuscicollis*, (d)–(g); the transition from alarm to mobbing calls in *C. jacchus*, (h)–(m); the mobbing calls of *S. fuscicollis*, (n)–(p).

of *S. fuscicollis* (Figure 1f and 1g). Alarm calls are typically made by both species in response to unexpected movements, such as that of a bird flying overhead. They may also follow unexpected noises. The response of other animals to the alarm call is usually immediate alertness and sometimes flight to a position of safety, such as a nest box, or under a platform or branch.

Again, the mobbing calls of both species are very similar; Figure 1h–m shows the transition from alarm call to mobbing call for *C. jacchus*, Figure 1n–p shows mobbing calls of increasing intensity in *S. fuscicollis*. It can be seen that Figure 1m and Figure 1o show considerable structural similarity. These calls are made in a variety of threatening situations, notably at the close approach of a human observer. A constant watch is maintained on the source of concern, and the calls are often accompanied by jerky, side-to-side movements of the upper part of the body. Other animals initially respond by looking around in order to locate the cause of the disturbance. They may then also make mobbing calls, or return to their previous activity, depending on the nature of the threat.

Parallels also exist between the species in the usage of excitement, fear and aggression calls, although in each case the structure of the call is different for each species. This suggests that either these calls are less "primitive" in the evolutionary sense, or that they have evolved in some way freer from constraints on their physical structure. The well-known arguments put forward by Marler (1955, 1957) concerning the relation between structure and function in the alarm and mobbing calls of birds, suggest that the latter is at least as important a consideration as the former.

Differences in call usage between the species are more difficult to assess than differences in call structure. This can be shown by a study of the contact call systems of both species. Contact calls occur in over 90 percent of all the calls made by captive isolated pairs of animals, and this makes them suitable for quantitative study in depth. The contact call systems of *S. fuscicollis* and *C. jacchus* are structured in a similar fashion. Both systems range from a short, soft single-call type (Figure 2a and 2g) to longer, louder multiple calls (Figure 2b–f for *S. fuscicollis* and Figure 2h–k for *C. jacchus*). All these calls occur during a wide range of normal activities. The louder calls in turn merge into the prolonged loud shrilling calls, made with the mouth wide open, which are characteristic of each marmoset species (Figure 3c and Figure 4c). The shrill call is made most frequently in three different

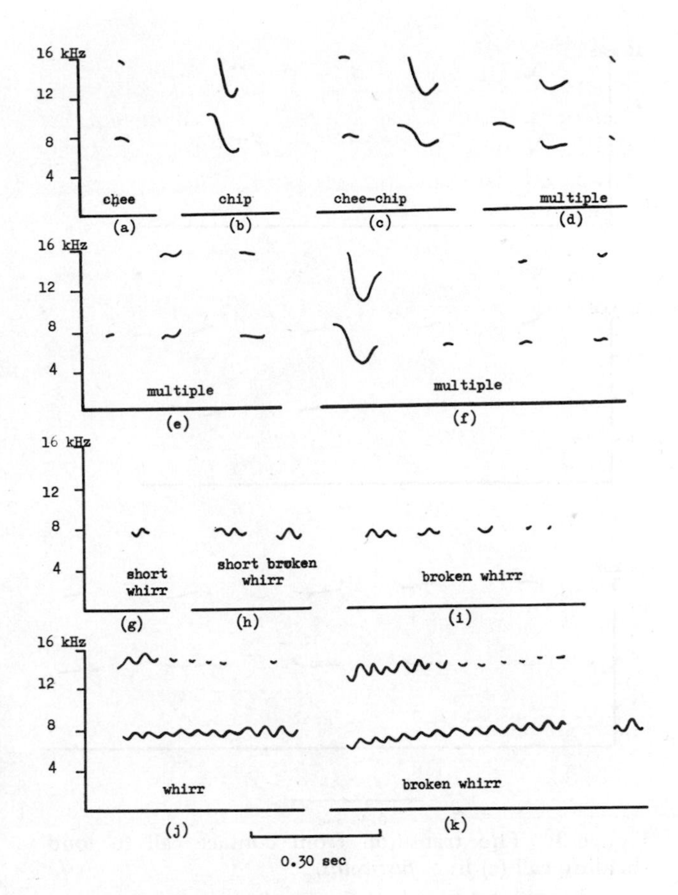

Figure 2. The contact calls of *S. fuscicollis*, (a)–(f), and *C. jacchus*, (g)–(k).

situations: when an individual is isolated, when two strange groups are placed together in close visual contact and at frequent intervals between groups that are widely separated.

It soon becomes obvious, however, even during preliminary investigations, that *S. fuscicollis* consistently makes more contact calls than *C. jacchus*. To discover why this should be so, and also why each species possesses such a variety of contact calls, a series of studies concerning the contextual use of these vocalizations was carried out.

Relationship Between Contact Calls and Activity Levels

The first hypothesis considered was that different types of contact call may be related to different activity levels. In this study, three activity states were defined: (1) Resting—the animal is in the nest box, hudding with its mate, or adopting a typical crouched resting posture; (2) Still—the animal is alert and looking around, or engaged in some activity such as feeding or grooming; (3)

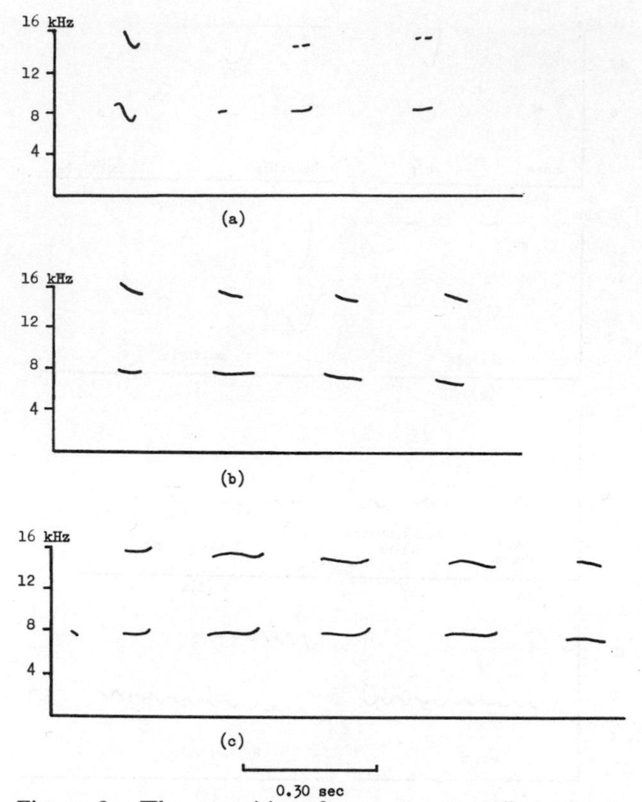

Figure 3. The transition from contact call to loud shrilling call (c) in *S. fuscicollis*.

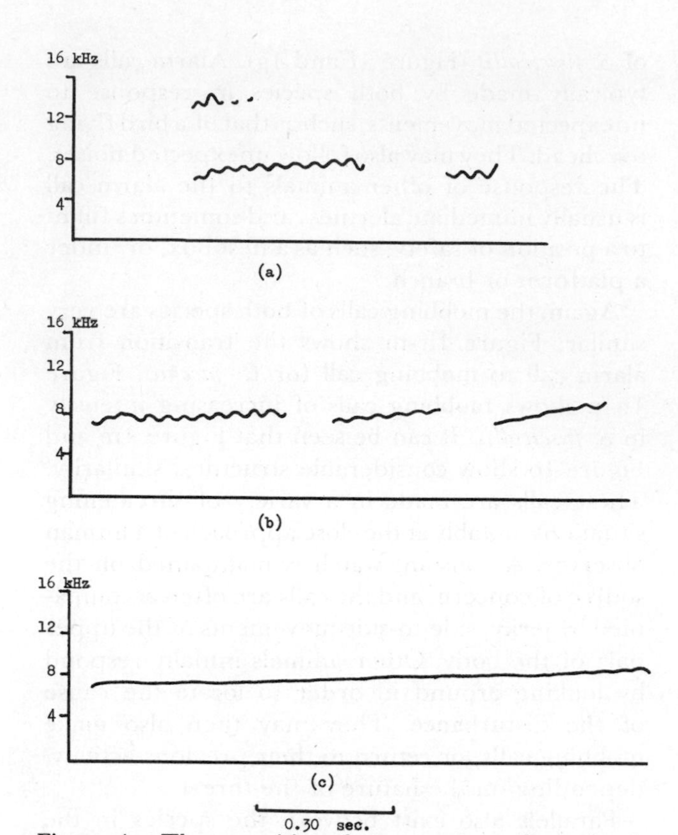

Figure 4. The transition from contact call to loud shrilling call (c) in *C. jacchus*.

Moving—the animal is engaged in any activity which involves the movement of all four limbs. At the onset of a call an animal's activity state and type of call were recorded. Observation sessions lasted for one hour during which time individual animals were observed for ten-minute periods.

The results are shown in Table 1. Both sets of contact calls are divided into four categories based on easily distinguishable acoustic features. The loud shrilling calls were ignored in these observations. In Table 1, the observed frequencies of use of the different call categories are shown and, in brackets, the expected frequencies. *C. jacchus* showed no significant preference for different call categories in different activity states. In *S. fuscicollis*, however, Chips were used significantly more by resting animals and Chee-chips and Multiples by active animals.

In an extension of this study, the relation between the vocalization patterns and activity scores, obtained by time sampling during forty-minute observation periods, was analysed. A similar trend toward the multisyllabic call types with increased activity score was found for *S. fuscicollis*, while no trends were seen in *C. jacchus*. For both species, there was a significant positive correlation between the activity score and the total calls for each observation session.

Thus, some of the different call types may be said to convey information about the current state of activity of *S. fuscicollis*, but not of *C. jacchus*.

Effect of Separation of the Pair on Call Use

The effect of physical separation on the pattern of call use was examined next. The vocalizations of pairs of animals were recorded under four conditions: (1) Both animals together in one cage; (2) The animals in separate cages with a distance of 45 cm between them; (3) The animals in separate cages, but with a visual screen placed between them; (4) One animal removed from the room and the other left in isolation.

The overall vocalization rates for both species over the four conditions are shown in Figure 5. The graph clearly shows how much more *S. fuscicollis* vocalize in all conditions. The table beneath the graph shows that for both species the proportion of contact calls is high in the first three conditions. When isolated, however, individuals of

Table 1. The relation between different contact-call types and activity state. * indicates p < .001 (Chi-square test).

Contact Calls	Resting	Still	Moving
Saguinus fuscicollis			
Chee	405 (369)	727 (756)	124 (131)
Chip	267 (188)*	338 (387)*	36 (67)*
Chee-chip	39 (120)*	309 (248)*	64 (43)*
Multiple	39 (64)*	151 (133)	39 (23)*
G-Test: p < .001			
Callithrix jacchus			
Short Whirr	33 (37)	169 (170)	10 (6)
Whirr	40 (40)	182 (181)	5 (6)
Short Broken Whirr	32 (26)	117 (120)	1 (4)
Broken Whirr	16 (18)	85 (83)	3 (3)
G-Test: Not significant			

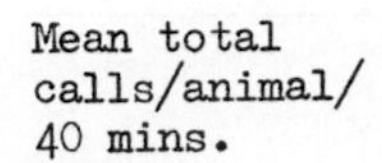

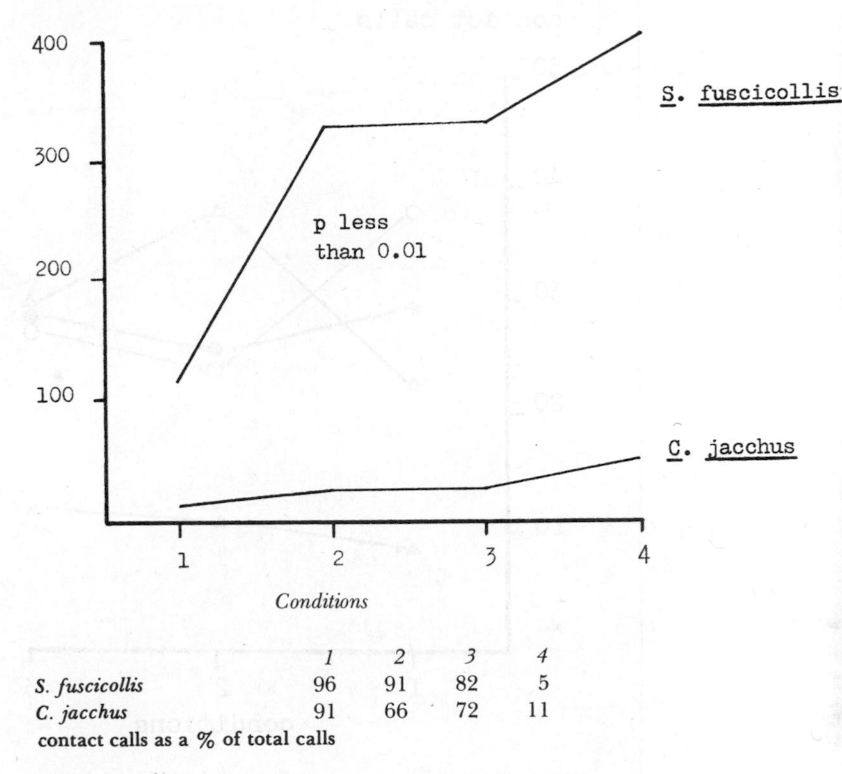

	1	2	3	4
S. fuscicollis	96	91	82	5
C. jacchus	91	66	72	11

contact calls as a % of total calls

Figure 5. The use of contact calls under different conditions of separation.

both species make very few contact calls, and shrilling calls are used most often.

The preference for different types of contact calls in the first three conditions is shown in Figure 6. Both species show significant fluctuations in call preference in the three conditions, partly related in the case of *S. fuscicollis* to the increased activity that occurs with physical separation. However, all the major changes in vocal pattern occur between conditions (1) and (2) for *S. fuscicollis*, but between conditions (2) and (3) for *C. jacchus*. In other words *S. fuscicollis* responds to physical separation with a change in vocal pattern (although this may be partly due to increased activity), but the loss of visual contact has no significant effect on this pattern. Conversely, the *C. jacchus* call pattern changes very little after separation (this again is in keeping with an increase in activity), but major alternations occur when visual contact is lost.

This suggests that *S. fuscicollis* can maintain contact by a purely vocal system which operates regardless of visual contact. In *C. jacchus*, however, the pattern of vocal communication would seem to be closely related to whether or not visual contact can be made. It is interesting to note that the changes induced by the loss of visual contact in *C. jacchus* (i.e., an increase in the use of the longer, louder multiple call types in preference to the short, soft single calls) are very similar to those which occur with increased activity in *S. fuscicollis*.

Interaction of Visual and Vocal Communication

Further observations were then designed to look more closely at the interaction of visual and vocal communication in both species. For this purpose, the animals of each pair were observed in separate cages, as in condition (2) of the last experiment. A record was kept of the vocalizations and concurrent behavior of each animal, including feeding, grooming, and gaze direction. Whilst only one animal could be watched at a time, the vocalization of both animals could be recorded as they occurred. Thus, the behavior of one animal could be related not only to its own vocalizations, but also to those of its mate.

For this analysis the contact call types were not treated individually, but as a single category for each species. The results are shown in Table 2. For example, while resting *S. fuscicollis* makes twenty percent of its contact calls. At the same time its mate, whose activity is unrecorded, makes twenty-two percent of its contact calls. As can be seen, there is little difference between the distribu-

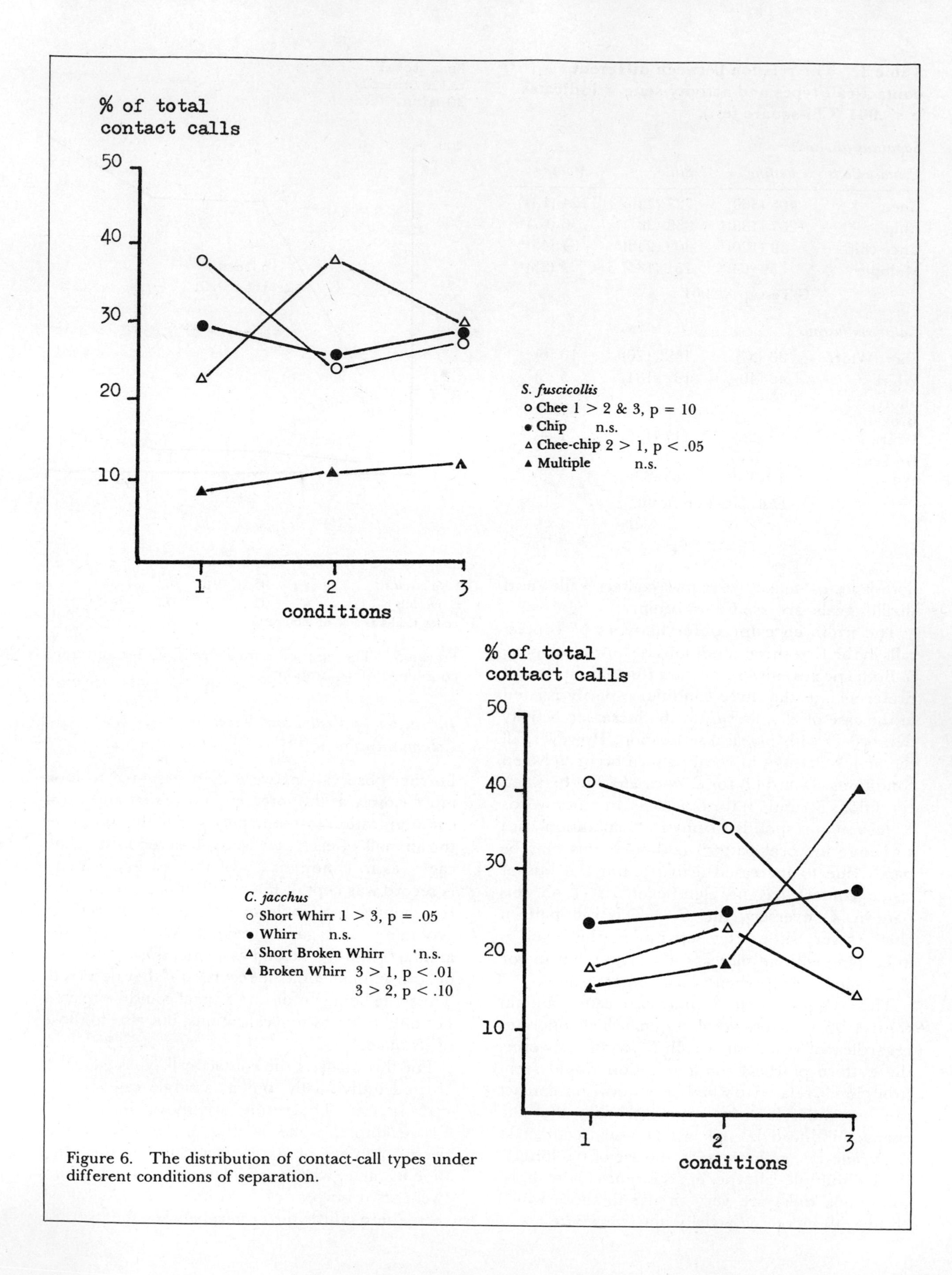

Figure 6. The distribution of contact-call types under different conditions of separation.

Table 2. The relation between the activity and contact calls of one animal and the contact calls of its mate during separation of the pair (in percentages).

Current Activity	*S. fuscicollis* Own Call	*S. fuscicollis* Mate's Call	*C. jacchus* Own Call	*C. jacchus* Mate's Call
In nest box	13	19	5	2
Resting	20	22	40	35
Still, looking around	33	25	31	30
Still, looking at mate	14	8	19	5
Running	7	7	0	4
Climbing on wire	1	2	0	1
Adjusting position	6	5	1	7
Grooming	4	5	3	8
Chewing wood			0	2
Marking			1	2
Defecating	0	1	0	0
Feeding	2	5	0	5
Total calls	7065	7056	1189	1041

tions of contact call use in the "own calls" and "mate's calls" columns for *S. fuscicollis*. The differences which do occur are to be expected from differences in activity level between the individuals. The results suggest that for *S. fuscicollis*, the use of contact calls by one animal is unrelated to the activity of its mate.

For *C. jacchus*, however, the distributions of contact-call use show a greater disparity. For example, an animal that is Still and looking at mate is almost four times more likely to vocalize than its mate, whereas an animal which is Still and looking around is no more likely to vocalize than its mate. Also, there appears to be a relation between one animal's movements and its mate's vocalizations. Summing all the categories which involve locomotion (i.e., Running, Climbing on wire, and Adjusting position), the data show that a moving animal is much less likely to vocalize than its mate, indicating either an inhibition from vocalizing in a moving animal, or an increased tendency for an animal to vocalize when its mate is moving. An earlier study, not discussed here, showed that for *C. jacchus*, the percentage of calls made while moving was approximately equal to the percentage of time spent moving which therefore discounts the inhibition hypothesis.

The results provide more information concerning the operational difference in the two contact-call systems, showing again how visual contact is an integral part of the *C. jacchus* contact communication system. For example, a *C. jacchus* individual, having seen its mate start to move, often responds with a contact call.

Call Sequences

The other major aspect of the use of contact calls which was investigated was their sequential use. The individuals of each pair were again kept in separate cages with a distance of 45 cm between them. The vocalizations of both animals were then recorded in sequence. An interval of more than five seconds between the beginning of any two successive calls indicated the end of one call sequence and the start of a new one. Numbers of calls per bout by the pairs ranged from 2 to 111 in *S. fuscicollis* and 2 to 9 in *C. jacchus*. Single calls were considered separately. The results of the subsequent sequential analysis are shown in Tables 3 and 4 for *S. fuscicollis* and *C. jacchus*, respectively. Pairs of consecutive calls separated by less than five seconds formed the "dyads" of this analysis. Thus, the second call of one dyad was frequently also the first call of another, and vice versa. From the analysis of the total number of call dyads, it can be seen that both species have a marked tendency to alternate vocalizations. In other words, a call by one member of the pair is likely to produce a rapid vocal response from the other.

Significant differences between the observed (O.) and the expected (E.) frequency of use of the individual contact-call types in certain positions in call bouts enable some judgments to be made concerning the role of each call type in such bouts. Both species possess a call which is used preferentially as the opening call after a period of silence. In *S. fuscicollis*, the Chee is common both as a single call and as the first call in a bout. The next call, which is likely to be made by the mate, is significantly more often a Chip (*see* "Preceded by Mate's Call" in Table 3). This is frequently followed by a Chip from the first animal (*see* "Followed by Mate's Call" in Table 3), and this alternation may continue for several calls. There is no call preferentially used to end a bout in the *S. fuscicollis* contact-call repertoire.

The Chee-chip call (which is highly associated with locomotion) is less likely to be involved in alternating call sequences. It tends to be made

Table 3. The sequential analysis of the contact calls of *S. fuscicollis*. O and E refer to observed and expected frequencies respectively. S-S refers to one animal calling twice in succession; S-Mate refers to one animal calling in response to the mate's call.

Single Calls (9% of total)	O.	E.	Diff.
Chee	272	225	Up 21%
Chip	263	237	n.s.
Chee-chip	106	121	n.s.
Multiple	24	59	Dn 66%
Other	28	51	Dn 45%

Chi-square: 45.7
$p < .001$

Total Call Dyads—5942	
S-S	2277 (39%)
S-Mate	3565 (61%)

Chi-square: 284
$p < .001$

	First Call in Bout			Last Call in Bout		
	O.	E.	Diff.	O.	E.	Diff.
Chee	540	431	Up 25%	432	431	n.s.
Chip	437	451	n.s.	492	451	n.s.
Chee-chip	226	233	n.s.	192	233	Dn 18%
Multiple	89	113	Dn 21%	110	113	n.s.
Other	36	100	Dn 64%	102	100	n.s.

First Call in Bout: Chi-square: 74.3, $p < .001$
Last Call in Bout: Chi-square: 11.1, $p < .05$

	Followed by Mate's Call			Preceded by Mate's Call		
	O.	E.	Diff.	O.	E.	Diff.
Chee	1197	1136	n.s.	1197	1136	n.s.
Chip	1311	1171	Up 12%	1382	1171	Up 18%
Chee-chip	494	655	Dn 25%	450	655	Dn 31%
Multiple	352	322	n.s.	324	322	n.s.
Other	211	281	Dn 25%	212	281	Dn 25%

Followed by Mate's Call: Chi-square: 79.8, $p < .001$
Preceded by Mate's Call: Chi-square: 122.5, $p < .001$

within continuous bouts of vocalization by a single animal. Multiple calls occur as a first call significantly less often than expected.

The *C. jacchus* contact-call system does not contain equivalents to these calls, in terms of their sequential use (Table 4). The Short Whirr is emitted as a single call significantly more than expected, but no call is used specifically to initiate a sequence. The Whirr call tends to follow a mate's call, but also tends to end bouts. The two types of Broken Whirr calls both promote call alternation, but also show an increased likelihood of ending bouts. Thus, although *C. jacchus* shows a trend toward alternation or contact calls, most of these sequences are rather short.

In summary, the *S. fuscicollis* contact-call system contains calls which tend to both start off and maintain alternating bouts of calling. The *C. jacchus* system has no sequence initiating calls and the "alternation maintenance" calls also tend to be the "end of sequence" calls.

Discussion

The results described above clarify the operational differences in the contact-call systems of the two species. The use of different types of *S. fuscicollis* contact call reflects the activity level of the vocalizing animal, but is independent of visual contact and unrelated to the general behavior of the other

Table 4. The sequential analysis of the contact calls of *C. jacchus*. Legend as in Table 3.

	Single Calls (45% of total)			Total Call Dyads—742	
	O.	*E.*	*Diff.*		
Short whirr	674	580	Up 14%	S-S	268 (36%)
Whirr	34	52	Dn 35%	S-Mate	474 (64%)
Short broken whirr	66	92	Dn 28%		
Broken whirr	23	48	Dn 52%		
Other	217	232	n.s.		
	Chi-square: 39.5 $p < .001$			Chi-square: 55.6 $p < .001$	

	First Call in Bout			Last Call in Bout		
	O.	*E.*	*Diff.*	*O.*	*E.*	*Diff.*
Short whirr	274	298	n.s.	270	298	n.s.
Whirr	24	26	n.s.	44	26	Up 69%
Short broken whirr	45	47	n.s.	65	47	Up 38%
Broken whirr	28	24	n.s.	37	24	Up 54%
Other	141	117	n.s.	96	117	n.s.
	Chi-square: 7.78 Not significant			Chi-square: 32.9 $p < .001$		

	Followed by Mate's Call			Preceded by Mate's Call		
	O.	*E.*	*Diff.*	*O.*	*E.*	*Diff.*
Short whirr	209	276	Dn 24%	216	276	Dn 22%
Whirr	30	24	n.s.	55	24	Up 129%
Short broken whirr	68	43	Up 58%	88	43	Up 104%
Broken Whirr	38	22	Up 73%	54	22	Up 145%
Other	129	109	n.s.	61	109	Dn 44%
	Chi-square: 47.6 $p < .001$			Chi-square: 167.7 $p < .001$		

animal in an adult pair. Sequential analysis shows, however, that the contact vocalizations of one animal are related to those of its mate. The *S. fuscicollis* contact-call system thus seems to operate independently from other modes of communication and seems capable of conveying enough information on its own to satisfactorily maintain contact among adult pairs of animals.

The pattern of use of *C. jacchus* contact calls is independent of the activity level of the vocalizing animal, but is affected by the presence or absence of visual contact and to some extent by the activity in which the other member of a pair is engaged. Sequential analysis shows that although there is a tendency to alternate vocalizations between an adult pair, there is little systematic use of the different call types in sequences. The *C. jacchus* contact-call system seems to be part of an integrated visual and vocal system for maintaining contact, with the vocalizations used mainly in a responsive fashion.

These findings, plus the fact that *S. fuscicollis* has a vocalization rate several times that of *C. jacchus* in most circumstances, suggest that *S. fuscicollis* inhabits denser forest than *C. jacchus*, according to Altmann's hypothesis (1967). Moreover, *C. jacchus* possesses several visual signals such as facial and genital displays which do not have parallels in *S. fuscicollis* (pers. obs.).

It can be seen that close cooperation between laboratory and field research is needed if the intricate relationships between ecology and social behavior in primate species are to be fully understood.

Acknowledgments

This research was carried out while the author was a research demonstrator at the Department of Psychology, University of Reading, England, and is a summary of part of a Ph.D dissertation.

I am very grateful to Dr. H. O. Box who was my supervisor at Reading University during this study.

Literature Cited

Altmann, S. A.
1976. *Social Communication Among Primates*, pp. 325–336. Chicago: University of Chicago Press.

Epple, G.
1968. Comparative studies on vocalization in marmoset monkeys (Hapalidae). *Folia primat.* 8:1–40.

Marler, P.
1955. Characteristics of some animal calls. *Nature* 176:6–8.
1957. Specific distinctiveness in the communication signals of birds. *Behaviour* 11:13–39.

Moynihan, M.
1970. Some behaviour patterns of platyrrhine monkeys, II *Saguinus geoffroyi* and some other tamarins. *Smithson. Contrb. Zool.* 28:1–77.

Muckenhirn, N. A.
1967. The behavior and vocal repertoire of *Saguinus oedipus* (Hershkovitz, 1966) (Callithricidae, Primates). Unpublished M. Sc. thesis, University of Maryland, College Park.

JENNIFER C. INGRAM
Department of Psychology
8-10 Berkeley Square
University of Bristol
Bristol BS8 1HH
England

Parent-Infant Interactions in the Common Marmoset (*Callithrix jacchus*)

ABSTRACT

Detailed observations of the interactions between parents and infants have been made on 5 adult pairs and 25 infant common marmosets (*C. jacchus*), housed in family groups. Intensive observations of parental behavior were taken up to the age of four months and comparisons made between infants in similar social situations. These comparisons searched for sex differences between infants, differences between singletons and twins, and differences within and between families. The frequencies with which older siblings carried infants were also compared within families. Correlations between the parental measures revealed that both parents (and particularly the mother) were responsible for initiating and regulating the increasing independence of their infants. Parents showed some differences in their responses to male and female infants, which in turn were dependent on the behavioral characteristics and individual differences of the two sexes.

Introduction

The behavior of marmosets and their social organization give rise to several interesting questions, particularly those concerned with mother-father cooperation in the care of young in a monogamous family unit, and the involvement of older siblings in these behaviors.

Methodology

Our colony was established four years ago with four adult pairs of common marmosets, *Callithrix jacchus*, (three of which were wild-caught), and we now have a total of 55 animals (July 1975), 10 of which are second generation infants. All the marmosets are housed indoors in family groups which are allowed to increase in size, to 10 or 12 in some cases. Most families live in large aluminum and wire cages, furnished with ropes, shelves, and swings, and two large families run free in small rooms (3 m × 2 m × 3 m) equipped with tree trunks, shelves, and ropes. The husbandry methods have been described elsewhere (Ingram, 1975a, 1975b).

The behavior studies have concentrated on parent-infant interactions and the development of the behavior repertoire in young who are living in family groups. The methods of recording and analysis for the parent-infant study are modified from those of Hinde, Rowell, and Spencer-Booth (1964) and all observations are made between 1000 hours and 1800 hours B.S.T.

Table 1 shows the 5 adult pairs and the 25 infants used in the study. When looking for differences between the sexes, between families and those of birth order within families, differences in social situations must be controlled for, and so comparisons are made between infants in similar situations only (*see* White and Hinde, 1975, for a similar approach).

Parents and infants were observed every weekday for 50 minutes each morning and afternoon from birth to 12 weeks old, and then for 50 minutes three times a week up to 18 weeks old. Several parental measures were recorded.

1. Percentage of time spent on both parents, and in larger families time spent on other siblings. When on the mother, it was noted whether the babies were in the "suckling position" or being carried.

2. When the babies started to be "rubbed off" or get off the parents, the percentage of time spent within a 15 cm radius of a parent ("Near") was recorded, and correspondingly observations were made of the time spent outside this 15 cm radius ("Far"). This distance was chosen since within it an infant could easily and quickly be retrieved, or climb on by itself. If other siblings were present in the family, the time spent near to a parent or older sibling (not twin) was also scored. Time spent "Far" was expressed as a percentage of the total time off the parents (and siblings).

3. During this 50-minute observation period, the number of approaches to and leavings from parents (and older siblings) was also scored. An approach is defined as each time the distance between the infant and another member of the family (particularly mother and father) was decreased from more than 15 cm to less (and vice versa for leavings), and when the boundary was crossed, the approacher (or leaver) was recorded. The difference between the percentage of approaches due to the infant and its leavings gives an indication of

Table 1. The identification of *Callithrix jacchus* parents and infants observed in this study.

	Parents				
	Wild-caught				*Captive-born*
Sets of Offspring	*Samson + Delilah (SD)*	*Pelham + Pippa (PP)*	*Andy + Alice (AA)*	*Nik + Vera (VN)*	*Billy + Bonito (BB)*
1st	SD1: ♀ (removed at 12mos)	PP1: ♀♀	AA1: ♀	VN1: ♂	BB1: ♂♂
2d	SD2: ♀♀	PP2: ♀♂	AA2: ♀♀ (3rd ♀ hand reared)	VN2: ♂♂	BB2: ♀♂
3d	SD3: ♀♂	PP3: ♀♂			
4th	SD4: ♀♀	PP4: ♀♀			

whether the infant or parent is primarily responsible for seeking proximity between them i.e. $Ap_i \times 100/(Ap_i + Ap_m) - L_i \times 100/(L_i + L_m)$ which is abbreviated to $\% Ap_i - \% L_i$ (Hinde and Spencer-Booth, 1967). The approach and leaving scores were recorded over a period of 20 minutes (during the 50-minute period), or until at least 30 "approaches" and "leavings" had been recorded. This was therefore dependent on the type of activity period and whether the infants were off frequently or not.

4. Every time an infant attempted to climb onto the mother or father, and to gain the nipple on the mother, the response of the parent (acceptance Mk_i or rejection R) was recorded. Also, the number of times that an infant went onto the nipple or was carried through the initiative of the parents (Mk_m or Mk_f) was recorded, so that a ratio of the number of rejections to the total number of occasions it attached or tried to attach to the nipple or be carried could be calculated: $R/(Mk_i + Mk_{m/f} + R)$(Hinde and White, 1974). This gives the relative frequency of rejections (R.F.R.). The contact initiation and rejection scores were recorded for the same 20-minute period as approaches and leavings.

Weekly means for all these parent-infant interaction measures were calculated up to the age of 12 weeks, and fortnightly means were used from the twelfth to eighteenth week. After this time, all infants were off their parents all the time and were usually completely weaned.

Graphs of median values for all infants were plotted to show overall developmental trends. Comparisons were made using the binomial test (Siegel, 1956) between individual infants in similar social situations to search for sex differences, and those of birth order within each family, in the amounts of interaction between parents and offspring. Other factors investigated include differences between singletons and twins, those between individual mothers and fathers (between family comparisons) and the ages of siblings involved in helping with infant care.

Relations between the age changes in the various parent-infant measures were assessed by calculating Spearman rank-order correlation coefficients (Siegel, 1956) between the values of these measures for all male and female infants and both parents. From these correlations, suggestions can be made about the roles of parent and infant in maintaining proximity, and responsibility for the development of infant independence.

Analysis of Results

Descriptive Method

Figures 1 and 2 show graphs of the median values of the parental measures from birth to 18 weeks for 12 female and 6 male infants. Table 2 is an example of the form of the results obtained from making comparisons between infants in similar social situations. The most reliable comparisons for investigating sex differences in behavior are those between heterosexual twins within families, since parental variation is eliminated. In Table 2, these are shown in the fourth row of results and include the results of 5 heterosexual twin comparisons from different families.

From the graphs and comparisons, mothers fed single female infants more than males, female twin infants suckled slightly more than males with a median level of about 35 percent during the first week (range 14 to 64 percent) dropping to 3 percent (0.5 to 5 percent range) during week ten (Table 2 and Figure 1a). Both sexes of singletons were carried for similar amounts of time by their mothers, but male twins were carried slightly more than female twins. Male infants, both singletons and twins, were carried more by their fathers. The single female infants were off their parents more than males, but both sexes of twins were off for similar amounts of time (Table 2). When alone the singletons spent similar amounts of time more than 15 cm from their parents (Far), but twin females tended to spend more time at a distance than males (Figures 1c and 2a). Infants approached their mothers more than their fathers, (two male and two female comparisons each with $p < 0.05$) and females tended to approach both parents more frequently than males (but none of these differences were significant at $p < 0.05$ level). Up to the end of the third week, the $\% Ap_i - \% L_i$ measure was negative for both parents and both sexes of infants, indicating that up to this age the parents were responsible for maintaining proximity with the infants (Figure 2b). During these early weeks, the infants were only off their parents from 2 to 8 percent of the time, and after being "rubbed off" they usually climbed up the wire mesh and hung on until their parents returned. Infant locomotor abilities were very limited at this age and so most approaches and leavings were performed by the parents, with most infant movements in the form of leavings. After the fourth week, the $\% Ap_i - \% L_i$ measure became positive, indicating that infants were responsible for seeking the prox-

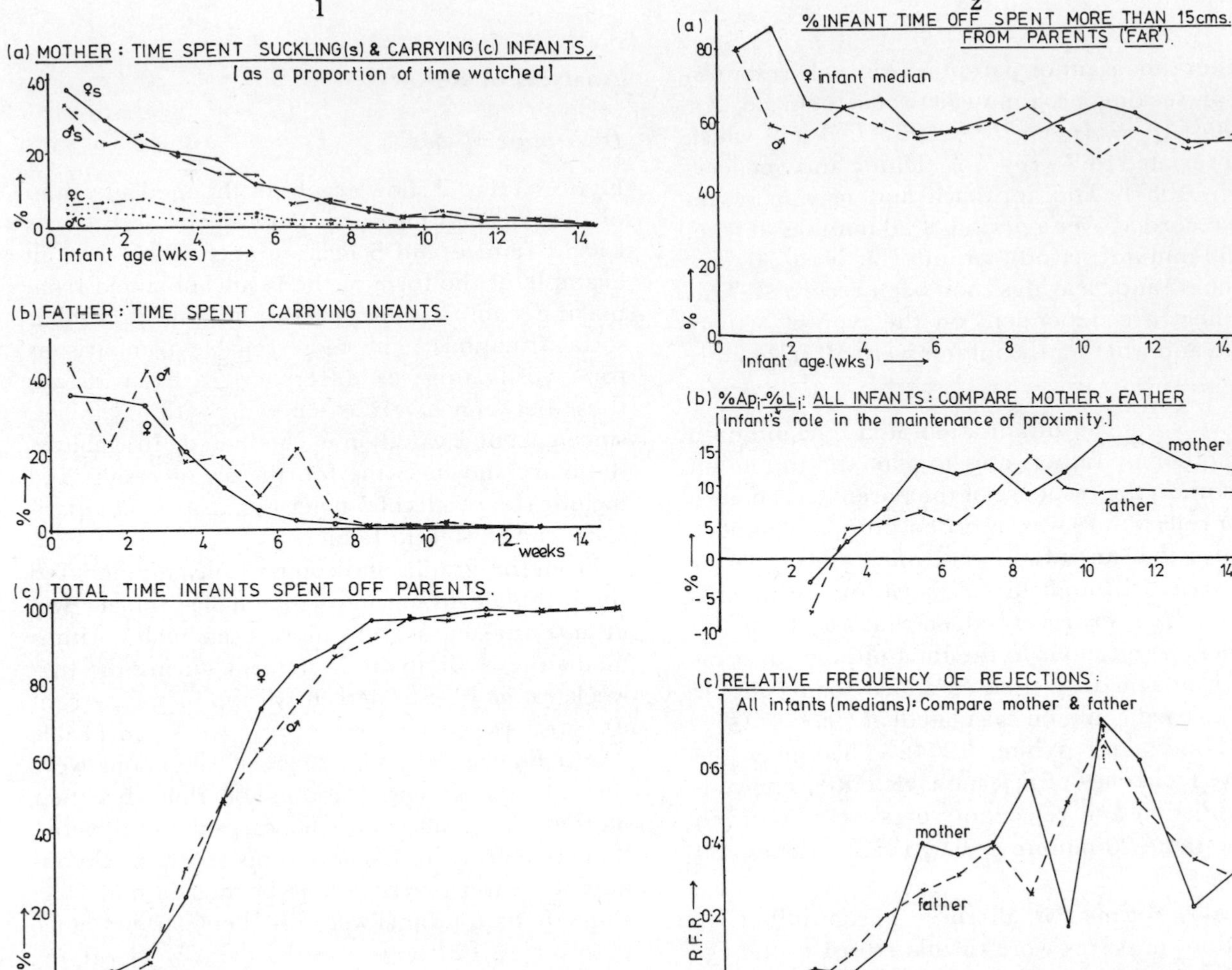

Table 2. Measures of parent-infant interaction: Comparisons made between male and female *C. jacchus*
Infants (male or female, singleton or twin) with consistently higher scores are shown, and the
Blanks = no data available; = approximately equal; N.S. = nonsignificant difference; M = mother;

	♂:♀ Comparisons	*Number of Comparisons*	*Number of Individuals*	*Parental Measures* Mother Feed	Mother Carry	Mother %Ap–%L
A.	Singletons	2	1 ♂ 2 ♀♀	♀ <.05	=	=
	Twin ♂♂:	2 (×2)	2 ♂♂		♂ .02 ♂ .02	
	Twin ♀♀		2 ♀♀	♀ .019 ♀ .002		♀ .02
	♂♀ Twins within families	5	5 ♂♂ 5 ♀♀	♂ N.S. ♀ N.S. (×3)	♂ .008 ♂ N.S. ♀ N.S.	♂ N.S. ♀ N.S. (×3)
	♂♀ Twins between families	2 (SD:PP)	2 ♂♂ (SD) 2 ♀♀ (PP)	.029 .006	.06	.05
B.	*Singletons(s): Twins*	3	3 ♂♂ 4 ♀♀	S < .001 =	= =	=

Figure 1. The course of parent-infant interactions in marmoset family groups. (a) and (b) The total time that infants spent on each parent and (c) the time spent off their carriers, as a percentage of the total time watched. In (a), the time spent in the suckling position or being carried by the mother is shown. Values plotted are the medians for 12 female and 6 male infants from four family groups up to 14 weeks old.

Figure 2. The course of parent-infant interactions in marmoset family groups. (a) The total time that infants were more than 15 cm from the parents as a proportion of the total time spent off the parents. (b) Infants' role in the maintenance of proximity. The number of approaches due to the infant (distance decreases from >15 cm to less) as a percentage of the total made, minus the number of leavings by the infant as a percentage of the total. (c) Relative frequency of rejections. The ratio of the number of occasions on which the infant attempted to climb onto the parent and was rejected (R) to the number of parent initiated contacts ($Mk_{m/f}$), infant initiated contacts (Mk_i) or rejections (R): $R/(Mk_i + Mk_{m/f} + R)$. Median values are shown for 18 infants from four families.

imity of their parents as they became more mobile.

The pattern of relative frequency of rejections by the parents complements that of the $\% Ap_i - \% L_i$ score (Figure 2c). Mothers rejected infants more frequently than fathers (one female comparison with $p < 0.05$), and three significant peaks of increasing strength of relative rejections were seen for the mother during the sixth, ninth and eleventh week for all infants. Two of these peaks correspond to higher $\% Ap_i - \% L_i$ scores and the eleventh week peaks are greatest for both measures. This suggests that the physical suckling tie is broken around this time. The peaks for relative rejections by fathers for male infants occur during weeks ten to twelve, and a smaller one around week fifteen, both of which coincide with high rejection levels by mothers. Fathers tended to reject females at a lower level over a longer period than their male offspring with the highest levels occurring between weeks eight and eleven, and another peak at week thirteen. From the comparisons (in Table 2) both parents tend to reject female infants more frequently than males. The pattern of relative rejection frequencies for all babies is therefore one of three peaks of increased rejection for mothers, and two peaks for fathers (weeks eight and eleven), but with fathers tending to be more receptive to the attempts of infants to be carried, particularly when they are frightened or anxious.

The effects of several social variations on these measures are now discussed briefly.

1. SINGLETONS COMPARED TO TWINS

Despite the limited data, the comparisons (in Table 2) show that single and twin females were fed and carried with similar frequencies by their mothers,

infants, and between singleton and twin infants, in similar social situations.
significance levels are for the binomial test.
F = father; %Ap–%L = infants role in proximity; R.F.R. = relative frequency of rejections.

Mother R.F.R.	*Father Carry*	*Father %Ap–%L*	*Father R.F.R.*	*Total Time Off*	*% Far (15 cm)*
	♂ .008	=		♀ .02	=
	♂ .02			♀ .002	
	♂ .02		♂ .016	♂ N.S.	
	♂ .02	=			
♀ N.S.				♀ .01	♀ N.S.
				♂ .035	♂ .05
♂ N.S.	♂ .008	♂ N.S.		♂ N.S.	
	♂ .016			♀ .008	
♀ N.S (×2)	♂ N.S.	♀ N.S. (×2)	♀ .03; ♀ N.S.	♀ N.S.	♀ N.S. (×3)
	.02				
	.002				
		N.S.		N.S.	N.S.
s N.S.	*s* < .001	*t* .02	*s* .05	*t* < .001	*t* .001
	s < .008			=	*t* .05

but the single male suckled more often than the twins. All singletons were carried more by their fathers and the difference was greater for the male comparison (Table 2). The female infants spent similar amounts of time off their parents, but the male twins were off much more than the single male. All twins spent more time at a distance (more than 15 cm) from their parents, and were usually with each other, whereas the singletons had no other companion and so spent more time close to their parents.

2. BIRTH ORDER WITHIN FAMILIES

The effects on the parental measures of whether an infant was one of a first, second, third or fourth set was investigated by comparing sets of twins of the same sex within each family. All infants, except for one, remained in the family group, so that these comparisons show the effects of the presence of older siblings on the behavior of parents with their latest infants.

Mothers fed their first and second female infants more than their third and fourth sets ($p < 0.05$ for three comparisons), and the first males suckled more than the second, which in turn fed more than the third set ($p < 0.05$). Similarly, first and second female and male infants were carried by their mothers more than those in the third or fourth sets ($p < 0.001$ for two comparisons). Fathers, on the other hand, carried all infants for similar amounts of time. Thus the presence of other siblings in the family, helping with infant care, seems to have most effect on the mother, suggesting an increasing tendency in the mothers to concentrate on the infants' essential food requirements, leaving transportation and comfort behaviors more to other family members.

When comparing the total time the infants spent off their carriers, there are some differences between the families. In general, first and second infants were off more than third and fourth infants for both males and females ($p < 0.05$ for four comparisons). For the time spent at a distance (Far), no clear distinctions were found, except that earlier infants tended to spend more time farther than 15 cm from their parents than those born subsequently.

Infants born in the second set tended to seek the proximity of both parents more frequently than those born third (nonsignificant differences). Relative frequency of rejection patterns could not be investigated for most comparisons (due to lack of data), but no striking differences were seen.

The first and second sets of babies tend to be fairly similar with respect to the parent-infant measures, as do the third and fourth sets to each other, with most differences occurring between these two groups. This may be due to the fact that by the time the third set of offspring are born, the first set are sub-adults and are much more able to handle and care for infants, than are juvenile siblings.

Thus these differences between successive sets of offspring, particularly for maternal measures, suggest that an increase in family size has most effect on the mother. Her responses to her infants are linked to the number of siblings able to share in their care, rather than a differential response to males and females, regardless of the family size.

3. INDIVIDUAL PARENTAL VARIATION BETWEEN FAMILIES

Comparisons were made (as before) between sets of twins in similar social situations in the five families. As would be expected, different parents show individual patterns for rearing their infants, but despite the differences the parental measures were not too dissimilar.

4. SIBLING INTEREST AND INVOLVEMENT

The involvement and interest of older siblings of different ages and sexes in the care of infants was examined by dividing them into three groups, with respect to the age gap between each sibling and infant. Group 1 contained siblings who were 5 to 10 months older than the infants in their families; Group 2 those 10 to 15 months older; and Group 3 siblings more than 15 months older than the infants. Each individual could be included in more than one group, as he or she entered a different age category with each successive set of infants born into the family. There were 13 siblings in Group 1 (9 females and 4 males), 6 in Group 2 (5 females and 1 male) and only 1 female (from PP1 twins) in Group 3. Thus the figures for Group 3 and some sections for Group 2 were not really extensive enough for many comparisons to be made, but even so, interesting results were obtained. The measures considered were the percentage of time spent carrying infants and $\% \; Ap - \% \; L$ due to the infants.

The results of the comparisons made within families of individuals in similar social situations can be seen in Table 3. These comparisons were made between male and female siblings (i.e., within

Table 3. Comparisons within and between sibling groups of siblings carrying infants and the maintenance of proximity of infants to siblings, within families.

Sibling Group 1 includes siblings which are 5 to 10 months older than the infants (adolescents); Group 2, those 10 to 15 months older than the infants (sub-adults); Group 3, those more than 15 months older than the infants (young adults).

The figures given are significance levels for the binomial test, and the infants with consistently higher scores are shown. Conventions as in Table 2.

	Within Families	*Number of Comparisons*	*Number of Individuals*	*Sibling Measures Carrying*	*%Ap–%L*
A.	Within groups ♂:♀ Group [1] (♂ + ♀ infants)	4	3♂♂ 3♀♀	♀ .008, .008	=
	Group [2] (♀ infants)	1	1♂ 1♀	♀N.S.	
B.	Between groups [1] : [2]—♀♀	2	2♀♀ (4 infants)	[2] .004	
	[1] : [2]—♂♂	2	2♂♂ (4 infants)	[2] .004 .016	
	[2] : [3]—♀	1	2♀♀ (2 infants)	[2] .016	
	[1] : [3] ♀	1	2♀♀ (2 infants)	[3] .035	

each group) and also between the groups.

Adolescent female siblings (Group 1) carried both sexes of infants more than males did and similarly females in Group 2 tended to carry infants more than male sub-adults, but the difference was much less than for the adolescents.

Comparisons of the carrying frequencies between the groups revealed that both male and female sub-adult siblings (Group 2) carried infants more than Groups 1 and 3, and the female in Group 3 carried infants more than female siblings in Group 1. The comparison of the *% Ap–% L* measures (indicating the relative role of the infant or sibling in maintaining proximity between them) showed that over the age of four weeks infants tended to approach all their siblings with similar frequencies.

The reasons for sub-adult siblings taking care of and carrying infants more than adolescents are probably twofold. Sub-adult animals are physically larger and so are more able to carry the weight of one or two infants than the half-grown adolescents are. Also, they are more experienced at handling babies (taking them correctly and "rubbing them off" efficiently) since they gained this experience from helping to a lesser extent with the previous infants. Young females are more attracted by infants than are males, and when the infants are very young, females touch them, try to take them and watch 'baby transfers' with great interest. From the results above it can be seen that this interest is reflected in the higher carrying frequencies for females than males, at all ages. The differences for adolescents are quite large, but they decrease to more similar levels by the sub-adult stage. The natural progression from this sub-adult level would be young adult females producing their own offspring which would then be carried much more by the young adult males.

5. SECOND GENERATION INFANTS COMPARED TO FIRST GENERATION INFANTS

Finally, a brief mention will be made of the differences seen between the development of first and second generation infants, with respect to the parental measures. Both sets of second generation infants were fed and carried less by the mother than first generation twins ($p < 0.001$ for two comparisons) but carried slightly more by their father. The most striking difference was that the BB1 twins (first set of second generation) spent consistently more time off their parents than all other male twins. The BB2 twins also spent more time off their parents than first generation twins, but the difference was not significant. There were no differences in the time spent more than 15 cm away from their parents between the two generations of infants. After the end of the third week, both sets of BB twins sought to maintain proximity with their parents more than first generation twins ($p < 0.01$ for three comparisons), and Bonito (mother) rejected them relatively more frequently than the wild-caught mothers did ($p < 0.05$ for two comparisons). The pattern of maternal relative rejection frequency differs quite markedly from that of first generation infants, with a peak for mother around the eighth week and another at week thirteen, which are both about two weeks earlier than those for wild mothers. The pattern for father also shows higher and earlier peaks for both sets of twins.

Thus these data (combined with behavior repertoire records) suggest that second generation infants become independent earlier than first generation infants. Some of these differences could possibly be related to experience, since these were Bonito's first two sets of infants, and the wild females may have had many previous infants. However, the differences are interesting and the development of other second generation infants is being studied at present to test these trends for other families.

Correlational Method

The second method of analysis was concerned with the relations between the age changes in the various parental measures for all infants. These relationships were studied by calculating Spearman rank order correlation coefficients between the values of the measures for each individual (ranked weekly to 12 weeks and then fortnightly). Selected results are shown in Table 4, with the 22 infants grouped together in families, so that familial characteristics are preserved. Positive and negative correlation totals are shown at the foot of each column to describe the overall relations between these parent-infant measures over the whole period. The correlations shown in the table are those between Total Time Off (T.T.Off) and Far, T.T.Off and Relative Frequency of Rejections (R.F.R.) and Far with $\% Ap_i$–$\% L_i$. For the two latter results, indications are made on Table 4 of the roles of parent (P) or infant (I) in increasing infant independence and in the maintenance of proximity, respectively. The results for the majority of infants will be discussed and the implications of these for the development of the young.

For two-thirds of infants, an increase in infant time off the parents up to 18 weeks old is negatively correlated with increased time spent more than 15 cm from the parents (nine significant with $p < 0.05$). When these infants were very young, they spent relatively more time far away from their parents. Later on, they spent more time near to them as interactive behaviors—such as grooming and playing—became more common, and the infants were less eager to climb onto their parents as soon as they approached. The results of this correlation for eight infants, however, are positive (two with $p < 0.05$) indicating that as they spent more time off their parents, they spent more time at a distance from them. Three of these infants are the sons in the VN family, and this may be a family characteristic.

If the increase in time off parents was due to changes in the behavior of the infant, the percentage of its approaches to the parents would decrease and its leavings would increase (i.e., a decrease in the infant's role in seeking proximity), and similarly the relative frequency of rejections by parents would decrease. Therefore, if changes in infant behavior were responsible, negative correlations would be seen between time off and these two measures. If, however, the increase in time off was due to changes in parental behavior, the correlations would be positive. From Table 4, it is apparent that three-quarters of the correlations for T.T.Off with maternal R.F.R. are positive (binomial test, $p < 0.05$), and thus changes in the behavior of mothers are important in stimulating the independence of most infants. For two infants only were changes in the time spent off the mother primarily due to changes in infant behavior, and for three others changes in both mother and infant were responsible (since correlations were very low or zero). A different pattern of correlations between the measures for fathers can be seen in Table 4. In only half of the relationships were fathers responsible for stimulating the increase in independence of infants by rejecting their attempts to be carried as they grew older. Those infants which took the initiative for becoming independent were all second-or third-born infants with older siblings present in the family group, and they may have spent more time with these siblings. Thus, mothers rejected many more infant attempts to gain the nipple or be carried than fathers did, and so encouraged infants to become more independent.

Similar arguments can be applied to the correlations between time spent at a distance and seeking the proximity of parents. For those eight infants with initial positive correlations between total time

Table 4. Spearman rank order correlations between four parental measures for individual *C. jacchus* infants up to 18 weeks of age.

Blanks = no data available; − negative correlations; + positive correlations; M = mother; F = father; %Ap−%L = infants' role in proximity; R.F.R. = relative frequency of rejections; P = parental influence in increasing independence or maintaining proximity; I = infants influence in increasing independence or maintaining proximity; * = $p < 0.05$; ** = $p < 0.01$.

Family	Infants		*Total Time Off:* *Far*	*R.F.R.* *M*	*F*	*Far:* *%Ap−%L* *M*	*F*
SD	SD2	♀	−.34				
		♀	−.25				
	SD3	♀	−.24	.28 P	.55 P*	−.52 I	.03 P
		♂	−.52	.32 P	.62 P*	−.05 I	.43 P
	SD4	♀	.56	−.30 I	.13 P	.07 I	.29 I
		♀	.11	.33 P	.03 P	.18 I	.08 I
PP	PP1	♀	.34				
		♀	.31				
	PP2	♂	−.64*			.12 P	−.18 I
		♀	−.77**			.46 P	0
	PP3	♀	−.79**	−.04 I	−.45 I	−.29 I	−.83 I*
		♂	−.25	.43 P	−.35 I	−.31 I	−.59 I
AA	AA1	♀	−.60*			.35 P	.50 P
	AA2	♀	−.56*	.40 P	−.56 I*	−.62 I*	.21 P
		♀	.26	.56 P*	−.04 I	.57 I	−.02 P
VN	VN1	♂	.68*	−.65 I	.58 P	.58 I	.60 I*
	VN2	♂	.76**	.42 P	−.14 I	.05 I	.21 I
		♂	.44	.35 P	−.17 I	.11 I	−.03 P
BB	BB1	♂	−.48*	.01 P	.27 P	−.39 I	.03 P
		♂	−.69**	.04 P	.04 P	−.08 I	−.21 I
	BB2	♀	−.63*	.46 P	−.35 I	−.54 I	.46 P
		♂	−.84**	.37 P	−.20 I	−.04 I	−.64 I
Totals	+		8	12	7	9	10
	−		14	3	8	9	7
Totals				12 P	7 P	3 P	8 P
				3 I	8 I	15 I (5 low)	9 I

off and time spent far, positive correlation between time spent more than 15 cm from the parents and the infants' role in maintaining proximity indicates that changes in the infants' behavior were involved, and a negative correlation that changes in parental behavior were responsible. For those infants with a negative correlation initially between T.T.Off and Time spent at a distance, the reverse situation will be true (positive correlation for changes in parental behavior, etc.). A low correlation again indicates that changes in both parental and infant behavior were involved in the age changes seen. From Table 4, it can be seen that 15 out of 18 of the infants were responsible for maintaining proximity with their mothers ($p < 0.05$), but five of these were low correlations indicating both mothers and infant were involved; in three, mothers were responsible (for one infant in AA family, and the PP2 twins). In only eight cases were infants primarily responsible for maintaining proximity with their fathers, in four, fathers were responsible (two of these results from AA family), and in six relationships both father and infant were involved in maintaining proximity. Thus, more infants tend to be involved in maintaining proximity with their mothers than with their fathers as they grow older, but again considerable individual differences are seen in the behavior of infants and in the relationships with their parents up to four months of age.

The preceding correlations elucidate some of the questions about the changes in the nature of the parent-infant relationship over the first 18 weeks of life.

Changes in the behavior of both parents, and particularly the mother, promote and regulate the increasing independence of infants. Fathers tend to be more willing to allow infants to initiate contact and are also responsible for maintaining proximity with some infants. Changes in the behavior of both parents and infant are involved in increasing the independence of yet others. Undoubtedly in all developing relationships the responses given by parent or infant, however subtle, initiate changes in the behavior of the other.

Further correlations have indicated that parents show some differences in their responses to male and female infants, which in turn are dependent on the behavioral characteristics of the two sexes. Differences in individual parent-infant interactions tend to be primarily due to parental behavioral variations for female infants, but differential male-infant behavior may be the cause of their varied relations with their parents.

The strength of some of these differences may be enhanced by the small number of parents involved and also the high proportion of twin infants used in the calculations.

As infants grow older, however, they rapidly show more interest in their physical and social environment. Parents encourage their independence by leaving them alone and allowing them to remain alone longer.

Discussion

When attempting to assess the effects of several independent variables on relationships as complex as those developing between parents and infants, White and Hinde (1975) have stressed the importance of equating sub-groups for all factors except the one under examination. This can be illustrated in the present study by the marked effect on parental behavior of the presence of different numbers of older siblings in the family group. In fact, several of the variables considered will most likely interact with one another to produce a more complex picture than that described above. However, several consistent differences were shown between maternal and paternal interactions with the young and also between male and female infants, when allowing for the variations in social situations.

The adaptiveness of the callitrichid parental and family system described above can be discussed in conjunction with the evolution of a monogamous pair bond.

The most plausible hypothesis for the involvement of the father in the care of young is related to the freeing of maternal energy for milk production. Since an adult female usually produces twin infants (in over 90 percent of cases), she must produce more milk than a primate feeding only one infant; thus her energy requirements to be met by foraging for food to support milk production will be high. In addition to this, she may conceive again during her postpartum estrus, and so two weeks after she has given birth to one pair of twins she could be supporting another pair "in utero." It seems unlikely that this high reproductive level could be maintained without the assistance of the male in transporting the rapidly growing infants, whose combined weight by the age of four weeks is about one third that of an adult.

The parent-infant relationship can also be considered from an evolutionary point of view in terms of the benefit and cost to the members involved with respect to their eventual reproductive success (Trivers, 1974). Parental investment can be defined as "anything done by the parent for the offspring that increases the offspring's chances of surviving to reproduce, at the cost of the parent's ability to invest in other offspring." The benefit to an infant is measured in terms of its chances of survival and future reproductive success, the cost in any reduction in the reproductive success of its unborn siblings. When considering the cost and benefit of producing twin infants the cost to the mother will be greater than for a single offspring, but she will increase her benefit by rearing twice as many offspring to maturity. In the family group situation, the continuous presence and help of a male will reduce the rearing cost to the female and increase the probability of the young reaching maturity. Thus it is also advantageous for the male to remain with the female in this situation to enhance the survival of his offspring. Therefore, if ecological factors favor dispersion into small coherent groups, the territorial behavior subsequently shown by the adults of such groups may serve as an adaptation to enhance adult survival, for the male to sire and rear several litters at once, and for a female to increase the success with which she rears her young.

Finally, evidence from Leutenegger's (1973) analysis of maternal-fetal weight relationships suggests that an evolutionary decrease in body size of the callitrichids has promoted the development of twinning as a secondary feature, rather than it being the retention of a primitive mammalian characteristic.

Thus, assuming that there is no strict breeding season, and that the female normally produces a set of twins every five or six months, it would appear to be advantageous for a male to remain with one female only. His contribution to the gene pool will be large since he sires four young each year. The female benefits from the continuous presence of the male by being able to concentrate her energy on foraging for milk production and having assistance with the successful rearing of the infants. Territorality is enforced by fierce aggression, especially between adult females, and a strong pair-bond develops between the adult male and female, enhancing the maintenance of a stable family unit, which is advantageous to all members.

Acknowledgments

I am very grateful to Dr. John Crook for support and encouragement throughout the project, and to Dr. Robin Dunbar for advice and discussion of the manuscript. I would like to thank Mr. M. Sherborne and Mr. C. O'Neill for their dedicated assistance with the maintenance of the marmoset colony, and Mrs. J. Sherborne for typing the manuscript. I would also like to thank the Medical Research Council for financial support.

Literature Cited

Hinde, R. A.; T. E. Rowell and Y. Spencer-Booth.
1964. The behavior of socially living rhesus monkeys in their first six months. *Proc. Zool. Soc. Lond.* 143:609–649.

Hinde, R. A. and Y. Spencer-Booth.
1967. The behavior of socially living rhesus monkeys in their first two and a half years. *Anim. Behav.* 15:169–198.

Hinde, R. A. and White, L. E.
1974. Dynamics of a relationship: Rhesus mother-infant ventro-ventral contact. *J. Comp. Physiol. Psych.* 86(1):8–23.

Ingram, J. C.
1975a. Parent-infant interactions in the common marmoset (*Callithrix jacchus*) and the development of young. Ph.D. dissertation, University of Bristol.
1975b. Husbandry and observation methods of a breeding colony of marmosets (*Callithrix jacchus*) for behavioral research. *Lab. Anim.* 9:249–259.

Leutenegger, W.
1973. Maternal-fetal weight relationships in primates. *Folia primat.* 20:280–293.

Siegel, S.
1956. *Non-parametric Statistics for the Behavioral Sciences.* London: McGraw-Hill.

Trivers, R. L.
1974. Parent-offspring conflict. *Am. Zoologist* 14:249–264.

White, L. E. and Hinde, R. A.
1975. Some factors affecting mother-infant relations in rhesus monkeys. *Anim. Behav.* 23:527–642.

R. J. HOAGE
National Zoological Park
Washington, D.C. 20008
and
Department of Anthropology
University of Pittsburgh
Pittsburgh, Pennsylvania

Parental Care in *Leontopithecus rosalia rosalia*: Sex and Age Differences in Carrying Behavior and the Role of Prior Experience

ABSTRACT

Between February 1974 and September 1975, an intensive investigation of infant care patterns in the golden lion tamarin (*Leontopithecus rosalia rosalia*) was conducted at the National Zoological Park. Eleven infants from seven litters were observed for the first twelve weeks postpartum. Two aspects of infant care were examined: (1) frequencies of infant carrying in adults and juveniles and (2) preferential carrying of infants of the same or opposite sex. Also, the degree of success in the delivery and care of neonates was compared in four sets of primiparous parents with differing previous experience in carrying and interacting with infants. The results showed that mothers remained the principal infant carriers through Week 3, but became secondary carriers from Week 4 to 12. In contrast, fathers, secondary carriers through Week 3, became principal carriers from Week 4 to 12. Juvenile females were tertiary carriers from the second to the seventh week, and after Week 8 were only incidental carriers. Juvenile males were also tertiary carriers but exhibited a concentrated period of carrying between Weeks 3 and 8, but were thereafter only incidental carriers.

The carrying data also indicated that in three of the four litters where heterosexual twins were born and survived the first 12 weeks, male and female parents preferred to carry infants of their own sex.

Finally, among primiparous parents those with the greatest previous exposure to infants had the greatest success with neonates. This suggests that a definite minimum time of exposure to infants is probably necessary for juveniles to ensure that they will be successful as primiparous parents.

Introduction

Parental behavior in the Callitrichidae (marmosets and tamarins) has been inadequately investigated, and most previous accounts were descriptive and based largely on short-term or intermittent observations (cf. Ditmars, 1933; Fitzgerald, 1935; Lucas et al., 1937; Altmann-Schönberner, 1965; Muckenhirn, 1967; Moynihan, 1970; and Coimbra-Filho and Magnanini, 1972). Also, these reports presented conflicting information and hypotheses on parental roles within and between species (see Epple [1975b] for a discussion). These differences can be attributed in part to the fact that nearly all studies have been conducted on captive animals in greatly varying environmental conditions. However, recent studies conducted on captive *Callithrix jacchus* (Box, 1975a and 1975b, and Ingram, 1977), *Saguinus fuscicollis* (Epple, 1975b) and *Callimico goeldii*[1] (Heltne et al., 1973) have contributed much to the establishment of quantitative methods for delineation of specific paternal and maternal roles (e.g., the differences found in infant carrying patterns). In the last few years, there have been several long-term field studies resulting in the accumulation of new data on the size, composition, and ecology of wild callitrichids. However, behaviors such as weaning, food sharing, and infant carrying are still difficult to observe consistently in the wild because of the small size and rapid movements of the animals and obscuring foliage. Clearly, until productive observation techniques can be developed for the field, the captive colony will remain the principal arena for quantitative research on the Callitrichidae.

In captivity, as well as in the wild, the social unit now thought typical of marmosets and tamarins is the monogamous family consisting of one adult mated pair and their offspring from successive litters (Epple, 1967, 1975a, 1975b; Hampton et al., 1966). Within this unit newborn infants become socialized and maturing juveniles begin to exhibit important aspects of parental care (e.g., infant carrying and food sharing). *L. rosalia* and *S. fuscicollis* juveniles, artificially separated from their families before or very soon after the arrival of newborn infants, have shown certain difficulties in successfully rearing their own young (Epple, 1975b; Coimbra-Filho and Mittermeier, 1976).

[1]Recently, *Callimico goeldii* has been raised to familial status (Callimiconidae) independent of but closely related to the Callitrichidae (Hershkovitz, 1970). Both families share a number of behavioral and physical characteristics.

Although the role of the father in infant care has received considerable attention, the degree of mother involvement, outside of nursing behavior, is less well known, and except for Ingram (1977) the role of older siblings has not been described in any detail (cf., Hill, 1957; Sanderson, 1957; Crandall, 1964; Napier and Napier, 1967; Mitchell and Brandt, 1972; Jolly, 1972; Snyder, 1972).

The purpose of this paper is to describe in detail the parental behavior patterns of golden lion tamarin (*L. r. rosalia*) adults and juveniles. To this end, two aspects of infant care-related behavior are examined: (1) frequencies of infant carrying in adults and juveniles, and (2) preferential carrying of infants of same or opposite sex. The importance of juvenile exposure to both parental and early infant behaviors is discussed.

Materials and Methods

As part of a more detailed investigation of social and sexual development in the golden lion tamarin, data on infant carrying were collected for 7 litters (11 infants) born at the National Zoological Park between February 1974 and May 1975. One pair was observed for three successive litters, another pair for two successive litters, and two pairs for one litter (Table 1). In those families where more than one litter was studied, family composition (except for the breeding pair) changed from one birth episode to the next as older juveniles and sub-adults were removed from the group. In all, 14 liveborn infants were observed in 5 different family groups (3 infants died within a week). Each litter was observed from birth to 12 weeks of age (see Table 1 for group histories).

Carrying occurred when an infant's weight was supported by another group member and it was off the substrate. This includes nursing or suckling positions while on the mother, in contrast to Epple (1975b) who separates nursing and carrying. Infants usually clung with the hands and/or feet when being carried. Infants were recorded as independent or "off" if most of the torso was in contact with the substrate and not supported by another animal. Active transport was not a criterion for carrying since a "carrier" could support an infant while lying down. Except when nursing or attempting to nurse, infants clung to the dorsum of the carrier.

Total observation time was 279.5 hours; observations were taken about five days per week, but without a strict schedule. Data were recorded between 0900 and 1800, with an emphasis on the afternoon hours. Beginning Day 1 (postpartum), individual carrying time was recorded in minutes: summaries were made every seventh day. Events occurring in Week 1 include Day 1 through Day 7 (all births occurred the night before Day 1); Week 2 includes Day 8 through Day 14 and so on (see Table 2).

In family groups with a single infant, carrying time plus independent locomotion ("off") time equaled total observation time, but in groups with twins, individual carrying time did not discriminate between the transport of one or two infants: if one infant was carried exclusively by animal X, and the second infant by animal Y, the sum of the carrying time for X and Y could be twice the total observation time. The combined carrying time for all members of a group can thus total more than the total observation time (see Table 2, Week 2 for example). Computations and statistical tests followed Ferguson (1959) and Siegel (1956).

All study groups were visually isolated and housed separately in either: (1) a three sectioned 6.1 m × 1.8 m × 2.5 m glass-fronted cage, or (2) a 4.6 m × 3.7 m × 4.6 m room. Cages were provided with natural branches, emphasizing the horizontal, and at least two nest boxes. Observers were in full view of the tamarins at all times.

This paper will cover two topics: (1) quantitative data on infant carrying within family groups, and (2) descriptions of the behavior of four sets of primiparous parents with differing previous infant experience.

Results

Infant Carrying Behavior: Weeks 1 to 12

Table 2 presents the average duration and percentage of time infants were carried by adults and juveniles of both sexes, as well as the percentage of time infants were independent of adults. Figure 1 shows the average percentages of carrying in four classes of family members during the first twelve weeks postpartum.

In six of the seven family groups, mothers were the sole carriers in Week 1; however, in Group D (Table 1), the mother attempted to strip the infant from her body two days after birth. Only the active relief and participation in carrying by her mate and his twin brother seemed to prevent the mother from harming the infant. This was the first birth for the pair. Thus, except for Group D, and a 10-minute steal of a dying infant by the father in Group A_1, fathers never carried infants

Table 1. Group composition, litters, and previous infant experience.

Nine litters born at National Zoological Park during the study period are listed. Litters A_1 through D survived the neonatal stage; litters in the two groups below the bar line died within the first three days. Capital letters plus numerals designate groups with multiparous parents (numbers indicate repeat litters in one group). Capital letters alone designate primiparous groups.

Group	*Size of Litters*	*Infant Birthdate*	*Father*	*Mother*	*Juveniles 1*	*2*	*3*
A_1	1 ♂*	11 II 74			♂	♂	
Age at Inf. Birth			> 5 yrs	6 yrs	9 mos	9 mos	
Previous Inf. Exper.			5 litters	4 litters	None	None	
B_1	1♀	15 III 74			♂	♀	
Age at Inf. Birth			> 8½ yrs	> 6¾	10 mos	10 mos	
Previous Inf. Exper.			3 litters	3 litters	None	None	
C	1♂ 1♀	5 IV 74					
Age at Inf. Birth			2¼ yrs	2 yrs			
Previous Inf. Exper.			22 days	159 days			
B_2	1♂ 1♀	16 VIII 74			♀		
Age at Inf. Birth			> 9 yrs	> 7¼ yrs	5 mos		
Previous Inf. Exper.			4 litters	4 litters	None		
A_2	1♂ 1♀	4 IX 74			♂		
Age at Inf. Birth			> 7⅓ yrs	6½ yrs	7 mos		
Previous Inf. Exper.			6 litters	5 litters	None		
B_3	1♂ 1♀**	26 III 75			♀	♂	♀
Age at Inf. Birth			> 9⅔ yrs	> 7¾ yrs	12½ mos	8 mos	8 mos
Previous Inf. Exper.			5 litters	5 litters	8 mos	None	None
D	1♂	19 V 75			(other ♂ adult)		
Age at Inf. Birth			2 yrs	2 yrs	2 yrs		
Previous Inf. Exper.			127 days	115 days	185 days		
E	1♂ 1♀	23 IV 75					
Age at Inf. Birth			3 yrs	2 yrs			
Previous Inf. Exper.			159 days	None			
F	1♂ 1♀	15 V 75					
Age at Inf. Birth			4¼ yrs	3½ yrs			
Previous Inf. Exper.			None	22 days			

*♂ twin died Day 3 postpartum.

**♂ triplet stillborn.

in the first seven days. Also, infants were always carried in Week 1.

In the second week, the mothers' carrying time decreased as fathers and juvenile females began to transport infants (see Table 3 for the timing of the first transfer to other family members by the mother). Mothers still carried 95 percent of the time observed (except in Group D where the mother carried only 25.5 percent of the time). Although juvenile females began to carry during the second week, male juveniles did not until the third week. Infants were always carried during the second week even though their activity levels and strength were increasing, thus preparing them for their first independent forays in Week 3.

Changes during the third week were dramatic; mothers carried less than 60 percent of time observed, but fathers increased to nearly 39 percent and the combined juvenile frequency increased to 15.5 percent. Infants moved independently for 2.8 percent of the time. On the average, infants were first observed more than 30 cm from a carrier on Day 20 (range: Day 17 to 25). In some instances, an infant independently left a carrier, but generally the mother was responsible for the first separation by scratching or rubbing the infant off and then

Table 2. Average carrying and independent locomotion ("off") time (in minutes) per week for *L. r. rosalia* infants. Percentages of total observation time are given in parentheses.**

Week	*Mothers (7)*	*Fathers (7)*	*Off time*	*Total Observ. time*	*Juvenile Males (4)*	*Juvenile Females (4)*	*Total Observ. Time*
1	1998.0 (88.49)	176.0 (7.81)		2258.0			1528.0
2	1478.5 (87.10)	278.5 (16.41)	.5 (.029)	1697.5		77.0 (5.74)	1342.5
3	967.0 (59.47)	630.0 (38.75)	35.5 (2.81)	1626.0	97.0 (7.94)	91.5 (7.49)	1221.0
4	557.5 (41.14)	722.5 (53.30)	141.5 (10.44)	1355.5	142.0 (13.87)	46.0 (4.49)	1023.5
5	583.0 (44.71)	555.5 (42.60)	325.0 (24.94)	1304.0	50.0 (4.98)	72.0 (7.17)	1004.0
6	270.0 (24.18)	552.0 (49.60)	357.0 (31.98)	1116.5	50.0 (5.13)	24.5 (2.54)	966.5
7	230.5 (20.15)	390.5 (34.14)	626.5 (54.76)	1144.0	40.0 (4.37)	41.0 (4.47)	916.0
8	168.5 (12.35)	388.5 (28.48)	942.0 (69.06)	1364.0	40.5 (3.81)	7.0 (0.66)	1062.5
9	169.0 (15.28)	241.5 (21.85)	733.5 (66.35)	1105.5	10.5 (1.15)	2.5 (0.28)	910.5
10	124.5 (9.63)	272.5 (21.08)	976.0 (75.48)	1293.0	8.5 (0.83)	3.5 (0.34)	1023.0
11	80.0 (5.99)	275.0 (20.65)	1086.0 (80.77)	1344.5	1.5 (0.14)	14.0 (1.30)	1074.5
12	39.0 (3.92)	127.0 (12.77)	836.0 (84.06)	944.4	3.5 (0.46)	1.0 (0.13)	754.5

**The sum of percentages may be greater than 100 since twins may be carried separately.

moving 30 cm beyond the infant. Thus, in the third week, a mother becomes more intolerant of infant weights and she actively begins to dislodge the infants. Rejected infants were soon retrieved by the father or a juvenile, but at this time juveniles also began rejecting infants, thereby making the father the principal refuge. As the mother's intolerance increased, the father's carrying grew more important, and, in the fourth week, adult males were the principal carriers (see Table 2).

During Week 4 both fathers and juvenile males carried infants more than at any other time (about 53 percent and 14 percent, respectively) while both mothers and juvenile females continued to exhibit declining interest. Infants were independent 10 percent of the time observed.

In the fifth week, when weaning is initiated, mothers and juvenile females exhibit slight increases in carrying time. At this point, infants were both more resistant to rejection by their mothers and more persistent in their attempts to board and nurse. Fathers and juvenile males exhibited their first decreases. The only trend truly constant from the fourth to the fifth week was a 14.5 percent increase in "off" time.

From Week 6 through 12 several patterns emerged. As the infants gained weight and exhibited greater independence, there was a steady decrease in carrying time by all family members (fathers declined at an average weekly rate of 4.26 percent and mothers at 5.83 percent). Carrying by male and female juveniles was less than 1.5 percent by Week 9. Yet, "off" time continued to increase at an average weekly rate of 8.45 percent achieving 69 percent by Week 8 and 84 percent by Week 12.

The following outline summarizes the carrying patterns discussed above:

Table 3. Day of first long-term, stable transfer of young in *L. r. rosalia*; i.e., the day postpartum when family members other than the mother first carried infants.

Group	*Father*	*Juvenile Female*	*Juvenile Male*	*Other Adult*
A_1	18		17	
B_1	14	15		
*C	32			
B_2	11	18		
A_2	9		16	
B_3	14	12		
*D	2			2
NZP average	**14.29**	**15.00**	**16.50**	
OCZ	2	(Oklahoma City Zoo; Blakely and Curtis, 1974)		
MJ (a)	7–9	(Monkey Jungle, Fla.; Snyder, 1972)		
*MJ (b)	2	(Monkey Jungle, Fla.; Snyder, 1972)		
*BZG	4	(Berlin Zoological Gardens; Altmann-Schönberner, 1965)		
RDJZ	11	(Rio de Janeiro Zoo; Coimbra-Filho and Magnanini, 1972)		
NYZS	8–10	(New York Zoological Society; Ditmars, 1933)		
Average	**6.33**			

*Known or probable inexperienced parents.

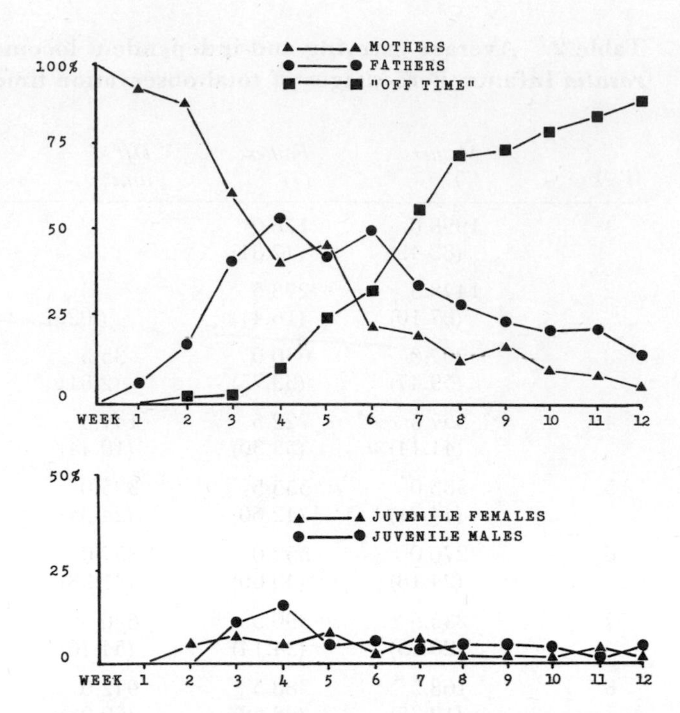

Figure 1. The change in carrying behavior by *L. r. rosalia* family members from birth to twelve weeks. Average percentages are given.

MOTHERS: Show a rapid decline in carrying from Week 1 through Week 4, but remain the principal carriers through Week 3; became secondary carriers from Weeks 4 to 12

FATHERS: Show a rapid increase in carrying for the first 4 weeks, but were still secondary carriers through Week 3; became principal carriers from Week 4 to 12.

JUVENILE FEMALES: From the second to the seventh week were tertiary carriers showing a weekly average of 6.5 percent; after Week 8 were only incidental carriers.

JUVENILE MALES: Also tertiary carriers, but show a concentrated period of carrying between Weeks 3 and 8 (averaging 8.0 percent per week); after Week 8 were incidental carriers.

"OFF" TIME: Occurs seldom in the first three weeks, but from Week 4 through 8, increased dramatically at an average of 13.2 percent weekly.

At the end of the twelfth week an infant weighs approximately 250 g (over one-third the weight of an average adult); it is alone about 84 percent of the time, and has been weaned. The size and agility of infants has increased and more complex social behaviors (especially play activities) have developed; this suggests that the transition from infant to juvenile stage begins about this time. By Week 12, juveniles have largely ceased to carry, and along with mothers act as refuge carriers for infants frightened by noises, low-flying birds, and other disturbing stimuli. Fathers, however, still will carry sleeping or resting infants as well as alarmed ones.

Sex and Age Class Differences in Carrying Frequencies

Since four of the seven study groups had twin infants and the remaining three had singletons, the analysis of carrying required that groups be separated into subdivisions based on the number of infants present. Thus, the only valid comparisons which could be made were between group members within each subdivision. Table 4 presents the data

Table 4. Tri-weekly average carrying percentages are presented for three classes of *L. r. rosalia* family members (mothers, fathers, and all others combined) in groups with twin infants.

The first twelve weeks postpartum are broken down into four three-week blocks: comparisons are made between classes in each block with the Mann-Whitney U-test.

Weeks	*Group*	*Mothers (4)*	*Fathers (4)*	*All Others (5)*	*Comparisons*	*U*	*P*	N_1
1–3	C	99.73	0.00		Mothers/Fathers	0	.014	4
	B_2	87.75	26.72	2.06 (1)	Mothers/Others	0	.05	3
	A_2	74.39	38.18	6.89 (1)	Fathers/Others	2	.20	3
	B_3	98.42	3.91	10.76 (3)				
4–6	C	85.43	4.03		Mothers/Fathers	6	.343	4
	B_2	45.85	73.92	1.16 (1)	Mothers/Others	0	.05	3
	A_2	22.86	68.25	11.35 (1)	Fathers/Others	0	.05	3
	B_3	70.02	35.32	18.67 (3)				
7–9	C	34.73	35.64		Fathers/Mothers	2	.057	4
	B_2	14.07	36.96	.74 (1)	Mothers/Others	1	.10	3
	A_2	7.40	32.54	2.24 (1)	Fathers/Others	0	.05	3
	B_3	20.78	21.71	8.56 (3)				
9–12	C	5.09	10.10		Fathers/Mothers	6	.343	4
	B_2	6.75	13.01	0.00 (1)	Mothers/Others	0	.05	3
	A_2	12.38	6.59	1.83 (1)	Fathers/Others	0	.05	3
	B_3	13.84	54.46	4.46 (3)				

and comparisons for the twin groups and Table 5 for the singleton groups. The Mann-Whitney U test was applied to the results.

Each table breaks the first twelve weeks postpartum down into four three-week blocks. Tri-weekly carrying percentages are presented in each block for three classes of family members (mothers, fathers, and all others[2]). Classes are compared with each other to determine which individuals are the major carriers in each block. It was assumed that if any consistencies or trends in carrying patterns persisted over several weeks, this analysis would delineate them.

In Weeks 1 to 3, mothers of twins carried significantly more than fathers and all others (in this case five juveniles) (Table 4). Weeks 4 to 6 saw no significant differences between mothers and fathers, but both parents carried significantly more than all others combined. For the seventh through the ninth weeks, the reverse of Weeks 1 to 3 occurred: fathers carried significantly more than both mothers and all others. Lastly, Weeks 10 to 12 showed the same condition as in Weeks 4 to 6: there was no significant difference between mothers and fathers, but both parental classes carried significantly more than all others combined.

From the first through the third weeks, there was only one noteworthy pattern in the singleton births (Table 5): mothers carried significantly more than all others (in this instance, three juveniles and one adult male), but not more than fathers. Weeks 4 to 6 saw fathers carrying significantly more than both mothers and all others. In Weeks 7 to 9, no significant carrying differences were determined. In the last three-week block, fathers carried significantly more than all others combined, but not more than mothers.

Overall, within the family group, it is clear that fathers and mothers are overwhelmingly the major carriers in the first twelve weeks postpartum, but mothers carried significantly more than fathers in Weeks 1 to 3 while fathers carried significantly more than mothers in Weeks 4 to 6 and 7 to 9. Juveniles acted as low-frequency incidental carriers.

[2]"All Others" is a class combining all juvenile males and females, and in one case, an adult male who was not a father.

Sex Preferences in Infant Carrying

In four litters where heterosexual twins were born and survived the first twelve weeks, recognition of sex and other distinguishing characteristics rarely occurred before the third week. By the end of Week 3, however, infants were individually

Table 5. Tri-weekly average carrying percentages in *L. r. rosalia* groups with single infants.
The first twelve weeks postpartum are broken down into four three-week blocks: comparisons are made between classes in each block with the Mann-Whitney U-test.

Weeks	*Group*	*Mothers* (4)	*Fathers* (4)	*All Others* (4)	*Comparisons*	*U*	*P*	N_1
1–3	A_1	88.26	8.13	4.27 (2)	Mothers/Fathers	1	.10	3
	B_1	74.63	18.11	7.26 (1)	Mothers/Others	0	.05	3
	D	35.66	48.34	14.34 (1)	Fathers/Others	1	.10	3
4–6	A_1	14.99	45.57	23.25 (2)	Fathers/Mothers	0	.05	3
	B_1	23.99	51.94	9.07 (1)	Mothers/Others	3	.35	3
	D	3.79	40.28	4.74 (1)	Fathers/Others	0	.05	3
7–9	A_1	13.85	18.65	11.61 (2)	Fathers/Mothers	2	.20	3
	B_1	14.26	47.47	1.50 (1)	Mothers/Others	2	.20	3
	D	0.56	2.89	0.33 (1)	Fathers/Others	1	.10	3
10–12	A_1	4.90	16.74	0.07 (2)	Fathers/Mothers	2	.20	3
	B_1	0.82	30.86	0.10 (1)	Mothers/Others	1	.10	3
	D	0.09	0.19	0.00 (1)	Fathers/Others	0	.05	3

identified which permitted the analysis of adult and juvenile preferences in the carrying of young of each sex (Tables 6 and 7).

Overall, mothers were observed to carry daughters more than sons, while fathers carried sons more than daughters. Juvenile females and males generally showed no distinct preferences, but in Group A_2 one juvenile male did slightly favor carrying the male twin over the female (Table 6). The values obtained in the proportional chi square analysis of each group shows that in three of the four groups the preference of parents for carrying infants of their own sex was significant (Table 7). The data for juveniles were insufficient for testing.

Parental Behaviors Exhibited by Juveniles

Much of the preceding has stressed the behavior of adults, particularly the differences between mothers and fathers, since juveniles generally exhibit low carrying frequencies and are incidental carriers compared to their parents. Yet there are some very important sex-related differences in the juvenile carrying which resemble the adult patterns.

Among juveniles, females begin to carry first (in Week 2) while mothers are still major carriers. Juvenile males, on the other hand, begin to carry in Week 3 when fathers are dramatically increasing their carrying time. Infant carrying peaks among juvenile females in the third week, the last week mothers are dominant in infant transport. Male juveniles and fathers both peak in the fourth week. In Week 3, both sexes of juveniles begin to show behaviors similar to patterns used by mothers (and later by fathers) for dislodging infants.

Table 6. Minutes of carrying infants of each sex by juvenile *L. r. rosalia*.

Groups	B_2	A_2	B_3	*Totals*
Juvenile ♀ + Infant ♀	7.0		4.5	11.5
Juvenile ♀ + Infant ♂	4.5		8.0	12.5
Juvenile ♂ + Infant ♀		54.0	2.0	56.0
Juvenile ♂ + Infant ♂		72.5	1.0	73.5

From the fourth through the twelfth week, juvenile males carried about 9 percent more than juvenile females, which mimics the trends shown by mothers and fathers over the same nine-week period (see Figure 1). Clearly, except for the first three weeks, both adult and juvenile males carry more than females in equivalent age classes.

In the fifth week, when weaning is initiated, both mothers and juvenile females, in a reversal from the previous week, show a slight increase in their carrying time. Fathers and juvenile males do not exhibit this pattern.

Another trend mimicking the behavior of parents was the preferential carrying of a male infant by a male juvenile (see above). In general, however, preferential carrying was not significant in juveniles (see Table 6).

Table 7. Within group analysis of preferences in carrying young of each sex in *L. r. rosalia* adults.**

		Minutes Infants Carried			Minutes Converted Into Proportions		Value of		
		♀	♂	Totals	♀	♂	Z	χ^2	P<
Group C:	Mother	132.5	36.0	168.5	.786	.214	12.869	165.609	.001
	Father	26.0	180.5	206.5	.126	.874			
	Totals	158.5	216.5	375.0	.423	.577			
Group B_2:	Mother	148.5	68.0	216.5	.686	.314	5.574	31.064	.001
	Father	127.5	164.5	292.0	.437	.563			
	Totals	276.0	232.5	508.5	.543	.457			
Group A_2:	Mother	89.0	40.5	129.5	.687	.313	4.814	23.178	.001
	Father	54.0	83.5	137.5	.393	.607			
	Totals	143.0	124.0	267.0	.536	.464			
Group B_3:	Mother	18.5	37.0	55.5	.333	.667	−1.853	3.345	.10
	Father	92.5	103.0	195.5	.473	.527			
	Totals	111.0	140.0	251.0	.442	.558			

**Chi-square applied to proportions in a 2 × 2 contingency table using a z-value for the normal deviate (adapted from Ferguson 1959:205).

The Role of Prior Experience in Parental Care

The following section will describe the behaviors observed in primiparous parents whose previous experience with infants differed.

In Group C, the adult male had only 22 days exposure to one newborn infant before being removed from his family group. Very probably his infant carrying phase had only just begun. The adult female, however, had had 159 days exposure to a litter of twins and probably had participated in carrying and interacting with infants.

Upon delivering twins, the mother in Group C was more protective toward the infants than were the two multiparous mothers previously observed. She vigorously prevented the father's investigatory attempts by screeching, jumping up, and slapping him away. The male became so wary of attempting to touch and/or take the infants that, even after two weeks, he had not performed the "Pressing Behavior" used by experienced animals for taking an infant off a carrier. "Pressing Behavior" includes pushing down or pressing the chest against an infant, whether carried or not, and tugging or nudging the infant, thus provoking a clinging response.

On the 32d day, this father began to carry infants, but even then, the infants initiated the activity by climbing on him after being dislodged by the mother. At first the male panicked and ran wildly over the branches in his cage, while one or both infants desperately clung to his back. After several minutes, he calmed down and quietly assumed his carrying duties. Within several days, his on-the-job training had given him enough experience to take and carry the infants normally.

Fortunately for the infants, the adult female in Group C did not harmfully reject the infants when the male failed to relieve her in the first few weeks. Although she became quite intolerant of her young by Day 16 postpartum, she did not bite or injure them even when vigorously rejecting them. When the father failed to carry the infants, she soon accepted them again.

In Group E, infant twins were stillborn. One infant, bearing no tooth marks, had fallen or been dropped onto the floor, while the other, with placenta still attached but with its tail and toes partially chewed and severed, was found in the nest box. In this pair, the adult male had 159 days exposure to infants but the female had no experience, having been removed from her family before another litter was born.

Group F had the least experience with infants and exhibited the most violent reaction to neonates. The mother had had only 22 days exposure to one infant and the father no experience. On the first day after birth, the mother was generally calm, except as in Group C, she was aggressively protective of the infants and attempted to prevent the father from contacting the young. This caused the male, as in Group C, to be wary of his mate and her young.

On Day 2, the mother began to run erratically

over the branches in the cage, making abrupt starts and stops, twisting, and attempting to dislodge the infants. The adult male, observing this behavior, made fumbling and inexpert attempts to contact the mother and infants. At one point, when one infant had dropped to the floor, the male apparently tried to retrieve it by manipulating the neonate with his forepaws, a behavior which did not permit the infant to cling. The mother and father then competed for possession of the infant and the mother was observed to kill it by crushing the face with severe bites. She was calm for the next 20 hours, but then repeated the violent and fatal pattern. The second infant was found dead the next midday—face and forehead crushed. Both infants were of average size and appeared healthy at autopsy. They had clung adequately and had been observed nursing. It appeared that the mother was extremely irritated by the weight and movements of the young on her dorsum; presumably the stimuli from their needle-sharp claws and the constant tugging of her fur were distressing.

Combined previous infant experience for this pair was only slightly over three weeks. Whether or not this was the variable which proved fatal for the twins remains to be demonstrated. But the events which subsequently transpired in another group, D, lend some support to this idea.

Group D was the first primiparous group for which data had been obtained on infant-carrying while the group members were juveniles. The father had 127 days of exposure to one infant, and the second male adult, the father's twin, had 185 days exposure to the same infant. The mother had the least previous experience, 115 days with one infant. With this background, the birth of a singleton was accepted calmly by the mother and the two males. The female exhibited moderate annoyance at the males' investigatory attempts by changing her position, avoiding approach, or, more rarely, screeching and slapping them away. This type of behavior was more typical of the multiparous mothers previously observed.

As in Group F, Day 2 brought a dramatic change. The female, despite her prior experience, began to exhibit erratic locomotion with abrupt twists and stops, and violent scratching and nipping of the infant, which produced facial lacerations and almost dislodged the neonate. At this point, the father and his twin brother began to carry the infant and ultimately relieved the female of 45.5 percent of her carrying duties during the first week. By the end of the second week, the period when the first infant transfers usually occur in multiparous groups, these two males were transporting the young tamarin 85 percent of the time. All three adults became less tolerant of the infant in Week 4, and it was left alone 30 percent of the time (compared to about 10 percent for most young in multiparous groups).

Discussion

Infant Carrying

Compared to other members of the Callitrichidae for which infant carrying data are available, the patterns in *L. r. rosalia* are quite distinctive for the most part, although a few features are shared by some other species. Among the golden lion tamarins at the National Zoological Park, fathers rarely began to carry infants in the first week postpartum, and, on the average, started to carry in the second or third week, becoming the major carrier during Weeks 4 to 12. In other groups of captive *L. r. rosalia,* fathers generally began to carry young in the second week, although variations, as in the National Zoological Park groups, appeared among inexperienced parents (see Table 3).

In captive *Callithrix jacchus* (Box, 1975b; Ingram, 1977), *Cebuella pygmaea* (Christen, 1974) and *Saguinus fuscicollis* (Epple, 1975b) fathers typically began carrying on the first day, and often carried infants more than any other family member in the early postpartum period. Within the first week after birth, captive *Tamarin tamarin* (= *Saguinus midas*) fathers have also been observed carrying infants (Christen, 1974). Only in *Callimico goeldii* was the timing somewhat like *L. r. rosalia;* Heltne et al. (1973) found that, although having just one offspring, the mother in one captive group remained the exclusive carrier through Week 2, but the father began to carry in Week 3, and became the principal carrier from that point through the last week of the study (Week 11).

In *C. jacchus* and *S. fuscicollis* both parents tended to show, generally, a gradual decrease in carrying (with limited fluctuations) after Week 1 (Box, 1975b; Epple, 1975b), while in the golden lion tamarin and Goeldi's monkey only the mother showed a similar general decline from the first week. In contrast, fathers in these two species exhibited a steady increase in transport frequency for three consecutive weeks after showing their first stable or long-term carries (Weeks 3 to 5 in *Callimico* and Weeks 2 to 4 in *L. r. rosalia*), and only after peaking in Weeks 4 or 5 did they begin to exhibit declining interest (Heltne et al., 1973; pers. obs.).

It was not uncommon in *S. fuscicollis* and *C. jacchus* for family members other than the mother and father (i.e., juveniles and subordinate adults) to take over and carry infants in the first few days postpartum (Epple, 1975b; Box 1975b). Among captive *L. r. rosalia,* however, experienced fathers rarely carried infants in the first week (see Table 3), and juveniles of either sex have similarly not been observed to transport neonates in the first seven days.

In *S. fuscicollis,* both fathers and mothers carried significantly more than all subdominant family members, and between adults, fathers or dominant males tended to carry the most over the first six weeks postpartum, although not significantly more than mothers (Epple, 1975b). In contrast, Box (1975b) found that in one *C. jacchus* group the females, including the mother and juvenile females, tended to carry more than males of equivalent ages; moreover, the older juvenile females, at times, exhibited a carrying frequency as high or higher than any other group member. In this respect, the *L. r. rosalia* pattern was more similar to *S. fuscicollis,* since after the third week golden lion tamarin males tended to carry more than females in both the adult and juvenile age classes.

There has been no attempt to distinguish the roles that male and female juveniles play in parental care; thus, little attention has been paid to sex differences, especially in regard to juveniles mimicking the behavior of the parent of the same sex. There is some indication that a greater sex-role differentiation exists in the process of maturation than has heretofore been reported.

The above comparisons suggest that former assumptions holding that callitrichid fathers, in general, carry infants most of the time beginning within a few days after birth may be incorrect for many species (cf., Napier, 1972; Jolly, 1972). Moreover, the involvement of family members in carrying may be dependent on age and sex variables and prior experience variables rather than on a species-specific difference.

Finally, the time when infants first left their carriers showed some uncharacteristic uniformity in three callitrichid species, *S. fuscicollis, C. jacchus,* and *L. r. rosalia,* and also in *Callimico goeldii.* Infants in these species moved off carriers between the second and fourth weeks (Epple, 1975b; Box, 1975b; pers. obs; and Heltne et al., 1973). In contrast, the rate of increase in "off" time was quite divergent. For example, in Week 9 postpartum, infants moved independently for 90 percent of the observed time in *C. jacchus* (Box, 1975b; Ingram, 1977) versus 66 percent in *L. r. rosalia* (pers. obs.). A *Callimico* infant matures even slower with "off" time reaching only 45 percent by Week 10 (Heltne et al., 1973).

Although there is no clear explanation for the interspecific similarity in the time when infants first left their carriers, the subsequent divergent rates of increasing independence seem likely to correspond to species-specific patterns of growth and weight gain. For *C. jacchus* and *S. fuscicollis,* adults range in weight from 160 to 360 g whereas *L. r. rosalia* and *C. goeldii* adults weigh between 400 to 700 g (Napier and Napier, 1967; pers. obs). In *C. jacchus,* sexual maturity is reached at 250 to 400 days (Hearn 1977), while for golden lion tamarins at the National Zoological Park, the earliest successful breeding occurred at 16 months (Kleiman, pers. comm.). Thus a slower rate of maturation appears likely for the larger species and, therefore, their infants probably remain dependent on adult and subadult carriers for longer periods.

Primiparous Parents and the Role of Prior Experience

During the course of this study, ten infants were born to multiparous parents. Of these, eight survived, one was stillborn, and one weakened and died within five days postpartum. None were killed, rejected, or, when dead, exhibited tooth marks. In the same period, seven infants were delivered to primiparous parents. Of these, two were stillborn, two were killed and three survived. One of the stillborn infants bore tooth marks and had the tip of its tail severed, while, due to aggressive parental rejection, one surviving infant bore facial lacerations for weeks. Whether or not a lack of prior infant experience was a variable causing much of the difficulty among these primiparous parents remains to be demonstrated. In mammals, first births classically cause difficulties, whereas subsequent births are usually less difficult and infants are successfully raised.

Nevertheless, in primates, where learned behavior is a substantial part of the adaptive strategy, inadequate preparation for parenthood may very well be an important component in causing the lack of success in the rearing of infants from the first births. The work of several investigators with rhesus macaques has demonstrated the importance of proper socialization of infants and juveniles for the subsequent expression of effective sexual and parental behaviors (Harlow and Harlow, 1965; Mitchell and Schroers, 1973). In the Callitrichidae, juvenile contact with and carrying of infants is probably a critical part of the socialization process. A report from Coimbra-Filho and Mittermeier (1976) lends some support to this conclusion. Three consecutive litters were observed in a pair of initially primiparous Brazilian *L. rosalia* (in this case a subspecies cross between *L. r. chrysomelas* and *L. r. rosalia*). Neither parent was known to have had any previous infant experience before being mated. In their first litter, both parents became overly excited and aggressive toward the newborn infants causing the deaths of both. Significantly, in their next two litters they repeated this pattern, although with some improvement in parental care before the demise of the infants.

The trial-and-error experiences of this pair did not seem to compensate for their lack of prior infant experience. If this pattern can be shown to be consistent among other groups of callitrichids, then the value of infant exposure previous to mating will be demonstrated.

Although infants of experienced multiparous parents at the National Zoological Park may have suffered mild mistreatment when rejected, multiparous parents never exhibited the apparent frustration and irritation observed in primiparous parents. Overall, no consistent pattern was observed in the rearing of first-born litters but, in general, all primiparous groups had certain difficulties with either the reaction of the mother or the father or both in some cases; yet interestingly, those with the most prior juvenile exposure to infants had the greatest success with neonates.

Though much more supportive data are needed, two tentative conclusions are perhaps worth proposing now, whether or not they are subsequently proved to be true.

1. Juveniles must very probably be exposed to newborn infants at least for the first 11 weeks postpartum if they are to be successful as primiparous parents. Table 1 shows that among the unsuccessful primiparous pairs at the National Zoological Park, 22 days of previous infant exposure was the most any one mother had as a juvenile, while among the successful primiparous parents, the mother in Group D, with 115 days, had the least prior exposure. However, without the timely assistance of other group members, she may not have been so successful in rearing her young.

Exposure to infants through the eleventh week postpartum is very likely most critical, since in groups with experienced parents, it is the period when male and female juveniles begin to exhibit specific paternal and maternal behaviors. Also, Weeks 3 and 4 appear to be especially important as the time when juveniles of both sexes first learn to reject infants without injury, and juvenile males, like their fathers, begin to dominate carrying in their age class.

2. The likelihood of a primiparous pair having success in rearing their first litter will be increased if, within their original family groups, they have had previous juvenile or sub-adult exposure to more than one birth and infant-rearing episode.

Acknowledgments

Without the generous support of Drs. D. G. Kleiman and J. F. Eisenberg of the National Zoological Park and Drs. M. I. Siegel and U. M. Cowgill of the University of Pittsburgh, this study would never have been conducted. I am especially indebted to the Smithsonian Institution for providing me with a predoctoral fellowship in support of the project. I would also like to express my thanks to Ms. J. Hitchcock, Ms. L. Dorsey and Mr. D. Mack for the stimulating discussions on various points of infant development, and to Mr. T. Davis, who supplied useful suggestions on data tabulation and analysis. Finally, I am grateful to my wife, Pat, and Ms. W. Holden for their help in preparing the manuscript, and to Dr. D. G. Kleiman who kindly critiqued the final draft.

Literature Cited

Altmann-Schönberner, V. D.
1965. Beobachtungen über Aufzucht und Entwicklung des Verhaltens beim Grossen Löwenäffchen, *L. rosalia. Zool. Gart.* 31(5):227–239.

Blakely, M., and L. Curtis
1974. Observations on birth in the golden lion marmoset (*L. rosalia*). *Proc. A.A.Z.P.A.* 50:5–7.

Box, H. O.
1975a. Quantitative behavior studies of marmoset monkeys (*Callithrix jacchus*). *Primates,* 16:155–165.

1975b. A social development study of young monkeys (*Callithrix jacchus*) within a captive family group. *Primates,* 16:419–436.

Christen, A.
1974. Fortpflanzungsbiologie und Verhalten bei *Cebuella pygmaea* und *Tamarin tamarin. Fortschritte der Verhaltensforschung.* Berlin and Hamburg: Paul Parey Verlag.

Coimbra-Filho, A. F., and A. Magnanini
1972. On the present status of *Leontopithecus*, and some data about new behavioral aspects and management of *L. r. rosalia*. In *Saving the Lion Marmoset*, edited by D. D. Bridgwater, pp. 59–69. Wheeling, W. Va.: Wild Animal Propagation Trust.

Coimbra-Filho, A. F., and R. A. Mittermeier
1976. Hybridization in the genus *Leontopithecus*, *L. r. rosalia* (Linnaeus, 1766) × *L.r. chrysomelas* (Kuhl, 1820) (Callitrichidae, Primates). *Rev. Brasil. Biol.* 36(1):129–137.

Crandall, L. S.
1964. *Management of Wild Animals in Captivity*. Chicago: University of Chicago Press.

Ditmars, R. L.
1933. Development of the silky marmoset. *Bull. N.Y. Zool. Soc.* 36:175–176.

Epple, G.
1967. Vergleichende Untersuchungen über Sexual- und Sozialverhalten der Krallenaffen (Hapalidae). *Folia Primat.* 7:37–65.

1975a. The behavior of marmoset monkeys (Callithricidae). In *Primate Behavior: Developments in Field and Laboratory Research*, vol. 4, edited by L. A. Rosenblum, pp. 195–235. New York: Academic Press.

1975b. Parental behavior in *Saguinus fuscicollis* (Callitrichidae). *Folia Primat.* 24:221–228.

Ferguson, G. A.
1959. *Statistical Analysis in Psychology and Education.* New York: McGraw-Hill.

Fitzgerald, A.
1935. Rearing marmosets in captivity. *J. Mammal.* 16:181–188.

Hampton, J. K.; S. H. Hampton; and B. T. Landwehr
1966. Observations on a successful breeding colony of the marmoset, *Oedipomidas oedipus*. *Folia Primat.* 4:265–287.

Harlow, H. F., and M. K. Harlow
1965. The affectional systems. In *Behavior of Non-Human Primates*, edited by A. M. Schrier, H. F. Harlow, and F. Stollnitz, pp. 287–334. New York: Academic Press.

Hearn, J.
1977. The reproductive endocrinology of the common marmoset, *Callithrix jacchus*. In *The Biology and Conservation of the Callitrichidae*, edited by Devra G. Kleiman, pp. 163–171. Washington, D.C.: Smithsonian Institution Press.

Heltne, P. G.; D. C. Turner; and J. Wolhandler
1973. Maternal and paternal periods in the development of infant *Callimico goeldii*. *Am. J. Phys. Anthro.* 38:455–459.

Hershkovitz, P.
1970. Cerebral fissural patterns in platyrrhine monkeys. *Folia Primat.* 13:213–240.

Hill, W. C. O.
1957. *Primates: Comparative Anatomy and Taxonomy, Volume III: Hapalidae.* Edinburgh: Edinburgh University Press.

Ingram, J.
1977. Parent-infant interactions in the common marmoset, *Callithrix jacchus*. In *The Biology and Conservation of the Callitrichidae*, edited by Devra G. Kleiman, pp. 281–291. Washington, D.C.: Smithsonian Institution Press.

Jolly, A.
1972. *The Evolution of Primate Behavior.* New York: MacMillan Company.

Lucas, N. S.; E. M. Hume; and H. H. Smith
1937. The breeding of the common marmoset, *Hapale jacchus*, in captivity. *Proc. Zool. Soc. Lond.* 107:205–211.

Mitchell, G., and E. M. Brandt
1972. Paternal behavior in primates. In *Primate Socialization*, edited by F. Poirier, pp. 173–206. New York: Random House.

Mitchell, G., and L. Schroers
1973. Birth order and parental experience in monkeys and man. In *Advances in Child Development and Behavior*, vol. 8, edited by H. W. Reese, pp. 159–184. New York: Academic Press.

Moynihan, M.
1970. Some behavior patterns of platyrrhine monkeys, II: *Saguinus geoffroyi* and some other tamarins. *Smithson. Contrb. Zool.* 28:1–76.

Muckenhirn, N.
1967. The social behavior and vocalizations of *Saguinus oedipus*. M. Sc. thesis, University of Maryland, College Park.

Napier, J. R., and P. H. Napier
1967. *A Handbook of Living Primates.* London: Academic Press.

Napier, P.
1972. *Monkeys and Apes.* New York: Bantam Books.

Sanderson, I. T.
1957. *The Monkey Kingdom: An Introduction to the Primates.* Garden City, N.Y.: Hanover House.

Siegel, S.
1956. *Nonparametric Statistics for the Behavioral Sciences.* New York: McGraw-Hill.

Snyder, P. A.
1972. Behavior of *Leontopithecus rosalia* and related species: a review. In *Saving the Lion Marmoset*, edited by D. D. Bridgwater, pp. 23–49. Wheeling, W. Va.: Wild Animal Propagation Trust.

SECTION V:

THE MANAGEMENT OF CALLITRICHID COLONIES

Introduction

The final section is concerned with problems of colony management, including the costs of developing and maintaining colonies under different conditions and the medical problems typically encountered. With several South American countries restricting exports of primates unless they are captive-bred, the pressure on the biomedical community is considerable to contribute to the development of alternatives to using wild-caught animals. Indeed, it is to their advantage to have specimens of known age, history, and genetic background. Hobbs, in the Panel Discussion, uses the term "sewer-rat monkey" to emphasize the fact that most callitrichids currently used in research are wild-caught individuals whose histories and biology are unknown. Therefore, there has been no standardization with respect to most variables which could affect the outcome of biomedical experiments. No researchers using rats or mice would fail to indicate the age, sex, weight, strain, and early experience of experimental subjects while with callitrichids these factors often are not reported, and the species may only be indicated as an afterthought.

The alternative methods available for supplying the biomedical community with research animals include harvesting from natural (or artificially created) populations on a regular basis, developing breeding colonies in the countries of origin, or developing breeding colonies in the user countries. The latter two alternatives are being pursued on a small scale, while the first remains unexplored and has numerous potential problems. In order for harvesting to be a successful technique, life tables must be developed for a species, based on prior long-term research where mortality and natality are known. Numbers of surplus animals cannot be estimated and quotas established without such information. And the effects of harvesting on population characteristics must be carefully monitored as long as harvesting continues. Although this approach is discussed by several participants, one of the biggest problems associated with the development of a harvesting program is its funding. Not only does an institution have to fund basic ecological studies over a number of years prior to the initiation of harvesting, but constant monitoring of a population must then be financed. It is not clear from Porter's chapter whether his organization or any commercial operation proposes to finance such monitoring. Harvesting or cropping programs unfortunately are also most susceptible to unethical behavior since numbers of animals can be inflated.

In the Panel Discussion, there was considerable disagreement about the usefulness of the export bans imposed by several South American countries, in part because the bans have created a black market in South American primates resulting in greater losses in transit of animals. There was also disagreement concerning what to do with illegal animals. There was some support for refusing to buy illegal animals in the hopes that the importers would ultimately cease operations, if there were no purchasers.

Some discussion was devoted to the choice of species for different kinds of studies. There appear to be only three or four callitrichids which are heavily used in biomedical research. In some cases, substitution of another species is apparently impossible; an example is the use of *S. oedipus* as a model for colonic cancer. The response of the biomedical community, should *S. oedipus* be declared an endangered species, is unknown. It is already illegal to export *S. oedipus* from Colombia due to the export bans. They are, however, still entering the United States in numbers (see Neyman, Section I), since it cannot be proved that they were illegally smuggled out of Colombia.

Where species can be substituted for each other, there is reluctance to change since researchers must begin anew to establish baselines. Kingston, in the Panel Discussion, suggests that researchers will refuse to buy a species, even if available in great numbers, if it is not the species which they have used previously.

The price of the common callitrichid species has increased dramatically since export bans have gone into effect. Yet, the cost of producing marmosets and tamarins is very low when compared with other laboratory primates. The reluctance of many biomedical researchers to breed their own animals still exists, especially with a view towards long-term propagation; however, attitudes are apparently changing as primate shortages become more acute.

A final problem area concerns the attitudes of users toward the producing countries and vice versa. Since the interests of both sometimes apparently conflict, the relationship is delicate. It is clearly necessary for biomedical researchers to be more sensitive to the interests of the South American countries from whom they get animals.

W. R. KINGSTON
Hillcrest House, Belton,
Loughborough, Leics. LE12 9TE,
England

The Cost of Developing and Managing a Marmoset Colony

ABSTRACT

The capital and running costs of initiating and developing a captive callitrichid breeding colony are discussed and analyzed. Estimates are made of the current cost of setting up a twenty breeding-pair colony under two alternative schemes and situations.

Introduction

The experience on which this paper is based stems mainly from a suggestion I made, as Chief Animal Technician to Fisons Ltd., in 1965, that a colony of marmosets should be established to provide a primate for their research program. Previously, private experience with the Callitrichidae, the poor survival records of even the best zoological gardens, and the failed attempt by a large pharmaceutical company to establish these animals had given them a reputation for nonviability under captive conditions. Several papers, however, had appeared in American journals by Drs. Levy, Artecona, and the Hamptons indicating considerable success with callitrichid colonies which prompted my suggestion to Fisons. We commenced with six pairs of *Callithrix jacchus,* housed in a small room with a number of stump-tailed macaques (*Macaca speciosa*).

The first young were born in 1966 and at this time we added small numbers of five other callitrichid species, all of which successfully bred, including *Saguinus oedipus, S. nigricollis, Callithrix argentata,* and *Cebuella pygmaea.* We maintained an average of some 40 breeding pairs until 1972 when the colony was disbanded. By then some 350 young had been reared and about 1,500 marmosets and tamarins imported directly for research had been handled. Our breeding success resulted in other institutions becoming interested in these animals and other colonies were established, often with our help. In the last year or two, this interest has greatly increased and there is now a large but unsatisfied demand for marmosets and tamarins resulting from export bans by South American countries for conservation purposes. Personally, I believe that the Callitrichidae have enormous research potential, but the supply of animals for research can and should arise from captive breeding. It is inexcusable and totally irresponsible to increase the already grave threats to the survival of wild populations by importations for research. The failure of the research community to breed adequate numbers of animals has greatly disappointed me and caused me to have regrets for having recommended and facilitated their use.

The costs of establishing a marmoset or tamarin breeding colony include capital items such as accommodation, caging, and initial breeding stock. The running or maintenance costs include labor, food, heating, lighting, ventilation, veterinary treatment, and standing charges such as depreciation, rates, and an appropriate proportion of the general overhead expenses of the home institution. The largest direct costs, labor and food, will probably not vary much, but the other costs, both capital and revenue, will vary considerably and will be dependent on the size and purpose of the breeding operation and the character and management of the institution. Fisons, a large commercial organization, had the benefit—or curse—of a large accounting section which prepared cost accounts in the utmost detail. I was amazed initially to find that my marmosets were expected to bear a proportion of the cost of such remote things as the managing-director's salary and the typing pool.

Capital Investment

Marmosets and tamarins are being successfully bred in accommodations ranging from fully air-conditioned, elaborate brick buildings costing upwards of $45 per square foot to light wood-framed and wire aviary structures with minimal shelter and heating at about $5 per square foot. I have bred six species for seven years in cages with a floor area of 0.25 sq m (3 sq ft) which were not too tall to be double-tiered; thus keeping two pairs on a floor area of 0.25 sq m. Using a double-tiered caging system and allowing for access to cages, food preparation and storage, a cage and utensil washing area, large pens for grouped young and a safety porch entrance to the building, a twenty pair colony in a single building or room would require 20 sq m (200 sq ft) or 1 sq m (10 sq ft) per pair. Single-tier cages or pens would increase this to 1.5 sq m (15 sq ft), with the service areas remaining the same. In this context, I consider vertical rather than horizontal space of greater value to these animals, if suitable perches and platforms are provided.

Cages may vary from heavy metal structures which would safely house a baboon to all-wire light-gauge weldmesh, and wood and wire bird-cage-like constructions. A heavily built and consequently expensive cage has little purpose, unless it must go through a mechanical cage washer or is likely to receive rough handling. Marmosets and tamarins are not very strong and their teeth are not suitable for gnawing, although they will sometimes chew wooden perches. Light metal, either aluminum or sheet steel, is hygienic and relatively inexpensive with 2.5 cm × 1.25 cm (1 in × 1/2 in) weldmesh of 19-gauge being perfectly adequate. Light plastic water tanks, a standard product, mounted in a metal or even wooden frame with a weldmesh front, and fitted so as to leave no exposed plastic edges appear to be a very promising alternative. The total cost of these including perches, cleaning trays, and nest boxes made of lengths of 17.7 cm (7 in) plastic piping is less than $35 each. Small wood and wire aviaries, with the wire fastened on the inside of the frames are another relatively cheap standard product. These can either be used indoors as the shelter section only (1 m sq and 2 m high) or also as an outdoor cage (2 m × 1 m × 2 m high) with a controllable communicating door between them. Present price (as of August 1975) of these in the United Kingdom is about $40. The more conventional all-metal cages, double-tiered in blocks of four, cost approximately $250 complete and mounted on castors.

The requirements for heating, lighting, and ventilation are to some extent determined by the local climate, density of stocking, and whether natural lighting is provided. Adult marmosets can tolerate freezing temperatures—even playing happily in the snow—and suffer no ill effects if they have access to a heated shelter. Young, however, should not be exposed to temperatures below 20° C during the first six weeks. Thus, even though the adults definitely benefit from access to all but the very worst weather, they must be confined indoors in the winter months when carrying young. The shelter should be maintained above 20° C regardless of how low the ambient temperature drops. Ventilation can be natural, subject to the heating requirements and stocking levels, but can be advantageously combined with heating by providing thermostatically controlled heating panels with forced fan driven ventilation. This can also provide cooling when heat gain from solar radiation is a problem. Lighting is important, and I suggest that 12 to 14 hours of good lighting should be provided daily. Thus, in temperate latitudes, additional artificial light is required for a few hours daily in winter even when good natural lighting is available. This can be on a timer and controlled by photo-electric cells.

The cost of the initial stock may or may not be regarded as a capital item and will depend on the source of the animals. With newly imported animals, losses will occur in acclimatization to captive conditions. When animals are obtained from commercial importers, the losses may vary from 2 or 3 percent to 50 percent, despite apparently identical treatment. One should aim for minimal time in any part of the supply chain, but this is of course difficult to control without personally trapping and shipping the animals. In the United Kingdom, marmosets and tamarins are subject to six months quarantine under specified conditions. Unless an organization can comply with these requirements, either captive-bred stock must be obtained or the institution must bear the high cost of maintenance by a licensed importer for this period. This is to some extent offset by better survival potential, but in my opinion it is so difficult to obtain animals free of quarantine restrictions that it is better to bear the extra cost of providing acceptable accommodations and having your own premises licensed as a quarantine center.

Maintenance Costs

Direct running costs include mainly labor and food. The technician must be intelligent, conscientious and, above all, sympathetic with the animals. In my experience these qualities are most likely to be found in women, but any capable person is perfectly able to cope with the day-to-day running of a colony of 100 animals, i.e., 20 to 30 breeding pairs and their young up to 12 months of age. Of course, additional support from an experienced senior technician and veterinary services are necessary. Since marmosets and tamarins require attention daily although only food and water is required at weekends, a part-time helper of an acceptable standard is required which may present financial difficulty if the colony or the organization is small. Direct wages would be about $5,000 annually at current United Kingdom rates, but may of course be higher in the United States.

Feeding costs have escalated drastically due to inflation, but for the currently accepted diet, about 42 cents (20 new pence) per animal per week is required. This includes two bananas, one apple, 250 g primate diet,[1] 70 g wholemeal bread, and one-half egg. Supplements include "Abidec," a water-soluble vitamin mix given twice weekly and "Cytacon," a B_{12} pediatric syrup given once weekly, both added to the drinking water, and wheat germ sprinkled on the food twice weekly. Regular administration of vitamin D_3 is essential to prevent a rickettic condition developing in both adult and young; 5,000 IU vitamin D_3 in one ml Arachis oil can be best administered orally at monthly intervals. Young will begin eating this diet at about six weeks of age and will consume a full ration at about ten weeks.

Heat, light, and ventilation charges will vary with location, design of buildings, and local energy costs and are difficult to quantify with any precision. Veterinary charges should not be a major item

[1]The primate diet used is manufactured by Coopers Nutritional Products Ltd.

once the colony is established and will, of course, depend on whether the veterinarian is employed by the institution or called in as a consultant. It is difficult to find a veterinarian with specialized knowledge of primates and most treatment will in fact be done by the technician in charge. Standing charges can also vary enormously and are specific to the individual institution, but standard accounting practice suggests an annual depreciation rate on the buildings and capital equipment of about 10 percent which should be included in the costs. Moreover, there must be an allowance for repairs and maintenance and either an agreed proportion or the actual cost of such items as stationary, telephone, postage, and transport according to the particular circumstances.

Total Costs for Two Types of Breeding Facilities

Marmosets and tamarins can be easily bred and their relatively high reproductive potential makes breeding them in captivity for research or conservation purposes a perfectly practicable proposition. Such an undertaking is, however, only of long-term value if the initial numbers are high enough to ensure an adequate genetic pool and subsequent generation matings are properly planned to avoid the undesirable effects of long-term inbreeding. The basic requirement is 20 unrelated breeding pairs, but these can be housed in a number of cooperating institutions thus avoiding the proverbial "eggs-in-one-basket." The following estimates are based not only on my own experience, but that of others and include the cost of establishing, developing, and running a colony of 20 breeding pairs. One of the estimates is based on a facility suitable for a public exhibit, and thus with revenue-earning potential. The second estimate is for a more intensive unit, suitable for an academic institution.

For an exhibit facility, I suggest a wooden prefabricated building—of the type available for agricultural purposes—to be erected on a good rodent-proof foundation. These buildings are made as standard modules, and the required size would be about 13 m × 3 m. They have enough windows for good natural lighting and heating and ventilation can be provided by a fan heater. Ten wood and wire pens of the type described earlier could be placed on each side and a larger divided 3 m × 2 m × 2 m high pen for groups of young animals erected at one end. The other end of the building could have a simple sink unit, and inner safety porch entrance, and cupboards with a good bench top for food preparation and storage. Artificial lighting would be "daylight" neon tubes and controlled by a photo-electric cell and time switch. Small outdoor runs could be fitted to each indoor pen with the divisions being double-wired to prevent fighting, and also to the pens for young. The animals would be on view to the public when outside and only the staff would enter the building itself. Once the animals have adapted, and especially when they have been born and reared in such a structure, they lose fear of human beings and make an excellent and entertaining exhibit, at least in the summer. This method of breeding should not be overlooked since it arouses public interest and may also help defray the cost, which is, after all, a major problem for a breeding colony. Current estimates for such a building, equipped as described, and assuming erection on a reasonably flat site with water, electricity, and drainage services within 30 m is about $5,000.

For a nonexhibit facility, I am assuming the use of a fully air-conditioned existing room of a floor area of 18 sq m, with services such as usually are found in a university. This room would be fitted with 20 metal double-tier cages and a pen for groups of young. Since such a room is unlikely to be built specifically for marmoset caging, I am costing it on a rental basis and assuming a building cost of $400 per sq m with an additional service charge to cover heat and light. Assuming that automatic cage-washing facilities exist in the building and are available to the staff of the unit, the work would involve only a half day and one hour at weekends. The annual costs of the two facilities may be compared below.

Costs	Unit "A" (Exhibit Facility)	Unit "B" (Intensive Facility)
Capital costs	$5,000	$1,250
Running costs		
Labor	5,000	3,000
Food	2,500	2,500
Rental charge		750
Heat, light, ventilation	700	400
Indirect labor, supervision, and veterinarian	1,500	1,500
Depreciation (10 percent on cost)	500	125
Standing charges and overheads	300	500
Total	$10,500	$8,775

The expected annual output of 20 established pairs of *Callithrix jacchus* after 12 months of undisturbed occupation of their cages, is at least 50 weaned young. Under these conditions *Saguinus* spp. are not as productive but, on the other hand, are apparently stronger and more disease resistant and with lower pre-weaning losses of young. I believe that with territorial security and stability for the animals and intelligent management of the colony, this output figure could be obtained for any species. With the birth of triplets and quadruplets which tend to increase in frequency under captive breeding, any excess young beyond twins (which may not be reared by the mother) can be hand-reared with very little difficulty including only one feeding outside of normal working hours. By the end of the second year, therefore, there will be a surplus of 50 young between 6 and 18 months of age. Apart from the replacement of breeding animals, these can be surplused since the original colony should certainly continue to breed successfully for a number of years. The first set of young will have cost around $400 each, but subsequent animals will be half this amount.

Proceeds from the sale of surplus young can, of course, reduce the cost of running the colony. At the current prices being offered for marmosets by the research community, this should result in a net profit. There should also be a ready sale to zoological gardens since captive-born animals are free of quarantine restrictions, tropical disease risks, as well as being tamer and very viable. Although the means of disposal is a matter for the colony owners to decide, surplusing must occur if the colony is to remain a moderate size.

During a recent visit to the large colony of *C. jacchus* developed by Imperial Chemical Industries, Ltd., which currently produces 1,000 young per annum, I had the opportunity to see their cost figures. Given the advantages of large-scale production which are, however, offset to some extent by overhead expenses of a very high order, their costs are quite close to the above-described estimates. It should be remembered, however, that my estimates are based on current British prices and salaries and, with so many possible variations of prices, based on local conditions, they cannot be regarded as more than a reasonably informed guide.

JAMES A. PORTER
South American Primates, Inc.
10525 S.W. 185th Terrace
Miami, Florida 33157

Planning Captive Breeding Programs in South America

ABSTRACT

Problems related to establishing breeding colonies of callitrichids in South America are discussed. These problems include training local personnel, combating indifference to the importance of medical research, making governments aware of the financial returns from the exportation of primates, working with officials who have different scientific backgrounds than corresponding officials in the United States, and combating publicity by some conservation groups that is unfavorable to the production and export of monkeys for research.

Three different methods of producing callitrichids for export are discussed, including harvesting from natural populations based on annual censuses, developing and harvesting from semifree breeding colonies, and developing caged breeding colonies. Methods of financing the operations are also discussed.

Introduction

Marmosets and tamarins are used in medical research such as virology, oncology, immunology, and dental studies. The South American countries that were the primary suppliers now forbid the export of wild-caught nonhuman primates; however, they may permit the export of monkeys for research if produced from breeding colonies. Other South American countries, not formerly exporting marmosets and tamarins, are now exporting a limited number; however, these countries may also place restrictions on the future export of wild-caught monkeys. Currently, sufficient numbers of marmosets and tamarins are not present outside South America to rapidly develop breeding colonies large enough to supply medical research needs. If callitrichids and other species of nonhuman primates are to be available in adequate numbers for medical research, they may have to be raised in breeding colonies located in the country of origin. South American Primates, Inc., is involved in planning such breeding farms.

Problems

Many biological, financial, political, and cultural problems must be confronted in planning South American breeding farms. Except for the effort of Tarpon Zoo, Inc., to establish a colony of squirrel monkeys (*Saimiri sciureus*) on an island in the Amazon River near Leticia, Colombia (Jerkins, 1972), there have been few efforts in this area. Reports (Bailey et al., 1974; Sponsel et al., 1974) describe some of the problems encountered by Tarpon Zoo and indicate some guidelines to be followed if success is to be obtained.

Persons with knowledge of primate colony management are scarce in South America. In most cases, local management personnel will have to be trained thus requiring the importation of experienced outsiders. Difficulties may be encountered in finding qualified persons able and willing to adjust to the changes in food, housing, language, and customs.

Also, the need for callitrichids in medical research is of little importance to most persons in South America. Many South American countries now place more importance on investigations to increase food production than on health research. Thus, an appeal for monkeys for medical research may not be enthusiastically received.

There is also a difference in perspective between officials in the United States and South American countries. In the United States such officials are often veterinarians because of the large laboratory animal colonies existing for medical research, while in South America most officials have a zoological background. Although this is not a major problem, the difference in perspective must be recognized.

In the United States and Europe, medical research may be attacked by the antivivisection groups. In South America, medical research also faces opposition, but often from conservation agencies with a strong voice in wildlife management. While not directly challenging the importance of medical research, they have influenced the restrictive export policy that has lumped the research and pet-trade needs and effectively contained both. One effort to equalize this situation was a meeting held in Lima, Peru, June 2–4, 1975, sponsored by the Pan American Health Organization, at which scientists described to South American regulatory officials the medical research uses of nonhuman primates.

In discussions with officials, an emphasis on the financial gain from exporting monkeys may result in officials citing figures indicating that a monkey valued for export at $10 may cost $75 in the United States, which suggests to them that a middle man makes an unreasonable profit. Actually, the quality of the monkey delivered to the purchaser from a breeding colony should be far different from that generally exported from South America. The wild-caught marmosets and tamarins purchased by dealers in the United States and Europe are usually in a poor nutritional state, heavily parasitized, and incubating diseases when received. A high percentage may die. A major part of the price difference is due to these factors. If animals were in good condition when shipped from South America, they would have a higher sale price, thus providing more exchange dollars to the country involved.

Harvesting from Wild Populations

The unrestricted capture and export of monkeys from South America will probably not be permitted in the future, thus different methods must be developed to acquire monkeys from there. One approach is the periodic harvesting of a percentage of monkeys from a limited area based on regular population counts; a second method is harvesting from semifree colonies, probably located on islands; and the third is the establishment of captive breeding in cages.

Primate censusing studies have been conducted in Colombia, Peru, Guyana, and Bolivia, in part by scientists working for the Pan American Health

Organization under contract with the United States National Academy of Sciences. Such studies should be continued longitudinally to gain more accurate information on primate numbers which can serve as the basis for establishing harvestable quotas which would not deplete the natural carrying capacity of a given area. Annual censuses would allow the adjustment of harvesting quotas based on the effects on the natural population of previous cropping. Trapping should not interfere with the breeding season, the first months after parturition, and the latter part of the gestation period. Some jungle areas in South America that should never be cleared for agricultural uses are large enough to support marmoset and tamarin populations from which harvesting could supply research needs. The harvesting of callitrichids and other nonhuman primates represents one way that an economic return could be obtained from these areas; however, since this has never been done, the governments, now restricting exports, are hesitant about permitting the introduction of these methods.

The most logical nations to initiate harvesting would be countries, such as Bolivia and Guyana, which still permit the export of monkeys. The censusing activities conducted in both countries should result in the data needed to start such a program. If successful in one country, other nations might later permit the use of this method. It is to be hoped that this approach would eventually gain general acceptance. The user would secure a cheaper monkey which, in many cases, would be adequate for his needs, and the host country would be receiving income from jungles with limited monetary value otherwise. Struhsaker et al. (1975) presented a program of this type for harvesting from the free-ranging primate populations of northern Colombia.

Harvesting from Semifree Populations

Semifree breeding programs are probably the most practical and acceptable approach for most South American countries. Colonies could be located on islands, in savanna, or in riparian forest. The size and number of designated areas will be based on the number of species and individuals to be harvested annually and the territorial and spatial requirements of each species of callitrichid. Unfortunately, it is difficult to estimate the social and spatial requirements of marmosets and tamarins.

Field observations of tamarins (*Saguinus* spp.) list most groups as consisting of 4 to 8 individuals (Moynihan, 1970; Thorington, 1968; Freese, 1975). Chapman and inhabitants of the Panama Canal Zone are reported as having observed groups of *Saguinus oedipus geoffroyi* as large as 12 or more (Moynihan, 1970), although Moynihan believes these were overestimates. Moynihan also noted that this species' territories may extend more than a quarter mile in one direction. Dawson's (1977) observations of *S. o. geoffroyi* support Moynihan's, with most groups having between 3 and 9 individuals. Geijesks reported *S. midas* as living in families of 5 to 20 individuals in Surinam (Thorington, 1968), which contrasted with the 8 groups Thorington studied in Brazil, numbering from 2 to 6 individuals per group. He reported a minimum density of one *S. midas* per 15 acres and a probable density of one per 7.5 acres in Amapá, Brazil, with the density being lower near Porto Platon and Santana.

My associate, Fontana, reports counts of 20 to 30 *S. midas* per group in Guyana. Castro and Soini (1977) counted 24 *S. fuscicollis* in one group at Campamento Callicebus in Peru, in contrast to the counts of 2 to 9 for other groups during the primate census in Peru. Freese (1975) reported densities ranging from 8.4 to 15.6 individuals per km^2 in different areas of Peru. Hernandez-Camacho and Cooper (1972) reported that *S. fuscicollis* ranged in group size from 9 to 20 in Colombia.

These reports suggest that there is a difference in the size of troops of the same species in different areas. These data contrast with the commonly expressed opinion that marmosets live as family groups with one mature male and female pair forming the nucleus of the unit. It may be that food availability is the major factor limiting the troop size of some species of marmosets and tamarins.

Unrelated marmosets often fight if caged together (Epple, 1970). In large free-ranging troops consisting of many breeding-age females, the females apparently live amicably together. If these large troops are indeed stable units, then more than the dominant female must be bearing young. Some species may never exist in large groups; thus, if such species are to be bred under semifree conditions, more territory must be provided. It is obviously more economical, however, to stock a colony with a species that is amenable to concentration.

To maximize production, competing species must be kept out of the colony area; however, *Aotus trivirgatus*, a nocturnal primate, might be stocked with marmoset or tamarin species because there may be minimal competition due to differences in activity rhythms. In this case, sufficient nesting areas might be the limiting factor.

An ecological survey should be made of selected areas before stocking. Predators, competing wildlife, vectors of diseases, available food, potential nesting areas, and water supply must all be considered. Caretakers should be housed in the area to prevent poaching. Supplemental plantings of fruit trees and selected vegetation must be made to increase the food supply of the area.

The specimens to be introduced will probably be captured elsewhere and transported to the designated site; however, before release, they should be conditioned, treated if necessary, and tattooed. When possible, an entire troop should be captured and later released together. In any case, compatible pairs or troops should be formed in a holding facility before the animals are released at the breeding site.

Large cages capable of housing troops of compatible individuals should be constructed at various locations in the release area. The introduced animals should be housed and fed within the cages for a few days so that the social group will be cemented. If the animals are used to and return to these cages for periodic supplemental feeding, the cages can then serve as capture sites when necessary.

Groups should be introduced at intervals and at different sites on the island or in the area. Close observation of groups during the stocking period should allow determination of the maximum carrying capacity of the area. When this point is reached, harvesting can begin. This must be constantly monitored by censusing to determine the optimum level of harvesting.

In some countries land is owned by and can be leased from the government for token payments. In other countries it may have to be purchased, but at low prices compared to prices in the United States. The other major costs would include the initial capture, selection, and transfer of monkeys to the release site, and the cost of management. I estimate that callitrichids from a semifree breeding colony will cost twice as much as animals harvested directly from the jungle which is considerably lower than comparably produced animals in the United States or Europe. As long as monkeys could be exported, the production from such colonies should have a ready market at these prices.

Cage-breeding Operations

The third and most expensive method of producing monkeys is in cages. Because of land costs and a harsher climate, cage breeding of callitrichids within buildings is the most common method in the United States and Europe. While labor and land costs would be cheaper in South America and the favorable climate would permit outside caging, there are advantages in having a breeding colony free from the whims of potential export or import restrictions which outweigh part of these cost benefits.

Cage breeding may be the preferred method of production due to high cost or nonavailability of islands close to a large city with good shipping connections and the necessary labor and supplies. Cage breeding might also be preferred to produce a high-quality monkey free from parasites and other undesirable pathogens and with a known pedigree. In the future this may become of prime importance, but only a small percentage of the medical researchers would now accept a cage-bred monkey if a lower priced animal was available.

Captive colonies might become a necessity if a particular species with a limited range were needed in medical research and the officials of the country decreed that they must be raised in cages to be exported for research. This would be an extension of the policy of some South American governments that require exports to come from approved captive-breeding colonies. These officials honestly believe that cage-bred monkeys rather than feral-derived monkeys would be more valuable in research and so justify the added cost. They possibly believe that conservationists would not protest the use of cage-bred monkeys for research whereas some conservationists do presently protest the use of feral monkeys. Further, officials might favor cage breeding because it is easier to regulate, i.e., it is easier to count monkeys in cages than on an island or in the jungle.

The conventional method used to breed marmosets in a laboratory environment is by pairing them in small cages (see Wolfe et al., 1972, 1975; Gengozian, 1969; Hampton et al., 1966; Kingston, 1969). However, Magnanini and Coimbra-Filho (pers. comm.) believe that large cages are necessary for each pair, if successful F_2 and F_3 generations are to be bred on a large enough scale to perpetuate the colony. Coimbra-Filho (1972) indicates that cramped enclosures permitting only limited movement cause serious psychological traumas prejudicial to normal reproductive behavior. Epple (1970) reports on a method of housing marmosets, 2 to 8 animals to a group in small rooms. She reported that no serious fighting occurred with the group, but that only the dominant female produced young. Because the cage size is going to be extremely important in determining the cost of the colony-produced marmoset, it will be necessary to get some definitive answers on the cage size for optimum production before a large commercial colony is started.

Financing Breeding Operations

Laws and regulations are interpreted by people. Because officials change positions or the same official may later change the interpretation of a law, a country by country review of governmental policies toward primate breeding colonies may not be useful; however, the officials in most countries I have visited indicate their country would provide assistance in the development of such programs. Most countries do have governmental lands that can be leased or purchased at a low price. In some countries this land is ideal; in others, the land offered would not be satisfactory. Brazil will provide loans of up to 50 percent of the cost of government-approved programs located in the Amazon River area to encourage development. Brazil will also permit the duty-free importation of goods into the Manaus area. Other governments have mentioned the granting of duty-free importation privileges to encourage breeding colonies. Of course, additional money will be needed, the amount varying according to the type and size of program. Since these types of commercial ventures are new, bankers cannot judge the soundness of individual requests for loans. With the current economic conditions, a group desiring to embark on such a project must have a substantial part of the necessary capital. Some countries may permit the exportation of monkeys from the jungle while a colony is being established to help defray costs; other countries will not.

In some countries it may be wise or necessary to involve local capital in a cooperative venture. Also, a government might become involved in the venture if they approve the proposal. The involvement of international banking groups, such as the World Bank or the Inter-American Bank, might strengthen the position of the developers, should a government later decide to confiscate the operation.

I have spent the past two years studying the feasibility of establishing primate breeding colonies in South America. This report is a summary of my conclusions based on these investigations. If some species needed for medical research are to be obtained, there is no alternative to the establishment of marmoset breeding colonies in South America. I further believe that these colonies can be successful and provide a quality monkey at a price considerably less than those currently produced in the United States and Europe.

Addendum. After this manuscript was completed there was a decision by the United States Department of Interior to declare several species of South American primates as endangered or threatened which may make any investment of private capital into breeding farms in South America a high risk.

Literature Cited

Bailey, R. C.; R. S. Baker; D. S. Brown; P. von Hildebrand; R. A. Mittermeier; L. E. Sponsel; and K. E. Wolf.
1974. Progress of a breeding project for non-human primates in Colombia. *Nature,* 248:453–455.

Castro, R., and P. Soini.
1977. Field studies on *Saguinus mystax* and other callitrichids in Amazonian Peru. In *The Biology and Conservation of the Callitrichidae,* edited by Devra G. Kleiman, pp. 73–78. Washington, D.C.: Smithsonian Institution Press.

Coimbra-Filho, A. F.
1972. Conservation and use of South American primates in Brazil. *Internat. Zoo Yearb.,* 12:14–15.

Dawson, G. A.
1977. Composition and stability of social groups of the tamarin, *Saguinus oedipus geoffroyi,* in Panama: Ecological and behavioral implications. In *The Biology and Conservation of the Callitrichidae,* edited by Devra G. Kleiman, pp. 23–37. Washington, D.C.: Smithsonian Institution Press.

Epple, G.
1970. Maintenance, breeding, and development of marmoset monkeys (Callithricidae) in captivity. *Folia Primat.,* 12:56–76.

Freese, C.
1975. A census of non-human primates in Peru. In *Primate Censusing Studies in Peru and Colombia,* pp. 17–41. Washington, D.C.: Pan American Health Organization.

Gengozian, N.
1969. Marmosets: Their potential in experimental medicine. *Ann. N. Y. Acad. Sci.,* 162:336–362.

Hampton, J. K., Jr.; S. H. Hampton; and B. T. Landwehr.
1966. Observations on a successful breeding colony of the marmoset, *Oedipomidas oedipus. Folia Primat.,* 4:265–287.

Hernandez-Camacho, J., and R. W. Cooper.
1976. The non-human primates of Colombia. In *Neotropical Primates: Field Studies and Conservation.* edited by R. W. Thorington and P. Heltne, pp. 35–69. Washington, D.C.: ILAR, National Academy of Sciences.

Jerkins, T.
1972. Free range breeding of squirrel monkeys on Santa Sofia Island, Colombia. In *Breeding Primates,* edited by W. B. Beveridge, p. 144. Basel: Karger.

Kingston, W. R.
1969. Marmosets and tamarins. *Laboratory Animal Handbook,* 4:243–250.

Moynihan, M.
1970. Some behavior patterns of Platyrrhine monkeys, II. *Saguinus geoffroyi* and some other tamarins. *Smithson. Contrbs. Zool.*, 28:1–77.

Sponsel, L. E.; D. S. Brown; R. C. Bailey; and R. A. Mittermeier.
1974. Evaluation of squirrel monkey ranching on Santa Sofia Island, Amazonas, Colombia. *Internat. Zoo Yearb.*, 14:233–240.

Struhsaker, T. T.; K. Glander; H. Chirivi; and N. Scott.
1975. A survey of primates and their habitats in northern Colombia, May–August 1974. In *Primate Censusing Studies in Peru and Colombia,* pp. 43–78. Washington, D.C.: Pan American Health Organization.

Thorington, R. W., Jr.
1968. Observations of the tamarin, *Saguinus midas. Folia Primat.*, 9:95–98.

Wolfe, L. G.; F. Deinhardt; J. D. Ogden; M. R. Adams; and L. E. Fisher.
1975. Reproduction of wild-caught and laboratory-born marmoset species used in biomedical research (*Saguinus* sp., *Callithrix jacchus*). *Lab. Anim. Sci.*, 25:802–813.

Wolfe, L. G.; J. D. Ogden; J. B. Deinhardt; L. Fisher; and F. Deinhardt.
1972. Breeding and hand-rearing marmosets for viral oncogenesis studies. In *Breeding Primates,* edited by W. B. Beveridge, pp. 145–157. Basel: Karger.

JEREMY J. C. MALLINSON
Zoological Director,
Jersey Wildlife Preservation Trust,
Les Augres Manor,
Trinity, Jersey,
Channel Islands.

Maintenance of Marmosets and Tamarins at Jersey Zoological Park with Special Reference to the Design of the New Marmoset Complex[1]

ABSTRACT

Sixteen callitrichid species have been maintained at the Jersey Zoological Park, of which seven have reproduced successfully. Over a ten-year period, experimentation with a variety of diets and patterns of accommodation has been carried out, resulting in the construction of a new marmoset complex design with the following factors in mind. Each indoor-cage unit has access to a connecting outdoor cage which the animals are allowed to use throughout the year. The inside area is not accessible to the public, thus acts as a "safe zone." Suggestions for the long-term environmental requirements of callitrichids in captivity are presented and the importance of coordinating data about captive populations for the establishment of breeding programs for rare species is discussed.

[1]Revised version of a paper presented at the *Conference on the Breeding of Endangered Species as an Aid to Their Survival* held May 1-3, 1972, in Jersey, Channel Islands.

Introduction

Zoological parks seldom have more than one or two species of marmosets and tamarins (Callitrichidae) in their collections, and those which do, rarely succeed in breeding a second generation. In the majority of cases this stems from the difficulties that zoos have experienced with the long-term maintenance of this highly specialized group. Some zoo managements, and indeed some research laboratories, do not take the time to read and digest the many excellent papers which set out the basic captive maintenance and general husbandry requirements for marmosets and tamarins. All too often, marmosets are maintained in standard primate caging with no regard to their special requirements.

Table 1. Checklist of species of Callitrichidae and Callimiconidae bred in captivity.[1]

Common Name	*Scientific Name*
*Common marmoset	*Callithrix jacchus*
White-eared marmoset	*Callithrix aurita*
*Black-pencilled marmoset	*Callithrix penicillata*
White-fronted marmoset	*Callithrix geoffroyi*
White-shouldered marmoset	*Callithrix humeralifer*
*Silvery marmoset	*Callithrix argentata*
Pygmy marmoset	*Cebuella pygmaea*
*Red-handed tamarin	*Saguinus midas*
*Saddle-backed tamarin	*Saguinus f. fuscicollis*
Moustached tamarin	*Saguinus mystax*
Red-mantled tamarin	*Saguinus fuscicollis illigeri*
*Black-and-red tamarin	*Saguinus nigricollis*
White-fronted tamarin	*Saguinus leucopus*
*Cotton-topped tamarin	*Saguinus oedipus*
Geoffroy's tamarin	*Saguinus oedipus geoffroyi*
Pied tamarin	*Saguinus bicolor*
Golden lion tamarin	*Leontopithecus rosalia*
*Goeldi's monkey	*Callimico goeldii*

[1]Editor's note: Hershkovitz (1975, *Folia primat.* 24: 137–172) considers *C. penicillata*, *C. geoffroyi*, and *C. aurita* to be subspecies of *C. jacchus* while Coimbra-Filho and Mittermeier (1973, *Folia primat.* 20: 241–264) still recognize them as distinct species.
*Eight of the 18 species listed have been successfully bred at the Jersey Zoological Park.

During the last 15 years, 16 callitrichid species have been represented in the Jersey Wildlife Preservation Trust's collection. During the first 10 years, the collection never comprised more than 25 specimens, although more recently (due to the construction of a new complex) the numbers have risen to over 50 individuals, 73 percent of which are captive bred; however, our population is small compared to the numbers now being used extensively in biomedical studies. Our success with their maintenance and breeding stems from our own experimentation with a variety of diets and patterns of accommodation.

As far as records show, several callitrichid species have only very rarely been kept in captivity and less than half of the species have so far successfully bred (Table 1), although only a small percentage of these reproduce frequently; however, various specific breeding programs are now underway, and the lifespan of some species in captivity has been greatly increased. As an example of the latter, a male Geoffroy's tamarin (*Saguinus oedipus geoffroyi*) which arrived in Jersey as an adult on July 6, 1961, died at an age of over 13 years (July 31, 1974). A male emperor tamarin (*Saguinus imperator*) which arrived in Jersey as a juvenile specimen on June 10, 1961, has lived at the zoo for over 15 years and is still thriving. Both of these specimens undoubtedly hold longevity records.

This paper presents data gathered from our experiences in Jersey, with special reference to and suggestions for the long-term environmental requirements of callitrichids in captivity.

Maintenance Requirements

Diet

Little has been published on the natural diets of marmosets and tamarins, but it is generally considered that they thrive on a diet high in animal protein. Callitrichids have been observed in the wild state to feed on a range of foodstuffs including sweet fruits, buds, berries, seed pods, spiders, grasshoppers, small lizards, freshly killed bats, and small birds (Bates, 1864; Moynihan, 1970).

In order to prevent captive animals from becoming bored by an unchanging diet, we vary the food as much as possible. We have found that a well-balanced diet comprised of mixed fruit (banana, orange, apple, pear, grape), vegetable matter (celery, carrot), animal protein (crickets, mealworms, hairless mice), boiled egg, raisins, brown bread, and a pellet mixture has sustained the collection well. However, like others we had to find out the importance of vitamin D_3 supplementation the hard way; osteomalacia frequently occurred with the adult specimens at intervals of 11 to 12 months, and with infants the condition developed during the first 3 to 5 months of life. It is interesting to note that the onset of this condition was recorded and at a similar period of time as long ago as 1885 when the traveler Hastings Dent brought two black-eared marmosets (*Callithrix penicillata*) back to England.

During the early 1960s the first successful remedy used by us to combat this condition was the injection of vitamin D_3 intramuscularly, giving 62,500 I.U. per adult at intervals of 10 months, and 20,000 I.U. to the young at approximately 4 months of age. Since the mid 1960s, however, we have found it preferable to administer vitamin D_3 orally and, as advised by Hampton et al. (1966), initially gave a supplement of 1,000 I.U. per specimen daily for a period of 11 weeks, which was decreased to 700 I.U., and then to a maintenance level of 500 I.U. Since the spring of this year (1975), all of our callitrichid collection has access to outside caging, and the sun's ultraviolet rays; some experimentation is being carried out to determine whether lower levels of vitamin D_3 will be adequate for marmosets and tamarins that are no longer accommodated totally indoors. (The vitamin product used is Rovimix D_3 400, manufactured by Roche Products Ltd., Switzerland.)

Accommodation

Initially, in the absence of any established guidelines, all the marmosets and tamarins in the Jersey collection were kept in indoor cages which measured 240 cm × 150 cm × 210 cm high, with the room temperature ranging between 17°–21° C (62°–70° F). This environment proved to be unsatisfactory, although it was found that the introduction of infrared lamps into each cage improved general conditions. Partly by eliminating the habitual huddling together of specimens, this improvement was apparently a result of the animals being able to periodically get to an area with a temperature of up to 29° C (85° F). It is now generally accepted that laboratories are better able to maintain marmosets successfully at a room temperature of 27°–28° C (80°–82° F) (see Hampton et al., 1966; Wolfe et al., 1972).

Each cage was furnished with a nest box and a frequently changed arrangement of sloping branches, for it had been found that alteration of cage environment, by regular renewal of the branching, helped to stimulate exercise and scent marking by the animals. It was also found to be important that at least one familiar scent-marked branch should be left for them when changing the majority of branches.

Since we were not only interested in captive propagation, but also in behavior, we realized that it was important to maintain some animals under conditions as close as possible to those present in the wild. For, as others have experienced (Epple, 1970), marmosets and tamarins confined to laboratory-type cages, which can be as small as 46 cm × 46 cm × 53 cm high (Wolfe et al., 1972), have only a limited repertoire of social responses and vocalizations.

In 1965, from mid May to mid September, we maintained some specimens of *Callithrix jacchus* outside on a small island in one of the zoo's duck ponds with only an unheated waterproof box for shelter. This exposure to the elements greatly benefited their condition, and thus, we decided to allow a family group of hybrids, black-penciled marmoset *Callithrix p. penicillata* × common marmoset *Callithrix jacchus*, access to a large outside aviary throughout the year, with the inside accommodation heated to 27° C (80° F). Even when the outside temperature fell as low as freezing, the five specimens maintained excellent physical condition.

We have found from these experiences that in order to study the general behavior of callitrichids, in addition to breeding them, the optimum accommodation includes a heated inside area with access to a large outside "aviary-type" enclosure throughout the year. At the same time, it is necessary to ensure that physical contact between neighboring pairs is impossible. In agreement with others (Epple, 1970; Hampton et al., 1966), we find that outdoor housing requires large cages with sufficient space for exercise in both cold and warm weather. When kept only indoors, marmosets and tamarins require temperatures of up to 29° C (85° F) at selected sites in the cages.

Taking account of these various considerations, a new marmoset and tamarin complex for the Jersey Zoo was designed and opened in the spring of 1975.

Marmoset Complex

The marmoset and tamarin complex consists of 16 indoor-cage units and 10 connecting outdoor cages (see Mallinson, 1975). The inside area is not accessible to the public, so acting as a "safe zone";

the animals are only on view in their outdoor enclosures.

The outdoor cages are richly furnished with sloping branches, vegetation, and granite boulders; the floor of the cages are earthen with some grass in each. The cages are constructed of hollow steel sections covered with a galvanized 2.54 cm "twil-weld" mesh which has been painted black in order to minimize any visual barrier. Due to the irregular shape of the outside enclosures, the cages vary in height from 2.1 m to 2.6 m and in floor area from 6.7 sq m to 8.9 sq m. The front of each cage has at least one panel of plate glass, which not only acts as a windbreak but provides extra security for the animals. The partitions between the cages are of an opaque corregated Perspex ("Plexiglass") so that physical contact between the groups is impossible.

The 16 indoor-cage units measure 122 cm × 91 cm × 152 cm high. The floor of the unit which slopes toward the front to facilitate cleaning is raised 76 cm above the floor of the house. The frames of the cage units are hollow steel sections, whilst the sides, floors, partitions, and shelves are made of opaque self-extinguishing ICI Darvic PVC.

Each unit has a slide on the back wall leading to the outside cage, which is operated from the front of the cage; there is also a slide between adjacent units. The front of each cage has two hinged doors covered with the 2.54 cm "twil-weld" mesh; these enable a person to enter the unit. The floor will support the weight of an adult human. Above the doors is a shelf running the whole width of the unit which is reached through sliding transparent Perspex partitions. The upper part of the cage has a single removable wooden nest box measuring 30 cm × 23 cm × 28 cm high in one corner, the rest of which is covered with the "twil-weld" mesh. The next-box design enables an attendant to service the nest box without disturbing the cage occupants or to catch an animal in it, either for examination through a glass sliding panel behind the outer wooden one, or for removal to another unit. Each unit has an infrared lamp to act as a heat locus, a wooden ladder, two standard perches, and a small natural branch. All cage embellishments can be removed for cleaning.

The temperature of the house ranges between 18°-27° C (65°-80° F) and is controlled by a thermostat. The electric lighting is controlled by a time clock, although a single 40-watt red-colored bulb is left on throughout the night. Two domed skylights with adjustable ventilation louvres are situated between the two rows of cages.

The wall and floor surfaces are painted with a chlorinated rubber paint (ICI). The wall skirting is coved and the floor well sloped with a 2.5 cm to 10 cm incline to facilitate cleaning and drainage. The house is hosed down twice daily and the normal relative humidity ranges between 60-75 percent.

Although the complex has only been in operation for a few months, it has been rewarding to witness how well the marmosets and tamarins have settled down to their new accommodation and how enthusiastic the public have been in seeing a variable collection of callitrichids in this more natural environment. Due to the spaciousness of the outside cages, and the fact that they are never shut out of their indoor "safe zone" units, they are exhibiting themselves well and are starting to reproduce successfully.

General Husbandry

Breeding Conditions

From our findings and those of others (Hampton et al., 1966; Kingston, 1970; Wolfe et al., 1972), we consider that satisfactory breeding results from keeping breeding pairs by themselves, as opposed to keeping marmosets in large family groups. We have found with both *Callithrix jacchus* and *Callithrix penicillata* that adult females will inhibit the young and prevent them from breeding, and unnecessary fighting amongst the family generally develops after the young have reached sexual maturity. Therefore, we leave young with parents until they are approximately twelve months old. For if the parents do produce further young within this time, the first generation offspring can assist in the postnatal handling of the new infants, and also benefit from the various other family interactions. With the current absence of wild-caught imported stock, this early social contact with the parents could prove important in the long-term breeding of marmosets and tamarins in captivity. The keeping of unrelated adult pairs together is seldom if ever successful, for fighting among animals of the same sex invariably occurs. We have experienced such social problems with *Callithrix jacchus*, *Callithrix penicillata*, hybrid *C. penicillata* × *C. jacchus*, and *Saguinus f. fuscicollis*.

Reproduction Data

Marmosets and tamarins have gestation periods ranging from 128-145 days, and many will breed throughout the year. Some of the specimens (*Callithrix jacchus* and hybrid *C. penicillata* × *C. jacchus*) in the Jersey collection have bred twice a year, and we have had cases of breeding three times within 12 months. The average interval between one birth followed by normal lactation and the next birth is 150–155 days.

Regarding development, infants have been observed to climb independently and eat soft fruit at 21 days of age; they may eat hairless mice at 25 days. Twin *Callithrix jacchus* who had lost their mother at 35 days of age survived on the rations provided, without further human intervention. Although the father plays an important part in carrying the young, passing them back to the mother during feeding periods, we have had one case where twin red-handed tamarins (*Saguinus midas*) lost their father at three days of age and were successfully reared by the mother alone.

Out of 57 litters born in Jersey (up to January 1976), 11 singletons, 40 sets of twins, 5 sets of triplets and 1 set of quadruplets have been recorded. It can be seen from Table 2 that of the recorded births approximately 19 percent are singletons, 70 percent twins, 9 percent triplets, and 2 percent quadruplets. In a census of 203 recorded parturitions in 6 different collections (Mallinson, 1971), the percentages were similar—20 percent singletons, 70 percent twins, 9 percent triplets and 1 percent quadruplets. It is also interesting to note that *Callithrix* spp. seem to have larger litters than *Saguinus* spp.and that with *Callimico,* singletons are the norm (Table 2).

It has been suggested that one of the major obstacles in the long-term breeding of callitrichids in captivity is the failure of successful breeding of second- and third-generation animals. There also appears to be a tendency for F_2 or F_3 breedings to produce both larger litters and in some cases fetuses. The three cases of quadruplets on record have all occurred with the common marmoset (Table 2). Within a two-year period, Mr. Frank

Table 2. Litter sizes in Callitrichidae and Callimiconidae at Jersey Zoological Park.[1]

Callitrichidae	*Singletons*	*Twins*	*Triplets*	*Quadruplets*
Common marmoset *Callithrix jacchus*	1	15	0	1
Black-pencilled marmoset *Callithrix penicillata jordani*	0	0	1	0
Hybrid marmoset *C. jacchus* × *C. p. penicillata*	1	1	1	0
Hybrid marmoset *C. p. penicillata* × *C. jacchus*	0	5	0	0
Hybrid marmoset *C. p. penicillata / C. jacchus* × *C. p. penicillata / C. jacchus* (F_2)	0	7	3	0
Silvery marmoset *Callithrix argentata*	1	3	0	0
Red-handed tamarin *Saguinus midas*	2	1	0	0
Black-and-red tamarin *Saguinus nigricollis*	0	2	0	0
Saddle-backed tamarin *Saguinus fuscicollis*	4	4	0	0
Cotton-topped tamarin *Saguinus oedipus*	2	2	0	0
Total births = 57	11	40	5	1
Callimiconidae				
Goeldi's monkey *Callimico goeldii*	3	0	0	0

[1]See editor's note, Table 1.

Table 3. Checklist of hybridization amongst Callitrichidae.[1]

Scientific Name	*Location*	*Reference*
Callithrix jacchus × *Callithrix p. penicillata*	Jersey 1966	
Callithrix p. penicillata × *Callithrix jacchus*	Jersey 1967, 1968, 1969	
Callithrix p. pencillata / Callithrix jacchus × *Callithrix p. penicillata / Callithrix jacchus* (F_2)	Jersey 1971, 1972, 1973, 1974, 1975	
Callithrix jacchus × *Callithrix geoffroyi*	Rio de Janeiro 1970	Coimbra-Filho, 1970
Callithrix jacchus × *Callithrix argentata*	London 1932	English, 1932
Saguinus midas × *Saguinus bicolor*	Cologne 1959	I.Z.Y. *Volume 1
Saguinus midas / Saguinus bicolor × *Saguinus midas*	Cologne 1961	I.Z.Y. *Volume 3
Saguinus midas × *Saguinus oedipus*	Washington 1964	I.Z.Y. *Volume 6
Saguinus oedipus × *Saguinus geoffroyi*[2]	Frankfurt 1970	Epple, 1970
Saguinus mystax × *Saguinus imperator*		Cruz Lima, 1945

[1]See editor's note, Table 1.
[2]Editor's note: often considered only subspecifically distinct.
**International Zoo Yearbook*, Zoological Society of London.

Rossi's father and daughter pair produced one set of twins, two sets of triplets and the one set of quadruplets (Rossi, pers. comm.). At Jersey, four out of the six instances of triplets and quadruplets have been to second-generation, captive-bred parents. The quadruplets born on March 10, 1975, at Jersey were from a partial F_3 breeding, for the grand-sire was born in Jersey on December 8, 1961.

Hybridization between *C. jacchus* and *C. penicillata* occurs frequently in captivity and as can be seen from Tables 2 and 3, hybrids have been bred with each other and produced viable offspring. Ten litters have been recorded to a Jersey pair within a four-year period.

Discussion

From our experience and others, the care and maintenance of marmosets and tamarins is now more fully understood, and the optimum conditions for viable breeding groups can be more easily established. It is Jersey policy to build a minimum population of four breeding pairs per species, and to cooperate with other conservation-minded zoological parks in specific breeding programs for these species. Already the Jersey Wildlife Preservation Trust is cooperating with the Twycross and London zoos in England and Wassenaar Zoo in Holland by having on loan and lending specimens of the silvery marmoset (*Callithrix argentata*). The Jersey Zoo has at present a nucleus of five pairs of this species.

Marmosets and tamarins can breed in the second year of life, can reproduce regularly at least twice a year, have a reproductive life of from 8 to 10 years, and a life expectancy of 10 to 15 years. As multiple births are the rule, callitrichids may produce 30 to 40 young during their lifetime (the Jersey F_2 hybrids have produced 23 infants within a 4-year period). I hope that zoos and research laboratories who have not previously had marmosets and tamarins in their collections, but wish to do so, will familiarize themselves with already published information so that the present large-scale wastage of animals will be reduced.

The survival of many nonhuman primates seems to depend on how much of their natural habitat will remain inaccessible. In captivity, a number of institutions, including Jersey, are recording second- and third-generation births, but it is with these that problems are being encountered. Therefore, it is of the utmost importance to coordinate data about the captive populations of the various species in captivity, to exchange ideas, and to have zoos and primate research centers combine efforts toward establishing self-sustaining breeding groups of the rare species. If careful attention is paid to both dietary and environmental requirements and sufficient numbers of different species can be brought together, with particular institutions being responsible for the long term interests of rare species, then the survival of the majority of callitrichids in captivity could be ensured.

Acknowledgments

My thanks to members of the Zoo staff who have noted their observations over the years.

Literature Cited

Bates, H. W.
1863. *The Naturalist on the River Amazon.* 2d edition. London: John Murray.

Coimbra-Filho, A. F.
1970. Acerca de um caso de hibridismo entre *Callithrix jacchus* (L., 1758) × *C. geoffroyi* (Humboldt, 1812). *Rev. Brasil. Biol.,* 30(4):507–517.

Cruz Lima, E. da
1944. *Mamiferos da Amazonia. Introdução Geral e Primates.* Museu Paraense Emilio Goeldi de Historia Natural e Etnografia: Belém do Pará.

Dent, H. C.
1886. *A Year in Brazil.* London: Kegan Paul Trench & Co.

English, W. L.
1932. Exhibition of living hybrid marmosets. *Proc. zool. Soc. Lond.*: 1079.

Epple, G.
1970. Maintenance, breeding and development of marmoset monkeys in captivity. *Folia primat.* 12:56–76.

Hampton, J. K.; S. H. Hampton; and B. T. Landwehr.
1966. Observations on a successful breeding colony of the marmoset, *Oedipomidas oedipus. Folia primat.* 4:265–287.

Kingston, W. R.
1970. On the breeding of marmosets and tamarins. *Lab. Primate Newsletter* 9:9–10.

Mallinson, J. J. C.
1971. Observations of the breeding of red-handed tamarin *Saguinus* (= *Tamarin*) *midas* with comparative notes on other species of Callithricidae (= Hapalidae) breeding in captivity. Jersey Wildlife Preservation Trust, Eighth Annual Report:19–31.

1975. Breeding marmosets in captivity. In *Breeding Endangered Species in Captivity,* edited by R. W. Martin, pp. 203–212. London: Academic Press.

Moynihan, M.
1970. Some behavior patterns of platyrrhine monkeys II *Saguinus geoffroyi* and some other tamarins. *Smithson. Contrbs. Zool.* 28:1–77.

Wolfe, L. G.; J. D. Ogden; J. B. Deinhardt; L. Fisher; and F. Deinhardt.
1972. Breeding and hand-rearing marmosets for viral oncogenesis studies. In *Breeding Primates,* edited by W. B. Beveridge, pp. 145–147. Basel:Karger.

JOHN L. CICMANEC
Litton Bionetics, Inc.
5516 Nicholson Lane
Kensington, Md. 20795

Medical Problems Encountered in a Callitrichid Colony

ABSTRACT

As a consequence of maintaining marmosets and tamarins in a laboratory environment for use in biomedical research, the principal disease problems of captive colonies have been defined. Of particular concern are highly fatal viral infections caused by *Herpesvirus platyrrhinae, H. hominis,* and measles (rubeola). Vaccines are available for control of herpesvirus infections, and rigorous personnel health programs are useful in controlling spread of *H. hominis* and measles to marmosets and tamarins. Among the group of bacterial pathogens isolated from callitrichids, *Pseudomonas aeruginosa, Klebsiella pneumoniae,* and *Salmonella* are most often associated with serious disease. Wild-caught marmosets and tamarins are affected with a wide variety of parasitic infestations; among these, *Prosthenorchis elegans,* an acanthocephalad, can create serious disease. Infestations due to nematodes and cestodes can be controlled with standard anthelmintics. Metabolic diseases which occur frequently are hypoglycemia and emaciation due to inadequate protein intake. Immediate corrective therapy can be applied, but precise etiologies remain obscure.

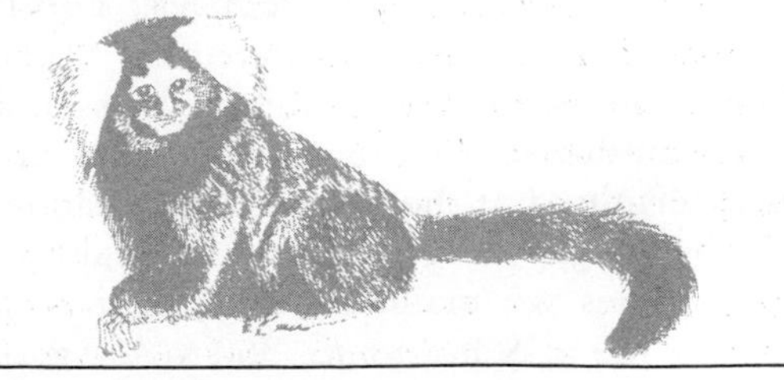

Introduction

Extensive experimental usage of marmosets and tamarins (particularly *Saguinus* spp. and *Callithrix* spp. has been made in biomedical research since 1960; one early report of their use appeared in 1936. During the period 1970 to 1972, *Saguinus* spp. ranked fourth among all nonhuman primates imported into the United States (Muckenhirn, 1975).

The use of callitrichids in research has stemmed primarily from their natural hematopoietic chimerism and its application to immunological studies and their susceptibility to human infectious hepatitis and oncogenic viruses, including *Herpesvirus saimiri* and *H. ateles.* Much of the description of clinical diseases of marmoset and tamarin species has detailed their high susceptibility to viral diseases associated with man and other primates; that is, measles (rubeola), *H. platyrrhinae,* and fatal bacterial infections.

This report provides a brief summary of reports of clinical diseases described in marmosets and tamarins as well as descriptions of clinical conditions encountered in our colony and of methods used in treating or attempted control of these diseases.

Clinical Examination and Clinical Procedures

Even within a laboratory environment, the small size and temperament of marmosets and tamarins places a distinct limit upon the extent and thoroughness of a physical examination. As is true with most other species of nonhuman primates, most disease processes are usually quite advanced before apparent clinical changes are noted.

As a routine practice, we examine each individual of the breeding colony at least once a month. This procedure involves manually catching the animal, and a visual and palpation examination is made. Particular attention is directed toward the oral cavity for paleness as an indication of anemia, ulcerations as possible signs of herpesvirus infection, and general condition of dentition. The abdomen is palpated for detection of nodule formation in response to prosthenorchis infestation, and detection of pregnancy in females. The limbs are palpated as a means of detecting broken bones and as an assessment of muscle mass and general body condition. Total body weight for each animal is determined at the time of examination.

The weight ranges of normal healthy adults of the species we maintain are: *Saguinus fuscicollis,* 340 to 400 g; *S. nigricollis,* 360 to 450 g; *S. oedipus,* 450 to 550 g; and *Callithrix jacchus,* 270 to 350 g.

Ten mg of ketamine hydrochloride[1] is injected intramuscularly to provide sedation for more thorough examination when indicated, and higher doses of up to 30 mg for a 350 g animal are used for minor surgical procedures. We have found the use of phencyclidine hydrochloride[2] to be much less predictable in marmosets and tamarins; hence, it is very infrequently used.

One-milliliter blood samples are often collected from the femoral vein for complete blood counts, and clinical samples of up to 4 ml of clotted blood are taken for clinical chemistry determinations. We have found normal values of marmosets and tamarins for glucose, phosphorus, calcium, and sodium to be at variance for normal values of humans and other primates. These are presented in Table 1 for *S. fuscicollis.*

Many drugs can be given orally, and if the animal will not readily swallow the medication, the preparation can be given via stomach tube. We prefer to give medications by the oral route rather than by injection whenever possible since many injectable medications are quite irritating when given into the small muscle mass of a callitrichid.

Table 1. Comparison of selected normal serum chemistry values of *Saguinus fuscicollis* with man.

Substance	*Saguinus fuscicollis*	*Homo sapiens*
Calcium (mg/dl)	9.6–12.1 $\bar{x}$ = 10.5	9.0–10.6
Glucose (mg/dl)	120–248 $\bar{x}$ = 172	60–100
Phosphorus (mg/dl)	4.1–8.3 $\bar{x}$ = 5.9	3.0–4.5
Sodium (mEq/l)	160–177 $\bar{x}$ = 168	136–142

Disease Problems

Herpesvirus platyrrhinae

H. platyrrhinae, also designated *Herpesvirus T* or *H. tamarinus,* infection has a rapid and highly fatal course in marmosets and tamarins. Experimental

[1]Ketaset®, Bristol Laboratories, Division of Bristol-Myers Company, Syracuse, New York.

[2]Sernylan®, Bio-Ceutic Laboratories, St. Joseph, Missouri.

studies with this agent have shown that it has a short incubation period of 7 to 10 days (Holmes et al., 1964; Melnick et al., 1964). Weakness, prostration, and rapid onset of death are characteristic clinical signs. Often ulcerative sores are seen on the lips and tongues of callitrichids affected with *H. platyrrhinae.* At postmortem examination, focal necrosis and hemorrhage within the liver, kidneys, and adrenal glands are observed, and the virus can be isolated from many organs.

It has been shown that squirrel monkeys (*Saimiri sciureus*) are the reservoir host for *H. platyrrhinae.* One method of controlling this disease is to keep marmosets and tamarins separated from squirrel monkeys, including the use of separate caretakers for callitrichid and squirrel monkey colonies within an animal facility. A modified live virus vaccine for *H. platyrrhinae* has been developed at the New England Regional Primate Research Center, and it is frequently used in laboratory colonies to control infections of *H. platyrrhinae.*

Herpesvirus hominis

H. hominis, the causative virus of human cold sores, also causes fatal infections in marmosets (Hunt and Melendez, 1969). While most humans have antibodies to this virus, the virus is only actively shed by humans during the time that actual sores exist. For this reason, persons with active cold sore lesions should be forbidden in the area of marmoset and tamarin colonies. Clinically and pathologically, this disease in marmosets is indistinguishable from *H. platyrrhinae,* so viral isolation or serology must be done for definitive diagnosis. A live virus vaccine for *H. hominis* for use in New World primates has also been developed at the New England Regional Primate Research Center.

Measles (Rubeola)

A third highly fatal viral disease of marmosets and tamarins that has been recognized is measles. The initial report of an epizootic of this disease was published by Drs. Levy and Mirkovic (1971) in which 326 animals died. Swollen eyelids and progressive lethargy were the only clinical signs noted. At postmortem examination, a giant cell pneumonitis was observed. Viral isolation and comparison of antibody titers in paired serum samples collected from marmosets and tamarins are two methods of making definite diagnosis of measles.

A possible source of measles infection to a callitrichid colony is by accidental infection from humans. Measles infection in newly received rhesus monkeys (*Macaca mulatta*) and other macaques is another source. Keeping a colony free of these two sources should do much toward preventing measles infection.

Human gamma globulin[3] preparations contain antibodies to measles, and for this reason all newly received marmosets and tamarins are injected with a human gamma globulin preparation at the beginning of the quarantine period. It is a useful practice to determine that personnel working with callitrichids have an antibody titer against measles; thus, they are not likely to acquire an active infection which could be transmitted to the colony.

In July 1973, we experienced an epizootic within our callitrichid colony which presented many similar but many dissimilar features of a measles infection. The clinical signs were muscular incoordination, apparent blindness, and rapid onset of death. Of the 88 animals at risk, 35 were affected (40 percent), and the disease was fatal to 70 percent of the affected individuals. Members of *Saguinus fuscicollis, S. nigricollis,* and *S. oedipus* were equally affected. Microscopic examination of tissues of affected animals did not show significant lesions for the most part; however, a few multinucleated giant cells were seen in the spleens, livers, and, occasionally, lungs of some of the animals. We then did hemagglutination inhibition tests for measles (rubeola) on some tamarins. The results of these tests are shown in Table 2.

Table 2. Results of hemagglutination inhibition tests for rubeola and rubella, July 1973, in *Saguinus fuscicollis* (S.f.), *S. nigricollis* (S.n.), and *S. oedipus* (S.o.)

Tamarin	*Number*	*Species*	*Rubeola**	*Rubella**
231-I	1	S.f.	32	20
231-I	2		32	20
249-I	1	S.n.	32	40
249-I	2		128	QNS**
403-I		S.o.	16	20
411-I		S.o.	32	20
582-I	1	S.o.	4	QNS
582-I	2		64	20
961-I		S.f.	32	20

*Reciprocal of titer. Numbers placed by tamarin number designate initial sample (1) or second sample collected 10 days later (2).

**QNS = Quantity not sufficient.

[3]Gammar®, Armour Pharmaceutical Company, Phoenix, Arizona.

Attempted viral isolation was unsuccessful. The conflicting serology data were not resolved, but we have found continuing high titers to rubeola in surviving tamarins, and on this basis we consider the disease outbreak to be measles, though atypical of the original epizootic described by Drs. Levy and Mirkovic.

Bacterial Diseases

Several bacterial infections are of particular importance in callitrichids. I would consider *Pseudomonas aeruginosa, Klebsiella pneumoniae,* and *Salmonella* sp. to be the most important. In a survey of fecal samples of newly received individuals, Deinhardt found a 23 percent incidence of *Pseudomonas* and a 19 percent incidence for *Salmonella.* Among throat swabs, a 64 percent incidence of *Klebsiella* was found (F. Deinhardt et al., 1967).

While it has been observed that a dramatic reduction in the incidence of bacterial infections occurs during the quarantine and conditioning period, each of these organisms can be a problem also within a conditioned colony. Even with extensive courses of antibacterial therapy, *Pseudomonas* will not be totally removed from a colony. It is important that separate isolations of bacteria and their antibiotic sensitivities be determined frequently. Carbenicillin,[4] gentamycin[5], and colistin[6] have been found to be effective antibiotics in the treatment of *Pseudomonas* infections. *P. aeruginosa* is a highly pathogenic organism, and natural-host defenses play a significant role in control of infection as well as administration of antibiotics.

We have used kanamycin,[7] chloramphenicol,[8] and ampicillin[9] to control *Salmonella* infections. Contamination of commercial feed with *Salmonella* seems to occur at least once a year. These infections generally occur in younger animals.

Klebsiella most frequently causes pneumonia in other primate species, but in callitrichids this organism often causes septicemia. Also, *Klebsiella* has been isolated from skin lesions in *Callithrix jacchus.* Chloramphenicol has been the most useful antibiotic in control of infections due to this organism in callitrichids.

Parasitic Diseases

At least three exhaustive studies have been made enumerating and classifying the parasites of marmosets (Cosgrove et al., 1968; F. Deinhardt et al., 1967; Porter, 1972). More than 40 species of parasites have been identified, and all classes of parasites, from protozoa through acanthocephalads, are well represented. From a clinical and colony-management point of view, infestation with the acanthocephalad, *Prosthenorchis elegans,* is most serious. The adult form of this parasite imbeds in the intestinal wall near the ileocecal junction; often the parasite penetrates through the intestinal wall, and deaths can occur due to peritonitis, enteritis, hemorrhage, or intussusception. No consistently effective anthelmintic treatment has been found for this parasite. Some authors have stated the infestation becomes self-limiting as marmosets and tamarins are held in the laboratory for more than 10 months providing that the facilities are free of cockroaches (J. B. Deinhardt et al., 1967). We are continuing to test experimental anthelmintics, including trichlorofon and tetrachlorethylene, in efforts to find a drug that will control infestation of this parasite. Tapeworm infestations which occur are due to *Raillietina* sp. and *Paratriotaenia.* Treatment with niclosamide[10] is effective in the control of these parasites. We routinely worm marmosets and tamarins with thiabendazole[11] during the quarantine period, and we have not seen serious infestations of nematodes in our colony.

Metabolic Conditions

The metabolic condition which occurs most frequently is hypoglycemia. With low blood-glucose levels, marmosets and tamarins can rapidly become unconscious if the condition is not detected in the early stages. Many species of New World primates are susceptible to hypoglycemia, although it most frequently occurs in callitrichids. Undoubtedly, their high metabolic rates contribute to the development of this condition. In truth, the development of hypoglycemia in most cases is caused by man-

[4]Geopen®, Roerig, Division of Pfizer Pharmaceuticals, New York, New York.

[5]Garamycin®, Schering Corporation, Kenilsworth, New Jersey.

[6]Coly-Mycin®, Warner Chilcott, Division of Warner Lambert Company, Morris Plains, New Jersey.

[7]Kantrex®, Bristol Laboratories, Division of Bristol-Myers Company, Syracuse, New York.

[8]Chloromycetin®, Parke, Davis & Company, Detroit, Michigan.

[9]Omnipen®, Wyeth Laboratories, Philadelphia, Pennsylvania.

[10]Yomesan®, Chemagro, Division of Baychem Corporation, Kansas City, Missouri.

[11]Omnizole®, Merck & Company, Rahway, New Jersey.

agement and relates to frequency of feeding of these species rather than a metabolic defect.

When severe hypoglycemia occurs, we administer 5 cc of 20 percent dextrose via stomach tube, and occasionally 1 to 2 units of insulin administered intraperitoneally in order to accelerate intracellular usage of the dextrose. If early symptoms such as muscle fasciculations and unsteadiness can be detected, immediate feeding of grapes or other fruit is often sufficient treatment.

Many of the details of callitrichid nutrition are still poorly understood. The natural diets of some marmoset and tamarin species have been defined, but in most laboratory situations it would be impractical to attempt to duplicate these natural diets. The basic problem with standard laboratory primate diets is that they are not consistently consumed by the animals; thus a serious deficiency in total protein intake can occur. Except in the specialized circumstance of feeding infants, the addition of fruit, which is readily consumed, is sufficient to meet daily caloric requirements. We feed suckling mice to adult callitrichids in order to increase protein intake. Unfortunately, not all members of the various marmoset and tamarin species consistently consume mice. Providing a diet of adequate protein composition that is consistently consumed by marmosets continues to be a problem.

Marmosets and tamarins specifically require that vitamin D be supplied in the form of vitamin D_3. Although most commercial laboratory primate feeds do supply vitamin D in this form, we supplement the prepared chow with 500 IU of a commercial vitamin D_3 preparation per animal weekly. It is also possible to provide sufficient vitamin D to marmosets by exposing them to ultraviolet light for a 15- to 30-minute period each day. Dosage of vitamin D_3 must be carefully calculated, especially for newborn animals, since overdosages of vitamin D can be fatal.

Summary of Necropsy Results

Table 3 summarizes the principal cause of death determined in 178 marmoset and tamarin necropsies. Among adults and juveniles, deaths during the initial three months in the laboratory are separated from those that occurred later. Infant deaths also are grouped separately.

Comparison with a similar report prepared by J. B. Deinhardt et al. (1967) indicates death due to *Prosthenorchis,* and *H. platyrrhinae* were common in their colony as well as ours. They also noted a significant number of deaths due to bronchopneumonia and septicemia. Despite extensive gross and microscopic examinations, a major cause of death could not be determined in 10 to 15 percent of the necropsies performed.

Mortality among newly received groups of marmosets and tamarins ranges between 15 and 35 percent and even within a conditioned breeding colony, up to 10 percent mortality can occur in one year. Among callitrichids receiving medical care for a defined clinical condition, the case fatality ratio is greater than 50 percent. Clearly, detailed studies of disease problems among marmosets and

Table 3. Primary cause of death among callitrichids (178 cases) in the Litton Bionetics colony, 1970–1975 (*Saguinus fuscicollis, S. nigricollis, S. oedipus, Callithrix jacchus*).

Cause of Death	*Number of Cases*		
	Under 3 Mos. in Lab	*Over 3 Mos. in Lab*	*New-borns*
Prosthenorchis sp.			
Enteritis	34	5	
Peritonitis	2		
Intussusception		2	
Other parasites (nematodes, tapeworms)			
Enteritis	6	2	
Peritonitis	4	1	
Lymphadenitis, vasculitis		2	
Bacterial enteritis	6	1	
Bacterial peritonitis		1	
Herpesvirus platyrrhinae	12		
Measles (rubeola)	6	18	1
Pneumonia	4	4	
Newborn pneumonia			5
Pleuritis	1		
Lungworms	1		
Anemia		1	
Trauma	3	4	
Thorns, fractures			
Newborn trauma			2
Nephritis	2	2	
Mineralization		2	
Newborn mineralization			5
Newborn acute tubular necrosis			1
Acute tubular necrosis	1		
Cystitis	1	1	
Starvation	3	2	
Parturition		1	
Bloat	1		
Undetermined	24	4	
Total	111	53	14

tamarins are urgently needed in order to make efficient use of these New World primates that are of unique value to biomedical research.

Literature Cited

Cosgrove, G. E.; B. Nelson; and N. Gengozian
1968. Helminth parasites of the tamarin, *Saguinus fuscicollis. Lab. Anim. Care,* 18:654–656.

Deinhardt, F.; A. W. Holmes; J. Define; and J. Deinhardt
1967. Marmosets as laboratory animals. IV. The microbiology of laboratory kept marmosets. *Lab. Anim. Care,* 17:48–70.

Deinhardt, J. B.; J. Define; M. Passovoy; R. Pohlman; and F. Deinhardt
1967. Marmosets as laboratory animals. I. Care of marmosets in the laboratory, pathology and outline of statistical evaluation of data. *Lab. Anim. Care,* 17:11–29.

Holmes, A. W.; R. C. Caldwell; R. E. Dedman; and F. Deinhardt
1964. Isolation and characterization of a new Herpesvirus. *J. Immunology,* 92:602–610.

Hunt, R. D., and L. V. Melendez
1969. Herpesvirus infections of nonhuman primates: A review. *Lab. Anim. Care,* 19:221–234.

Levy, B. M., and R. R. Mirkovic
1971. An epizootic of measles in a marmoset colony. *Lab. Anim. Sci,* 21:33–39.

Melnick, J. L.; M. Midulla; I. Wimberly; J. G. Barrera-Oro; and B. M. Levy
1964. A new member of the Herpesvirus group isolated from South American marmosets. *J. Immunology,* 92:596–601.

Muckenhirn, N. A.; C. H. Southwick; R. W. Thorington, Jr., et al.
1975. *Nonhuman Primates. Usage and Availability for Biomedical Programs.* Washington, D.C.: National Academy of Sciences.

Porter, J. A., Jr.
1972. Parasites of marmosets. *Lab. Anim. Sci,* 22:503–506.

Panelists:
K. Hobbs (chairperson),
J. Cicmanec, N. Gengozian,
S. Hampton, W. R. Kingston,
J. Porter.

**PANEL DISCUSSION:
AUGUST 19, 1975**

Problems of Management of Large Marmoset Colonies

HOBBS: There is a good deal of detail to be covered, e.g., in just the disease aspect alone; this is one of the major problems of management. Perhaps a lot of the more factual detail we can go through quickly, since I would like to expand this discussion to include the far more controversial aspects and take advantage of the fact that we have people interested in conservation here together with those interested in the availability and use of marmosets. I think it would be a great pity to limit this discussion just to problems of management. Having said that, perhaps I can just run through the major problems of management. First of all, we have the problems associated with the use of imported animals, the disease as a result of stress and trauma, of trapping and transportation which of course are common throughout all the primate species, not just marmosets. There is the whole problem of doing research using a wild-trapped animal. In the United Kingdom, a pet phrase has been "sewer-rat monkey." Nobody would dream of using a sewer rat, but here we are using monkeys almost like that. Secondly, there are the problems which occur as a result of housing such animals in laboratory environments, finding the best sorts of caging, and the behavioral needs of the particular species linked with the need for good reproduction. Environmental control is perhaps a little more specific for some of the New World monkeys than it is for other monkeys. And then there are problems associated with the improved well-being of the animal once it is established in a captive condition. For example, some say that the F_2 animals are producing larger offspring, and we are getting problems of distocia. In some colonies this doesn't seem a problem at all. Also in certain colonies, triplets and quads become more common, which in turn requires the hand-rearing of those rejected young. Then there is the cost aspect. The obvious problem is the lack of financing, especially in the last couple of years. But it is also a problem of commitment, of getting the research user committed to breeding programs because of the cost. This is where contributions such as Mr. Kingston's and the ICI (Imperial Chemical Industries) detailed costing are very valuable. They put the situation into a sort of realistic aspect. In fact, ICI says that they can breed marmosets for less than they can breed beagles. Therefore, the problems of convincing the research user are at last coming to a head. This, of course, is helped by the lack of availability from the user countries.

I think the other major point that we want to cover is where the interests of conservationists can clash or coincide with the interests and the needs of research workers. Very briefly, I think that conservation per se obviously causes problems for the user in that it creates a lack of availability in the first place. On the other hand, it can help to promote the establishment of breeding colonies for that very reason. The dilemma at the moment, I think, is that those who are able to establish self-sustaining breeding colonies are having difficulties in getting the stocks now in order to do it. I am very disappointed that we are not able to have direct access to national representatives from the countries of origin to whom we could say that we have the money, we have the knowledge to breed, and could they help us get the nuclei stock out over the next one or two years.

WILSON: I acknowledge that one of the major obstacles is acquiring more marmosets, but I think in order to justify doing so, a very concerted effort has to be made to pair up the rarer species. For example, who knows where the different species of marmosets are? *L. rosalia* has a very extensive *Studbook* as does *Callimico*, but for some of the other rarer marmosets, before we can justify importation under the United States Endangered Species Act, we must show that efforts have been made to locate the appropriate animals here in the United States. I think it would be very good if everybody could list what they have and are willing to send elsewhere to help establish breeding programs.

HOBBS: There are two distinct needs for captive colonies: for the conservation of the species and for the research needs.

Another point that was already mentioned by Hearn is that we have as much as 60 percent losses when we bring in marmosets, especially into Europe. This is a direct consequence of the importation bans that exist, especially for *C. jacchus* from Brazil. These are not really bans, because the authorities are not able to enforce them completely. There is a black-market trade, so that these populations are in fact probably showing more losses than ever before. So, we have a very unsatisfactory situation where a lot of animals are lost because of bans imposed to justify conservation.

And lastly, there is the question about whether we should breed for research use in the user country or in the countries of origin. This is, of course, very controversial. Many people might feel that to invest time, money, and resources in countries which are developing quickly and have potentially unstable political and even religious situations could destroy an awful lot of investment.

PORTER: Part of the problem is—and I am all in favor of getting all the breeding colonies started in the user country—getting the initial stock to set up colonies. These governments do not see eye to eye with what we're thinking. They want breeding farms to be set up in South America so you may have to move the animals out of breeding farms down there to breeding farms up here. As long as you are not confiscated, you can produce a cheaper monkey down there. But you would have to have a breeding farm set up here at least so you will never be behind—we're being black-mailed right now. They tell us what we have to do if we want to get monkeys. They had a meeting in Lima June 2-4, 1975, in which the users presented their needs to the regulatory officials of three different South American countries. I don't think any monkeys are going to be exported in a hurry as a result, but it certainly was the thing to do.

THOMAS: Why must the facility be constructed in the country of origin?

PORTER: Because they say so.

THOMAS: Is that the only reason?

HEARN: I think there is a sound reason for having at least a staging point in the country of origin.

THOMAS: There is a great difference between a staging point and a breeding facility in the country of origin.

HEARN: The greatest loss seems to occur between leaving the country and arriving in the user country. If we could cut down that loss, then the overall demand that we have at the moment would probably evaporate.

PORTER: To give you one example. Coimbra is not here right now, but in Brazil, if you want to export monkeys from there legally, they must come from a breeding farm unless you get a special dispensation.

CICMANEC: I think a part of it is that they would like to be involved in development, to gain the technology, and practice some of these more sophisticated techniques, to work more toward an end product rather than just to be a continual supplier of raw products. There is a lot of orientation and thinking in this direction in South America. If the situation was reversed, we would be motivated in the same direction ourselves.

PORTER: Yes, I think they are entirely honest in their convictions and actually they have been forewarning us for a number of years that this was going to occur. I think we should have been making preparations in setting up breeding colonies. But we didn't. Now they essentially say, if you want monkeys you set up breeding farms down here. I think they are justified in saying this, and I have no objections to raising monkeys in South America as long as the doors will stay open. But I think we will have a better negotiating position with the South American governments if we set up breeding colonies in the user countries to supplement what is down there. Right now we haven't got a strong position to negotiate from.

HAMPTON: I agree with Porter in the sense that even if we could get marmosets now as in the past, it would not solve the problem. I think there is the economic problem of who will pay for these breeding programs in the user countries.

KINGSTON: Well, they will pay in the United Kingdom. There is a ready market in the U.K. for *C. jacchus*, and users will pay an economic price. Thus, it is commercially viable to breed *jacchus* in the U.K. and to make a quite substantial profit. As regards ICI, they are breeding 1,000 *jacchus* a year regularly. They are in the F_4 generation. And there is no fall-off in breeding potential. They are having a higher incidence of triplets and quads. They are producing the marmosets at around £70 per animal at 12 months of age and that is with a considerable loss of animals because for economic reasons they do not rear the third or fourth young which the parent will not rear; they argue that to have staff on overtime is just not economically viable. As a result, they have lost about 400 to 500 young out of a production of something like 900 weaned. Some of those were obviously preweaning deaths (they wean at about 12–16 weeks) of the triplets and the quadruplets born which have to be hand reared. But there is a market. In fact, *C. jacchus* are still arriving in the U.K., I don't know how. I was asked to deal with 170 only the other day. They are still being offered quite freely at quite a reasonable price, considering that they must be smuggled over the borders. The appalling fact is that, if they are caught in Brazil, they have to be transported by canoe or ox cart or whatever they use in Brazil, resulting in tremendous loss even prior to the actual shipment.

COOPER: Since these are patently illegal animals, why are we talking about the means to improve their survival? I really question the ethics of even receiving those animals in violation of the regulations.

EISENBERG: They are probably coming in from Holland, but not one single European Common Market country has ratified the International Treaty. Only the United States and several Latin American countries have. The idea was that all the European Common Market countries would ratify the treaty at one time because of their ties; but, because of a few holdouts, they haven't ratified it. That's why this anomolous situation exists. Hopefully, this is to be ratified in the next couple of months.*

KINGSTON: Yes, but if the animals actually arrive, surely on any grounds, you wish to improve their survival. Since they are actually there, you can't allow them just to die.

COOPER: I think that's not true. In fact, as long as those animals are purchased, as long as they're not allowed to die or be kicked out of the country, the market continues. The illegal trade can't be sustained without a market and the market is research.

KINGSTON: Well, they are being offered. A dealer in Holland is offering 100 *C. jacchus* a week at £26 each which is quite a reasonable price. We

*Editor's note. EEC countries finally decided to ratify the International Treaty separately. By summer 1976, England and Germany had ratified.

were offered very recently 100 per week, for as many weeks as we liked. I would emphasize I am not acting as an importer, I am acting as a kind of freelance consultant on marmoset survival in the United Kingdom.

COOPER: I would recommend that, if these offers come in writing from known dealers, you mail those offers with a letter to the Ministry of Agriculture in Brazil, informing them that these animals are being offered to you.

HOBBS: We have, but we have had no response from the authorities in Brazil.

HEARN: The dealers quote their sources as being the Guianas and Venezuela.

COOPER: Then tell Prince Bernhard about it; he's chairman of the Survival Service Commission and should have some control over the country and the airlines which have been notorious for this kind of trade for ten years.

HOBBS: I think what we're saying is that it is impractical for any exporting country to stop this sort of black market, especially a country the size of Brazil. This must be an enormous, impossible task.

COOPER: Not at all. The marmosets only occur in a small area. They're only shipped out of a few places and probably by a very few people. It only takes a few truckloads.

HOBBS: I think that's very encouraging. Because people like myself have set up to receive animals and then at the last minute we have said no, we can't be part of this. One reason is because of the ethics, and the second because of the financial losses. It is a real dilemma. I would rather see a real ban; this is no ban at all.

LORENZ: I would hope that in spite of the unsettled legal issues that exported animals, whether legal or illegal, once they arrive outside their country of origin, could be confiscated and put to good use. I think if we breed enough animals in captivity then we can probably help the wild population much more. This is, after all, the central issue of this conference.

KINGSTON: There are only three or four species so far in demand for research purposes. I would think it was in the interests of conservation to encourage breeding of those animals because researchers are going to obtain them one way or the other. Obviously not for *L. rosalia* or anything like that.

PORTER: As long as there is a demand for them, they are going to get in legally or illegally. I think the best thing for us to do is to consider how we can get legal animals. Right now, we are not producing enough. We should be considering ways that we can work with these governments to get animals legally. Right now, emphasis is on breeding operations down there.

HOBBS: I think many people would like to see that qualified. If you are going to try to get animals out legally, then it should specifically be to set up breeding nuclei, not for research per se. I would sympathize with the conservationists there. This is a justified qualification, because the research community has been irresponsible. There have been written warnings for years.

KLEIMAN: If we are thinking about setting up breeding colonies in user countries and having viable populations, we ought to be thinking about what species we are using. Mr. Kingston mentioned that there are only three or four species that are being used. I have gotten the impression that there is a significant difference in the breeding of *C. jacchus* as compared with *Saguinus* species. I would like people's opinions on that because no one in the United Kingdom seems to have trouble breeding *jacchus* in great numbers, with good survivorship and into the F_4 and F_5 generation. While for the *Saguinus* colonies, I think that there have been difficulties. Now, if that is indeed the case, we ought to think about whether or not it is wise in the United States—where *Saguinus* is the most commonly used genus—to maintain large colonies of *Saguinus* for research when, in fact, we may never be able to overcome possible breeding problems in that genus.

PORTER: One marmoset species is not the same as another; one may have uses for one purpose, another has uses for other purposes. We have to decide which are most important for medical reasons.

KLEIMAN: But the question is how many researchers in biomedical research, or within animal behavior, need particular species for specific reasons. Researchers seem to because they start a program with a particular species. But how many do need *Saguinus* and could they switch over to *Callithrix* and still get similar results?

GENGOZIAN: There are two basic areas of study in which I could easily go to another species, such as *C. jacchus*. This would mean a lot of backtracking in genetics that we have already done. Also, I mentioned in my talk that the cotton-top has the colonic cancer. Now we have not seen that in other species. We have not had access to as many *Callithrix,* that's true, but that would be one area in which we would be limited to cotton-tops. But other types of problems could be done with other marmoset species, at times with great personal expense and professional expense.

BAILEY: Has there been an attempt to take an inventory of all the different marmosets in this country and also to project the needs of the biomedical research community over the next decade or two, so that projections can be made as to how many animals we have to breed?

WHITNEY: There are several species of South American monkeys that are irreplaceable as models for human diseases. One is colon cancer. There was no good animal model for spontaneous colon cancer until Dr. Gengozian came up with the high incidence in *Saguinus oedipus.* There is no animal model for human hepatitis A except the chimpanzee and *S. mystax.* The development of a vaccine against hepatitis A which is one of the prevalent infectious diseases in the world and also the testing of diagnostic procedures is entirely dependent upon *S. mystax.* This is the single species that biomedical research is most interested in and, in fact, Delta Regional Primate Center has a contract now to breed 1,000. We have only been able to get 200 of those up to now. Merck, Sharp and Dohme wants another 2,000 for breeding colonies so they can raise their own, but have not been able to get them. The other South American primate that seems to be irreplaceable in biomedical research is the owl monkey, particularly for malaria. It is one of the few primates that is susceptible to all of the species of human malaria, so that all of the therapeutic regimes that are being developed against malaria are tested in the owl monkey. So we are really talking about the owl monkey and *S. mystax* as being irreplaceable and another good model being *S. oedipus* for colon cancer. When *S. mystax* became unavailable, they then started screening other species of *Saguinus* and, in fact, initially *S. nigricollis* showed some susceptibility to hepatitis A, but not clearly as positive as *S. mystax* had been. Speaking in the field of oncogenic virus research, initially susceptibilities were found for *Herpes saimiri* and *Herpes ateles* in *Saguinus,* but in fact more emphasis is now going to Epstein-Barr, a human virus, and *Ateles* and *C. jacchus* are susceptible to tumor formation with this virus. Perhaps we don't know our specific species needs for a predictable period of time simply because a broad-range survey never was initially made. It was just the initial species to be found susceptible and everyone said, that's the one we want. The others were never really looked at.

KLEIMAN: Can I get the second half of my question answered as to the differential reproduction?

EPPLE: I think it is really important that we try to find out whether this difference is real or not. Is it a species difference or is it a difference in maintenance. I have had no difficulties in breeding *S. fuscicollis* into the fifth generation. My generations have not been becoming smaller or showing a negative survival rate. I think *S. fuscicollis* is a good candidate to try for multigeneration breeding.

HAMPTON: I can't help but think that there are species differences, because in my situation there are several different species under the same conditions and one species will do well while another species won't.

KINGSTON: Well, as far as the United Kingdom is concerned, *C. jacchus* was used because it was the only easily available marmoset in the 1960s when we first started. We had a very awkward time and made all sorts of errors; we lost young, they got ricketts, there were epizootics, lots of trial and error. But I think we can truthfully say that we can rear three viable fully healthy *C. jacchus* per breeding pair per year as regularly as clockwork, year after year. But I feel, if we had had *S. mystax* or any of the others and gone through the same process of trial and error, of changing diets and cage size, and all the 101 things which make this work, I see no reason why we should not have done equally well with *mystax* in the U.K. I can see no reason why you shouldn't do equally well with any other species. I think you just have to get your methods right for the different species.

COOPER: I appreciate your feelings about this. I doubt, however, that that is the case. If you had gotten *S. mystax,* I don't think you would be in the same place that you are now. The two species are quite different. That is kind of a gestalt feeling. There is not a lot of data, but the experience that people have had with *mystax* is that they breed in captivity but there are problems with diet, and they are much more nervous than *jacchus.* Their evolutionary history is different and their means of survival have been different. It looks to me like *jacchus* is the more generalized form, more adaptive in that there is a range of habitat in which it is found. Not only that, but you go back to breeding in the 1930s and *jacchus* was always successful. There is a history of *jacchus* pairs breeding immense families for 40 years. That is not true for *mystax,* and it is not true with a number of other *Saguinus.* It is not true with *Leontopithecus,* and it never has been. And I think you may be right that you could do something, but it is really dangerous to assume that you could. It would be worth trying, but I wouldn't make that assumption.

MITTERMEIER: That is a good point. *Jacchus* even among the other *Callithrix* is a very adaptable species. It is the only one that isn't disappearing. In fact, it has been introduced into a number of places in southern Brazil and in the ranges of other *Callithrix* and it seems to be replacing them. You can even find them within the city limits of Rio de Janeiro.

KINGSTON: *Saguinus* is healthier in a broad sense. We have had five species of marmoset and tamarin in the same room under the same conditions. We have had all kinds of disease outbreaks and the ones that died were invariably the *jacchus,* you could have *Saguinus* in the next cage and the *jacchus* would die and *Saguinus* would not turn a hair. We have never even lost a young *Saguinus.* We have only bred 40 or 50—I'm not talking in hundreds—but we lost dozens of *jacchus* young and adults for various reasons, but we have never lost a *Saguinus;* they do seem to be, under our conditions, much stronger and disease-resistant animals. We have had *S. nigricollis, S. oedipus,* and *S. illigerii.*

GENGOZIAN: In our colony, we have found that *S. oedipus* seems to be a very hardy animal. I paid no attention to *oedipus* whatsoever and there were no problems.

KINGSTON: Are we not overdoing it? Is *jacchus* in any way an endangered species? Are not these bans which are being imposed in blanket fashion because of the all-goodness of administering a ban? Now I can see that in practical terms, if you don't have a total ban, you could have anything coming out described as *jacchus,* but is there any harm in reasonable numbers of *jacchus* being allowed out?

THOMAS: You cannot say if an animal is endangered or not until you know what the demand drain is. It doesn't make any difference if you've got five million, if the annual toll taken is six million, then it is endangered. It's the numbers game.

EISENBERG: I think there is another point. These countries that have a raw material which is desired by large politically organized blocks that control lots of money are going to force you to do something. That's the whole name of the game. There is no logic here, this is about politics. I think the countries are being very smart about it too. We would do the same thing, if we had a resource that we sold for a nickel and someone else turned around and sold it for five dollars.

HAMPTON: If we can breed *jacchus* so well, why do we need to get them?

KINGSTON: Because there is no large commercial breeder of *jacchus* in the United Kingdom yet. The big colony is in the hands of a large pharmaceutical company. There are a number of companies with small colonies hoping to supply their own needs, but there is obviously quite a large demand from other smaller firms who have not yet decided to breed their own. And there is the practicality for research. Research is done on groups of animals and, in the pharmaceutical field, they want 50 males and 50 females of approximately the same age at the same time. Now, if you have a colony of 20 or 30 pairs, you are never going to get a viable research group at any one time of comparable animals. So that you've got to have a very big colony indeed to supply even a modest pharmaceutical research need. There is a need for a large commercial effort.

HOBBS: In fact, there is a need to start doing what we have been doing with other common laboratory species.

KINGSTON: I still say that it is conservationally sensible, wise and ethical to encourage the commercial breeding on a large scale of this species which is predominantly required for pharmaceutical research purposes. At the same time, I am in total agreement with an absolute ban on the other species.

HOBBS: The climate is so right now to get the user to breed in any country, that it is in the interests of conservationists to help encourage this outlook right now.

COOPER: Encourage it, how?

HOBBS: By helping to clarify the situation as regards the legality of these bans that really aren't bans. Because of this, it makes the situation worse instead of better. And to encourage the legality of exporting from user countries with the specific qualifications that these are for breeding nuclei.

COOPER: I don't agree with that. I think that primates have been used in research for many many years and there is not a single incident regardless of the numbers available in the wild, that setting up breeding colonies has done anything but increase the need. It's just talk, we will build and build but when a wall comes up, we will push it down. The talk about the research community not being responsible has gone on since 1962. They are not going to be responsible and so why should conservationists work hard to help them.

HOBBS: I agree with you entirely, but I am saying that I sincerely believe that the time is right now simply because of shortages. The other point I would just like to make is that I would disagree with your point of proliferation of use. If you look at the rodents and lagomorphs in the U.K., the use for any particular project has in fact decreased because you've got improved standards. You've got an improved yardstick on which to base your results.

COOPER: But, there are very obvious differences in reproductive capacity and lifetime, and I don't think it is comparable at all.

KINGSTON: I remain convinced that a ban which is unworkable, and it is obvious that the Brazilian one at this moment is unworkable, is doing more harm than a properly controlled ban in which a proper licensed export is permitted. I still think that you will not get a ban to work in this political climate and economic times.

THORINGTON: I would disagree. I think that the ban is working very effectively; it is making you breed *C. jacchus* which you would not be doing if the ban did not exist. The price is £25 per animal which is beginning to compare favorably with the £70 for which you can breed them. If we can just make the ban work a little bit better—you could still get the black-market animals, but they would cost £75.

KINGSTON: I am for the ban entirely. I am for conservation. I am not a research worker. I am saying that, if you let enough come out to get these commercial colonies or individual-user colonies going, you are still serving the long-term interests of conservation. They will get them out anyway.

THORINGTON: But at £75, they won't be using so many. At £150 they will be using fewer. Yes, they will come out all right, but somebody is going to pay for them.

EISENBERG: There won't be any more marmosets coming from Holland in about six or seven weeks if the Common Market countries ratify the treaty and that will drive the price up a little bit higher.

KINGSTON: Well, in actual fact, there are enough *jacchus* in the U.K. at this moment. I don't think you need to import any more, but they are in the hands of a commercial pharmaceutical company which is offering them to other commercial pharmaceutical companies. But, what if their program changes and they don't want marmosets any more?

HOBBS: I would just like to raise one more controversial point, and that is the difficulties which were discussed yesterday of taking censuses in various areas and by varying methods. I am just wondering how much damage such censusing could do by giving falsely pessimistic information to national authorities?

MITTERMEIER: I don't think any of the information that is coming from the censuses is false or pessimistic. For species that are common, we are finding them to be pretty common and species that are rare are found to be pretty rare. I think the censuses indicate very clearly what the situation is. One of the points we made yesterday was that it is not absolute, but relative data that we are getting, and relative densities are important in determining marmoset conservation.

PORTER: Of course, I am interested in what those numbers mean, but what can we do with them, can you harvest a certain percentage of monkeys out of those areas? For instance, in Peru, they have white-lipped marmosets; presumably they are going to survive, they survive well in the outskirts of Iquitos. Does the Peruvian government now have enough information that they could, if they wished, permit the exportation of so many marmosets per year to the research laboratories?

MITTERMEIER: What are they going to get out of it, if they permit it?

PORTER: Well, it is up to them to negotiate with the United States. I think they are negotiating with PAHO right now. But, does this information have any validity after we accumulate it? For instance, I work in Guyana and there is a group down there from the National Academy of Sciences right now accumulating information on monkeys. When they turn this over to the government, what is the Guyana government going to do, if there isn't an interpretation of the information? Are they going to put a ban on the export of monkeys? If nobody can give the relative value of the information when it is presented to the Guyana government, presumably a conservation group would say, "Well, here they only have 6 or 10 members of this species per square kilometer. Probably this isn't enough, let's put a ban on it."

GREEN: I am disturbed these past two days, I think there is a fair proportion of "researchers" and a fair proportion of "conservationists," and I hear the terms researcher and conservationist used very harshly. Are not researchers interested in conservation?

HOBBS: We certainly are.

GREEN: So why make the distinction?

PORTER: We want to have monkeys available for research 20 or 50 years from now. So we have to be conservation-minded, but at the same time, since I am interested in setting up breeding colonies on islands, I want to know what these population figures mean. Do they mean you can harvest so many animals out of that area? I don't know how to evaluate these figures. I'm asking for information.

EISENBERG: Well, a little more money is going to have to be put into research before that can be appropriately interpreted. There may be enough data for two or three marmoset species to give you some reasonably hard answers to the questions you ask, but certainly not for more than that. What this requires is an estimation of carrying capacity for a given habitat type and the extent of that habitat type has to be mapped. Then, the population currently supported by that habitat type has to be estimated. All of that requires the support of field workers over a period of time in the areas of choice. And if funds will ever be graciously offered—and they are not expensive pieces of research—to accomplish this in a professionally competent manner, you will have reasonable answers. But up until now, it has been a struggle to get money for this kind of research. Because the so-called "researchers" that I have been hearing about don't consider it to be very important research. They have other priorities and money for them. We are at a crunch about money now.

PORTER: If we do develop these so-called semi-free breeding operations in South America, naturally we are going to have to formulate our answers as we go along.

EISENBERG: Right. So when requests come through for funding from various agencies concerning this sort of ground-level, game-management-type research project, these will have to be funded. I rather think that Guyana will be the testing ground for this kind of work, and it will be with *Saguinus midas.*

KINGSTON: But then you come up against a snag in that the pharmaceutical research worker is a very conservative animal and, if papers appeared using a specific species of marmoset, it's no good producing, with all respect to Mr. Porter, thousands of *midas.* You won't get researchers to use them because the comparative work has been done on *jacchus,* and it is the only species of any importance in the U.K. You could bring in thousands of *midas* and they wouldn't buy one, even if they are only one-quarter the price of *jacchus.*

CICMANEC: John [Eisenberg], what type of research is it that you are asking for? Is it management preserves or is it just basic data about these species?

EISENBERG: The funding of research concerning basic ecological data for the species in question.

CICMANEC: For what purpose?

THORINGTON: For an experimental situation in which you can start cropping and see what the effect is on the population, knowing what your basic population is, what the habitat is like, what is going on before you start manipulating. So you know whether it is your manipulation that is causing the population to drop or not.

EISENBERG: We have had two field studies reported on here, one on *Saguinus oedipus oedipus* and one on *S. oedipus geoffroyi.* To the best of my knowledge, those are the only two field studies on any species of marmoset that have been conducted in a given area over a period of at least a year. Now, this is really remarkable, isn't it? Because you have to do a study at least over several annual cycles before you can even begin to get an idea of your natality and mortality and the carrying capacity for a defined habitat type for a species. Now, one project was funded at a reasonable level because of a tie-in with some medical importance for the species in question; the other study was carried out by sheer grit and dedication with an absolute minimum of funding, outside of some money to live on. If you want solid answers, you have to give a little to these people to do this. We're not talking about mega bucks, this is chicken feed, cigarette money.

CICMANEC: What are the groups presently supporting this type of work and what do you see as potential groups who might be solicited to support this research?

EISENBERG: Work on primates traditionally has been supported by the National Science Foundation and the National Institutes of Health. It is becoming increasingly difficult for NIH to provide funds in a lot of areas, not just in field research. In any case, I don't think NIH's contributions toward field research were all that large proportionate to other expenditures. But yet the very questions that now seem so important are the ones that should have been addressed 5 and 10 years ago with very modestly funded research projects. You don't have to build a building, you don't have to hire permanent staff and cost out the pensions. That is the tragedy from my standpoint.

CICMANEC: Are there other groups able to support or be enticed to support this work, or is more money simply going to have to come from these two sources.

EISENBERG: The NIH funding pool is shrinking; I don't know if the end is in sight. NIH unfortunately became saddled with welfare programs that everyone is getting fed up with and so its budget is being curtailed. The original intent of NIH has been completely lost in the shuffle of administering all sorts of programs that should have been handed to some other agencies. It's only the National Science Foundation that is left, although people can always talk about Ford and these other foundations. I have never seen them take very much interest in an individual investigator. A lot of field work originates in the minds of one or two individuals; there is no monolithic organization of field workers, they are not unionized and organized, there is no spokesman for them. Field workers are very independent, and most really hate any sort of associations, even involvement with other people. They are only interested in what they are doing and so they suffer from this; it's an occupational handicap. I don't see any magic sugar daddy coming out of the sky with his bag of gold that is going to make it happen! (applause)

KINGSTON: Can't we take another line here? If research is prepared to spend the money if viral hepatitis is of world wide importance, if ICI can spend as they have done £271,000 on a marmoset breeding colony, if research wants these animals, why can't they be asked? It is chicken feed to a research company, they could support a group of scientists for months in the heart of the Brazilian jungle for petty cash. I mean I worked in a large firm for many years and I know, it's nothing.

THORINGTON: They have been asked. Years ago, they said, "Well, we can get our marmosets for $30 a piece."

KINGSTON: But in view of the current bans and if the commercial pharmaceutical firms want these animals to make money—and this is what they do want—then if it could be stated that they won't get anymore animals unless and until the status of the species is established in the wild, I would be very amazed if the comparatively very small sums were not forthcoming, providing it is put to them hard enough.

PORTER: I think we commercial people and Thorington know what company we're talking about and what marmosets they are interested in. The way they put it to me was that they have financed these things before and haven't got a cent out of it. They have plenty of money and no objection to it, but they don't think they are going to get anything back. They don't trust the countries down there.

THORINGTON: They put a mere $2,000 into some research at one point.

HOBBS: I think everybody agrees that the timing is better now than it has ever been. Are there any positive constructive recommendations? This conference is a rare opportunity to really get something from this meeting to the research community or, in fact, to the press. It strikes me that both sides are really talking the same language now. The commercial users want pressure, they want a total ban to promote commercial breeding. I think maybe this is where we should be involving the World Health Organization that is already very concerned indeed and is making moves to set up a foundation trust. The WHO approach may take time and their influence may come to bear in a few years, but perhaps this is a good opportunity to give that a thought. I mean, after all, they are the people who are recommending that you use monkeys in your work, whether it is hepatitis or polio vaccine testing or whatever. We should be twisting their arms now.

HEARN: The problem seems to revolve around time and the interests of research both in the field and the laboratory. In the field, we have seen how difficult it is to get good data on population and habitat measures, tree species, ways of management, shelters, and so forth. Although this is bound to take a lot of time, when we are dealing with endangered species this time isn't available. On the other hand, the medical problems are urgent and require being settled as soon as possible. Isn't it possible to separate these two things? If one considers the endangered species, perhaps approaches for funding can be channeled much more toward conservation funding societies and a more positive management approach be taken, such as providing shelters by boring holes in a nest tree or setting up nest boxes. This might at the site increase survival chances of endangered species. On the other hand, when we get to the laboratory, we're dealing with a few species, most of them unendangered. Here, regulated export will help the system along, and science will probably fund the project. The danger is that we get the two interests crossed. I think that if we get a report from the field that marmosets are in danger and the government is developing a conservation policy, then one gets the system stirred up and all marmosets are banned from export. This is in no one's interest and creates a black market. So if we can separate these two issues and have two funding approaches, perhaps we can come to a more reasonable conclusion.

MARTIN: (Litton Bionetics) I think that at a minimum everyone would agree that it should be resolved that we should not purchase animals which are clearly or even marginally illegal. If a gathering such as this can put that out, then people who may not be here or may not be in the mainstream of the work will realize that this is going forth and that this will certainly help to lower the black market in primates of all sorts.

HOBBS: I vote for that. Is there anyone here who can tell us what needs to be done to get this ratification of the Washington Treaty applied to these EEC countries? What is holding it up?

EISENBERG: Well, the fur dealers are holding it up as I understand it, in Spain and Italy. There is a lot of hide trade that goes on there for leather goods, and it is a political boondoggle at this point. The idea was to bring the entire Common Market in in a block ratification. You cannot get a concensus because of this special interest pressure in the two aforementioned countries. I think that marmosets coming through Holland is just an incidental thing. I don't think the Dutch animal dealers have any clout with the Dutch parliament. I'm sure it will happen. There is no way any endangered marmosets can come through Holland, for example, and come into the United States without the receiver being liable to prosecution. It could still happen but, if we catch them, we can put the screws on them. On the other hand, technically it is not illegal for these marmosets to move from Holland to England. It may be unethical but it is not, by the letter of the law, illegal at this moment.

HOBBS: I am sure it is not illegal for Americans to receive imports from England. I am told on very good authority that batches of animals, in fact, have been coming via various back doors to England and then being shipped off back to Miami.

EISENBERG: The only way they can come in is if the country of origin documents have in some way been forged. It is possible for this to happen but, if the appropriate footwork was done, the receivers of this could be taken to court.

KINGSTON: How would they respond to the fact that ICI are offering *jacchus* (F_3 and F_4 *jacchus*) to people in the United States, their own associations, of course? Is that illegal?

THOMAS: No. Permits can be issued if you can establish iron-clad proof of origin.

KINGSTON: Perhaps I am biased in this, since I have been concerned with research, but are we not exaggerating the effect on conservation of endangered species of the research demand? Is it not very small compared with the destruction of the habitat? There are statistics of imports into the United Kingdom of all animals including by family, but it is a comparatively small number; we're talking about 6,000 to 10,000 marmosets. Although I know that a lot more have died, surely that is comparatively small compared with blanket deforestation which is surely far more dangerous to the survival of the species in the long term.

KLEIMAN: But we have more control over what we personally do, as Bob [Cooper] brought up yesterday, than what some group of colonizers in Colombia do, so we really have to start with ourselves.

COOPER: Speaking of starting with ourselves. Of the four countries of interest as far as marmosets are concerned, Brazil, Guyana, Colombia, and Peru, a project is getting under way to breed animals in captivity (including *S. mystax*) in Iquitos, Peru. It seems to me that the focus on Peru is really most appropriate. A few years ago in Iquitos at the Amazonian Biology Congress, there were representatives of these countries, and it was then apparent to all of us there that the Peruvian government officials responsible and the Peruvian academics were really ready to and were doing something. They had already gotten actual data on the consumption of monkey meat by indigenous people. Soini is now working for the government along with Castro and through PAHO, surveys have been conducted in Peru. We have people here at the conference from PAHO, Melendez who is the head of this project, and Blood from the United States Primate Steering Committee who is very concerned with this project and its funding. Also Bob Whitney who is from the Division of Research Services from which much of the present funding is coming. They are trying to locate someone to operate the breeding efforts in Peru, to concentrate on *Saguinus mystax*, *Aotus*, and squirrel monkeys. They will certainly be breeding animals in cages and it looks as though they will be trying to look at some semi- or free-ranging situations as well. One concrete thing that we could do, since the species *Saguinus mystax* is picked out, is individually to offer what we know to those people involved. How much knowledge is there about *mystax*? I have had very little experience; other people have had more. Anyone who has had experience could benefit these people by saying something about *S. mystax*. It would be a shame to leave this meeting without trying to focus what knowledge there is about this species.

THORINGTON: We've heard that *C. jacchus* is really needed in the U.K. We have heard that *S. oedipus* is an important animal and that's crucial in that the populations are really low in the wild. We've heard that something is being done about *S. mystax*. Gengozian has expressed why he needs *S. fuscicollis illigeri* for his work. One animal that has been used a bit but I have not heard any expression of need for it, is *S. nigricollis*. Is there any real reason for people to be trying to breed *S. nigricollis?*

KINGSTON: No one that I know of, certainly not in the U.K. or in Europe.

THORINGTON: Now there are a lot of others that nobody really needs, because they don't know if they are worth anything. That would seem to limit us to four basic marmosets that people seem to feel are really needed. U.K. seems to have enough *C. jacchus* at this point and, if they don't get any more they are probably going to work out some way to breed them in spite of the protestations. There are a few *S. oedipus* in the country here. In fact, I worry a little bit that there may be more represented in colonies owned by people here than there are in the forests that are being studied. It's almost as if we have got the basic nucleus for *S. oedipus*, and people had better start thinking about their breeding. It seems to me that there are large colonies of *S. fuscicollis*. We know where a good many of the animals are at present through the ILAR surveys that Nancy Muckenhirn made recently. And so we have animals in the United States and in Great Britain that could be used constructively for the colonies. One question is how many people who are breeding marmosets are thinking in the 20- to 30-year range and are actively considering breeding strategies. How are they outbreeding, are they keeping good records of their breeding? How about the genetics, are they trying to select for something? Are these being considered?

HOBBS: Well, I don't think so. This is one of the points that we are concerned about because they are in commercial setups.

KINGSTON: Well, you see, with *jacchus*, there are a lot, possibly 5,000 or 10,000, but everyone of those, except for small colonies, is in the hands of commercial concerns. Now the emphasis is on commercial and if they decide they don't want them, I can assure you they would shut the colony down tomorrow afternoon.

THORINGTON: And you'd better be there to pick them up.

KINGSTON: Yes, quite. It happened to me with Fisons; that colony was producing 100 to 150 young a year as regularly as clockwork, and I was called in one afternoon and told: "We can't see any foreseeable use for these, arrange to get rid of them."

THORINGTON: It is not unique to the United Kingdom, I assure you. It is common in the United States as well. It is frequently simply the result of a panel review of a research program; they decide not to fund the research program for another year, and the colony is gone.

KINGSTON: So, they are not safe although there

are a lot at the moment. There are enough *C. jacchus* in the U.K. to insure a continuity of the supply, and if genetics and the like are watched reasonably intelligently, there is no reason why they shouldn't go on indefinitely. But I think it would be very unwise to assume that they will.

THORINGTON: I would just remark that if that's the way medical research is going to continue to run things, in the face of the obvious problems, then a pox on your town.

KINGSTON: No. Well, it's either one of two things, either you want a government-sponsored breeding center which is not subject to commercial pressures in the same way or you want a good commercial supplier who will keep going, for which he has to be making money. At the moment, the climate is such that he could make money in the foreseeable future and for a very long time. But he can't get any animals; they haven't got any. There are commercial firms ready and waiting to breed them at this moment with cages, staff, every damn thing, and they can't obtain a single marmoset to do it with.

THORINGTON: Bob [Whitney], would you or Ben [Blood] care to comment on the thoughts that you are having at NIH on this, about not letting the animals go down the drain in the future?

WHITNEY: I was just thinking I'd rather have a pox than measles! Ben, where are you?

BLOOD: There is a very great concern that opportunities are being lost with colonies of primates that have, for one reason or another, lost their funding or lost their interest to the sponsors. We do not have a mechanism that is established to handle this quickly, mainly because we do not have the mechanism to fund, which is exactly what is always needed. Somebody needs some money in a hurry. But at least we are facing the problem and are trying to find a way to pass the hat should important opportunities like this come along, to the various agencies of the federal government, be they in the health sector or NSF or others that might have an interest. There are a number of such cases right now; none, I think, that are concerned with marmosets, but several other primates, particularly macaques and baboons. The Primate Steering Committee is grappling with this problem, and I would hope that they would have some kind of an answer soon. It is, of course, part of a much bigger picture to fund any kind of a national primate program, both in this country and for work with foreign countries. The first thing that we really must come up with is a program that will merit funding so that we can give this top priority.

HOBBS: You must agree with me that we are exhausted in every sense of the word, although we haven't covered the title of the panel discussion.

INDEX